Lecture Notes in Physics

367

Y. Osaki H. Shibahashi (Eds.)

Progress of Seismology of the Sun and Stars

Proceedings of the Oji International Seminar
Held at Hakone, Japan, 11–14 December 1989

Springer-Verlag
Berlin Heidelberg GmbH

Editors
Y. Osaki
H. Shibahashi
Department of Astronomy, Faculty of Science
University of Tokyo, Bunkyo-ku, Tokyo 113, Japan

ISBN 978-3-662-13772-7 ISBN 978-3-540-46645-1 (eBook)
DOI 10.1007/978-3-540-46645-1

Originally published by Springer-Verlag Berlin Heidelberg New York in 1990
Softcover reprint of the hardcover 1st edition 1990

2153/3140-543210 – Printed on acid-free paper

Preface

It has been known for several years that the sun is oscillating with very low amplitudes in millions of global eigenmodes. By using information from these eigenmodes we are now able to explore the internal structure of the sun, just as we probe the interior of the earth by using seismic data. This new field of research is called "helioseismology." The rapid progress of the study of helioseismology is outstanding. Helioseismology was born in 1975 when the solar five-minute oscillations that were discovered in 1960 were definitely identified as a superposition of many eigenmodes of the sun. By the early 1980s it was found that the sun is oscillating in many modes with a wide range of wavenumbers. The oscillating region is dependent on the mode. The oscillations at the photospheric level are seen directly, and it is certain that even the region as deep as the nuclear reacting core is oscillating in some modes. The relative accuracy of frequency measurements of solar oscillations is now as good as 10^{-5}. By using such highly accurate observational information, we are now able to "see" the inside of the sun fairly well. In the last decade, we have begun to "see" the sound speed distribution and rotation in the sun and to investigate other physical processes in the sun. It is evident that helioseismology is very successful in studying the internal structure of the sun.

In recent years, pulsation and oscillation-related phenomena have been discovered not only in the sun but also in many stars that were hitherto regarded as non-pulsating. They include white dwarfs, Ap stars, and early-type O and B stars with slow and rapid rotation. The most important characteristics in these stars are that their oscillations are usually multi-periodic with several modes of oscillation involved. The same method used in helioseismology may in principle be applied to stellar oscillations. — This is called "asteroseismology." Asteroseismology is still in its infancy but has the potential to develop into a major field of stellar physics, since it is the only observational method that probes the interiors of stars, which otherwise cannot be seen directly.

The Oji International Seminar on "Progress of Seismology in the Sun and Stars" was held at Hakone Prince Hotel in the beautiful surroundings at the foot of Mt. Fuji, at Hakone (near Tokyo), 11–14 December 1989. The prime purpose of this seminar was to bring together solar and stellar astronomers (both theoreticians and observers) who have been actively working in the field of solar and stellar oscillations to discuss their latest progress in helio- and asteroseismology. There were 60 participants from 16 countries, who discussed subjects ranging from the Kamiokande II measurements of solar neutrinos to comparisons of various inversions of solar internal rotation, to the long-term variation of solar oscillation properties, and to Ap and white dwarf oscillations. There were very lively discussions during the conference and the session chairmen had difficulty in keeping the program on schedule. There was also an informal exchange of ideas and discussions in a more relaxed atmosphere after the formal sessions, since all participants stayed in the same hotel during the seminar.

The workshop was sponsored by the Fujihara Foundation of Science and the Japan Society for the Promotion of Science. We are very grateful to the Fujihara Foundation of Science for the generous financial support and to the Japan Society for the Promotion of Science for their administrative assistance. We wish also to thank Miss Miyuki Kudo for her secretarial assistance with miscellaneous problems during the workshop; her help made the workshop really enjoyable both for participants and those that accompanied them.

The editorial work of these proceedings was done during the stay of one of us (H.S.) at the Institute for Theoretical Physics, University of California, Santa Barbara, during the seminar "Helioseismology — Probing the Interior of a Star" which was held from January to June 1990 just after the Oji International Seminar. H.S. thanks all the staff of the Institute for their kind hospitality. We also thank Mrs Keiko Sakurai for her assistance in the editorial work.

Finally we should note that these proceedings were processed using the TeX macro package from Springer-Verlag.

March 1990

Yoji Osaki
Hiromoto Shibahashi

Contents

Part IV Observations of Solar Oscillations

LIST OF PARTICIPANTS

Al-Khashlan, A.	King Abdulaziz City for Sci. & Tech.	Saudi Arabia
Ando, H.	National Astronomical Observatory	Japan
Balona, L.A.	SAAO	South Africa
Belmonte, J.A.	Inst. de Astrofísica de Canarias	Spain
Belvedere, G.	Citta Universitaria	Italy
Braun, D.	NOAO/NSO	U.S.A
Cacciani, A.	University of "La Sapienza"	Italy
Damé, L.	L.P.S.P.	France
Däppen, W.	ESA/ESTEC Space Science Dept.	The Netherlands
Demarque, P.	Yale University	U.S.A.
Domingo, V.	ESA/ESTEC Space Science Dept.	The Netherlands
Dziembowski, W.	N. Copernicus Astronomical Center	Poland
Foing, B.H.	ESA/ESTEC Space Science Dept.	The Netherlands
Fröhlich, C.	Physik.-Meterolog. Observatorium	Switzerland
Gabriel, M.	Universite de Liège	Belgium
Goode, P.R.	New Jersey Institute of Tech.	U.S.A.
Gough, D.O.	University of Cambridge	U.K.
Harvey, J.W.	NOAO/NSO	U.S.A.
Hiei, E.	National Astronomical Observatory	Japan
Hill, F.	NOAO/NSO	U.S.A.
Hirayama, T.	National Astronomical Observatory	Japan
Ichimoto, K.	National Astronomical Observatory	Japan
Isaak, G.R.	University of Birmingham	U.K.
Jefferies, S.M.	University of Delaware	U.S.A.
Kambe, E.	University of Tokyo	Japan
Kawaler, S.D.	Iowa State University	U.S.A.
Korzennik, S.G.	UCLA	U.S.A.
Kosovichev, A.G.	Crimean Astrophys. Observatory	U.S.S.R.
Kotov, V.A.	Crimean Astrophys. Observatory	U.S.S.R.
Kuhn, J.R.	Michigan State University	U.S.A.
Kumar, P.	HAO	U.S.A.
Kurtz, D.W.	University of Cape Town	South Africa
Lee, U.	University of Tokyo	Japan
Leibacher, J.W.	NOAO/NSO	U.S.A.
Libbrecht, K.G.	Caltech	U.S.A.
Matthews, J.M.	University of British Columbia	Canada
Nakahata, M.	University of Tokyo	Japan
Narasimha, D.	Tata Inst. Fund. Research	India
Nishikawa, J.	University of Tokyo	Japan
Osaki, Y.	University of Tokyo	Japan
Pallé, P.L.	Inst. de Astrofísica de Canarias	Spain

Paternó, L.	Universitá di Catania	Italy
Rhodes, E.J., Jr.	University of Southern California	U.S.A.
Roxburgh, I.W.	Queen Mary College	U.K.
Saio, H.	University of Tokyo	Japan
Sakurai, T.	National Astronomical Observatory	Japan
Schmider, F.-X.	Queen Mary College	U.K.
Sekii, T.	University of Tokyo	Japan
Shibahashi, H.	University of Tokyo	Japan
Smeyers, P.	Katholieke Universiteit Leuven	Belgium
Stein, R.F.	Michigan State University	U.S.A.
Suematsu, Y.	National Astronomical Observatory	Japan
Takeuti, M.	Tohoku University	Japan
Unno, W.	Kinki University	Japan
van der Raay, H.B.	University of Birmingham	U.K.
Vauclair, G.	Observatoire Midi-Pyrénées	France
Vauclair, S.	Observatoire Midi-Pyrénées	France
Vorontsov, S.V.	Institute of Physics of the Earth	U.S.S.R
Weiss, W.W.	Universität Wien	Austria
Woodard, M.F.	Caltech	U.S.A.

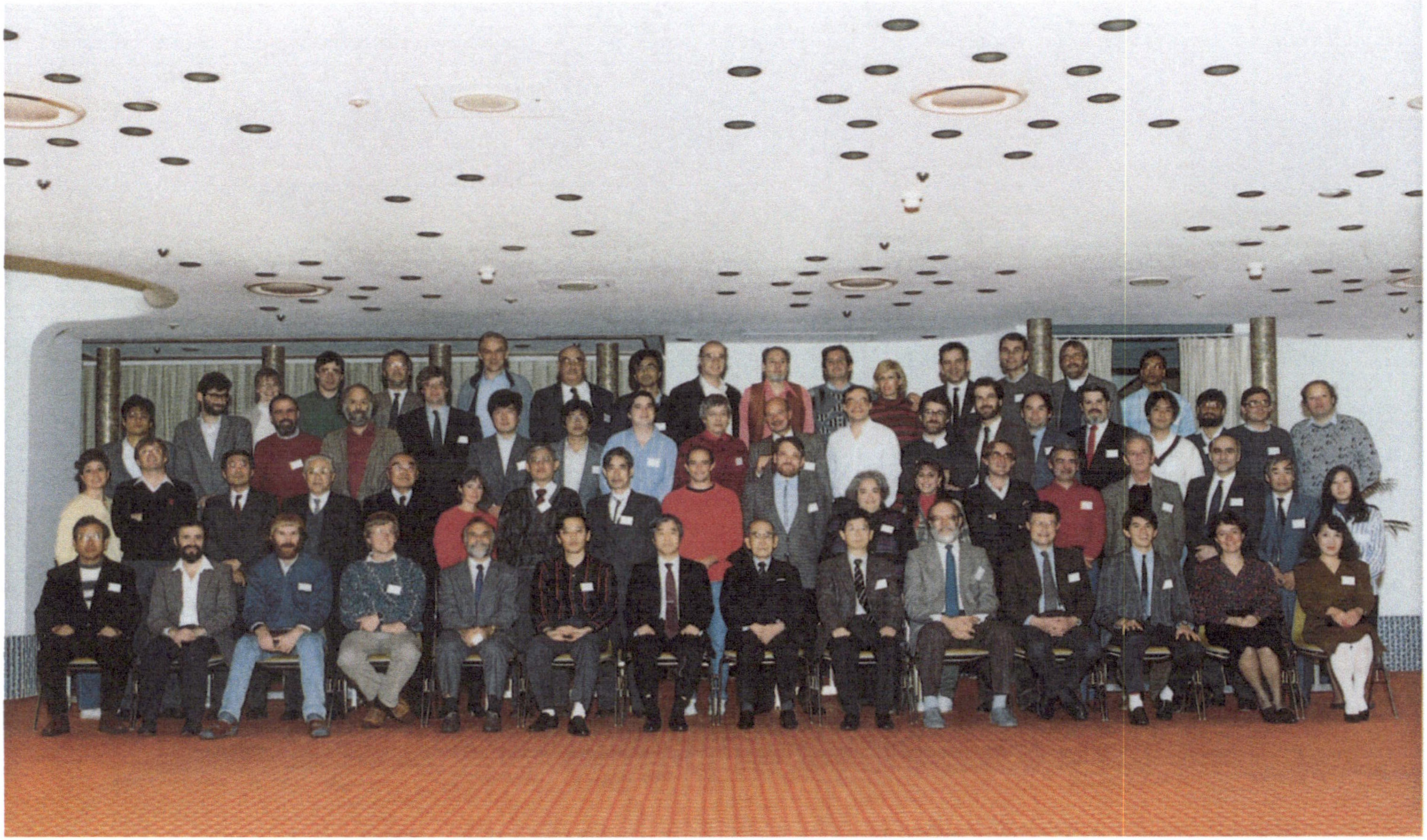

OJI INTERNATIONAL SEMINAR
PROGRESS OF SEISMOLOGY OF THE SUN AND STARS
11–14 DEC. 1989, HAKONE, JAPAN

1.Takeuti ; 2.Balona ; 3.Vorontsov ; 4.Libbrecht ; 5.Gough ; 6.Shibahashi ; 7.Hori (JSPS) ; 8.Fukushima (Fujihara Foundation) ; 9.Osaki ; 10.Leibacher ; 11.Demarque ; 12.Foing ; 13.S.Vauclair ; 14.Mrs.Shibahashi ; 15.Mrs.Goode ; 16.Goode ; 17.Misawa (Yomiuri Shinbun); 18.Sakaizumi (Fujihara Foundation); 19.Fukushima (Fujihara Foundation); 20.Mrs.Gough ; 21.Hirayama ; 22.Unno ; 23.Jefferies ; 24.Weiss ; 25.Mrs.Stein ; 26.Mrs.Kawaler ; 27.Belmonte ; 28.G.Vauclair ; 29.van der Raay ; 30.Isaak ; 31.Ando ; 32.Kudo (secretary) ; 33.Kambe ; 34.Korzennik ; 35.Domingo ; 36.Fröhlich ; 37.Damé ; 38.Lee ; 39.Ichimoto ; 40.Braun ; 41.Roxburgh ; 42.Paternò ; 43.Matthews ; 44.Pallé ; 45.Kawaler ; 46.Kotov ; 47.Rhodes ; 48.Sekii ; 49.Narasimha ; 50.Kosovichev ; 51.Däppen ; 52.Mrs.Kuhn ; 53.Kuhn ; 54.Kurtz ; 55.Gabriel ; 56.Smeyers ; 57.Saio ; 58.Cacciani ; 59.Stein ; 60.Bervedere ; 61.Mrs.Belvedere ; 62.Dziembowski ; 63.Harvey ; 64.Hill ; 65.Kumar

I

Introduction to Seismology of the Sun and Stars

Introductory Review of Solar and Stellar Oscillations

Hiromoto Shibahashi

Department of Astronomy, University of Tokyo, Bunkyo-ku, Tokyo 113, Japan

Abstract: Fundamental properties of seismological approach to the sun and stars are reviewed. I present also some results of helio- and asteroseismology to provide a common background for the more specialized reviews and the contributed papers in these proceedings.

1 Introduction

Any oscillation phenomenon of a system is useful to understand the system itself. When solar five-minute oscillations were definitely identified in 1975 as a superposition of many eigenmodes of the sun, we became able to use them as a diagnostic tool of the sun's interior, and a new field of research, helioseismology, became open. The rapid progress of the study of helioseismology is outstanding. Comparing (k, ω)-diagrams of the solar oscillations observed in the 1970's with the latest ones (e.g., Fig. 1 of Libbrecht and Woodard in these proceedings), we can easily realize how accurate observations became during the last decade. The relative accuracy of frequency measurements of solar oscillations is nowadays as good as 10^{-5}. In this sense, helioseismology is a quite rare field of precise science in astrophysics. By using such highly accurate observational information, we are now able to "see" the inside of the sun fairly well. The detection of solar neutrinos was a powerful method of examining the physical state of the central region of the sun. The observed neutrino flux is only about 1/3 of the theoretically expected value, and this problem has cast a shadow on our understanding solar structure. Helioseismology will, hopefully, be able to conclude whether the cause of the neutrino problem is a deficiency in the standard solar models.

It is the sun itself which powers and controls the oscillations. As compared with an aspect of helioseismology as a precise science, the study of the excitation mechanism of the solar oscillations is still order-of-magnitude science. We are still struggling to explain theoretically even to within orders of magnitudes the observed amplitudes and lifetimes of modes. The observed amplitudes, phases, and lifetimes of oscillations provide us with further information about the sun, but these quantities have not yet been utilized so efficiently as the accurately measured eigenfrequencies. In the meantime, highly accurate observational data of solar oscillations have been accumulated over a time interval as long as a solar activity cycle. Observations show that there are slight changes of the frequencies, the amplitudes, and the lifetimes of the oscillation modes of the sun. These tell us something about the source of the solar activity. We expect that information other than the eigenfrequencies will be utilized in helioseismology in the next decade.

In this review, I present some fundamentals of the seismological approach to the sun and stars to provide a common background for the more specialized reviews and the contributed papers in these proceedings. Detailed discussions and reviews are outside the scope of this paper. For more detailed introductory reviews, readers should consult, for example, Deubner and Gough (1984), Brown *et al.* (1986), Unno *et al.* (1989), and Vorontsov and Zharkov (1989).

2 Oscillation Properties of Stars

2.1 Basic Properties of Adiabatic Oscillations

Let us consider oscillations of a non-rotating, non-magnetic, spherically symmetric gaseous star. Small-amplitude oscillations of such a star can be separated into normal modes, each of which can be expressed in terms of $\exp(i\omega t)$ and a spherical harmonic $Y_l^m(\theta, \phi)$ with respect to time t and co-latitude θ and longitude ϕ, respectively. Since we are now considering a three-dimensional body, its normal modes are characterized by three quantum numbers: the spherical degree l which corresponds to the number of total nodal lines on the surface, the azimuthal order m which corresponds to the number of nodal lines in the longitudinal direction on the surface, and the radial order n which is the number of nodal surfaces intersecting the radial direction. For example, the displacement eigenvector $\boldsymbol{\xi}$ can be written in the spherical coordinates (r, θ, ϕ) as

$$\begin{aligned}\boldsymbol{\xi} &= \Re\left\{ \boldsymbol{\xi}_{nlm}(r,\theta,\phi)\exp(i\omega_{nl}t)\right\} \\ &= \Re\left\{ \left[\xi_{r,nl}(r), \xi_{h,nl}(r)\frac{\partial}{\partial\theta}, \xi_{h,nl}(r)\frac{\partial}{\sin\theta\partial\phi}\right] Y_l^m(\theta,\phi)\exp(i\omega_{nl}t)\right\}. \qquad (1)\end{aligned}$$

Here the dependence of the mode on the distance from the stellar center, r, is described by the functions of ξ_r and ξ_h which are functions of only r, and $\Re$ means the real part. As the star is supposed to be spherically symmetric, there is no special preferred choice of the axis $\theta = 0$, and hence the eigenfrequencies and the physics of oscillations must be independent of the choice of the axis. Since the azimuthal order m is measured with respect to a specified axis, the eigenfunctions $\xi_r(r)$ and $\xi_h(r)$ and the eigenfrequency are independent of m.

There are two kinds of restoring force; gaseous pressure and buoyancy. Waves whose restoring force is mainly the gaseous pressure are acoustic waves, while those whose main restoring force is buoyancy are internal gravity waves. The normal modes of acoustic waves and gravity waves are called p modes and g modes, respectively. Besides p modes and g modes, there is a unique f mode for each l, which is essentially a surface wave.

In most stars, the dynamical timescale is very much shorter than the thermal timescale, and hence the oscillations are well approximated as adiabatic ones. The equation governing these oscillations for $l = 0$ is a Sturm-Liouville type with an eigenvalue of ω^2. However, the equation for $l \neq 0$ oscillations is a fourth-order differential equation, but it is reduced to simpler forms under some approximations. Hereafter, let us follow Gough's (1986) approximation and compare it with Unno *et al.*'s (1989) treatment. Gough (1986) derived

$$\frac{d^2\Psi}{dr^2} + k_r^2(\omega^2, r)\Psi = 0, \tag{2}$$

where $\Psi(r)$ is the radial part of $\tilde{\Psi} \equiv \rho^{1/2}c^2\nabla \cdot \boldsymbol{\xi}$, that is,

$$\tilde{\Psi} = \rho^{1/2}c^2\nabla \cdot \boldsymbol{\xi} = \Psi(r)Y_l^m(\theta, \phi)\exp(i\omega t) \tag{3}$$

and

$$k_r^2(\omega^2, r) = \frac{\omega^2 - \omega_{\rm ac}^2}{c^2} - \frac{l(l+1)}{r^2}\left(1 - \frac{N^2}{\omega^2}\right). \tag{4}$$

Here, ρ and c are the density and the sound velocity, respectively, and $\omega_{\rm ac}$ and N are the acoustic cut-off frequency and the Brunt-Väisälä frequency defined as

$$\omega_{\rm ac}^2(r) \equiv \frac{c^2}{4H_\rho^2}\left(1 - 2\frac{dH_\rho}{dr}\right) \tag{5}$$

and

$$N^2(r) \equiv -g\left(\frac{d\log\rho}{dr} - \frac{1}{\Gamma_1}\frac{d\log p}{dr}\right), \tag{6}$$

respectively, where H_ρ denotes the density scale height. As clearly seen in Eqs. (2) and (4), there are two wave propagation zone, in which ω^2 is positive. These are a region in which

$$\omega^2 > \omega_+^2(r) \tag{7}$$

and a region in which

$$\omega^2 < \omega_-^2(r), \tag{8}$$

where $\omega_\pm^2(r)$ are the solutions of $k_r^2(\omega, r) = 0$ $(\omega_+ \geq \omega_-)$. In regions satisfying the condition (7) (P-zone), waves propagate as acoustic waves, and in regions satisfying the condition (8) (G-zone), waves propagate as gravity waves. Otherwise, waves are evanescent.

The characteristic frequencies $\omega_\pm^2$ are well approximated, as far as $L_l^2 \gg N^2$ and $L_l^2 \gg \omega_{\rm ac}^2$, by

$$\omega_+^2(r) \approx L_l^2(r) + \omega_{\rm ac}^2(r) - N^2(r) \tag{9}$$

and

$$\omega_-^2(r) \approx N^2(r), \tag{10}$$

respectively, where $L_l^2(r)$ is the Lamb frequency defined by

$$L_l^2(r) \equiv \frac{l(l+1)c^2(r)}{r^2}. \tag{11}$$

Unno *et al.* (1989) used a slightly different dispersion relation

$$k_r^2(\omega^2, r) = \frac{1}{\omega^2 c^2}(\omega^2 - L_l^2)(\omega^2 - N^2). \tag{12}$$

Though the physical causes of the acoustic cut-off frequency and the Brunt-Väisälä frequency are different, their values are accidentally very close except for the case of $N^2 \leq 0$.

Hence, the usage of the Lamb frequency and the Brunt-Väisälä frequency as the critical frequencies are practically convenient in massive stars. Indeed, in Unno *et al.*'s (1989) approximation, $\omega_{\rm ac}$ is replaced by N in Eq. (4), and $\omega_\pm^2$ are given by

$$\omega_+^2(r) = \max\Big[L_l^2(r), N^2(r)\Big] \tag{13}$$

and

$$\omega_-^2(r) = \min\Big[L_l^2(r), N^2(r)\Big], \tag{14}$$

respectively.

In the case of an early-type star having a radiative envelope, these two approximations are consistent because $\omega_{\rm ac}^2 \approx N^2$ reduces Eq. (9) to $\omega_+^2(r) \approx L_l^2(r)$. However, we should distinguish the acoustic cut-off frequency from the Brunt-Väisälä frequency in the case of stars having a convective envelope. If we would approximate the critical cut-off frequencies by Eqs. (13) and (14) in analyzing oscillation properties of a wholly convective polytrope, we would not have any outer turning points near the surface. This approximation is obviously inappropriate in this case. Equation (9), on the other hand, gives the critical frequency $\omega_+^2(r) \approx L_l^2(r) + \omega_{\rm ac}^2(r)$ in the convective envelope where $N^2 = 0$, while it gives $\omega_+^2(r) \approx L_l^2(r)$ in the deep radiative interior. Hence, Eq. (9) is a better approximation in treating stars having a convective envelope like the sun. The distribution of ω_+^2 and ω_-^2 in a solar model is drawn in Fig. 1.

If $\omega_- < \omega < \omega_+$ at the boundaries $r = 0$ and $r = R$, Eq. (2) gives an eigenvalue problem with a discrete eigenvalue ω^2. [As for details of the boundary condtions, see Unno *et al.* (1989).] Eigenmodes oscillating in a P-zone are p modes, and those trapped in a G-zone are g modes. In the case of main-sequence stars, distinction of p, f, and g modes is simple. However, as a star evolves, some of modes become to have a mixed character of an acoustic wave and a gravity wave. This is because the acoustic wave cavity and the gravity wave cavity in a star become to overlap in some frequency ranges (cf. Unno *et al.* 1989).

The quantization condition based on the WKBJ approximation of Eq. (2) gives a relation between the radial order n and the eigenfrequency $\omega(n,l)$:

$$(n+\epsilon)\pi = \int_{r_1}^{r_2} k_r(\omega^2, r)dr, \tag{15}$$

where r_1 and r_2 are the turning points at which $k_r = 0$.

In the case of high-order p modes with low degrees ($n \gg l \approx 1$), the quantization condition (15) leads to

$$\nu_{nl} \equiv \frac{\omega_{nl}}{2\pi} \approx \nu_0\left(n + \frac{l}{2} + \epsilon\right), \tag{16}$$

where

$$\nu_0 \equiv \left[2\int_0^R \frac{dr}{c}\right]^{-1} \tag{17}$$

is the inverse of twice the sound travel time between the center and the surface (Vandakurov, 1967, Tassoul 1980, Smeyers *et al.* 1988). This formula is used to search for stellar oscillations of high-order p modes. In the higher order approximation, the frequencies of p modes are approximately given by

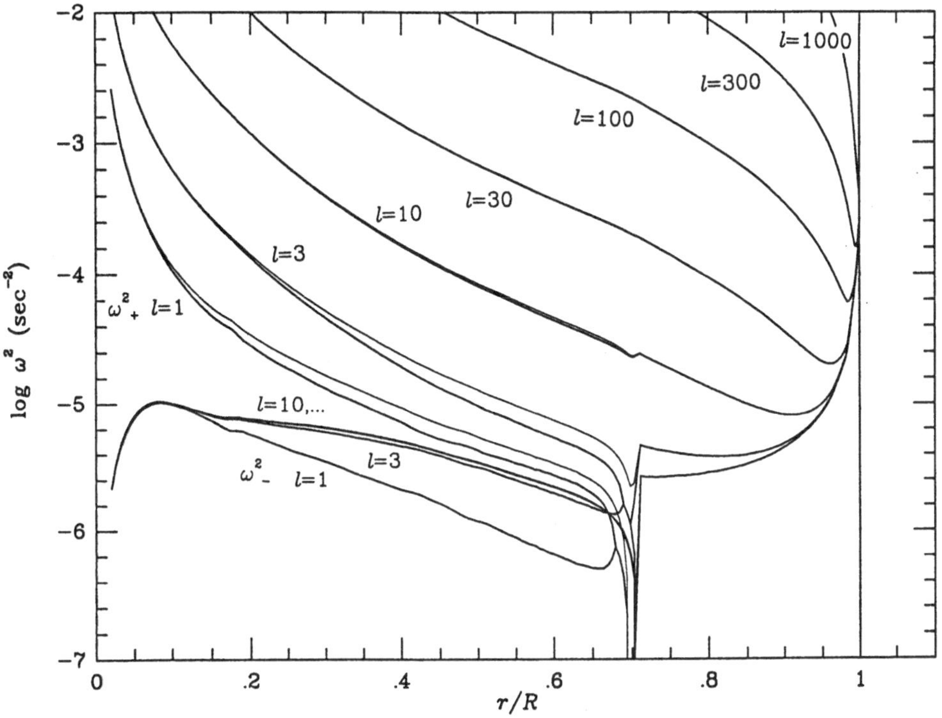

Fig. 1. Propagation diagram for a solar model. The squared critical frequencies ω_+^2 and ω_-^2 for $l = 1$, 10, 30, 100, 300, and 1000 are drawn as functions of r/R. The distribution of $L_l^2 + \omega_{ac}^2$ are also shown by thin lines.

$$\nu_{nl} = \nu_0\left(n + \frac{l}{2} + \epsilon\right) + \frac{[l(l+1) + \delta]A}{\nu_0(n + \frac{l}{2} + \epsilon)}, \tag{18}$$

where δ and A are constants dependent on the structure of the star (Tassoul 1980).

The quantization condition (15) leads to an analogous relation in the case of high-order g modes of low degree (Vandakurov 1967, Tassoul 1980):

$$\Pi_{nl} \equiv \frac{2\pi}{\omega_{nl}} \approx \frac{\Pi_0}{\sqrt{l(l+1)}}\left(n + \frac{l}{2} + \epsilon\right). \tag{19}$$

Here Π_{nl} denotes the period of the g_n mode of degree l and

$$\Pi_0 \equiv 2\pi^2\left[\int_0^R \frac{N}{r}dr\right]. \tag{20}$$

This formula is used in the search of g modes of the sun.

The quantization conditions are useful in the asymptotic approach in the inverse problem, in which functional forms of certain physical quantities are determined from the observed frequency spectrum. With increasing accuracy of the observations, more accurate asymptotic treatments are being required. Vorontsov (1990) and Smeyers and Ruymaekers (1990) present such treatments in these proceedings.

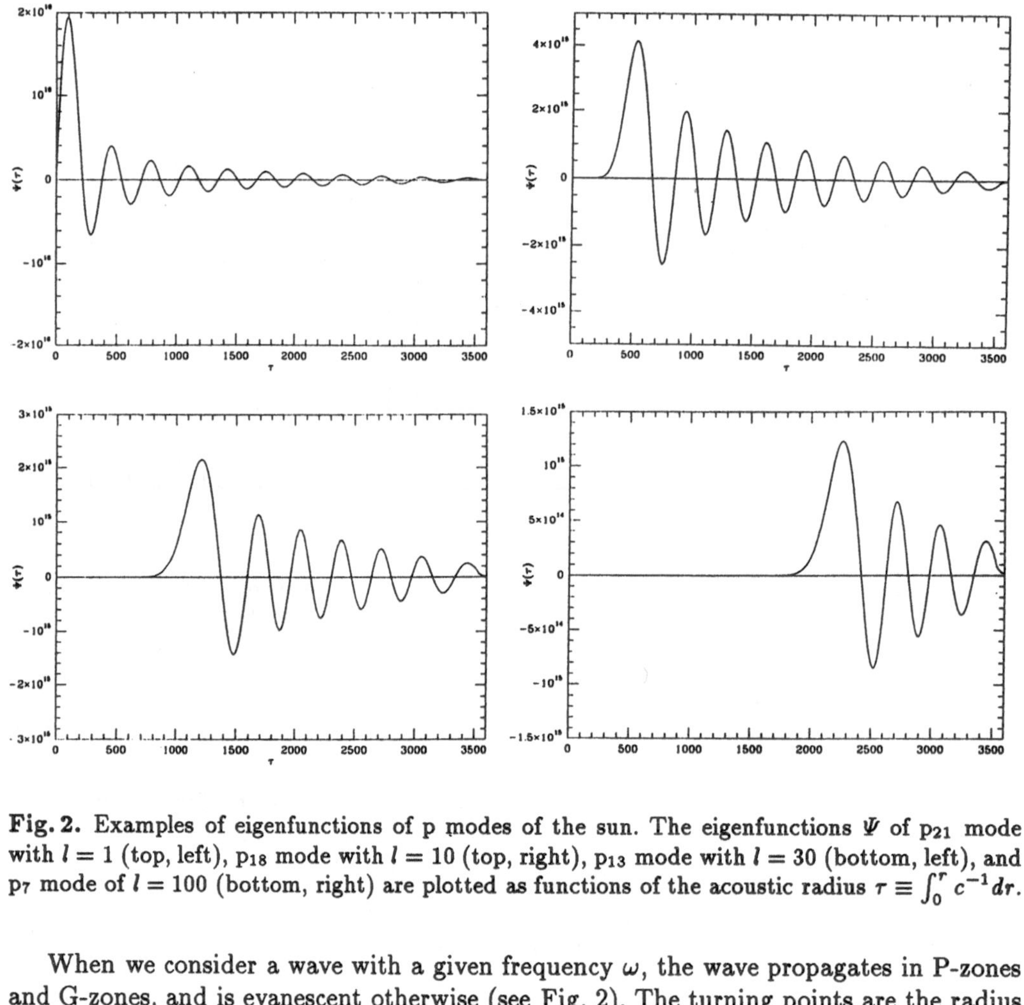

Fig. 2. Examples of eigenfunctions of p modes of the sun. The eigenfunctions Ψ of p_{21} mode with $l = 1$ (top, left), p_{18} mode with $l = 10$ (top, right), p_{13} mode with $l = 30$ (bottom, left), and p_7 mode of $l = 100$ (bottom, right) are plotted as functions of the acoustic radius $\tau \equiv \int_0^r c^{-1}dr$.

When we consider a wave with a given frequency ω, the wave propagates in P-zones and G-zones, and is evanescent otherwise (see Fig. 2). The turning points are the radius at which $\omega^2 = \omega_+^2(r)$ given by Eq. (9) for p modes and the radius at which $\omega^2 = \omega_-^2(r) = N^2(r)$ for g modes. As for turning points of p modes, since the Lamb frequency dominates over both the critical cut-off frequency and the Brunt-Väisälä frequency in the deep interior, the inner turning points occur where

$$l(l+1)c^2(r)/r^2 = \omega^2. \tag{21}$$

We can expect to extract from each of the p modes some information of the stellar structure as deep as the inner turning point given by Eq. (21).

Figure 3 shows the so-called (l, ν)-diagram [or (k_h, ω)-diagram] of a solar model, in which eigenfrequencies of p modes $\nu_{nl} \equiv \omega_{nl}/2\pi$ calculated as eigenvalues of a full fourth-order differential equation are plotted against the spherical degree l. In this figure we also show some critical frequencies whose inner turning points are $r = 0.05, 0.10, \ldots$ We can see from this diagram that we may use low degree p modes of the sun in the five-minute range as a probe for the region as deep as $r = 0.05$, in which nuclear reactions occur and most of the ^{8}B neutrinos are generated.

As for the outer turning points of p modes, they are almost functions of only frequency ω^2 and independent of the spherical degree l. The frequencies corresponding to the outer turning points at $r/R = 0.980, 0.990, 0.995$, and 0.999 are shown in the right hand side of Fig. 3.

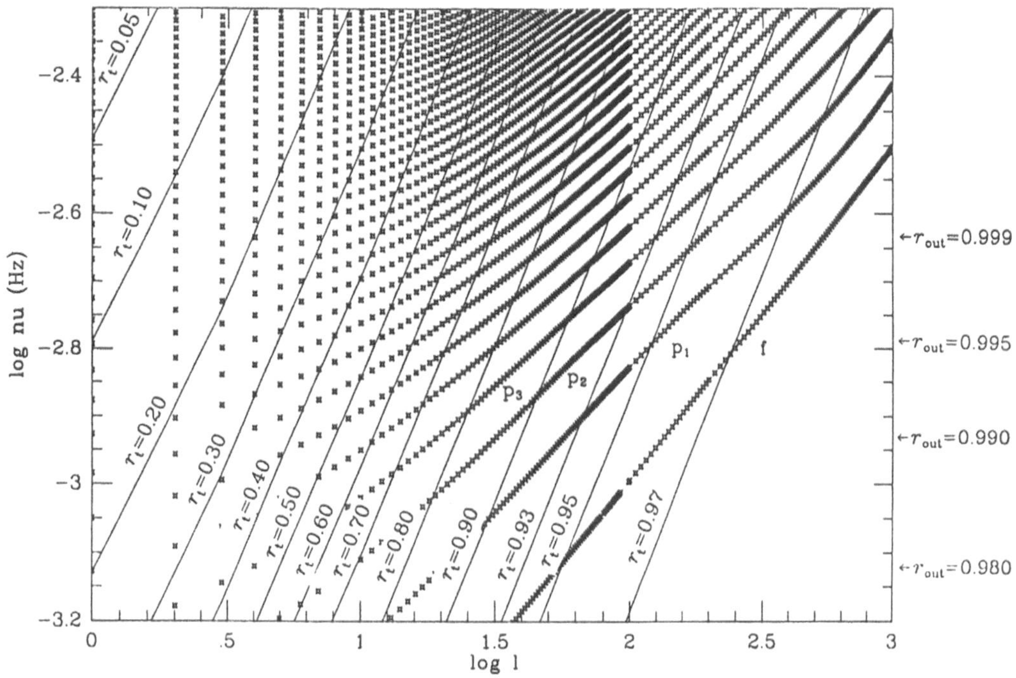

Fig. 3. (l, ν)-diagram of a solar model. The abscissa and the ordinate are $\log l$ and $\log \nu$, respectively. Straight lines show the critical (l, ν)-relation for the inner turning points $r/R = 0.05$, 0.10, 0.20, 0.30, 0.40, 0.50, 0.60, 0.70, 0.80, 0.90, 0.95, and 0.97. The frequencies corresponding to the outer turning points at $r/R = 0.980, 0.990, 0.995$, and 0.999 are shown in the right hand side.

In the wave propagation zone, the wave energy flux is constant, while it exponentially decreases outside. Since the extent of the wave propagation zone differs from mode to mode, even if the same energy is input into different modes, the observable surface amplitudes are different. Figure 4 shows amplitudes of the radial displacement eigenfunctions $\xi_{r,nl}(r)$ of a solar model at the photosphere as functions of frequency ν in the case that the wave energy $\frac{1}{2}\omega^2 \int_0^M |\xi|^2 dM_r$ is equal to GM/R. As seen in this figure, for a given radial order n (i.e., for a series of p_n modes), the surface amplitude of ξ_r is the largest at $\nu \approx 3\,\mathrm{mHz}$, and decreases rapidly with decreasing frequency. On decreasing the frequency, the outer turning point becomes deeper so that the evanescent zone near the surface becomes thicker, and hence the surface amplitude is attenuated more. On the other hand, with the increasing of the frequency, the inner turning point becomes deeper so that the wave propagation zone extends deeper, and hence the surface amplitude becomes relatively smaller. The global feature of Fig. 4 is determined by a balance of the attenuation effect and the depth of the wave propagation zone. The amplitudes of g modes are invisible in the same figure. Comparing this diagram with the observed amplitudes of modes of solar oscillations, we can estimate the kinetic energy of each mode. This information is used to investigate the excitation mechanism of the oscillations.

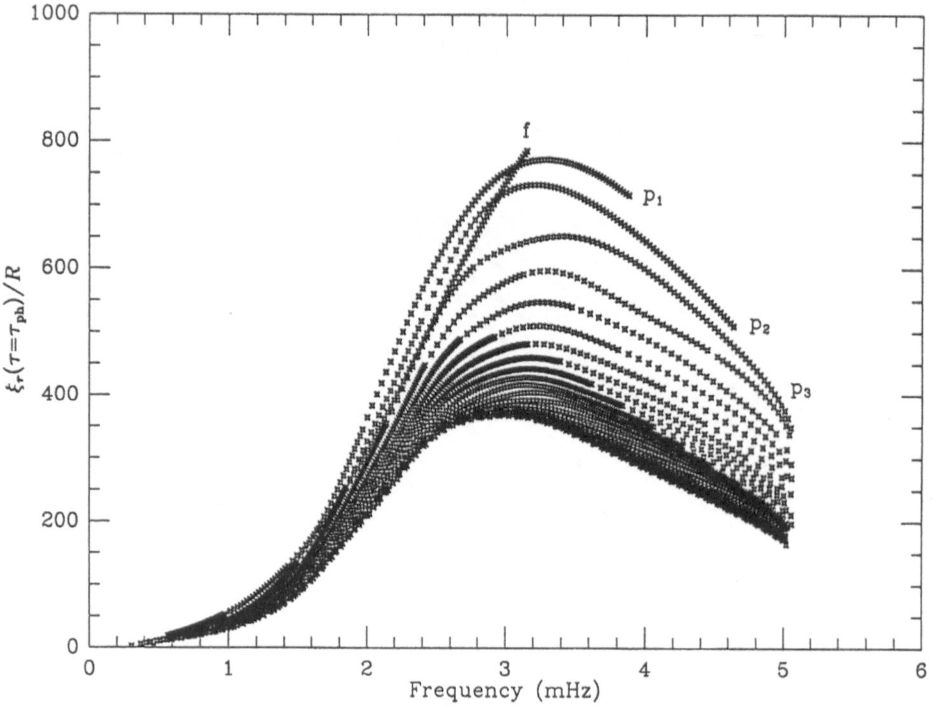

Fig. 4. Amplitudes of the eigenfunctions $\xi_{r,nl}$ of a solar model at the photosphere as functions of frequency for a given kinetic energy $\frac{1}{2}\omega^2 \int_0^M |\xi|^2 dM_r = GM/R$.

The critical cut-off frequency ω_+ for p-mode oscillations is finite at the surface. If we assume there is an isothermal atmosphere far above the photosphere, the cut-off frequency is given by

$$\omega_{\mathrm{cut}} = \frac{c}{2H_\rho}. \tag{22}$$

Above this frequency, waves propagate outward with energy leakage, and there should be no standing wave. However, in cases of the sun and some Ap stars, observations report some oscillations with frequencies higher than the theoretically estimated cut-off frequency (Jefferies *et al.* 1988, Kurtz 1990b). In the solar case, the frequency spectrum of these high-frequency components is fairly wide, but in the case of Ap stars, the frequency spectrum of the components higher than the theoretical critical cut-off frequency is very sharp. Oscillations higher than the theoretically estimated cut-off frequency may provide us some hints on the excitation mechanisms or the atmospheric structure. As for Ap stars, since the frequency spectrum of these modes is so sharp, observed oscillations seem standing waves. To keep them standing waves, there should be a barrier of $\omega_+(r)$ somewhere in the atmosphere, and the oscillation may be leaky waves outside of it. There should be some energy supply to overcome the energy leakage. As for the barrier, existence of a low temperature atmosphere (Shibahashi and Saio 1985), a density jump (Balmforth and Gough 1990), and a barrier of the mean molecular weight (Vauclair and Dolez 1990) have been suggested.

2.2 Effects of Rotation and Magnetic Fields

Rotation and global magnetic fields lift the $(2l+1)$-fold degeneracy of eigenfrequency with respect to the azimuthal order m. In the case of rotation, the splitting of the eigenfrequency observed from an inertial frame is expressed, to first order, in the form

$$\omega_{nlm} - \omega_{nl0} = \int_0^R \int_0^\pi K_{nl}(r,\theta)\Omega(r,\theta)d\theta dr, \tag{23}$$

where $\Omega(r,\theta)$ is the rotational frequency of the star and K_{nl} is a function dependent on the star and the mode. Figure 5 shows examples of $K_{nl}(r,\theta)$ for a solar model. Modes with low azimuthal orders m have sensitivity extended over all latitudes. With increasing $|m|$, modes become sensitive only near lower latitude zones. Since the rotational frequency splittings of low-azimuthal-order modes are smaller than those of high-azimuthal-order modes, the observational errors of the rotational frequency splittings are smaller for high-$|m|$ modes. Equation (23) is used to infer the internal rotation of the sun from inversion of the rotational frequency splittings.

Effects of magnetic fields in Ap stars on the oscillations seem to dominate over those of rotation. In most of these stars, magnetic fields seem roughly to be dipole and the symmetry axis is oblique to the rotation axis. Observed oscillations in Ap stars look like dipole p modes which are axisymmetric with respect to the magnetic axis. It is expected that oscillation data of these stars provide us some information on magnetic fields of Ap stars (see Kurtz 1990a, b).

3 Observations of the Solar Oscillations

A detailed review of the observations of solar oscillations is outside the scope of this introduction. Here we focus ourselves on one of the hot topics: long term variation of oscillations. Various observing teams reported first the observed long-term variation of frequencies at the Aahus meeting in 1986 (Woodard and Noyes 1985, 1988, Gelly *et al.* 1988, Henning *et al.* 1988, Isaak *et al.* 1988, Jiménez *et al.* 1988). There was, however, inconsistency between observers; one group of teams reported that the eigenfrequencies

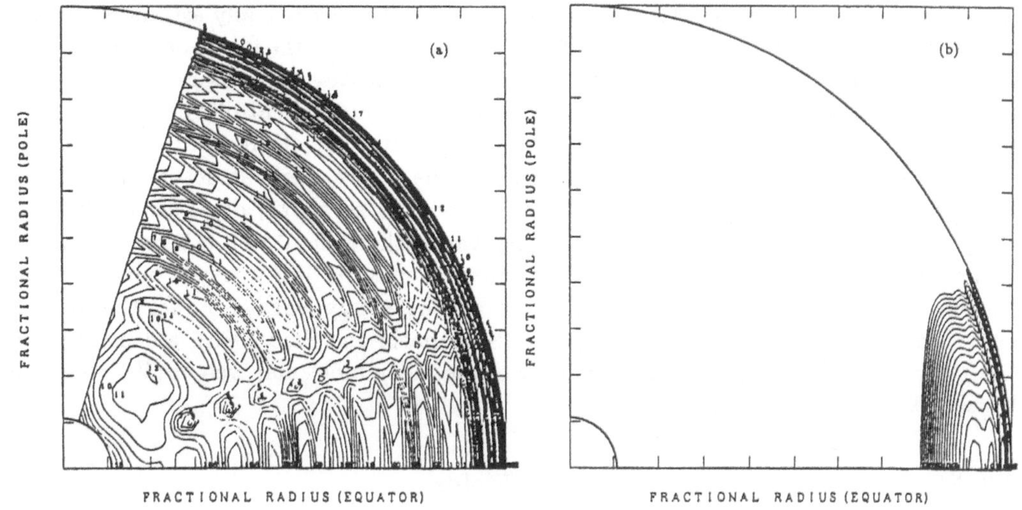

Fig. 5. (a)Contour map of the kernel $K_{nl}(r, \theta)$ for p_{23} mode with $l = 5$ and $m = 3$ of a solar model. The contour lines are drawn for 27 levels from $\log K = -5$ to 2 with a step of 0.25. (b)The same as (a) but for p_3 mode with $l = 60$ and $m = 60$.

were higher in the solar active phase than in the inactive phase, while another group claimed the opposite tendency. With the increase of the observational time spans, statistical accuracy becomes better. Furthermore, since the solar activity is now going to reach its maximum phase, more reliable correlation between long-term variation of oscillation properties and the solar activity will be obtained in a few years. In these proceedings, Jefferies *et al.* (1990), Libbrecht and Woodard (1990), and Pallé *et al.* (1990) report that the frequencies of the solar p modes are slightly higher in solar active phases than in the inactive phases. The variation of the frequencies, the amplitudes, the line-widths of eigenmodes, and the rotational splittings of frequencies will provide us some hints to understanding the solar activity and the excitation mechanism of oscillations. This information is of a new kind, which will be useful for diagnosing both the thermal structure of the sun as a heat engine, and the physics of convection. A new aspect of helioseismology will be opened and activated in the 1990's.

With the upgrade of the quality of CCD cameras, two-dimensional observations over the solar disk came to be carried out. These observations make it possible to identify each of individual mode patterns, and much valuable information is obtained through these observations. Here, I wish to make a remark on the analysis of these observations. Solar oscillations have been observed as the oscillation of the velocity fields on the solar surface or as the oscillation of the brightness. As for the former, only the oscillations of the line-of-sight velocity are measurable from the Doppler shift of photospheric lines. Both the radial and the horizontal components of the velocity fields of eigenmodes contribute to the line-of-sight velocity. Due the boundary condition at $r = R$ which requires the Lagrangian pressure perturbation to be zero there, the ratio of the horizontal displacement $(l \times \xi_{h,nl})$

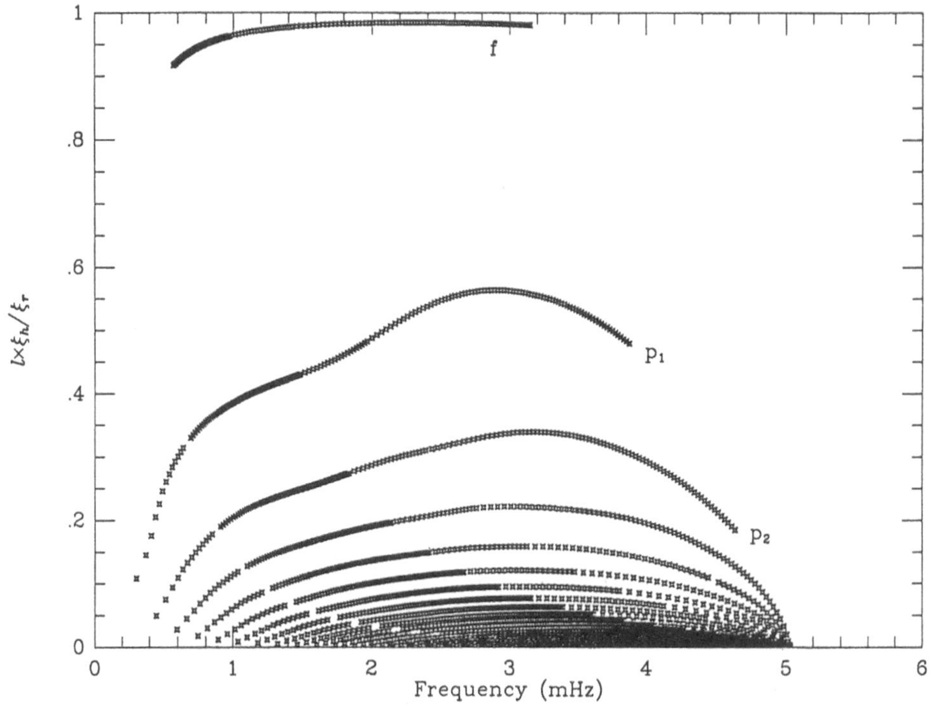

Fig. 6. The ratios of $l \times \xi_h(R)/\xi_r(R)$ for f modes with $30 \leq l \leq 1000$ and for p modes with $1 \leq l \leq 1000$ for a solar model as functions of the frequency $\nu \equiv \omega/2\pi$.

to the radial displacement $\xi_{r,nl}$ at the surface is given by

$$\frac{l\xi_{h,nl}(R)}{\xi_{r,nl}(R)} = \frac{l \times (GM/R^3)}{\omega_{nl}^2}, \tag{24}$$

where the factor l comes from the spatial derivative operator in Eq. (1). Figure 6 shows this ratio for p modes of a solar model for $1 \leq l \leq 1000$. Obviously, this ratio is of the order of unity for high-degree f modes, and becomes smaller with increasing radial order for p modes while it becomes larger with increasing radial order for g modes. Therefore, the contribution of the horizontal component is substantially larger with increasing the spherical degree l and decreasing the radial order n in the case of p modes. Furthermore, the θ component of the velocity eigenfunction of a single mode has an angular dependence which is different from the angular dependence of r component. Therefore special attention should be paid to analyses of the Doppler shift measurement of high-degree p modes. In order to investigate the excitation mechanism from observational results, we have to analyze the amplitudes and the line-widths in the power spectrum. If we ignore the effect of the horizontal component of velocity fields, we would overestimate the amplitude of each mode and therefore the excitation efficiency.

4 Forward Problem

As soon as the solar five-minute oscillations became recognized as a superposition of eigenmodes of the sun, attention was paid to the possibility of using this tremendous number of eigenmodes as a diagnostic of the internal structure of the sun. If we have some solar models with different parameters, we can search for the model whose eigenfrequencies fit the best to the observed frequencies. This forward problem approach has been used by various authors within the frame of the so-called standard solar models. In this framework, we construct solar models so that they have the solar luminosity and the solar radius at a reasonable solar age by adjusting the chemical composition and the mixing length. There are discrepancies between the theoretical eigenfrequencies of standard models thus constructed and the observed frequencies, in the sense that theoretical frequencies of low-degree modes are lower than the observed ones while those of high-degree modes are higher than the observations. Uncertainties of some input physics to construct a solar model are, however, larger than the observational errors of frequencies. The problem with this approach is, hence, to make ambiguities in input physics small. Most stellar evolution codes were developed earlier than the opening of helioseismology, and numerical accuracy as high as the requirement of helioseismology was not required in stellar evolution code at that time. Therefore, to define quantitatively the discrepancies between the real sun and the solar models, we have to examine carefully the accuracy of the evolutionary code itself as well as the input physics. The uncertainties in input physics are (1) treatment of convection, (2) equation of state, (3) opacity, and (4) nuclear reaction rate. The problem is reviewed by Gabriel (1990), and recent investigation of the equation of state is described by Däppen (1990).

5 Inverse Problem

There is no definitely established guideline for constructing models other than the procedure for the standard solar model. Hence, even if we succeed in constructing a non-standard solar model providing a better fit to the observation of oscillations, we cannot exclude other possibilities for the solar structure. On the other hand, functional forms of certain physical quantities can be directly determined by solving integral equations whose known functions are provided by observational data. This is the inverse approach. The inverse problem is the only method of determining the solar internal rotation since there is no standard model of the stellar internal rotation.

The inverse problem is divided into two categories: the linear problem and the non-linear problem. If we have a reasonable model, we can solve integral equations to obtain differences of certain physical quantities between the model and the real sun by analyzing the differences between the observed oscillation quantities and the model. For example, by treating the differences between the theoretical frequencies of a model and the observed frequencies of p modes as known quantities, we can derive from the variational principle an integral equation to obtain differences of the sound speed and the density between the model and the real sun:

$$\frac{\delta\omega_{nl}^2}{\omega_{nl}^2} = \int_0^R K_{\rho,c^2}^{nl}\frac{\delta\rho}{\rho}dr + \int_0^R K_{c^2,\rho}^{nl}\frac{\delta c^2}{c^2}dr, \tag{25}$$

where $K_{\rho,c^2}(r)$ and $K_{c^2,\rho}(r)$ are kernels which depend on the mode. Another example of a similar integral equation is that for the rotational splittings given as Eq. (23). In that equation, the rotational frequency splittings should be regarded as known quantities and the rotational profile $\Omega(r,\theta)$ is the unknown function to be determined. The integral equation thus derived is of the Fredholm type, and various methods of solving helioseismological inverse problems have been developed: optimized averaging kernel method, constrained least-squares method, spectral expansion method, and so on.

Nonlinear inverse problems of helioseismology have so far been attacked by asymptotic inversion. Let us consider a quantization condition given by Eq. (15) and treat $n(l,\omega)$ as a continuous and differentiable function of continuous variables l and ω. Differentiation of Eq. (15) with respect to ω and l leads to

$$\frac{\partial n}{\partial \omega}\pi = \int_{r_1}^{r_2} (\partial\omega/\partial k_r)^{-1} dr \tag{26}$$

and

$$\frac{\partial n}{\partial l}\pi = \int_{r_1}^{r_2} (\partial\omega/\partial k_h)/(\partial\omega/\partial k_r) r^{-1} dr. \tag{27}$$

The left-hand sides of Eqs. (26) and (27) are obtained from observational data, and the integral function in the right-hand side of Eq. (26) is the radial component of the wave group velocity and that of Eq. (27) shows how the wave is diffracted during its propagation. Hence, we can recognize Eqs. (26) and (27) as integral equations to let us know how wave packets propagate. Since the propagation of waves is determined by the physical structure of the medium, we can infer the physical structure of the medium itself from Eq. (15).

Since the frequencies of p modes are mainly determined by the sound speed distribution in the sun, the sound speed distribution in the sun is the easiest information to be extracted. Various inversion techniques have been used to extract information of the solar internal structure from the p-mode frequency spectra. The general tendency of the various inverted results is in good agreement (Christensen-Dalsgaard *et al.* 1985, 1988, Gough and Kosovichev 1988, Kosovichev 1988, Shibahashi and Sekii 1988, Vorontsov 1988, Dziembowski *et al.* 1990, Vorontsov and Shibahashi 1990). If we compare the inverted results with the model 1 of Christensen-Dalsgaard (1982), we can conclude that the sound speed in the real sun is a few per cent higher in the range of $0.3 \le r/R \le 0.7$ and it seems several per cent higher in the deep interior in $r/R < 0.2$ than the model. The inverted results of the sound speed $c(r)$ show clearly a kink near $r/R \approx 0.70$, which is a manifestation of the sharp change of the temperature gradient at the base of the convective envelope.

The other physical quantities are less sensitive than the sound speed to the p-mode frequency spectra. Some attempts have been made to extract information of the density structure from the p-mode frequency spectra by inversion. The results obtained by Gough and Kosovichev (1990) and Kosovichev (1990) show that the density in $r/R < 0.2$ seems about ten per cent smaller than the standard solar models. This may indicate some mixing had occurred in the solar core.

As stated previously, helioseismological inversion is the only method of inferring the solar internal rotation. The first attempt to invert the frequency-splitting data was made by Duvall *et al.* (1984). At that time, the observational data were limited to the frequency splittings of sectoral modes ($m = \pm l$). Since those modes have large amplitudes in the equatorial zone, the inverted result indicated the rotational profile near the equatorial plane. Resolving individual modes differing in m with the same value of (n, l) is important in order to investigate the solar internal rotation as a function of both the depth and the latitude. Fully two-dimensional observations over the solar disk made it possible to resolve individual modes and to infer the solar internal rotation as a function of both the radius and the co-latitude. As described in Section 2, the observational errors of the rotational frequency splittings are smaller for high $|m|$ modes. Hence the internal rotation rate inferred by inversion is still more reliable in lower latitudes than in higher latitudes.

Observational data obtained at different phases of the solar activity cycle indicate possible long-term variation of rotation and of the envelope structure of the sun. These items are discussed by Libbrecht and Woodard (1990), Kuhn (1990), and Goode *et al.* (1990) in these proceedings.

6 Excitation Mechanism of Solar Oscillations

Though helioseismology is very successful in diagnosing the solar internal structure by using very accurate frequencies both in observation and theory, the study of the excitation mechanism of solar oscillations is still in an undeveloped stage. In fact, though we often have to discuss the physics of the sun with the same accuracy of the observed eigenfrequencies which is as accurate as 10^{-5}, we are struggling to explain theoretically the observed amplitudes of oscillations and the lifetimes of modes even to within an order of magnitude.

There exist two different models for the excitation mechanism of solar oscillations. One of the possibilities is the linear overstability of the eigenmodes, which is thought to be responsible for the pulsation of Cepheid variables (Ando and Osaki 1975). However, whether or not p modes are found to be overstable depends on the treatment of convection, and it is still quite difficult to estimate the effects of turbulent convection on the thermal stability of p modes. If we assume the p modes are overstable, the amplitudes of these modes will grow until some nonlinear effects suppress them. However, the observed amplitudes of p modes seem to be too small to induce such nonlinear effects. Nonradial f modes should be stable, because the κ mechanism does not operate for these modes due to their solenoidal nature. On the other hand, observations show that the amplitudes of f modes with high degree l are comparable with those of p modes with the same range of l. These facts cast a serious suspicion on the justification of the hypothesis of self-exciting overstability of oscillations.

The alternatively proposed excitation mechanism is the stochastic excitation of eigenmode oscillations by turbulent convection (Goldreich and Keeley 1977a, b). This mechanism is essentially similar to the phenomenon in which the oscillations of a system are excited at resonant frequencies and maintained at some level by the Brownian motion of individual particles that compose the system. In this model, the eigenmodes are supposed to be thermally stable, and the amplitudes of oscillations are determined by the balance of the excitation rate due to turbulent convection to the damping rate. Theoretical problems in this model are then how to estimate the excitation rate by turbulence and how to determine the damping rate. The stochastic excitation model is reviewed by Osaki (1990) in these proceedings.

Observations provide us with the dependence of the oscillation amplitudes and of the line widths of peaks in the power spectrum on ω and l. These quantities should be directly related with the excitation mechanism of oscillations, and we should carefully compare these data and theoretical models. Observations show that these quantities reveal long-term variation related with the solar activity as well as eigenfrequencies. These observational quantities yield us some information about the interior of the sun, and should be used as helioseismic data.

Whatever the excitation mechanism is, the treatment of convection is one of the main problems in examining the mechanism. Recent supercomputers make it possible to simulate realistic, three-dimensional, compressible convection. Stein and Nordlund (1990) report their simulation results in these proceedings.

7 Seismology of Stars in General

The success of helioseismology encourages seismological study of stars in general. Nevertheless, since the amplitudes of oscillations in sun-like stars are expected to be very small, observation is a challenging task. Observable modes in stars in general are limited to those with $l < 4$ in the case of observations of the luminosity variation and of the Doppler shift measurement of spectroscopic lines. In the case of high-order modes of low degree, the asymptotic formula given by Eq. (18) indicates that the frequency spectrum of p modes is equally spaced with a spacing of ν_0. Observers are trying to find such equally spaced frequency spectra of oscillations. To do so, Gelly *et al.* (1986) made a Fourier spectrum of the frequency spectrum of the oscillation data of α Cen itself, and found equi-distant peaks in the frequency spectrum. This technique is widely used now. So far marginally positive detection of sun-like oscillations of α Cen, α CMi, and ϵ Eri have been reported (Gelly *et al.* 1986, Noyes *et al.* 1984). They were detected by means of the Doppler velocity measurement or the spectroscopic line intensity measurement. Belmonte *et al.* (1990) report the discovery of acoustic oscillations in a sun-like star by means of high-speed photometry in this conference.

According to Eq. (18), the frequency difference between the (n,l) mode and the $(n-1,l+1)$ mode is given by $(4l+6)A\nu_0^2/\nu_{nl}$. This quantity A is sensitive to the central structure of the star, while the spacing ν_0 is strongly dependent on the mass and the radius of the star. Then the combination of the spacing of ν_0 and the minute spacing of frequencies of (n,l) modes and $(n-1,l+1)$ modes provide us information of the mass and the age of the star (Christensen-Dalsgaard 1984, Ulrich 1986, Gough 1987).

In addition to the purpose of diagnosing individual stars, through the investigation of the dependence of the amplitudes of oscillations on various stellar parameters, seismological study of stars in general may be useful for understanding the excitation mechanism of the sun-like oscillations.

So far, stars showing evidently multi-modes of oscillations are δ Scuti stars, rapidly oscillating Ap stars, and white dwarfs. These stars seem excellent candidates for seismically testing the basic assumptions and physics of stellar evolution theory. As for δ Scuti stars, one of the theoretical problems is the mode-selection mechanism in dense oscillation spectra. The pulsation and the chemical peculiarity had been thought to be exclusive of each other. The discovery of the rapidly oscillating Ap stars has raised a suspicion of this hypothesis. The investigation of oscillations of the rapidly oscillating

Ap stars and the δ Scuti stars will be helpful to understanding the physics of these stars. The rapidly oscillating Ap stars are the unique objects in the sense that the oscillations in these stars are influenced by strong magnetic fields. Seismic investigation of these stars is expected to provide us with some information about the magnetic fields of these stars. The oscillations in white dwarfs are influenced by degeneracy, convection, the past nuclear reactions, cooling rate, and diffusion of elements. None of these physical processes have been well established. The oscillations of these stars will be helpful to understand these basic processes. Seismology of the rapidly oscillating Ap stars and δ Scuti stars are reviewed by Kurtz (1990a) and Dziembowski (1990), respectively, in these proceedings.

Observations of variation in profiles of spectroscopic lines are also useful for detecting oscillations and identifying modes, and high-degree modes as high as $l \approx 10$ are detectable if stars are rapidly rotating so that the line shapes are wide enough. This method has been applied to early-type stars. Studies of these stars have not yet reached the level of seismology. However, the investigations of long-term variation in oscillations and its relation to stellar activity such as the Be-phenomena will soon become another aspect of asteroseismology.

References

Ando, H., and Osaki, Y. 1975, *Publ. Astron. Soc. Japan*, **27**, 581.

Balmforth, N., and Gough, D.O. 1990, submitted to *Astrophys. J.*

Belmonte, J.A., Pérez Hernández, F., and Roca Cortés, T. 1990, in these proceedings.

Brown, T.M., Mihalas, B.W., and Rhodes, E.J., Jr. 1986, in *Physics of the Sun, Vol. I.*, eds. P.A. Sturrock, T.E. Holzer, D.M. Mihalas, and R.K. Ulrich (Reidel, Dordrecht), p.177.

Christensen-Dalsgaard, J. 1982, *Monthly Notices Roy. Astron. Soc.*, **199**, 735.

Christensen-Dalsgaard, J. 1984, in *Theoretical Problems in Stellar Stability and Oscillations*, eds. A. Noels and M. Gabriel (Université de Liège, Liège), p.155.

Christensen-Dalsgaard, J., Duvall, T.L., Jr., Gough, D.O., Harvey, J.W., and Rhodes, E.J., Jr. 1985, *Nature*, **315**, 378.

Christensen-Dalsgaard, J., Gough, D.O., and Thompson, M.J. 1988, in *Seismology of the Sun and Sun-like Stars*, ed. E. Rolfe (ESA SP-286, Noordwijk), p.493.

Däppen, W. 1990, in these proceedings.

Deubner, F.-L., and Gough, D.O. 1984, *Ann. Rev. Astron. Astrophys.*, **22**, 593.

Duvall, T.L., Jr., Dziembowski, W.A., Goode, P.R., Gough, D.O., Harvey, J.W., and Leibacher, J.W. 1984, *Nature*, **310**, 22.

Dziembowski, W. 1990, in these proceedings.

Dziembowski, W.A., Pamyatnykh, A.A., and Sienkiewicz, R. 1990, preprint.

Gabriel, M. 1990, in these proceedings.

Gelly, B., Fossat, E., Grec, G., and Pomerantz, M. 1988, in *Advances in Helio- and Asteroseismology, IAU Symp. No. 123*, eds. J. Christensen-Dalsgaard and S. Frandsen (Reidel, Dordrecht), p.21.

Gelly, B., Grec, G., and Fossat, E. 1986, *Astron. Astrophys.*, **164**, 383.

Goldreich, P., and Keeley, D.A. 1977a, *Astrophys. J.*, **211**, 934.

Goldreich, P., and Keeley, D.A. 1977b, *Astrophys. J.*, **212**, 243.

Goode, P.R., Dziembowski, W.A., Rhodes, E.J., Jr., and Korzennik, S. 1990, in these proceedings.

Gough, D.O. 1986, in *Seismology of the Sun and the Distant Stars*, ed. D.O. Gough (Reidel, Dordrecht), p.125.

Gough, D.O. 1987, in *Nature*, **326**, 257.
Gough, D.O., and Kosovichev, A.G. 1988, in *Seismology of the Sun and Sun-like Stars*, ed. E. Rolfe (ESA SP-286, Noordwijk), p.195.
Gough, D.O., and Kosovichev, A.G. 1990, in *Inside the Sun*, eds. G. Berthomieu and M. Cribier (Kluwer, Dordrecht), in press.
Henning, H.M., and Sherrer, P.H. 1988, in *Advances in Helio- and Asteroseismology, IAU Symp. No. 123*, eds. J. Christensen-Dalsgaard and S. Frandsen (Reidel, Dordrecht), p.29.
Isaak, G.R., Jefferies, S.M., McLeod, C.P., New, R., van der Raay, H.B., Pallé, P.L., Régulo, C., and Roca Cortés, T, 1988, in *Advances in Helio- and Asteroseismology, IAU Symp. No. 123*, eds. J. Christensen-Dalsgaard and S. Frandsen (Reidel, Dordrecht), p.201.
Jefferies, S.M., Duvall, T.L., Jr., Harvey, J.W., and Pomerantz, M.A. 1990, in these proceedings.
Jefferies, S.M., Pomerantz, M.A., Duvall, T.L., Jr., Harvey, J.W., and Jaksha, D.B. 1988, in *Seismology of the Sun and Sun-like Stars*, ed. E. Rolfe (ESA SP-286, Noordwijk), p.279.
Jiménez, A., Pallé, P.L., Pérez, J.C., Régulo, C., Roca Cortés, T., Isaak, G.R., McLeod, C.P., and van der Raay, H.B. 1988, in *Advances in Helio- and Asteroseismology, IAU Symp. No. 123*, eds. J. Christensen-Dalsgaard and S. Frandsen (Reidel, Dordrecht), p.205.
Kosovichev, A.G. 1988, in *Seismology of the Sun and Sun-like Stars*, ed. E. Rolfe (ESA SP-286, Noordwijk), p.533.
Kuhn, J.R. 1990, in these proceedings.
Kurtz, D.W. 1990a, in these proceedings.
Kurtz, D.W. 1990b, *Ann. Rev. Astron. Astrophys.*, **28**, in press.
Libbrecht, K.G., and Woodard, M.F. 1990, in these proceedings.
Noyes, R.W., Baliunas, S.L., Belserene, E., Duncan, D.K., Horne, J., and Widrow, L. 1984, *Astrophys. J. Letters*, **285**, L23.
Osaki, Y. 1990, in these proceedings.
Pallé, P.L., Régulo, C., and Roca Cortés, T. 1990, in these proceedings.
Shibahashi, H., and Saio, H. 1985, *Publ. Astron. Soc. Japan*, **37**, 245.
Shibahashi, H., and Sekii, T. 1988, in *Seismology of the Sun and Sun-like Stars*, ed. E. Rolfe (ESA SP-286, Noordwijk), p.471.
Smeyers, P., Briers, R., Tassoul, M., Degryse, K., Polfliet, R., and Van Hoolst, T. 1988, in *Seismology of the Sun and Sun-like Stars*, ed. E. Rolfe (ESA SP-286, Noordwijk), p.623.
Smeyers, P. and Ruymaekers, E. 1990, in these proceedings.
Stein, R.F., and Nordlund, Å. 1990, in these proceedings.
Tassoul, M. 1980, *Astrophys. J. Suppl.*, **43**, 469.
Ulrich, R.K. 1986, *Astrophys. J. Letters*, **306**, L37.
Unno, W., Osaki, Y., Ando, H., Saio, H., and Shibahashi, H. 1989 *Nonradial Oscillations of Stars (2nd Edition)* (University of Tokyo Press, Tokyo).
Vandakurov, Y.V. 1967, *Astron. Zh.*, **44**, 786.
Vauclair, S., and Dolez, D. 1990, in these proceedings.
Vorontsov, S.V. 1988, in *Seismology of the Sun and Sun-like Stars*, ed. E. Rolfe (ESA SP-286, Noordwijk), p.475.
Vorontsov, S.V. 1990, in these proceedings.
Vorontsov, S.V., and Shibahashi, H. 1990, in these proceedings.
Vorontsov, S.V., and Zharkov, V.N. 1989, *Sov. Sci. Rev. E. Astrophys. Space Phys. Rev.*, **7**, 1.
Woodard, M.F., and Noyes, R.W. 1985, *Nature*, **318**, 449.
Woodard, M.F., and Noyes, R.W. 1988, in *Advances in Helio- and Asteroseismology, IAU Symp. No. 123*, eds. J. Christensen-Dalsgaard and S. Frandsen (Reidel, Dordrecht), p.197.

II

Physical Processes in Modeling the Sun

Solar Equilibrium Models and Physical Processes Governing the Solar Internal Structure

M. Gabriel

Institut d'Astrophysique de l'Université de Liège,
5, avenue de Cointe, B-4200 LIEGE, Belgium

Abstract: In the first part of the paper, the uncertainties on the fundamental data and on the input physics relevant for the computation of the internal structure of the Sun are reviewed. In the second part, I present several tests done in order to compute a solar model with a relative accuracy of 3×10^{-5}. I also propose a modification of the difference method which allows to achieve the required accuracy with a relatively low number of mesh points.

1 Introduction

Solar seismology has produced a need for theoretical models able to predict oscillation frequencies over a large domain with the same accuracy as the observations (of the order of 3×10^{-5} for 5 min. p-modes).

This raises many numerical problems but even if they can be overcome, it remains that the basic astronomical data and the physics are not known with such a high precision. Worst, parts of the physics are still poorly understood.

As a result, our objective can only be to produce a solar model which has a high mathematical coherency. It will be through the resolution of the inverse problem that we will know how it diverges from the real sun. Then, it will be possible to search for physical processes suitable to solve the discrepancies.

Therefore, we will discuss only the problems arising in the computation of an accurate standard solar model. We assume that the initial sun was chemically homogeneous and that its mass remained constant (no accretion, no mass loss). We also suppose that matter is transported only by convection and we neglect both turbulent diffusion (Schatzman *et al.* 1981) and gravitational settling (Gabriel *et al.* 1984, Cox *et al.* 1989, Bahcall and Loeb 1990). We also neglect magnetic field and rotation (see Pinsonneault *et al.* 1989). Of course we ignore WIMPS and possible variations of the gravitational constant.

We first review the main uncertainties on the physics. Then we discuss the mathematical difficulties encountered in the computation of simplified solar models.

2 Uncertainties in the Physics of the Standard Solar Model

2.1 Basic Astronomical Data

$GM_\odot = 1.32712438 \times 10^{26}$ c.g.s. is well known. Cohen and Taylor (1989) give $G = (6.67259 \pm 0.00085) \times 10^{-8}$ c.g.s. This implies that the uncertainty on $M_\odot$ is of 8×10^{-4}.

$R_\odot = 6.9599 \times 10^{10}$ cm corresponds to a standard semi-diameter of 959.63″ (Parkinson 1983). However, Wittman (1977) and Laclare (1983) give respectively (960.00 ±0.09)″ and (959.37 ± 0.02)″. This means that $R_\odot$ is known with an accuracy of 2×10^{-4}.

For the solar constant S, I took the average of the observations since 1983 given by Crommelynck (1989a). I obtain $S \simeq 1365.25\,\mathrm{W\,m^{-2}}$ or $L_\odot = 3.8395 \times 10^{33}$ c.g.s. The comparison of this value with that obtained by Wilson and Hudson (1988) from ACRIM ($S = 1367.72\,\mathrm{W\,m^{-2}}$, $L_\odot = 3.846 \times 10^{33}$ c.g.s.) is compatible with the error of 1.5 to 2×10^{-3} on the solar constant given by Crommelynck (1989b).

The solar age is close to that of the oldest meteorites (4.53 Gyr). Values of 4.6 and 4.7 Gyr are most often used. However Guenther (1989) has recently given arguments for an age of 4.49 ± 0.04 Gyr. He also showed that the two extreme values lead to solar models with initial hydrogen abundance different by 2×10^{-3} which is not negligible.

Given the uncertainty on the solar age we may start its evolution on the ZAMS rather than during the gravitational contraction.

Anders and Grevesse (1989) have recently reviewed the solar and meteoritic abundances. There remains a large discrepancy between the solar (log Fe =7.67) and meteoritic (log Fe = 7.51) abundances of iron. This leads to Z/X values respectively of 0.02746 and 0.02668 ± 8.5 %. The error on Ne (which contributes to 10 % of Z) is close to 125 % and Demarque (1989) has shown that it affects significantly the p modes frequencies. The error on C, N and O is still of the order of 10 %. They also pointed out that elements with high first ionization potentials (among which C, N, O and Ne) are depleted in the solar wind relative to the photosphere. If these observations are confirmed, the observed abundances might not be representative of these in the radiative core.

2.2 The Input Physics

The MHD equation of state (Hummer and Mihalas 1988, Mihalas *et al.* 1988, Däppen *et al.* 1988) is the best presently available. It also provides analytic expressions for the first and second order derivatives of the free energy and therefore allows an accurate computation of all the required thermodynamic quantities. MHD is now available for $Z = 0.02$. However the C and O conversion into N by the CNO cycle increases Z by 3 % (Bahcall and Ulrich 1988). This introduces an error of about 5×10^{-4} on the pressure in the central regions which can be easily taken into account.

The only source of accurate opacities is the Los Alamos Opacity Library (Huebner *et al.* 1977). According to Huebner (1986) its accuracy is better than 30 %.

The uncertainties on the chemical composition of the Sun have non-linear effects on opacities. Turck-Chièze *et al.* (1988) find differences which may reach 8 % between opacities for 3 different compositions. Bahcall and Ulrich (1988) have also pointed out that the C and O conversion in N increases the opacity by up to 7.6 % for $r < 0.2R_\odot$.

The problem with the opacity tables is to know how to interpolate. Christensen-Dalsgaard (1982) found differences of up to 3 % on ρ, T and p between two solar models obtained with spline and linear interpolations. Morel *et al.* (1989) find differences of up to 1 % on T and 5 % on ρ and p. They also point out that because of the course mesh of the tables, it is impossible to tell which interpolation formula is the best. A good solution is perhaps to use an analytic formula with coefficients obtained by fitting with the tables (see Bahcall and Ulrich 1988, Cox *et al.* 1989).

Nuclear reactions are discussed in Parker (1986), Bahcall and Ulrich (1988) and Turck-Chièze *et al.* (1988). The uncertainty is of 2 % for the p-p and He^3–He^3 reactions, of 4 % on He^3–He^4 and of 7 % on Be^7–p. This means that in the Sun the energy generation rate is known within 3 %.

Though the mixing length theory is very elementary, it is generally considered that convection is not a critical problem for the Sun because the temperature gradient departs from the adiabatic value for $T \leq 2 \times 10^4$ K only and because the mixing length is calibrated on the solar model. For this second reason, the choice of the $T(T_{\rm eff}, \tau)$ law is also considered as noncritical. However, the outer layers contribute significantly to $\int dr/c$. If for convection there is little we can do, we should at least use a good $T(\tau)$ law and a good pressure at the outermost point.

Another problem related to convection is the amount of overshooting at the bottom of the convective envelope. Several theories have been developed (Shaviv and Salpeter 1973, Ulrich 1976, Xiong 1980, Langer 1986) but they all contain at least one free parameter. The study of overshooting by Schmitt *et al.* (1984), Unno *et al.* (1985) and Pidatella and Stix (1986) suggests that it is moderate (a few tenths of a pressure scale height).

3 Numerical Problems

Nearly all stellar evolution codes use the 30 years old Henyey method. It has been described in details by Henyey and Levee (1965), Kippenhahn *et al.* (1967) and also by Larson and Demarque (1964), Bodenheimer *et al.* (1965), Iben (1965), Paczynski (1969), and Eggleton (1971).

The Henyey method is a second order one. More recently fourth order codes have been developed by Smith and Fuchs (1976), Budge (1987) and Morel (1988). Morel is an active member of the GONG solar model team and, if he is cautious, his models should be the best. The problem could be to see how close they can be approach with second order codes.

Models used for the discussion of this section have been computed with the simplified physics defined for the level 1 GONG models (Christensen-Dalsgaard 1988).

3.1 Difference Equations

The difference equations are usually written in one of the following forms

$$\frac{y_{j+1} - y_j}{x_{j+1} - x_j} = f_{j+1/2} + \frac{1}{24} f''_{j+1/2}(x_{j+1} - x_j)^2 \tag{1}$$

$$\frac{y_{j+1} - y_j}{x_{j+1} - x_j} = \frac{1}{2}(f_j + f_{j+1}) - \frac{1}{12} f''_{j+1/2}(x_{j+1} - x_j)^2 \tag{2}$$

In our models with the largest number of points, the relative difference between $f_{j+1/2}$ and $1/2(f_{j+1} + f_j)$ is of the order of 10^{-4}. Therefore, we may expect differences of a few times 10^{-4} between models computed with the Henyey method.

We have computed ZAMS models using Eq. (1) (type I) and Eq. (2) (type II), without and with Taylor expansion (subscripts 1 and 2) and with a criterion on $\Delta L/L$ or $\Delta L/L(R)$ for the mesh spacing (subscripts a and b) and we have compared them with a model (SEARS) computed with the shooting method. The global parameters of some models are given in table I as well as these of Morel (1989), Christensen-Dalsgaard (1989) (JCD), and Berthomieu and Provost (1989) (B-P). As for the definition of β, see Christensen-Dalsgaard (1988). Figures 1 and 2 show the comparison of models I_{1a} and II_{1a} with SEARS. They show that errors are about 4 times smaller for model II_{1a} than for I_{1a}.

Table 1. Global parameters of ZAMS models — $X = 0.733$, $\beta = 7.22$.

Model	$R(10^{10})$	$L(10^{33})$	$T_c(10^6)$	ρ_c	$p_c(10^{17})$
Morel	6.152450	2.808212	12.97423	83.56032	1.488211
JCD	6.152486	2.808254			
B-P	6.152519	2.808035	12.97438	83.55824	1.488192
SEARS	6.152420	2.808205	12.97402	83.56253	1.488227
I_{2b}	6.152120	2.807774	12.97358	83.55946	1.488120
II_{2b}	6.152310	2.808053	12.97395	83.55984	1.488168
II_{2bc}	6.152457	2.808248	12.97422	83.56056	1.488213

3.2 Choice of the Variables and Strategy near the Center

For most choices of variables, it is necessary to start the integration from the center with a Taylor expansion. For two choices this is however not necessary. As far as the behaviour near the center is concerned, the best variables are r^2, $m^{2/3}$, $L^{2/3}$, p, ρ, T. Then the solution of the difference equations is equivalent to a two terms expansion. Henyey's variables also allow to start the integration right from the center, however only the first term of the expansion is correct.

To test the influence of the strategy near the center, we have computed models with a Taylor expansion up to $m/M = 1.25 \times 10^{-4}$ and $r/R \simeq 10^{-2}$. The comparison of models II_{1a} and II_{2a} shows that the Taylor expansion improves the solution close to the center (mainly on L). However farther out p and ρ are marginally worst by 1 to 1.6×10^{-5}.

3.3 Strategy for the Outer Layers

Interpolation in a triangle or a square of envelopes produces discontinuities at the fitting point with the interior. For this reason some have preferred to use the Henyey method up to the surface. We have chosen to keep the classical procedure. It allows the use of a 4th order method in the outer layers where ionization takes place and also to reduce the size of all vectors. We find that the discontinuities at the fitting point are all smaller than 10^{-5} when the mass fraction in the envelope is 10^{-2} and when L and $T_{\rm eff}^4$ vary by 0.25 % over the sides of the square.

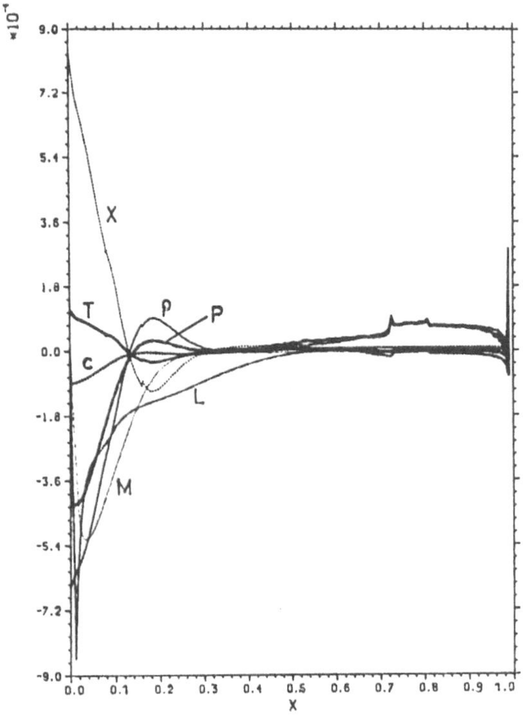

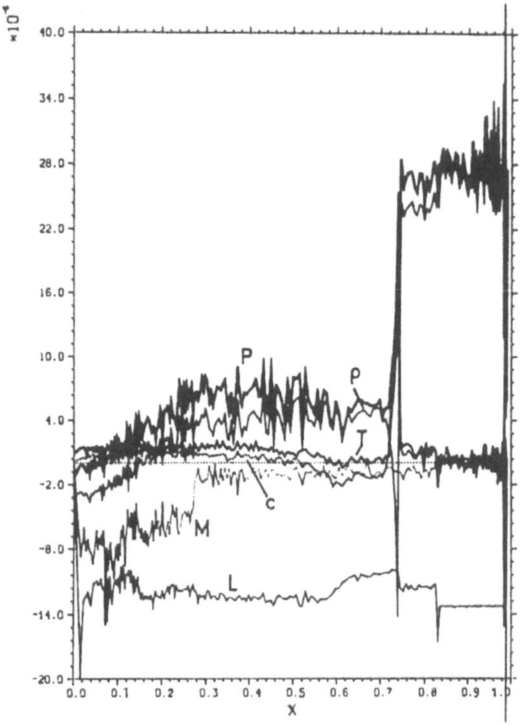

Fig. 1 (left) and 2 (right). Relative differences between SEARS and ZAMS models of type I_{1a} and II_{2a} respectively. The letter near each curve shows the variable it refers to (c is the sound speed).

3.4 Definition of the Grid Points

Eggleton (1971) (see also Press *et al.* 1986) has introduced a method of automatic mesh points allocation. It has the advantage of having a constant mesh spacing but it makes it difficult if not impossible the positioning of a mesh point just at the bottom of the convective envelope.

This is easily done with the classical method used in our code. It requires that the variations $\Delta y = |y_{j+1} - y_j|$ fulfill the following two conditions

$$\max\left[\frac{1}{AF(1)}\frac{\Delta T}{T}, \frac{1}{AF(2)}\frac{\Delta\rho^{1/3}}{\rho^{1/3}}, \frac{1}{AF(3)}\frac{\Delta p^{1/4}}{p^{1/4}}, \frac{1}{AF(4)}\frac{\Delta L}{L_t}, \frac{\Delta X}{AF(5)}\right] < 1 \qquad (3)$$

$$\min\left[\frac{2.1}{AF(1)}\frac{\Delta T}{T},\frac{2.1}{AF(2)}\frac{\Delta y^{1/3}}{\rho^{1/3}},\frac{2.1}{AF(3)}\frac{\Delta p^{1/4}}{p^{1/4}},\frac{2.1}{AF(4)}\frac{\Delta L}{L_t},\frac{2.1}{AF(5)}\Delta X\right]\leq 1 \quad (4)$$

where X is our independent variable and $AF(1) = AF(2) = AF(4) = 0.03$, $AF(3) = 0.0075$, $AF(5) = 0.006$.

L_t is either the local luminosity or the surface value. We have computed models for these two choices of L_t. They have respectively 600 and 450 points in $0 \leq m/M \leq 0.99$. The comparison of these two type II models shows that model II_{2a} improves over II_{2b} by less than 2.2×10^{-5} on ρ and p and by less than 10^{-5} on T.

3.5 Gravitational Energy Release

In all codes this term is computed to the first order. This leads to an error of 0.5 to 1 % for time steps of 2.5×10^{8} yr. However during the main sequence evolution of the Sun, the gravitational energy release amounts only to 5×10^{-4} of the luminosity. The error on L is therefore smaller than 10^{-5} and we may keep the first order approximation.

3.6 Last Improvement

We have introduced the truncation error term of Eq. (2) in the code. This gives a model (II_{2bc} in table I) which is in very good agreement with Morel's (1989) (see Fig. 3) in the radiative zone and shows somewhat larger differences on ρ and p in the convection zone.

3.7 Chemical Composition and Energy Generation

In the Sun, energy is produced by the p-p chain and the CNO cycle. The later contributes to 1.75 % of the luminosity and to 8 % of the energy generation ε at the center. In these reactions He^3, C^{12}, N^{14} and O^{16} reach an equilibrium abundance within a time which is not everywhere short compared to the solar age. When He^3 is maximum it is no longer at equilibrium. However 12.5 % of the luminosity is still produced farther out in a ZAMS model ($t = 4.93 \times 10^7$ yr) and 3.6 % in the Sun. This clearly shows that we may not assume that He^3 is at equilibrium to compute the energy generation rate. It is easy to see when C^{12} is at equilibrium for then its abundance increases with temperature. From this it can be seen that equilibrium is reached nowhere in the ZAMS model and only for $m/M < 0.12$ in the Sun. At this point the CNO cycle contributes by 0.1 % to ε. As expected O^{16} is hardly depleted in the sun. For the CNO cycle it would be safer to follow the detail of the reactions (see also section 2.2).

To compute the abundances with an accuracy of 10^{-5} we must use an accurate and asymptotically stable difference scheme. These used in evolution codes are not appropriate for they are stable but of the first order (Bodenheimer *et al.* 1965), of the second order but unstable (Kippenhahn *et al.* 1967) or of the first order and unstable (Arnett and Truran 1969). We propose to use a stable higher order scheme obtained from the Taylor expansion of the abundance X around the value in the model being computed:

$$X_j = X - \left(\frac{dX}{dt}\right)(t - t_j) + \sum_{i=1}^{I} C_i(t - t_j)^{i+1} \quad . \quad (5)$$

The C_i are computed from the $(I + 1)$ last computed models and we obtain

$$X = \left[X_n - \sum_{i=1}^{I} CORX_i (X_{n-i} - X_{n+1-i}) \right] + \frac{dX}{dt} (1 - CORXP)(t - t_n) \ . \tag{6}$$

CORXP and $CORX_i$ are functions of the time at the $(I + 2)$ last models only and they are the same at all mesh points.

We have used that scheme to compute a calibrated "solar" model with the same parameters as for model II_{2bc}. The global properties of several models are given in table II. Figure 4 shows the comparison with Morel's model. The differences are quite large close to the center. They probably result from the discrepancy on $X_c = 3 \times 10^{-4}$.The comparison with the 2400 points model by Christensen-Dalsgaard (1989) shows a better agreement. The maximum difference are of the order of 10^{-5} on T and the sound speed, of 2×10^{-5} on X but still of the order of 10^{-4} on ρ and p. The causes of these discrepancies are still to be understood.

Table 2. Global parameters of calibrated models.

Model	$R(10^{10})$	$L(10^{33})$	X_S	β	$T_c(10^6)$	ρ_c	X_c
Morel	6.959899	3.845999	0.7314562	6.429864	14.94495	165.9609	0.368746
JCD	6.959891	3.845963	0.7314525	6.432985			
B-P	6.959923	3.845967	0.7320910	6.448275	14.95746	166.4970	0.368601
present work	6.959897	3.845983	0.731449	6.42903	14.94627	166.0696	0.368437

4 Conclusions

Given the uncertainties in the physics our objective can only be to compute a solar model with a high mathematical coherency. For this reason it is essential that codes be compared to the GONG level 1. If this is not done it will be impossible to disentangle the mathematical and the physical causes of differences between models.

Though the choice of the difference scheme influences the accuracy of the model, with the Henyey method it is necessary to have at least 2400 points to reach the required accuracy. For this reason it is perhaps better to develop higher order codes.

Finally, special attention should be given to the computation of the chemical composition. It requires a high order asymptotically stable numerical scheme.

Acknowledgments

I wish to thank Drs. G. Berthomieu, P. Morel, and J. Provost for helpful discussions and for making their models available to me.

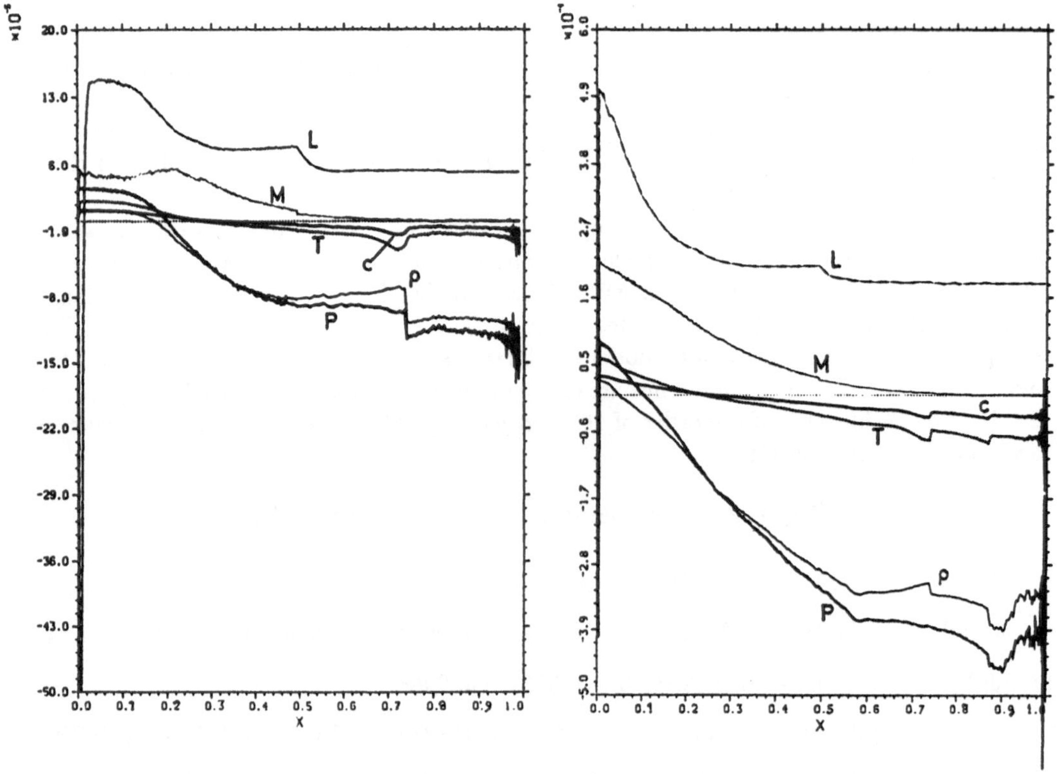

Fig. 3 (left) and 4 (right). Relative differences between Morel's (1989) models and ours. They compare the ZAMS and the calibrated "solar models" respectively.

References

Anders, E., and Grevesse, N. 1989, *Geochemica and Cosmochimica Acta*, **53**, 197.

Arnett, W.D., and Truran, J.W. 1969, *Astrophys. J.*, **157**, 339.

Bahcall, J.N., and Loeb, A. 1990, *Astrophys. J.*, in press.

Bahcall, J.N., and Ulrich, R.K. 1988, *Rev. Mod. Phys.*, **60**, 297.

Berthomieu, G., and Provost, J. 1989, private communication.

Bodenheimer, P., Forbes, J.E., Gould, N.L., and Henyey, L.G. 1965, *Astrophys. J.*, **141**, 1019.

Budge, K.G. 1987, *Astrophys. J.*, **312**, 217.

Christensen-Dalsgaard, J. 1982, *Monthly Notices Roy. Astron. Soc.*, **199**, 745.

Christensen-Dalsgaard, J. 1988, Computational procedures for GONG solar model project.

Christensen-Dalsgaard, J. 1989, private communication.

Cohen, E.R., and Taylor, B.N. 1989, *Physics Today*, **42**, No. 8, BG8.
Cox, A.N., Guzik, J.A., and Kidman, R.B. 1989, *Astrophys. J.*, **342**, 1189.
Crommelynck, D. 1989a, *Adv. Space Res.*, **9**, No. 7, 51.
Crommelynck, D. 1989b, (private communication).
Däppen, W., Mihalas, D., Hummer, D.G., and Mihalas, B.W. 1988, *Astrophys. J.*, **332**, 261.
Demarque, P. 1989, Communication to the 1989 Annual GONG meeting.
Eggleton, P.P. 1971, *Monthly Notices Roy. Astron. Soc.*, **151**, 356.
Gabriel, M., Noels, A., and Scuflaire, R. 1984, *Mem. Soc. Astron. Italiana*, **55**, 169.
Guenther, D.B. 1989, *Astrophys. J.*, **339**, 1156.
Henyey, L.G. and Levee, R.D. 1965, in *Methods in Computational Physics, Vol. 4*, eds. B. Alder, S. Ferhbach, and M. Rotenberg (Academic Press, New York), p.333.
Huebner, W.F. 1986, in *Physics of the Sun, Vol. 1*, ed. P.A. Sturrock (Reidel, Dordrecht), p.33.
Huebner, W.F., Mertz, A.L., Magee, N.H., and Argo, M.F. 1977, Astrophysical Opacity Library, Los Alamos Sci. Lab. Rept. LA-6760-M.
Hummer, D.G. and Mihalas, D. 1988, *Astrophys. J.*, **331**, 794.
Iben, I. 1965, *Astrophys. J.*, **141**, 993.
Kippenhahn, R., Weigert, A., and Hofmeister, E. 1967, in *Methods in Computational Physics, Vol. 7*, eds. B. Alder, S. Ferhbach, and M. Rotenberg (Academic Press, New York), p.129.
Laclare, F. 1983, *Astron. Astrophys.*, **125**, 200.
Langer, N. 1986, *Astron. Astrophys.*, **164**, 45.
Larson, R.B., and Demarque, P. 1964, *Astrophys. J.*, **140**, 524.
Mihalas, D., Däppen, W., and Hummer, D.G. 1988, *Astrophys. J.*, **331**, 815.
Morel, P. 1988, *Code de structure interne*, p. 56.
Morel, P. 1989, private communication.
Morel, P., Provost, J. and Berthomieu, G. 1989, *Solar Phys.*, in press.
Paczynski, B. 1969, *Acta Astron.*, **19**, 1.
Parker, P.D. 1986, in *Physics of the Sun, Vol. 1*, ed. P.A. Sturrock (Reidel, Dordrecht), p.15.
Parkinson, J.H. 1983, *Nature*, **304**, 518.
Pidatella, R.M., and Stix, M. 1986, *Astron. Astrophys.*, **157**, 338.
Pinsonneault, M.H., Kawaler, S.D., Sofia, S., and Demarque, P. 1989, *Astrophys. J.*, **338**, 424.
Press, W.H., Flannery, B.P., Teukolsky, S.A., and Vetterling, W.T. 1986, *Numerical Receipes* (Cambridge Univ. Press, Cambridge), p.608.
Schatzman, E., Maeder, A., Angrand, F., and Glowinski, R. 1981, *Astron. Astrophys.*, **96**, 1.
Schmitt, J.H., Rosner, R., and Bohm, H.U. 1984, *Astrophys. J.*, **282**, 316.
Shaviv, G., and Salpeter, E.E. 1973, *Astrophys. J.*, **184**, 191.
Smith, R.L., and Fuchs, H. 1976, *Astrophys. Space Sci.*, **50**, 63.
Turck-Chièze, S., Cahen, S., Cassé, M., and Doom, C. 1988, *Astrophys. J.*, **335**, 415.
Ulrich, R.K. 1976, *Astrophys. J.*, **207**, 564.
Unno, W., Kondo, M., and Xiong, D.R. 1985, *Publ. Astron. Soc. Japan*, **37**, 235.
Wilson, R.C., and Hudson, H.S. 1988, *Nature*, **332**, 210.
Wittman, A. 1977, *Astron. Astrophys.*, **61**, 225.
Xiong, D.R. 1980, *Chinese Astron.*, **4**, 234.

Progress Towards a Unified Equation of State

Werner Däppen

Space Science Department of ESA, ESTEC,
2200AG Noordwijk, The Netherlands
and
Institute for Theoretical Physics, University of California,
Santa Barbara, CA 93106, U.S.A.

Abstract: A recent comparison of thermodynamical quantities, computed in the chemical and physical picture, has revealed a remarkable agreement in the hydrogen and helium ionization zones (on an isochore of $\log \rho = -5.5$, with ρ in $\mathrm{g\,cm^{-3}}$). This agreement is due to an unexpectedly dominating (classical) Coulomb pressure term. The analogous comparison at a somewhat higher density ($\log \rho = -3.5$) shows still striking similarities, despite the different treatment of bound states in the two formalisms. The results suggest use of a relatively simple parametrized equation of state for solar purposes.

1 Introduction

Observed solar oscillation frequencies probe, among other things, the equation of state of the solar interior. Mihalas, Hummer, and Däppen (Hummer and Mihalas, 1988; Mihalas, Däppen, and Hummer, 1988; Däppen *et al.*, 1988) have recently developed a new treatment of the partition functions in the equation of state (hereinafter MHD), which is part of an ongoing effort to recompute opacities for stellar envelopes ("Opacity Project"; see e.g. Seaton, 1987). The MHD equation of state was used by Christensen-Dalsgaard, Däppen and Lebreton (1988) for solar models, where it significantly reduced the difference between observed and computed p-mode frequencies.

The MHD formalism is realized in the so-called chemical picture in which bound systems are interpreted as autonomous species, with reactions between each other. The equation of state developed at Livermore (Rogers, 1986; Iglesias, Rogers, and Wilson, 1987; hereinafter Livermore equation of state), is realized in the physical picture where explicitly only fundamental species (*i.e.* electrons and nuclei) are treated.

While a more systematic comparison of results from the two approaches is underway, Däppen, Lebreton, and Rogers (1990) have begun to examine some selected cases of immediate solar interest. The first case studied was that of the H and He ionization zones of the Sun, because of their importance for p modes (Christensen-Dalsgaard, Däppen, and Lebreton, 1988).

After a presentation of the MHD and Livermore equations of state, I will discuss the physical cause of the rather surprising agreement found in a previous, first comparison. Then I will show new results obtained at a higher density ($\log \rho = -3.5$).

2 MHD and Livermore Equations of State

2.1 The MHD Equation of State

In the chemical picture, perturbed atoms must be introduced on a more-or-less ***ad-hoc*** basis to avoid the familiar divergence of internal partition functions (see for instance Ebeling, Kraeft, and Kremp, 1976). In other words, the approximation of unperturbed atoms precludes the application of standard statistical mechanics, *i.e.* the attribution of a Boltzmann-factor to each atomic state. The conventional remedy of the chemical picture against this is a modification of the atomic states, e.g. by cutting off the highly excited states in function of density and temperature of the plasma. Such cut-offs, however, have in general dire consequences due to the discrete nature of the atomic spectrum, *i.e.* jumps in the number of excited states (and thus in the partition functions and in the free energy) despite smoothly varying external parameters (temperature and density).

The MHD equation of state avoids these discontinuities (in the free energy) by introducing 'soft' cut-offs in the form of occupational probabilities. These occupation probabilities have the same function as the 'hard' cut-offs, that is ***they indicate how we imagine that atomic states in a plasma are modified.*** Once this description of states is adopted, one applies statistical mechanics as usual, using Boltzmann factors.

The MHD equation of state is further characterized by detailed internal partition functions of a large number of atomic, ionic, and molecular species. Full thermodynamic consistency is assured by analytical expressions of the free energy and its first- and second-order derivatives. This not only allows an efficient Newton-Raphson minimization, but, in addition, the ensuing thermodynamic quantities are of analytical precision and can therefore be differentiated once more, this time numerically. Reliable third-order thermodynamic quantities are thus calculated.

In the occupation probabilities, perturbations by charged and neutral particles are taken into account. Correlations between the two effects are neglected (for lack of knowing how to describe them); thus the occupation probabilities due to charged and neutral perturbers are simply multiplied. The resulting weighted internal partition functions Z_s^{internal} are (for species s)

$$Z_s^{\text{internal}} = \sum_i w_{is} g_{is} \exp\left[-\frac{E_{is} - E_{1s}}{kT}\right] \quad . \tag{1}$$

The coefficients w_{is} take into account charged and neutral surrounding particles. In physical terms, w_{is} gives the fraction of all particles of species s that can exist in state i with an electron bound to the atom or ion, and $1 - w_{is}$ gives the fraction of those that are so heavily perturbed by nearby neighbours that the state is effectively destroyed. Perturbations by neutral particles are based on an excluded volume treatment and perturbations by charges are calculated from a fit to a quantum-mechanical Stark-ionization theory. Hummer and Mihalas's (1988) choice has been

$$\ln w_{is} = -\left(\frac{4\pi}{3V}\right)\left\{\sum_{\nu} N_\nu (r_{is} + r_{1\nu})^3 + 16\left[\frac{(Z_s+1)e^2}{\chi_{is} k_{is}^{1/2}}\right]^3 \sum_{\alpha \neq e} N_\alpha Z_\alpha^{3/2}\right\} . \qquad (2)$$

Here, the index ν runs over neutral particles, the index α runs over charged ions (except electrons), r_{is} is the radius assigned to a particle in state i of species s, χ_{is} is the (positive) binding energy of such a particle, k_{is} is a quantum-mechanical correction, and Z_s is the net charge of a particle of species s. Note that $\ln w_{is} \propto -n^6$ for large principal quantum numbers n (of state i), and hence provides a (density-dependent) cutoff for Z_{is}^{internal}. Finally, the MHD equation of state also includes a Debye-Hückel term for the Coulomb-pressure correction, partially degenerate electrons, and radiation pressure.

2.2 The Livermore Equation of State

It is clear from the preceding subsection that the advantage of the chemical picture lies in the possibility to model complicated plasmas, and to obtain numerically smooth and consistent thermodynamical quantities. Nevertheless, the heuristic method of the separation of the atomic-physics problem from that of statistical mechanics is not satisfactory, and attempts have been made to avoid the concept of a perturbed atom in a plasma altogether. Thus in the physical picture only fundamental particles (electrons and nuclei) enter. Since no chemical and ionization reactions have to be controlled (that would be very difficult in terms of the unwieldy chemical potential), there is nothing that prevents the use of the otherwise practical grand-canonical partition function, and to try to build a theory of partially ionized plasmas similar to well-know cluster expansions for real gases (Rogers, 1981; for an introduction into cluster expansions see Huang, 1963).

To explain the advantages of this approach for partially ionized plasmas, it is instructive to discuss the activity expansion for gaseous hydrogen. The interactions in this case are all short ranged and the pressure is determined from a self-consistent solution of the equations (Hill, 1960)

$$\frac{p}{kT} = z + z^2 b_2 + z^3 b_3 + \ldots \qquad (3)$$

$$\rho = \frac{z}{kT}\left(\frac{\partial p}{\partial z}\right) \qquad (4)$$

where $z = \lambda^{-3}\exp(\mu/kT)$ is the activity, $\lambda \equiv h/\sqrt{2\pi m_e kT}$ is the thermal (de Broglie) wavelength of electrons, μ is the chemical potential, and T is the temperature. The b_n are cluster coefficients such that b_2 includes all two particle states, b_3 includes all three particle states, etc. The second cluster coefficient for hydrogen includes the formation of H_2 molecules as well as scattering states in the ${}^1\Sigma_g$ potential. It also includes scattering states in the ${}^3\Sigma_u$ potential and all excited electronic state potentials. The third cluster coefficient includes H_3 bound states, $H - H_2$ and $H - H - H$ scattering states. Equation (3) demonstrates that the equation of state for associating gases can be obtained without an explicit knowledge of the occupation numbers of associate pairs. Further details can be found in the review by Däppen, Keady, and Rogers (1990).

To illustrate how the physical picture allows avoiding the divergences that plague the chemical picture, I note that b_2 is convergent, because the bound state part of b_2

is divergent but the scattering state part, which is normally omitted in the chemical approach (e.g. in MHD), has a compensating divergence. Consequently the total b_2 does not contain a divergence of this type (Ebeling, Kraeft, and Kremp, 1976, Rogers, 1977). A major advantage of the physical picture is that it incorporates this compensation at the outset. As a result, the Boltzmann sum appearing in the atomic (ionic) free energy is replaced with the so called Planck-Larkin partition function (PLPF), given by (Ebeling, Kraeft, and Kremp, 1976)

$$\mathrm{PLPF} = \sum_{nl}(2l+1)\left[\exp\left(-\frac{E_{nl}}{kT}\right) - 1 + \frac{E_{nl}}{kT}\right] \tag{5}$$

The PLPF is convergent without additional cut-off criteria as are required in the chemical picture. I stress, however, that despite its name the PLPF is not a partition function, but merely an auxiliary term in a virial coefficient (see Däppen, Anderson, and Mihalas, 1987).

3 Results

3.1 Previous Low-density Comparison

For convenience, a representative result from Däppen, Lebreton, and Rogers (1990) is shown in Fig. 1, which compares MHD and Livermore with the simple Eggleton, Faulkner, and Flannery (1973) equation of state (EFF). For our physical conditions, EFF closely corresponds to the Saha equation with ground-state-only partition functions. It does not contain the Coulomb-pressure correction.

The absolute representation of part *a* is merely able to show the difference between MHD (or Livermore) and EFF results. The difference between MHD and Livermore is alone visible in the magnified part *b*, which shows the *relative* differences between, on the one hand, MHD and Livermore values and the EFF values (serving as reference) on the other hand. This relative plot now not only allows to see the difference between MHD and Livermore results, but also to realize their striking similarity.

In order to find the *physical reason* of this agreement, I varied parameters of the MHD equation of state. I found out that on the selected isochore of $\log\rho = -5.5$ all thermodynamical quantities are mainly dominated by the Debye-Hückel term for the Coulomb pressure correction, and thus the excited states (differently treated in the MHD and Livermore approach) do not matter much. Nevertheless, the influence of the Coulomb term in causing the deviation from the EFF values is somewhat hidden, because it is not the thermodynamical derivative of the Debye-Hückel term itself (it would be 1-2 orders of magnitude too weak), but rather the shifts in the ionization equilibria due to the Debye-Hückel term, which are responsible for the non-ideal part of the result. It follows that the agreement shown in Fig. 1 (Däppen, Lebreton, and Rogers, 1990) is merely a manifestation of the same treatment of the Coulomb pressure and has nothing to do with excited states.

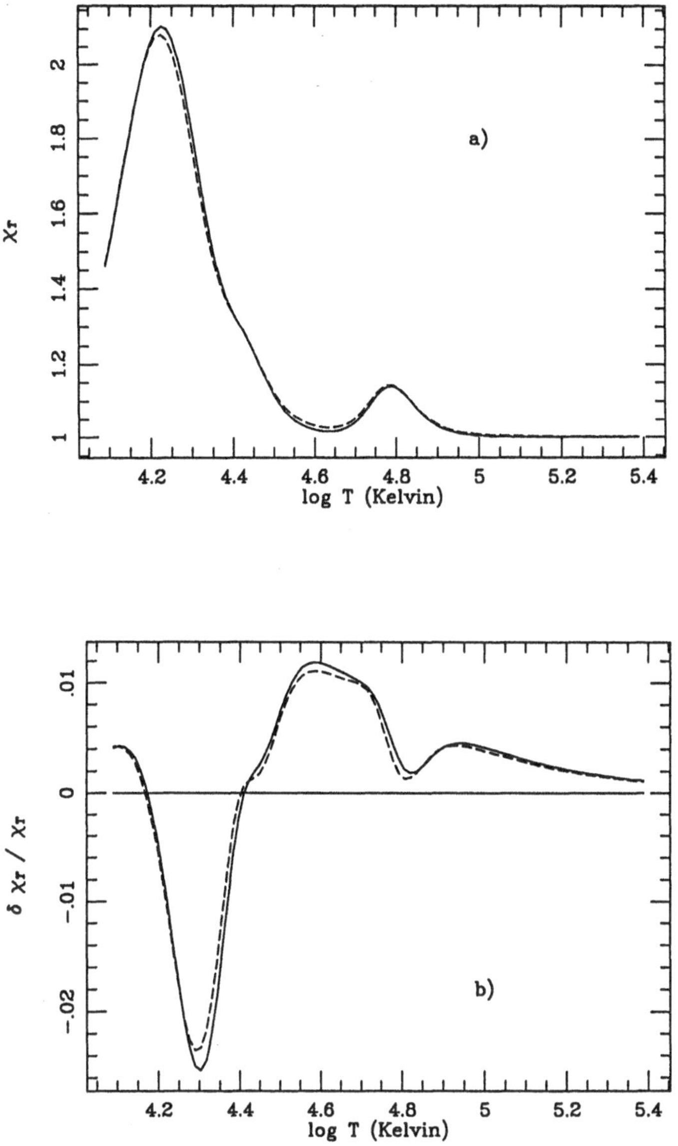

Fig. 1. Comparison of logarithmic pressure derivative $\chi_T = (\partial \ln p/\partial \ln T)_\rho$ on an isochore with $\rho = 10^{-5.5}$ g cm^{-3}. Part *a* shows absolute values; the solid line representing EFF, and the dashed line MHD. The chemical composition is hydrogen and helium only, with number abundances of 90% H and 10% He. The Livermore result would lie indistinguishable on the MHD curve. Part *b* shows the relative differences between Livermore and EFF values, *i.e.*, $(\chi_T^{\text{Livermore}} - \chi_T^{\text{EFF}})/\chi_T^{\text{EFF}}$ (solid line) and between MHD and EFF values, *i.e.*, $(\chi_T^{\text{MHD}} - \chi_T^{\text{EFF}})/\chi_T^{\text{EFF}}$ (dashed line). From Däppen, Lebreton, and Rogers (1990).

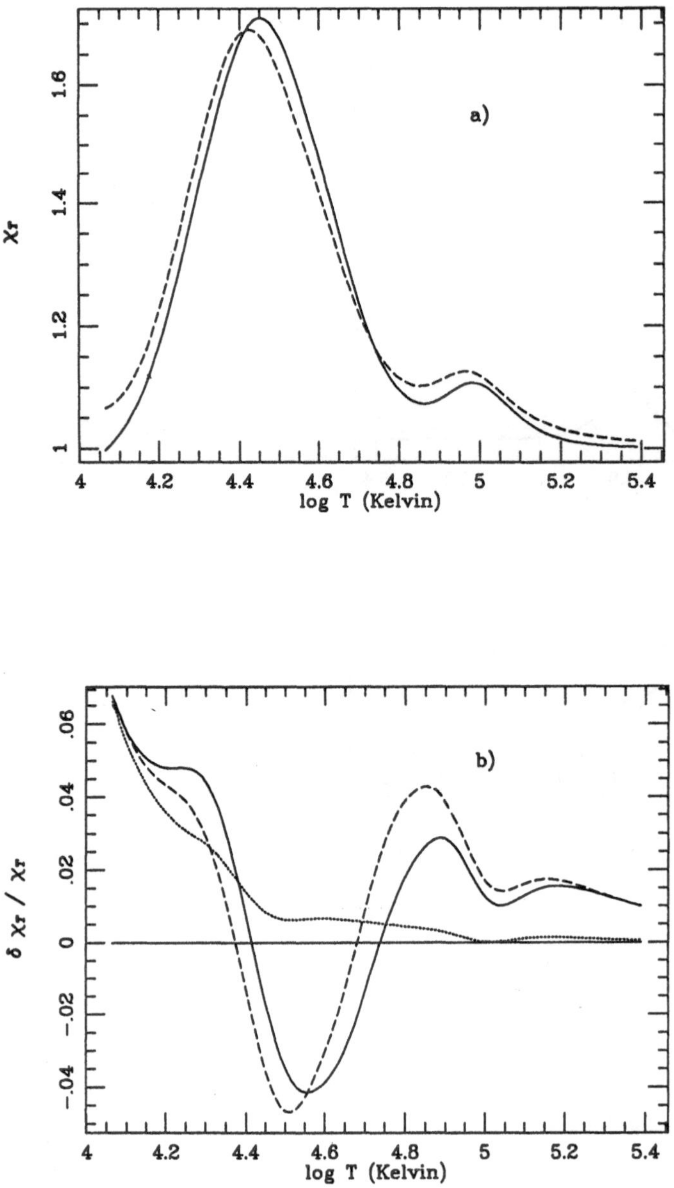

Fig. 2. Same as in Fig. 1, but for $\rho = 10^{-3.5}$ g cm^{-3}. The only addition is the dotted line in part *b*, which shows the result obtained from MHD with a disabled Debye-Hückel term (therefore representing the deviation due to internal partition functions alone).

3.2 Present Higher-density Comparison

In search for manifestations of the influence of internal partition functions, we have repeated the comparison of Däppen, Lebreton, and Rogers (1990) for $\log \rho = -3.5$. Figure 2 shows the result for χ_T.

4 Conclusions

Several conclusions can be made. First, the difference between MHD and Livermore values on the one hand and the EFF value on the other hand has increased by a factor of about three. Second, the difference between MHD and Livermore results is now clearly visible, *though it is still significantly smaller than the amount by which both differ from EFF*. Third, at the low temperature end, the deviations from EFF are effectively due to the internal partition functions, not to the Debye-Hückel term. We do not yet understand why this is so. Fourth, the rather marked domination of the Coulomb pressure term in the thermodynamical quantities suggests that for many solar applications relatively simple equations of state (e.g. an 'extended' EFF including the Debye-Hückel term), might be adequate, provided they are carefully checked against more realistic formalisms, such as the MHD or Livermore equation of state.

Acknowledgements

I thank Forrest Rogers for the Livermore results of Figs. 1-2. Computing of the MHD results was supported by the CCVR of the Ecole Polytechnique at Palaiseau (France). This work was supported in part by the National Science Foundation under Grant No. PHY82-17853, supplemented by funds from the National Aeronautics and Space Administration.

References

Christensen-Dalsgaard, J., Däppen, W., and Lebreton, Y. 1988, *Nature*, **336**, 634.

Däppen, W., Anderson, L.S., and Mihalas, D. 1987, *Astrophys. J.*, **319**, 195.

Däppen, W., Keady, J., and Rogers, F. 1990, in *Solar Interior and Atmosphere*, eds. A.N. Cox, W.C. Livingston and M. Matthews (Space Science Series, University of Arizona Press), in press.

Däppen, W., Lebreton, Y., and Rogers, F. 1990, *Solar Phys.*, in press.

Däppen, W., Mihalas, D., Hummer, D.G., and Mihalas, B.W. 1988, *Astrophys. J.*, **332**, 261.

Ebeling, W., Kraeft, W.D., and Kremp, D. 1976, *Theory of Bound States and Ionization Equilibrium in Plasmas and Solids* (Akademie Verlag, Berlin, DDR).

Eggleton, P.P., Faulkner, J., and Flannery, B.P. 1973, *Astron. Astrophys.*, **23**, 325.

Hill, T.L. 1960, *Statistical Thermodynamics* (Addison-Wesley), Chapt. 15.

Huang, K. 1963, *Statistical Mechanics* (John Wiley), Chapt. 14.

Hummer, D.G., and Mihalas, D. 1988, *Astrophys. J.*, **331**, 794.

Iglesias, C.A., Rogers, F.J., and Wilson, B.G., 1987, *Astrophys. J. Letters*, **322**, L45.

Mihalas, D., Däppen W., and Hummer, D.G. 1988, *Astrophys. J.*, **331**, 815.

Rogers, F.J. 1977, *Phys. Lett.*, **61A**, 358.

Rogers, F.J. 1981, *Phys. Rev.*, **A24**, 1531.

Rogers, F.J. 1986, *Astrophys. J.*, **310**, 723.
Seaton, M. 1987, *J. Phys. B: Atom. Molec. Phys.*, **20**, 6363.

Effects of Overshooting and Magnetic Field at the Base of the Solar Convection Zone on the 5-Minute p-Mode Eigenfrequencies

Lucio Paternó

Istituto di Astronomia, Universitá di Catania, Italy, and
CNR - Gruppo Nazionale di Astronomia, UdR di Catania, Italy

Abstract: The effect of an overshooting layer at the base of the solar convection zone, where a magnetic field can be intensified to largely exceed the equipartition value, on the p-mode eigenfrequencies is investigated and the results are compared with observations.

1 Introduction

Recent measurements of p-mode rotational splittings (Morrow 1988; Rhodes *et al.* 1988, 1989; Woodard and Libbrecht 1988) have pointed out that the surface differential rotation persists through the solar convection zone, at the base of which there is the tendency towards a rigid rotation. The inversion of helioseismic data indicates that the abrupt rotational change occurs in a thin layer located between $0.7\,R_\odot$ and $0.6\,R_\odot$.

This is the region where the convective overshooting may take place (Pidatella and Stix 1986) and dynamo generated magnetic field can be intensified (Belvedere *et al.* 1990; Spiegel and Weiss 1980; Spruit and Van Ballegooijen 1982).

We investigate separately the perturbations introduced by the overshooting and magnetic field on the p-mode eigenfrequencies as calculated using a standard solar model and compare them with observations.

2 Solar Equilibrium Models

In order to carry out the present analysis, we use three different solar equilibrium models: i) standard; ii) overshooting; iii) overshooting with magnetic field.

The first is a classical standard model which makes use of the standard mixing-length theory of convection. This model has initial abundance $Z/X = 0.027$ and $Y = 0.25$ and reaches the present luminosity and radius after $4.75\ 10^9$ years with a convection zone about 191,000 km deep; the model is very similar to the Model 1 of Christensen-Dalsgaard (1982).

The second is one which uses a non local treatment of convection based on the original work of Shaviv and Salpeter (1973) and its extension to the solar case (Pidatella and Stix 1986). This model shows approximately the same overall characteristics of the first one, except for the fact that now the convection penetrates below the layer determined by the Schwarzschild criterion of instability, extending its almost adiabatic stratification to a depth of about $0.65\,R_{\odot}$. In the overshooting layer sound speed is in average a few per cent larger than that in the corresponding region of the standard model.

The third model includes the presence of a magnetic field in the overshooting region. The strength of the field is estimated by making use of the instability analysis of Spruit and Van Ballegooijen (1982) for flux tubes in a spherical shell, which includes the stabilizing effect of magnetic tension produced by the shell curvature. It comes out from this analysis that modes with large azimuthal order number, corresponding to small scale magnetic structures, can be stable in the weak buoyancy overshooting region for field strengths greatly exceeding the equipartition value. For high order modes the limit field strength can be estimated to be:

$$B^2 \simeq 16\pi P |\nabla - \nabla_{\rm ad}| \tilde{k}^2 \tag{1}$$

where P is the gas pressure and $\tilde{k}$ is the dimensionless wave number. For $\tilde{k}$ corresponding to the dimensions of the active regions, values of B ranging from 10^6 to $3.5\ 10^6$ G can be attained through the overshooting layer. In this equilibrium model the hydrostatic balance then modifies as follows:

$$\frac{d}{dr}\left(P + \frac{B^2}{8\pi}\right) = -\frac{\rho G M_r}{r^2} \tag{2}$$

This model saves the same characteristics of the second one; however, owing to the presence of the magnetic field through the overshooting layer, the sound speed is from 0.1% to 0.4% larger than that of the second model.

3 Eigenfrequencies

The eigenfrequencies of non radial adiabatic oscillations are computed, in the Cowling approximation, for the three equilibrium models using a standard procedure as that described by Unno *et al.* (1979). The linearized equations are integrated outwards, using a fourth order Runge-Kutta algorithm, matched onto the adiabatic solutions for a plane isothermal atmosphere at $T = T_{\rm eff}$, and the eigensolutions found by Newton-Raphson iteration.

Since our analysis is limited to modes with the turning point located in the proximity of the overshooting layer, namely with ℓ large, we use the Cowling approximation. Also we do not fit the solutions at the temperature minimum, as usually done, because our equilibrium models do not have a separately calculated atmosphere. Moreover the influence of the magnetic field has been taken into account only in the equations for the equilibrium model, with the confidence that this is the dominant effect.

These approximations affect the eigenfrequencies in the sense that their values could deviate a little from those obtained with more accurate procedures. However, since the present analysis is based on mutual comparisons of trends, the above considered approximations are likely to be irrelevant.

4 Results

The results of the present analysis are summarized in Fig. 1, where the frequency differences of the models with overshooting (O) and overshooting plus magnetic field (O+B) with those of the standard model (S) are shown for modes with $\ell = 20$, 40, 50 and 60. For comparison also the frequency differences between the observed mode frequencies (OBS), as identified by Duvall *et al.* (1988), and those of the standard model are shown. The horizontal line on the top of each square of the figure indicates the turning points of the considered modes in terms of solar radius. Horizontal and vertical axes display respectively the radial order of modes and frequency differences.

The remarkable frequency differences between observations and the standard model, of the order of 10 μHz or more, can be attributed to the above mentioned approximations, at least for the part exceeding 5 μHz.

It is clear that the effect of the overshooting dominates the effect of the presence of the magnetic field, this latter being smaller by an order of magnitude, giving frequency differences not larger than a few tenths of μHz. As expected, both effects are especially important for the modes whose turning point is just below, as for some modes with $\ell = 20$, or inside the overshooting layer, as for most modes with $\ell = 40$. In all of the examined cases the trend is to shift the frequencies towards the observed values. On the other hand, this substantial shift, which in some cases leads the frequencies of the overshooting model to match the observed ones, does not necessarily imply that this model is more correct than a classical standard model, until we are not able to produce very accurate solar standard models. Nevertheless overshooting can help. The small magnetic effect is near the present limits of the observational detectability and certainly much smaller than the uncertainty of the present models. This excludes the possibility of verifying by means of mode frequency analysis whether or not the solar dynamo is really located in the overshooting layer at the base of the convection zone.

References

Belvedere, G., Proctor, M.R.E., and Lanzafame, G. 1990, these proceedings.

Christensen-Dalsgaard, J. 1982, *Monthly Notices Roy. Astron. Soc.*, **199**, 735.

Duvall, T.L., Jr., Harvey, J.W., Libbrecht, K.G., Popp, B.D., and Pomerantz, M.A. 1988, *Astrophys. J.*, **324**, 1158.

Morrow, C.A. 1988, in *Seismology of the Sun and Sun-Like Stars*, ed. E. Rolfe, ESA SP-286 (ESA Publication Division, Noordwijk), p.91.

Pidatella, R.M., and Stix, M. 1986, *Astron. Astrophys.*, **157**, 338.

Rhodes, E.J., Jr., Cacciani, A., Korzennik, S., Tomczyk, S., Ulrich, R.K., and Woodard, M.F. 1988, in *Seismology of the Sun and Sun-Like Stars*, ed. E. Rolfe, ESA SP-286 (ESA Publication Division, Noordwijk), p.73.

Rhodes, E.J., Jr., Cacciani, A., Korzennik, S., Tomczyk, S., Ulrich, R.K., and Woodard, M.F. 1989, *Astrophys. J.*, in press.

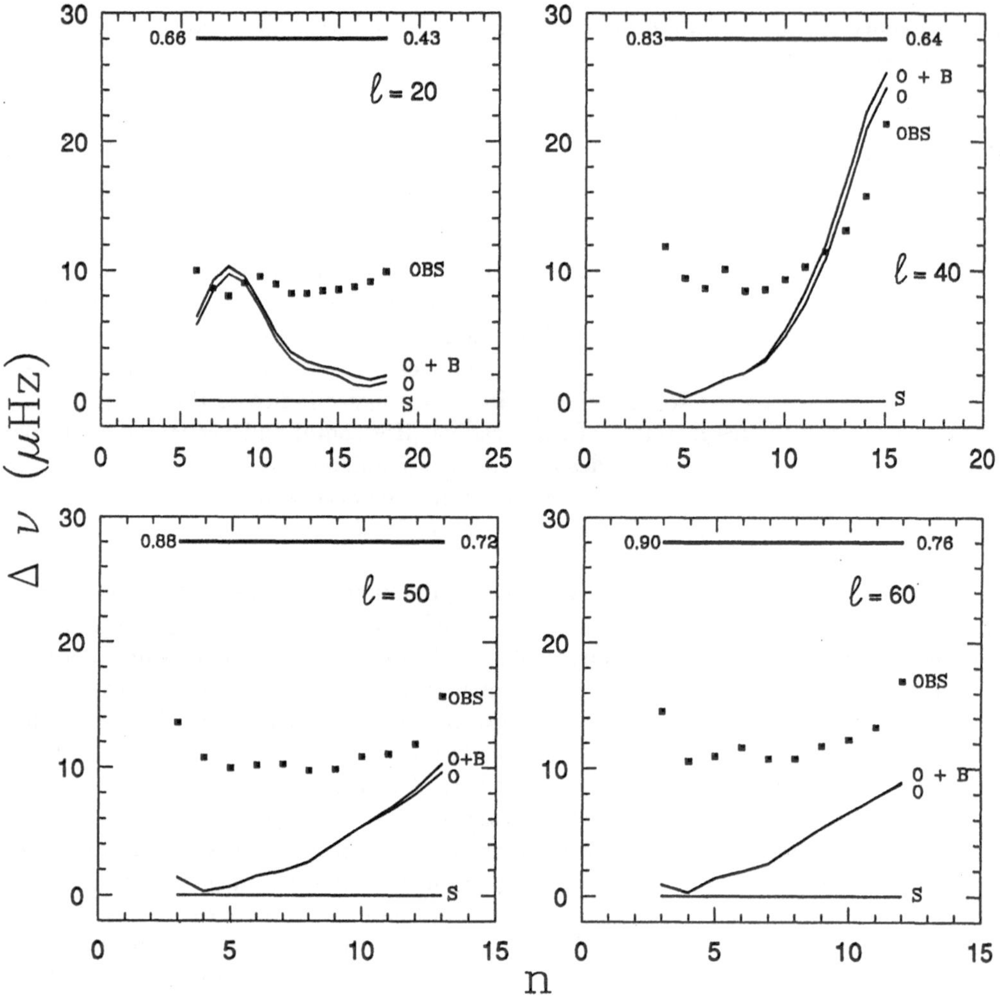

Fig. 1. The eigenfrequencies of a solar standard model (S) are compared with those of a model with overshooting (O), a model including a dynamo generated magnetic field in the overshooting region (O+B) and observations (OBS), for p-modes with ℓ = 20, 40, 50, 60. Horizontal and vertical axes display respectively the radial order of modes (n) and frequency differences (μHz) relative to the standard model. The horizontal line on the top of each square indicates the turning points of the considered modes in terms of solar radius.

Shaviv, G., and Salpeter, E.E. 1973, *Astrophys. J.*, **184**, 191.

Spiegel, E.A., and Weiss, N.O. 1980, *Nature*, **287**, 616.

Spruit, H.C., and Van Ballegooijen, A.A. 1982, *Astron. Astrophys.*, **106**, 58.

Unno, W., Osaki, Y., Ando, H., and Shibahashi, H. 1979, *Nonradial Oscillations of Stars* (University of Tokyo Press, Tokyo).

Woodard, M.F., and Libbrecht, K.G. 1988, in *Seismology of the Sun and Sun-Like Stars*, ed. E. Rolfe, ESA SP-286 (ESA Publication Division, Noordwijk), p.67.

Convective Nonovershooting in Stellar Cores

D. Narasimha and H. M. Antia

Theoretical Astrophysics Group, Tata Institute of Fundamental Research, Homi Bhabha Road, Bombay 400 005, India

Abstract: Mixing-length approximation appears to be consistent with the normal mode analysis of the transport of heat flux due to turbulent convection, in the stellar core as well as in the envelope of stars of a range of spectral types and luminosity classes that we have investigated. However, in spite of demonstrating the self-consistency of the mixing length approximation we do not see any justification for accepting any constant multiple of the mixing-length as a measure of the scale length for convective overshooting into the radiative zones. The convective velocity field in the interior of model of a star of ZAMS mass $10M_{\odot}$ has been examined at three representative epochs during the main sequence phase, using the linear convective modes. The extent of overshooting is found to be less than $0.1H_p$ in all the cases while the mixing-length within the convection zone is typically around $1/3H_p$.

The structure of even an idealized nonrotating star can be computed only if the profile of its chemical composition is known. Except possibly at the onset of the hydrogen burning in the stellar core, the run of the chemical composition cannot be determined from observations in view of the various mixing processes during the evolution of the star. At present turbulent convection is believed to be the main agent for mixing both in the convection zone and the adjacent radiatively stable layers. The aim of the present paper is to study the extent of convective overshooting in the stellar core. The stars crossing the Cepheid instability strip are natural choice for this investigation because, (i) their mass can be estimated by various methods (cf., Simon 1986) and (ii) convective overshooting has been suggested as a possible solution to the discrepancy in the value of mass obtained by various techniques (Böhm-Vitense 1986). Consequently, we have examined the convective velocity field in the core of a stellar model of ZAMS mass $10M_{\odot}$, at three representative epochs during its main sequence lifetime.

For this purpose we use a model of convection based on a superposition of linear convective modes. Narasimha and Antia (1982) have demonstrated that it is possible to find a superposition of linear convective modes which reproduces the convective flux in the mixing-length model. Using this superposition we can estimate the convective velocity or temperature field in the stellar core. To find the linear superposition we use the following approach. First a stellar model is constructed using some version of mixing-length approximation. The linear convective modes in this model are studied, including the mechanical and thermal effects of turbulent convection via the turbulent transport coefficients. The linear stability theory itself does not give the amplitude of the modes, which is determined by demanding that the convective flux due to a linear superposition of unstable convective modes reproduces the convective flux assumed in the mixing-length model over the entire convection zone. This problem is nontrivial because the number of unstable convective modes is finite and most of them are localized in some part of the convection zone. In fact if the turbulent transport coefficients are neglected then it turns out that it is impossible to find such a superposition of linear convective modes. Further, as pointed out by Antia, Chitre and Narasimha (1984) the mixing length at a given depth can be identified with the width of convective flux profile of the dominant convective mode at that depth. Consequently, in a self-consistent model the mixing length used in the construction of the initial stellar model should agree with the equivalent width of the convective flux profile of the dominant mode at corresponding depth. As a result such a self-consistent model is essentially parameter free. It is found that for a variety of stellar convection zone models that we have investigated the consistency is satisfied for a choice of the mixing-length given by

$$\frac{1}{L} \approx \sqrt{\frac{1}{H_T^2} + \frac{1}{d^2}} \tag{1}$$

where H_T is the temperature scale height and d is the distance from the nearest boundary of the convection zone. In the case of solar convection zone the analysis has additional merit of explaining the observed variation of the granular and supergranular velocity field in the atmosphere.

The zero age main sequence model was constructed for a non-rotating star of mass $10M_\odot$, and chemical composition $X = 0.70$ and $Z = 0.03$. The model was evolved up to the core-helium burning phase. During the evolution no mixing was invoked outside the convection zone, but mass loss was included using a semi-empirical relation given by Tarafdar(1988). The linear convective modes were computed at the three representative phases given in Table 1.

Table 1. Properties of models studied

Model	Age (10^7 yrs)	X_c	Mass ($M_\odot$)	Luminosity ($L_\odot$)	$T_{\rm eff}$ (K)	Radius ($R_\odot$)	Radius of C.Z. ($R_{\rm star}$)
ZAMS	0	0.700	10	4889	22986	4.406	0.232
MS1	1.28	0.283	9.764	7193	20420	6.772	0.136
MS2	1.60	0.020	9.557	7443	20391	6.908	0.039

These representative models were refined prior to using them for the stability analysis. The extent of the superadiabaticity in the convective core was determined by appealing to the mixing-length approximation and the grid size was redefined as needed for the numerical accuracy of the analysis of the convective modes. This requires spacing of finer than a few thousandth of the pressure scale height in the top of the convection zone and the adjacent radiative layers because of steep gradients in the eigenfunctions in this region. Because of these steep gradients the eigenfunctions fall off very rapidly outside the convection zone giving very little overshooting. We believe that the main reason for this is the large ratio of the subadiabatic temperature gradient in the radiative layer to the near zero superadiabatic gradient in the convective part.

As in the case of the red giant (cf. Antia, Chitre and Narasimha, 1984) only one mode ($l = 1$, fundamental) contributes to the convective flux transport in almost the entire convection zone. In spite of the high Reynolds number the velocity field within the convection zone, away from the boundary appears almost like that of a laminar flow. The flux transport due to the dominant mode is displayed in Fig. 1 along with the model flux. The amplitude of this mode has been fixed by fitting the convective flux profile due to the mode with the model flux. The self-consistency is apparent from a comparison between the two profiles. Note that the radial length scale of the two profiles is also a measure of the mixing-length. All the other unstable modes contribute very little to the convective heat transport, and hence their amplitude is expected to be very small.

In Figs. 2, 3 and 4 we have displayed the radial and horizontal component of the velocity eigenfunction of the dominant mode as well as the convective velocity as estimated by the mixing-length theory (MLT) for the three models. The horizontal velocity has a discontinuity at the boundary of convection zone as we switch from a viscous to inviscid layer. The velocity falls off rapidly as we move away from the convection zone, but after some distance the velocity profile shows some spatial oscillations. This is due to the fact that we have moved from a region of fine mesh size to a coarser one. Because of the limitations imposed by the computer time and memory it is not possible to maintain a small mesh spacing over large intervals. But the radial velocity drops off smoothly to negligible value within a distance of approximately 0.03 times the pressure scale height for all the models investigated. Consequently we expect negligible mixing due to the overshooting of the eddies responsible for energy transport inside the convection zone; if there exists mixing induced by convective eddies of smaller size, it should be a diffusive process rather than dynamical. Hence it could depend only on the scales of small eddies which do not see the local pressure scale-height.

To ensure that the results are not affected by the discontinuity at the convection zone boundary, we have repeated the calculations by assuming that there is some overshooting in the equilibrium model. The velocity and flux profile in this model is assumed to fall off exponentially with a suitable scale height. With this assumption the discontinuity in horizontal velocity is removed and both the horizontal and vertical components of velocity fall off at essentially the same rate as shown in the Figs. 2–4.

We have also investigated the stability of the oscillatory modes. Some of the gravity modes trapped in the region of steep composition gradient are only marginally stable, with life time comparable to the nuclear time scale of the star. Possible mixing induced by these gravity modes cannot be ruled out, but if such a process of mixing exists it could be important even in stars which do not have a convective core.

In conclusion we find that there is very little convective overshooting beyond the boundary of convection zone. We believe this is mainly because of the large ratio of the subadiabatic gradient in the radiative layer to the superadiabatic gradient in the convection zone. If the same analysis is applied to the top of solar convection zone then we get a velocity scale height in the atmosphere which is comparable to the pressure scale height thus giving significant overshooting. It may be noted that in the solar convection zone the superadiabatic gradient is very large near the top.

Acknowledgements

It is a pleasure to thank Professors W. Unno and J.-P. Zahn for valuable discussions and advice.

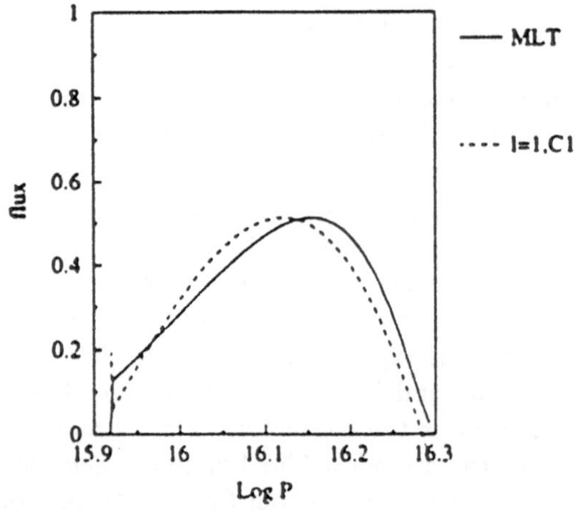

Fig. 1. The convective flux profile for model MS1.

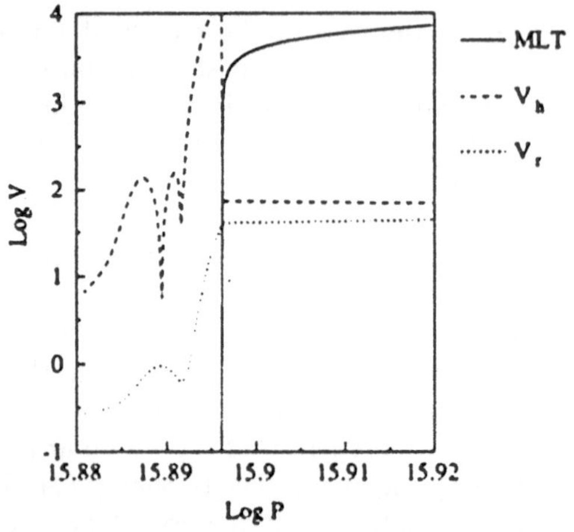

Fig. 2. The velocity field in the model ZAMS.

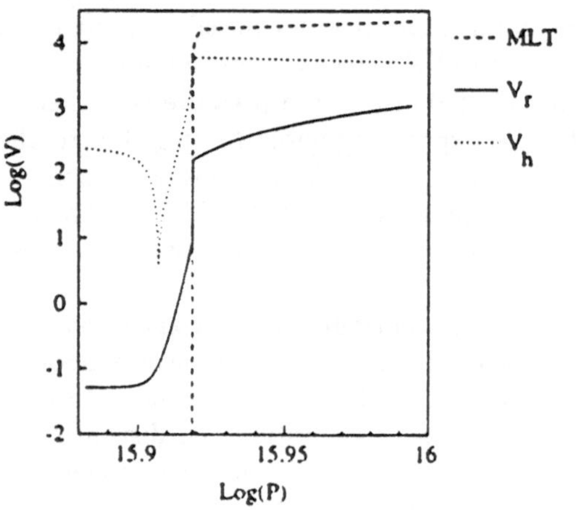

Fig. 3. The velocity field in the model MS1.

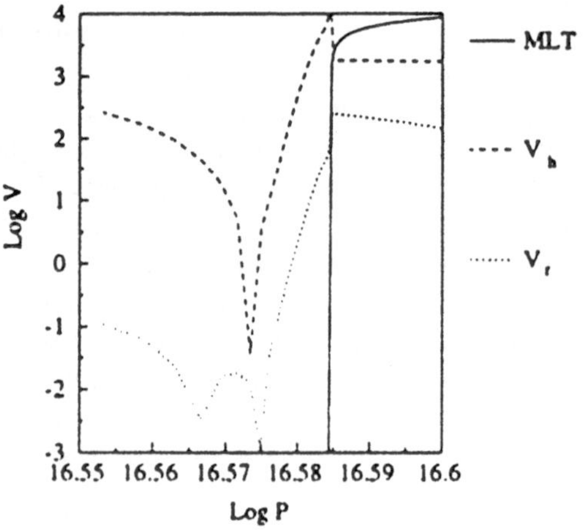

Fig. 4. The velocity field in the model MS2.

References

Antia, H.M., Chitre, S.M., and Narasimha, D. 1984, *Astrophys. J.*, **282**, 574.

Böhm-Vitense, E. 1986, in *Stellar Pulsation*, eds. A.N. Cox, W.M. Sparks, and S.G. Starrfield (Springer, Berlin), p.159.

Narasimha, D., and Antia, H.M. 1982, *Astrophys. J.*, **262**, 385.

Simon, N.R. 1986, in *Stellar Pulsation*, eds. A.N. Cox, W.M. Sparks, and S.G. Starrfield (Springer, Berlin), p.148.

Tarafdar, S.P. 1988, *Astrophys. J.*, **331**, 932.

Solar Neutrino Experiments

Masayuki Nakahata

Institute for Cosmic Ray Research, University of Tokyo,
Tanashi-shi, Tokyo 188, Japan

Abstract: The currently running solar neutrino experiments, the ^{37}Cl radiochemical experiment and the Kamiokande-II experiment, indicate the deficit of ^{8}B solar neutrinos. The Gallium solar neutrino experiments(GALLEX and SAGE) are under way and are expected to give first results in near future. The present status of these experiments and future solar neutrino experiments are discussed.

1 Introduction

It is widely accepted that the fusion of light nuclei is the main energy source of the sun. The solar core is believed to be in the hydrogen burning phase, a sequence of nuclear reactions which convert four hydrogen nuclei into a helium nucleus, two positrons and two neutrinos. The standard solar model (Bahcall and Ulrich 1988) predicts that 98.5 % of the energy is generated by a series of reactions called p-p chain and the other 1.5 % is generated by the CNO cycles. Figure 1 shows the predicted energy spectra of solar neutrinos of various origins of the nuclear reactions; pp, pep, ^{7}Be, and ^{8}B neutrinos from p-p chain, and ^{13}N, ^{15}O, and ^{17}F from the CNO cycle. Also shown are the threshold energies of the neutrino captures on ^{37}Cl and ^{71}Ga, together with the analysis threshold of the Kamiokande-II experiment.

The ^{37}Cl experiment performed by R. Davis *et al.* over 20 years has revealed a surprising result, namely a deficit of solar neutrinos, what is called "solar neutrino problem". Recently this problem was confirmed by the Kamiokande-II experiment using a method completely different from the ^{37}Cl experiment. Various possible solutions of the solar neutrino problem are proposed, ranging from errors in the input parameters for the calculations, to particle-physics solutions such as neutrino oscillations (Gribov and Pontecorvo 1969; Bilenky and Pontecorvo 1978; Wolfenstein 1978; Bethe 1986; Mikheyev and Smirnov 1986), neutrino magnetic moments (Voloshin *et al.* 1986; Voloshin and Vysotskii 1986), or exotic massive particles at the core of the sun (Spergel and Press 1985; Faulkner and Gilliland 1985).

The ^{71}Ga radiochemical experiments are under way and are expected to give first results in near future. Their energy threshold is sufficiently low that they are sensitive to the pp neutrinos.

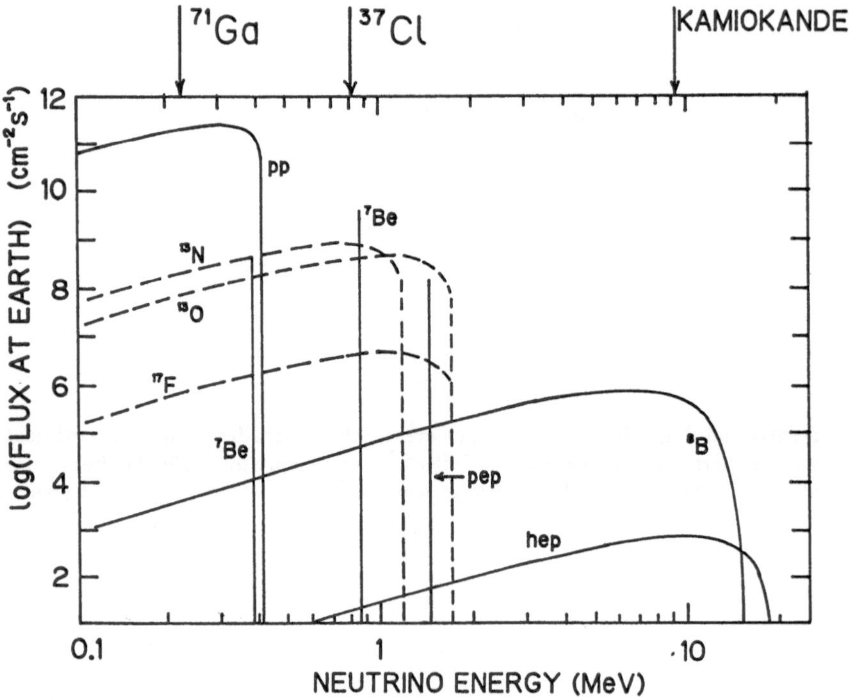

Fig. 1. Expected energy spectra of solar neutrinos. Adopted from Bahcall and Ulrich (1988). Also shown are the threshold energies for ^{37}Cl and ^{71}Ga capture reactions and the analysis threshold of the Kamiokande-II experiment.

Several future solar neutrino experiments are proposed, which plan to measure the ^{8}B neutrino spectrum precisely, flux of ^{7}Be neutrinos, or integrated neutrino flux over several million years.

In this report the present status of the ^{37}Cl, Kamiokande-II, and ^{71}Ga experiments are reviewed and future solar neutrino experiments are briefly presented. In the final section, the solutions of the solar neutrino problem are discussed based on the results of the present experiments and future experiments.

2 ^{37}Cl Radiochemical Experiment

The ^{37}Cl radiochemical experiment has been operating since 1968 at a level of 1480 m underground in the Homestake Gold Mine in U.S.A. The neutrino target is 2.2×10^{30} atoms (133 tons) of ^{37}Cl in the form of 3.8×10^{5} liters of liquid perchloroethylene, C_2Cl_4. Neutrinos are captured by ^{37}Cl by the reaction, $^{37}\mathrm{Cl} + \nu_e \rightarrow {}^{37}\mathrm{Ar} + e^-$ with the threshold energy of 0.814 MeV. In the standard solar model the ^{37}Cl detector is sensitive to a linear contributions of $\nu_{^8\mathrm{B}}(77\,\%)$, $\nu_{^7\mathrm{Be}}(14\,\%)$, $\nu_{^{15}\mathrm{O}}(4\,\%)$, $\nu_{\mathrm{pep}}(3\,\%)$. The observed capture rate averaged from 1970 to 1985 is 2.1 ± 0.3 SNU ($1\,\sigma$) (Rowley *et al.* 1985; Davis 1987), where SNU $= 10^{-36}$ captures/atom/sec. On the other hand, the standard solar model (SSM) predicts 7.9 ± 2.6 SNU (3σ) (Bahcall and Ulrich 1988), which is about 4 times larger than the measured rate.

The ^{37}Cl detector was stopped for about one year and a half since 1985 because of its pump failure. Davis *et al.* resumed the experiment in October, 1986 and the result till May, 1988 was reported to be 4.2 ± 0.8 SNU (Davis 1988), which is a much higher value compared with the earlier result. Figure 2 shows the observed capture rate as a function of time. High rates around 1977 and in 1986-1988 are apparent. Davis claims that there is an apparent change in the capture rate which is greater than the statistical fluctuation. Figure 3 shows the time variation of 5 extraction running averages of measured solar neutrino flux with number of sunspots plotted upside down. The neutrino capture rate seems to have anti-correlation with the solar activity. If it is indeed real, one can predict the capture rate to decrease till 1990 in which the solar activity will become maximal. The solar neutrinos are produced deep in the core of the sun. Therefore, it should be constant for periods of about Kelvin time ($\sim 3 \times 10^7$ years). Hence, if the correlation with the solar 11 years' cycle is real, it should be due to the property of neutrinos, such as the magnetic moment of neutrinos. Of course, one has to wait until 1990 to conclude the anti-correlation.

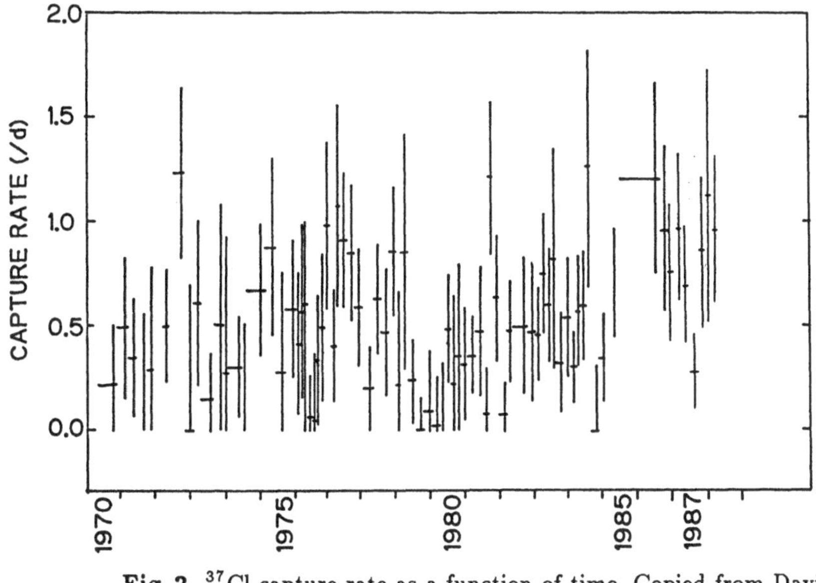

Fig. 2. ^{37}Cl capture rate as a function of time. Copied from Davis (1988).

3 Kamiokande-II Experiment

Kamiokande-II experiment is a counter experiment that detects energies and directions of the electrons scattered by solar neutrinos, $\nu_e e^- \rightarrow \nu_e e^-$. These electrons peak in a narrow forward direction ($\sigma_\theta = 28°$ at 10 MeV). Therefore, the solar neutrino signal and backgrounds are clearly separated using the directional information. The energy information of the scattered electrons is useful for the test of possible distortion of the neutrino

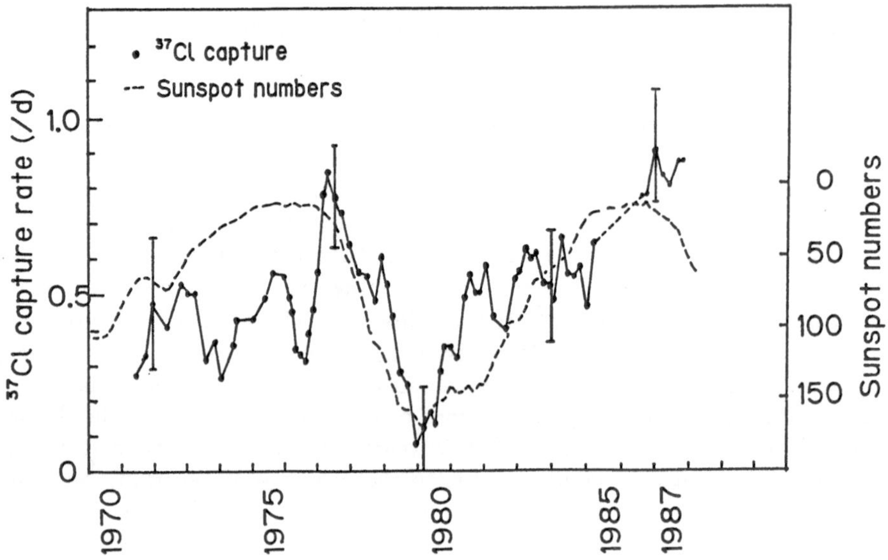

Fig. 3. Running averages of the capture rates with sunspot numbers. Each data point of the data corresponds to the average of the adjacent 5 runs.

spectrum which might be due to neutrino oscillations. Furthermore, Kamiokande-II is a real time experiment, which is essential for the test of short time variation, such as day/night difference or seasonal variation of the flux.

The Kamiokande-II detector is a 3000 ton water Cherenkov detector (Hirata *et al.* 1988), the inner volume (2140 tons) of which is viewed by 948 20-inch ϕ photomultipliers. The detector is located 1000 m underground in the Kamioka mine in Japan. The large counter experiment that has a threshold energy as low as several MeV suffers from enormous backgrounds. It mainly consists of external γ's from the surrounding rocks, β-rays from radioactive elements dissolved in the water and β-rays from unstable nuclei produced by cosmic ray muons. External γ-rays are reduced by the anti-counter layer which surrounds the whole inner detector and the remaining γ-rays are eliminated by limiting the fiducial volume to only 680 ton near the center of the detector. Water has been constantly purified to minimize the radioactive contents of ^{238}U, ^{226}Ra, ^{222}Rn, etc. The backgrounds induced by cosmic rays are eliminated by taking a timing and spatial correlation with preceding cosmic rays.

Kamiokande-II published the first result of the ^{8}B solar neutrino measurement using 450 days' data taken from January 1987 through May 1988 (Hirata *et al.* 1989a). Figure 4 shows an example of the $\cos\theta_{sun}$ distribution with an energy threshold of 10.1 MeV, where θ_{sun} is the angle between the scattered electron and the direction from the sun. An enhancement near the forward direction indicates the evidence for solar neutrinos. The observed electron energy spectrum is shown in Fig. 5 with the expectation from the standard solar model. The flux of ^{8}B solar neutrinos is obtained by fitting the electron energy spectrum with a scaled spectrum of the expectation. The obtained flux using the data with $E_e \geq 9.3$ MeV is;

$$\text{flux} = (0.46 \pm 0.13(\text{stat}) \pm 0.08(\text{sys})) \times \text{SSM} , \tag{1}$$

where SSM is the standard solar model prediction, SSM $= 5.8 \times 10^6\,\text{cm}^{-2}\text{s}^{-1}$.

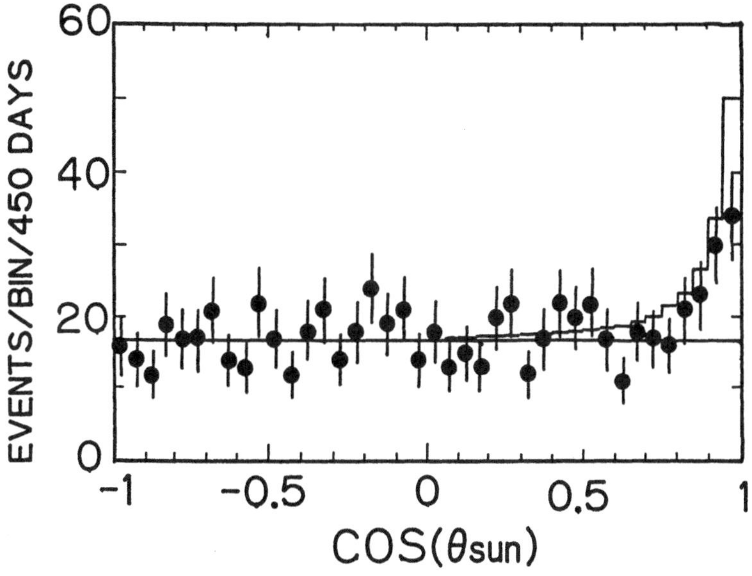

Fig. 4. $\cos\theta_{\text{sun}}$ distribution with an energy threshold of 10.1 MeV, where θ_{sun} is the angle between the scattered electron and the direction from the sun. This figure is made based on the data taken from January 1987 through May 1988.

In June 1988 the detector and the analysis methods were improved and the background rate has been reduced by a factor of three since then. The energy threshold of the data analysis has been lowered to 7.5 MeV since the improvements. The preliminary results of the 288 days' data from June 1988 through April 1989 is reported in Hirata *et al.* (1989b). A $\cos\theta_{\text{sun}}$ distribution is shown in Fig. 6 with the energy threshold of 7.5 MeV. The flux obtained from the improved data is;

$$\text{flux} = (0.39 \pm 0.09(\text{stat}) \pm \sim 0.06(\text{sys})) \times \text{SSM} \quad (\text{preliminary}) . \tag{2}$$

The result is statistically consistent with the previous result.

The Kamiokande-II result is the second positive observation of solar neutrinos. The ^{8}B solar neutrino flux is about $40 \sim 50\,\%$ of the SSM prediction. The Kamiokande-II result is statistically consistent with the result of the ^{37}Cl experiment during the same time period. The analysis relevant to the time variation of the solar neutrino flux, such as day/night effect, seasonal effect, or correlation with solar activities are under study. However, the observational period is too short to make a meaningful test of the time variation, especially the variation with solar activity. Continuous operation of 2 - 3 years is needed to discuss it.

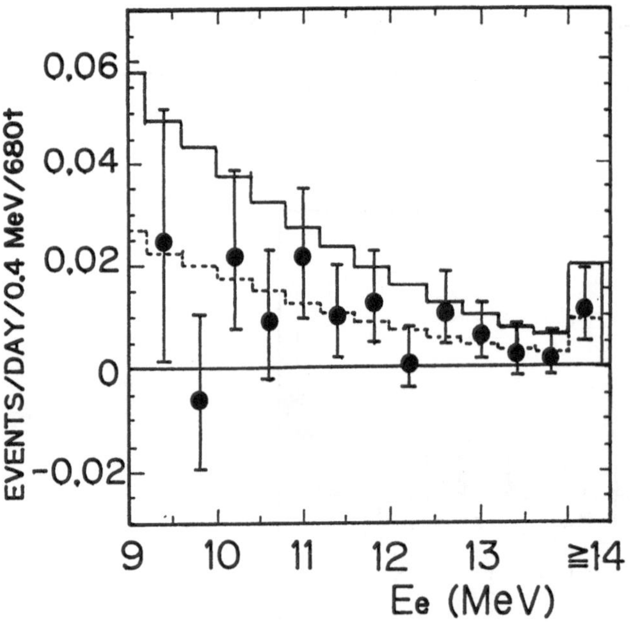

Fig. 5. Energy distribution of the solar neutrino signal (January 1987 - May 1988). The histogram is the distribution predicted by the standard solar model. The highest bin corresponds to $E_e \geq 14$ MeV. The dotted line shows the best fit to the data (0.46×SSM).

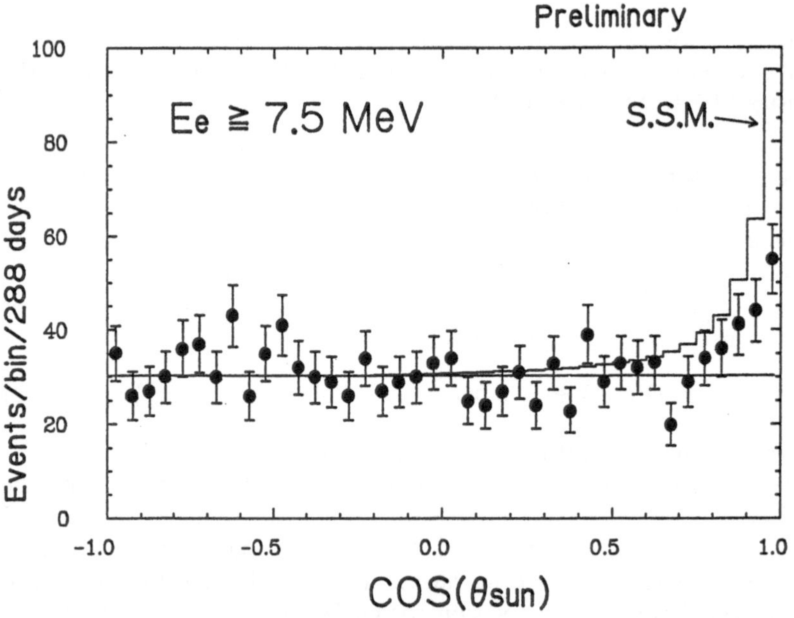

Fig. 6. $\cos\theta_{sun}$ distribution with an energy threshold of 7.5 MeV based on the data taken from June 1988 through April 1989.

4 Gallium Radiochemical Experiment

Two Gallium radiochemical experiments, GALLEX and SAGE, are under way (Kirsten 1989; Gavrin 1989). They are summarized in Table 1. The basic process of the Gallium experiment is $^{71}\text{Ga} + \nu_e \rightarrow {}^{71}\text{Ge} + e^-$ with the threshold energy of 0.233 MeV, thus sensitive to pp neutrinos. In the standard solar model the ^{71}Ga detectors are sensitive to a linear contributions of $\nu_{pp}(56\,\%)$, $\nu_{^7\text{Be}}(26\,\%)$, and $\nu_{^8\text{B}}(11\,\%)$. The expected total capture rate is 1.09 captures per day per 30 tons' Gallium.

Table 1. Gallium solar neutrino experiments.

	GALLEX	SAGE
site	Gran Sasso (Italy)	Baksan (USSR)
depth	3100 m.w.e.	4700 m.w.e.
source	30 ton Ga in the form of $GaCl_3$ solution	60 ton metallic Ga
vessel	single tank	10 pieces of 6 ton tank
extraction	air purge	chemical procedure
start of experiment	February 1990	already started with 30 ton Ga in 1989

These experiments expose the detector to solar neutrinos for a given time (14-30 days), after which ^{71}Ge atoms are extracted and the electron capture decays of the ^{71}Ge ($\tau_{1/2} = 11.4$ days) are counted in a low background proportional counter.

The SAGE detector is located in the Baksan Neutrino Observatory in USSR. 60 ton metallic Gallium is used. The experiment is already started with 30 ton Gallium. At present an effort to lower the background level is being made now. It is reported that the background rate is reduced to the level of < 1 event/day (Gavrin 1989).

The GALLEX detector is located in the Gran Sasso laboratory in Italy. 30 ton Gallium in form of concentrated aqueous Galliumchloride solution is used. The start of the experiment is planned to be February 1990.

5 Future Solar Neutrino Experiments

Several proposed solar neutrino experiments are briefly described. They are summarized in Table 2.

Table 2. Proposed solar neutrino experiments. In the calculation of the expected event rate, the flux of ^{8}B neutrinos is assumed to be 0.46 times the prediction from the standard solar model.

experiment	reaction	detection threshold	ν source	rate (/yr)
SNO	$\nu_e + D \to e^- + p + p$	5 MeV	^{8}B	4340
1000 ton D_2O	$\nu + D \to e^- + n + p$	2.2 MeV	^{8}B	2370
	$\nu + e^- \to \nu + e^-$	5 MeV	^{8}B	490
Super-Kamiokande	$\nu + e^- \to \nu + e^-$	5 MeV	^{8}B	8400
BOREX	$^{11}B + \nu_e \to {}^{11}C + e^-(\gamma)$	3.5 MeV	^{8}B	1060
(200 t natural B)	$\nu + {}^{11}B \to \nu + {}^{11}B + \gamma$	4.45 MeV	^{8}B	60
	$\nu + e^- \to \nu + e^-$	3.5 MeV	^{8}B	690
Indium detector (10 ton In)	$\nu_e + {}^{115}In \to e^- + {}^{115}Sn + 2\gamma$	0.74 MeV	^{7}Be, pep	200
ICARUS I	$\nu_e + {}^{40}Ar \to e^- + {}^{40}K + \gamma$	11 MeV	^{8}B	100
(200 ton liq. Ar)	$\nu + e^- \to \nu + e^-$	5 MeV	^{8}B	100
^{98}Mo	$\nu_e + {}^{98}Mo \to e^- + {}^{98}Te$	1.7 MeV	^{8}B	
^{205}Tl	$\nu_e + {}^{205}Tl \to e^- + {}^{205}Pb$	43 keV	pp	

5.1 SNO (Sudbury Neutrino Observatory)

The SNO project (Canada-USA-UK collaboration) consists in 1000 tons of heavy water D_2O surrounded by purified light water H_2O (Aardsma *et al.* 1987; Ewan *et al.* 1987). The experiment detects the Cherenkov radiation of electron and gamma from the charged current reaction ($\nu_e + D \to e^- + p + p$), the neutral current reaction ($\nu + D \to e^- + n + p$ ($n + Cl \to Cl + \gamma$; 2.5 ton NaCl is planned to be added into D_2O)), and $\nu + e^- \to \nu + e^-$. The sensitive neutrino source is ^{8}B neutrinos. This detector will be able to measure the total neutrino flux (including ν_μ and ν_τ) as well as the neutrino energy spectrum.

5.2 Super-Kamiokande

Super-Kamiokande project (Japan) plans to build a 50 kilo-ton water Cherenkov detector (22 kilo-ton fiducial mass) at the Kamioka mine in Japan (Totsuka 1987). It will detect ^{8}B neutrinos by $\nu + e^- \to \nu + e^-$ scattering with high statistics. The detector will have 40 % of its surface covered with the cathodes of photomultipliers, which is twice as dense as the present Kamiokande-II detector.

5.3 BOREX

The BOREX project (USA-Italy) consists in 2000 tons Boron-loaded liquid scintillator (200 ton Boron) viewed by photomultipliers (Raghavan and Pakvasa 1988). It will detect the ^{8}B neutrinos by the charged current reaction, $\nu_e + {}^{11}B \to e^- + {}^{11}C^*$ ($\to {}^{11}C + \gamma$), by the neutral current reaction, $\nu + {}^{11}B \to \nu + {}^{11}B^*$ ($\to \gamma + {}^{11}B$) and also by electron scattering, $\nu + e^- \to \nu + e^-$.

5.4 Indium Experiment

The reaction of ^{115}In and neutrinos, $\nu_e + {}^{115}\mathrm{In} \rightarrow e^- + {}^{115}\mathrm{Sn}^*$, is followed by two gamma rays (or electron + gamma) within a short time interval ($\tau_{1/2} = 3.3\,\mu$sec). The threshold energy of this reaction is so low (0.128 MeV) as to be sensitive to pp neutrinos. However, the decay of Indium itself prevents constructing a realistic experiment for pp neutrinos. An Indium experiment for the measurement of ^{7}Be and pep neutrinos is proposed (Bellefon *et al.* 1989).

5.5 ICARUS Project

ICARUS project (Italy-USA) is a 3000 ton liquid Argon time projection chamber, which will be installed in Gran Sasso (Bahcall *et al.* 1986). The present problem consists in the feasibility to drift ionization electrons over long distances. A small project, ICARUS I, which use 200 tons of liquid Argon, is being developed as a first step.

5.6 Geochemical Experiments

Geochemical experiments will give integrated flux over several million years. There are two proposed geochemical experiments.

The Molybdenum experiment counts the number of ^{98}Te atoms produced by the $\nu_e + {}^{98}\mathrm{Mo} \rightarrow e^- + {}^{98}\mathrm{Te}$ reaction ($\tau_{1/2}$ (^{98}Te)= 4.2 million years), which has a threshold energy of 1.7 MeV. Therefore, it can measure the integrated ^{8}B neutrino flux. The Los Alamos group (Wolfsberg *et al.* 1985) is measuring the present abundance of ^{98}Te in about 2600 tons of a Molybdenum ore, in which 10^7 ^{98}Te atoms are expected, using a dedicated mass spectrometer. The analysis is in progress and should give results in near future.

The ^{205}Tl experiment (Henning *et al.* 1985) counts the number of ^{205}Pb atoms ($\tau_{1/2}$ (^{205}Pb)= 15 million years) produced by the $\nu_e + {}^{205}\mathrm{Tl} \rightarrow e^- + {}^{205}\mathrm{Pb}$ reaction using the technic of accelerator mass spectroscopy. Because of the very low threshold (43 keV), this experiment can measures the integrated flux of pp neutrinos.

6 How to Solve the Solar Neutrino Problem

The experimental procedure to solve the solar neutrino problem is discussed in this section.

The possible solutions of the solar neutrino problem can be classified into three categories.

1. Neutrino properties, such as neutrino oscillations or neutrino magnetic moments.
2. Errors in input parameters or in assumptions in the standard solar model, such as errors in nuclear cross sections, in opacities, or in the assumption that the initial matter density is homogeneous.
3. WIMPs (Weakly Interacting Massive Particles) in the core of the sun.

If at least one of the following items were observed by solar neutrino experiments, the solution would be the neutrino properties.

a) Very small capture rate in Gallium experiment. This condition would be met, if the capture rate is much smaller than 132−20−14 = 98 SNU, where 132 SNU is the estimated total capture rate, 20 SNU is the 3σ error of the estimation and 14 SNU is the contribution of the ^{8}B neutrinos to the total capture rate (Bahcall and Ulrich 1988).
b) Possible distortion of the neutrino spectrum. SNO, Super-Kamiokande, and BOREX would test this possibility.
c) Possible short time variation, such as correlation with 11 years' cycle or day/night difference, of neutrino fluxes. Further operation of the ^{37}Cl experiment and Kamiokande-II and future solar neutrino experiments would test this possibility.
d) Possible difference between ν_e flux and total ν flux ($\nu_e + \nu_\mu + \nu_\tau$). SNO and BOREX would test it.

If none of the above possibilities are met, can we separate the solutions of the second and the third categories, i.e. error in the standard solar model vs. WIMPs ? The precise measurement of the flux of each neutrino source as well as the precise measurement of the internal temperature of the sun by the helioseismology would separate these possibilities.

7 Conclusion

The recent result of the ^{37}Cl experiment in 1986-1988 gives a high capture rate, 4.2 ± 0.8 SNU, in contrast to the earlier (1970-1985) result of 2.1 ± 0.3 SNU. The capture rate seems to change with time more than expected from the statistical fluctuation. The data till the next solar maximum ($\sim$1990) would further test this possibility.

Kamiokande-II has been observing ^{8}B solar neutrinos. The observed flux obtained from 450 days' data in the time period of January 1987 through May 1988 with $E_e \geq$ 9.3 MeV is (0.46 ± 0.13(stat) $\pm$ 0.08(sys)) $\times$ SSM. The background rate to the solar neutrino measurement was reduced by a factor of three since June 1988 because of the improvements of the detector and of analysis methods. The preliminary result of the flux after the improvement is (0.39 ± 0.09(stat) $\pm \sim$ 0.06(sys)) $\times$ SSM, which is obtained using 288 days' data from June 1988 through April 1989. The two results of the Kamiokande-II is statistically consistent with the result of the ^{37}Cl experiment during the same data-taking period.

Both ^{37}Cl and Kamiokande-II experiments present the deficit of the solar neutrino flux. We hope this problem will be solved by further operation of the currently running experiments and future solar neutrino experiments.

References

Aardsma, G., Allen, R.C., Anglin, J.D., Bercovitch, M., Carter, A.L., Chen, H.H., Davidson, W.F., Doe, P.J., Earle, E.D., Evans, H.C., Ewan, G.T., Hallman, E.D., Hargrove, C.K., Jagam, P., Kessler, D., Lee, H.W., Leslie, J.R., MacArthur, J.D., Mak, H.-B., McDonald, A.B., McLatchie, W., Robertson, B.C., Simpson, J.J., Sinclair, D., Skensved, P., and Storey, R.S. 1987, *Phys. Lett. B*, **194**, 321.

Bahcall, J.N., Baldo-Ceolin, M., Cline, D.B., and Rubia, C. 1986, *Phys. Lett. B*, **178**, 324.

Bahcall, J.N., and Ulrich, R.K. 1988, *Rev. Mod. Phys.*, **60**, 297.
Bellefon, A., Barlouaud, R., Borg, A., Ernwein, J., and Mosca, L. 1989, "Progress Report: Feasibility Study of an Indium Scintillator Solar Neutrino Experiment," DphPE 89-17 (Commissariat A l'energie Atomique, Centre d'etudes Nucleéaires de Saclay, France).
Bethe, H.A. 1986, *Phys. Rev. Letters*, **56**, 1305.
Bilenky, S.M., and Pontecorvo, B. 1978, *Phys. Rep.*, **41**, 225.
Davis, R., Jr. 1987, in *Proc. of the Seventh Workshop on Grand Unification, ICOBAN '86*, ed. J. Arafune (World Scientific, Singapore), p.237.
Davis, R., Jr. 1988, in *Proc. of the 13th Int. Conf. on Neutrino Physics and Astrophysics, Neutrino '88*, ed. J. Schneps, T. Kafka, W.A. Mann, and Pran Nath (World Scientific, Singapore), p.518.
Ewan, G.T., Mak, H.B., Robertson, B.C., Allen, R.C., Chen, H.H., Doe, P.J., Sinclair, D., Davidson, W.F., Storey, R.S., Earle, E.D., Jagam, P., Simpson, J.J., Hallman, E.D., Mcdonald, A.B., Carter, A.L., and Kessler, D. 1987, *Sudbury Neutrino Observatory Proposal*, SNO 87-12, October (Queen's University, Canada).
Faulkner, J., and Gilliland, R.L. 1985, *Astrophys. J.*, **299**, 994.
Gavrin, V.N. 1989, Talk in the *Inside the Sun, IAU Colloq. No. 121.*
Gribov, V., and Pontecorvo, B. 1969, *Phys. Lett.*, **28B**, 493.
Henning, W., Kutschera, W., Ernst, H., Korschinek, G., Kubik, P., Mayer, W., Morinaga, H., Nolte, E., Ratzinger, U., Müller, M., and Schüll, D. 1985, in *Proc. of Conf. on Solar Neutrinos and Neutrino Astronomy* eds. M.L. Cherry, K. Lande, and W.A. Fowler, AIP Conference Proc. No.126 (American Inst. Phys., New York), p.203.
Hirata, K.S., Kajita, T., Kifune, K., Kihara, K., Nakahata, M., Nakamura, K., Ohara, S., Oyama, Y., Sato, N., Takita, M., Totsuka, Y., Yaginuma, Y., Mori, M., Suzuki, A., Takahashi, K., Tanimori, T., Yamada, M., Koshiba, M., Suda, Y., Miyano, K., Miyata, H., Takei, H., Kaneyuki, K., Nagashima, Y., Suzuki, Y., Beier, E.W., Feldscher, L.R., Frank, E.D., Frati, W., Kim, S.B., Mann, A.K., Newcomer, F.M., Van Berg, R., and Zhang, W. 1989a, *Phys. Rev. Letters*, **63**, 16.
Hirata, K.S., Kajita, T., Koshiba, M., Nakahata, M., Oyama, Y., Sato, N., Suzuki, A., Takita, M., Totsuka, Y., Kifune, T., Suda, T., Takahashi, K., Tanimori, T., Miyano, K., Yamada, M., Beier, E.W., Feldscher, L.R., Frati, W., Kim, S.B., Mann, A.K., Newcomer, F.M., Van Berg, R., Zhang, and Cortez, B.G. 1988, *Phys. Rev. D*, **38**, 448.
Hirata, K.S., Kajita, T., Kifune, T., Kihara, K., Nakahata, M., Nakamura, K., Ohara, S., Sato, N., Suzuki, Y., Totsuka, Y., Yaginuma, Y., Mori, M., Oyama, Y., Suzuki, A., Takahashi, K., Takei, H., Tanimori, T., Koshiba, M., Suda, T., Tajima, T., Miyano, K., Miyata, H., Yamada, M., Fukuda, Y., Kaneyuki, K., Nagashima, Y., Takita, M., Beier, E.W., Feldscher, L.R., Frank, E.D., Frati, W., Kim, S.B., Mann, A.K., Newcomer, F.M., Van Berg, R., and Zhang, W. 1989b, "Recent Solar Neutrino Data from the Kamiokande-II Detector," contributed paper to the XIV International Symposium of Lepton and Photon Interactions, Stanford, 1989, ICR Report 195-89-12.
Kirsten, T. 1989, Talk in the *Inside the Sun, IAU Colloq. No. 121.*
Mikheyev, S.P., and Smirnov, A.Yu. 1986, *Nuovo Cimento*, **9C**, 17.
Raghavan, R.S., and Pakvasa, S. 1988, *Phys. Rev. D*, **37**, 849.
Rowley, J.K., Cleveland, B.T., and Davis, R., Jr. 1985, in *Proc. of Conf. on Solar Neutrinos and Neutrino Astronomy* eds. M.L. Cherry, K. Lande, and W.A. Fowler, AIP Conference Proc. No.126 (American Inst. Phys., New York), p.1.
Spergel, D.N., and Press, W.H. 1985, *Astrophys. J.*, **294**, 663.
Totsuka, Y. 1987, in *Proc. of the Seventh Workshop on Grand Unification, ICOBAN '86*, ed. J. Arafune (World Scientific, Singepore), p.118.
Voloshin, M.B., and Vysotskii, M.I. 1986, *Soviet J. Nuclear Phys.*, **44**, 544.
Voloshin, M.B., Vysotskii, M.I., and Okun, L.B. 1986, *Soviet J. Nuclear Phys.*, **44**, 440.

Wolfenstein L. 1978, *Phys. Rev. D*, **17**, 2369.
Wolfsberg, K., Cowan, G.A., Bryant, E.A., Daniels, K.S., Downey, S.W., Haxton, W.C., Niesen, V.G., Nogar, N.S., Miller, C.M., and Rokop, D.I. 1985, in *Proc. of Conf. on Solar Neutrinos and Neutrino Astronomy* eds. M.L. Cherry, K. Lande, and W.A. Fowler, AIP Conference Proc. No.126 (American Inst. Phys., New York), p.196.

Improving the Asymptotic Approximations of Higher-Order Radial Oscillations in Stars

P. Smeyers and E. Ruymaekers *

Astronomisch Instituut, Katholieke Universiteit Leuven, Belgium
*Research Assistant of the National Fund for Scientific Research (Belgium)

Abstract: Since the initial work of Ledoux (1962) on the asymptotic representation of higher-order radial oscillation modes, it has been customary to use the square of the angular frequency as the large parameter in the governing second-order differential equation. In the present investigation, the introduction of a different large parameter is shown to improve the accuracy of the first and second asymptotic approximations of the eigenfrequencies of the radial oscillation modes, for the equilibrium sphere with uniform mass density and the polytropes with index $n = 2$ and $n = 3$. The definition of the new large parameter is based on the use of the second-order differential equation in the divergence of the radial displacement.

1 Introduction

Asymptotic representations of higher-order linear isentropic oscillation modes of stars have become very important in connection with the 5-min oscillations detected at the surface of the Sun. They are useful tools for extracting significant parameters from the frequencies measured and offer potentially large reductions in the computation of eigenfrequencies of solar models if a sufficiently high degree of approximation can be reached (see, e.g., Bahcall and Ulrich 1988, Mullan and Ulrich 1988).

Since the initial work of Ledoux (1962) on the asymptotic representation of radial oscillation modes, it has been customary to use the square of the angular frequency σ as the large parameter. However, in her paper on second asymptotic approximations of higher-order non-radial oscillation modes of stars, Tassoul (1980) observed that a choice of the large parameter other than σ^2 and $1/\sigma^2$ leads to closer asymptotic approximations for the eigenfrequencies of non-radial modes of the equilibrium sphere with uniform mass density.

Therefore, we set ourselves the task of determining a large parameter that would improve the asymptotic approximations of the eigenfrequencies of radial oscillation modes with lower orders κ, which are of interest for the solar 5-min oscillations. The orders of the radial oscillations identified up to now on the solar surface range from $\kappa = 12$ to $\kappa = 33$ (Duvall *et al.* 1988, Pallé *et al.* 1989).

We concentrate on purely radial oscillation modes since, in this case, the Eulerian perturbation of the gravitational potential must not be neglected and as a result the errors of the eigenfrequencies stem only from the approximations made in the course of the asymptotic treatment.

2 Asymptotic Approximations of the Radial Oscillation Modes of a Star

Linear isentropic radial oscillations of a star are governed by the second-order differential equation of the canonical form

$$\frac{d^2\zeta}{dr^2} + \left[\frac{\sigma^2}{c^2} + M_0(r)\right]\zeta = 0\,, \tag{1}$$

where

$$\zeta = r^{-3}\left(\rho c^2\right)^{-1/2}\delta r\,, \tag{2}$$

$$M_0(r) = -\frac{2}{r^2} - \frac{1}{2\rho c^2}\frac{d^2(\rho c^2)}{dr^2} + \frac{1}{4}\left[\frac{1}{\rho c^2}\frac{d(\rho c^2)}{dr}\right]^2 + \frac{1}{r\rho c^2}\frac{d(\rho c^2)}{dr} + \frac{4g}{rc^2}\,. \tag{3}$$

In this equation, r is the radial distance from the center, σ the angular frequency in the time dependency of the form $\exp(i\sigma t)$, ρ the mass density, g the gravity, c the isentropic sound velocity, and δr the radial displacement.

One generally assumes that the mass density can be expanded near $r = R$ in a power series as

$$\rho = \rho_0 (R-r)^{n_e}\left[1 + \frac{\rho_1}{\rho_0}(R-r) + \cdots\right], \tag{4}$$

where n_e is the effective polytropic index characterizing the star's surface layers. Associated series expansions are derived for the pressure P and the square of the isentropic sound velocity c. It follows that the second and third terms of the coefficient $M_0(r)$ have a double pole at $r = R$ and that the coefficient c^{-2} of σ^2 and the last two terms of the coefficient $M_0(r)$ have a simple pole at $r = R$.

In the usual constructions of first asymptotic approximations from $r = R$ by means of the methods devised by Langer (1935) and Olver (1974), the last term in Definition (3) of the coefficient $M_0(r)$, $4g/(rc^2)$, is neglected in comparison to the terms that display a double pole at $r = R$. Note that this term contains c^2 in its denominator as does the term σ^2/c^2 and that, at $r = R$, $4g/r$ becomes equal to 4 times the inverse of the square of the star's dynamic time scale $(GM/R^3)^{1/2}$. For higher-order radial oscillation modes, σ^2 is very large in comparison to $4GM/R^3$ so that the term $4g/(rc^2)$ can be neglected. However, when σ^2 is not so large, the term $4g/(rc^2)$ may no longer be negligible in comparison to the term σ^2/c^2 near $r = R$, and it may be suitable to incorporate the term $4GM/R^3$ into the large parameter.

The question then arises whether no other terms of the coefficient $M_0(r)$ that display a simple pole at $r = R$ must be included in the large parameter. After several indecisive attempts, we considered the second-order differential equation in $\alpha = \nabla \cdot \delta \boldsymbol{r}$

$$\frac{d^2 w}{dr^2} + \left[\frac{\sigma^2}{c^2} + \frac{4}{r}\frac{g}{c^2} + \frac{1}{2}\frac{1}{\rho}\frac{d^2\rho}{dr^2} - \frac{3}{4}\frac{1}{\rho^2}\left(\frac{d\rho}{dr}\right)^2 + \frac{1}{\rho}\frac{d\rho}{dr}\left(\frac{1}{r} - \frac{1}{2}\frac{1}{z}\frac{dz}{dr}\right)\right.$$
$$\left. + \frac{1}{r}\frac{1}{z}\frac{dz}{dr} + \frac{1}{2}\frac{1}{z}\frac{d^2 z}{dr^2} - \frac{3}{4}\frac{1}{z^2}\left(\frac{dz}{dr}\right)^2\right] w = 0\,, \tag{5}$$

where

$$w = \rho^{1/2} r\, z^{-1/2} c^2 \alpha\,, \tag{6}$$
$$z = \sigma^2 + 4 g r^{-1}\,. \tag{7}$$

We defined the large parameter involving all terms of Eq. (5) that display a simple pole at $r = R$

$$\lambda^2 = \sigma^2 + \frac{GM}{R^3}\left[4 - \Gamma_{1,R}\frac{n_e}{n_e+1}\left(1 + \frac{n_e}{2}\right)\right] \tag{8}$$

and observed that the use of this large parameter in Eqs. (1) and (5) leads to identically improved first and second asymptotic approximations of the eigenfrequencies of lower-order radial oscillation modes of the equilibrium sphere with uniform mass density and of the polytropes $n = 2$ and $n = 3$. In Definition (8), $\Gamma_{1,R}$ is the value of $\Gamma_1 = (\partial \ln P/\partial \ln \rho)_S$ at $r = R$ and is set equal to 5/3 in our computations.

The first asymptotic approximations σ_1 of the eigenfrequencies are determined by means of an eigenvalue equation identical to that of Ledoux (1962) except that σ^2 is replaced by λ^2. Consequently, these approximations are given by

$$\sigma_1 = \left\{\left[\left(\int_0^R \frac{dr}{c}\right)^{-1}\left(2\kappa + n_e + \frac{1}{2}\right)\frac{\pi}{2}\right]^2 - \frac{GM}{R^3}\left[4 - \Gamma_{1,R}\frac{n_e}{n_e+1}\left(1+\frac{n_e}{2}\right)\right]\right\}^{1/2}, \tag{9}$$

where κ is equal to the number of nodes of the eigenfunction δr between $r = 0$ and $r = R$.

The eigenfunctions associated with the first asymptotic approximations σ_1 of the eigenfrequencies are identical to those associated with Ledoux' (1962) first asymptotic approximations σ_L.

The second asymptotic approximations σ_2 of the eigenfrequencies are determined by an eigenvalue equation with a form similar to that of the eigenvalue equation of Tassoul and Tassoul (1968)

$$\lambda \sin\left(\lambda \int_0^R \frac{dr}{c} - n_e\frac{\pi}{2} - \frac{5\pi}{4}\right) + \gamma \cos\left(\lambda \int_0^R \frac{dr}{c} - n_e\frac{\pi}{2} - \frac{5\pi}{4}\right) = 0\,, \tag{10}$$

where γ is a constant determined as follows:

$$\gamma = \frac{1}{2} R \left[\frac{1}{\rho c}\frac{d^2(\rho c^2)}{dr^2}\right]_{r=0} - \frac{1}{2}\frac{GM}{R^3}\left[4 - \Gamma_{1,R}\frac{n_e}{n_e+1}\left(1+\frac{n_e}{2}\right)\right]\int_0^R \frac{dr}{c}$$
$$+ \frac{1}{4}\int_0^R \frac{1}{(R-r)^{1/2}}\frac{d}{dr}\left\{(R-r)^{3/2} c \left\{\frac{2}{\rho c^2}\frac{d^2(\rho c^2)}{dr^2} - \left[\frac{1}{\rho c^2}\frac{d(\rho c^2)}{dr}\right]^2\right\}\right\} dr$$
$$+ \frac{1}{2}\int_0^R \left[\frac{1}{r\rho}\frac{d(\rho c^2)}{dr} - \frac{1}{8}\frac{d^2 c^2}{dr^2} - \frac{1}{r}\frac{dc^2}{dr} + \frac{4g}{r}\right]\frac{dr}{c}\,. \tag{11}$$

The second term in the expression for γ does not appear when σ^2 is used as the large parameter, as was done by Tassoul and Tassoul (1968).

3 Results

Table 1 contains the exact eigenfrequencies and their second asymptotic approximations determined successively by means of the eigenvalue equation of Tassoul and Tassoul (1968) and our Eq. (10) for the radial oscillation modes with order $\kappa = 10$ and $\kappa = 30$ of the equilibrium sphere with uniform mass density and the polytropes $n = 2$ and $n = 3$. The eigenfrequencies are denoted as σ_{ex}, σ_T, and σ_2, respectively, and are expressed in the unit $(GM/R^3)^{1/2}$. For the second asymptotic approximations σ_T and σ_2, the relative errors with respect to the corresponding exact eigenfrequencies are given.

For the equilibrium sphere with uniform mass density, the relative errors of the second asymptotic approximations σ_2 of the eigenfrequencies are smaller than 2×10^{-6} in absolute value from the radial oscillation mode with order $\kappa = 3$. The improvement of the accuracy is substantially smaller for the second asymptotic approximations σ_2 of the eigenfrequencies of the polytropes $n = 2$ and $n = 3$. In these cases, the relative errors are reduced by a factor somewhat larger than 4 in comparison to the errors of the eigenfrequencies obtained by means of the eigenvalue equation of Tassoul and Tassoul (1968).

From eigenvalue Eq. (9), it follows that, even in the first asymptotic approximation, the separation between the eigenfrequencies of radial oscillation modes with successive orders κ and $\kappa - 1$ is not constant. Since λ^2 is larger than σ^2 for the models considered, the separation decreases as the order κ increases and leads to a concave-left curvature in a usual echelle diagram (see Mullan and Ulrich 1988). For large values of κ, the separation tends towards the constant value derived with the use of σ^2 as large parameter.

Table 1.

	κ	σ_{ex}	σ_T	rel. error	σ_2	rel. error
$n = 0$	10	20.4368947413	20.4376979690	3.93 (−5)	20.4368942412	−2.45 (−8)
	30	57.0175411606	57.0175787339	6.59 (−7)	57.0175411734	−4.07 (−10)
$n = 2$	10	16.4077	16.4223	8.9 (−4)	16.4106	1.8 (−4)
	30	43.7542	43.7551	2 (−5)	43.7544	5 (−6)
$n = 3$	10	15.4250	15.4451	1.3 (−3)	15.4297	3.0 (−4)
	30	40.3678	40.3699	5 (−5)	40.3683	1 (−5)

References

Bahcall, J.N., and Ulrich, R.K. 1988, *Rev. Mod. Phys.* **60**, 297.

Duvall, T.L., Harvey, J.W., Libbrecht, K.G., Popp, B.D., and Pomerantz, M.A. 1988, *Astrophys. J.* **324**, 1158.

Langer, R.E. 1935, *Trans. Amer. Math. Soc.* **37**, 397.

Ledoux, P. 1962, *Bull. Acad. Roy. Belg., Cl. Sci., 5e Série,* **48**, 240.

Mullan, D.J., and Ulrich, R.K. 1988, *Astrophys. J.* **331**, 1013.
Olver, F.W.J. 1974, *Asymptotics and Special Functions* (Academic Press, New York).)
Pallé, P.L., Pérez Hernández, P.L., Roca Cortés, T., and Isaak, G.R. 1989, *Astron. Astrophys.* **216**, 253.
Tassoul, M. 1980, *Astrophys. J. Suppl.* **43**, 469.
Tassoul, M., and Tassoul, J.-L. 1968, *Astrophys. J.* **153**, 127.

Second-Order Asymptotic Theory of Solar Acoustic Oscillations

S. V. Vorontsov

Institute of Physics of the Earth, Moscow 123810, USSR

Abstract: The second-order asymptotic theory is developed for low- and intermediate-degree acoustic oscillations of the Sun and Sun-like stars, in a form convenient to use for helioseismic inversions.

The asymptotic theory of high-frequency acoustic oscillations is widely used in solar seismology, in particular as a basis for constructing techniques of nonlinear inversion of observational data. It usually assumes Cowling approximation and asymptotic expansion limited by terms of $1/\omega$. Both sound-speed inversions (e.g. Vorontsov and Zharkov 1989) and direct computations (Kosovichev 1988) show however that gravity perturbation and higher-order terms must be taken into account to study the structure of the solar core. First attempt to include the effects of gravity perturbation (Vorontsov 1988) used the variational principle. The present report describes more general asymptotic theory developed without any recourse to Cowling approximation. Effects of gravity perturbation (and buoyancy forces) are included together with all other terms of order $1/\omega^2$.

Although the second-order asymptotics can be written down directly for a full fourth-order system of differential equations [such problems are known in geoseismology, e.g., Woodhouse (1978)], somewhat different approach was found to be more convenient. By the appropriate choice of dependent variables, the equations of linear adiabatic oscillations of a self-gravitating star are reduced to the form

$$\frac{d}{dr}\begin{pmatrix}\omega\xi\\ \eta\\ P\\ \omega S\end{pmatrix} =$$

$$\omega\begin{pmatrix} 0 & \frac{h_2}{h_1}\left[\frac{l(l+1)}{\omega^2}-\frac{r^2}{c^2}\right] & -\frac{1}{h_1}\frac{l(l+1)}{\omega^2} & 0\\ \frac{1}{r^2}\frac{h_1}{h_2}\left[1-\frac{N^2}{\omega^2}+\frac{4\pi G\rho}{\omega^2}\right] & 0 & 0 & \frac{1}{h_2}\frac{l(l+1)}{r^2\omega^2}\\ \frac{4\pi G\rho h_1}{r^2\omega^2} & 0 & 0 & \frac{l(l+1)}{r^2\omega^2}\\ 0 & -\frac{4\pi G\rho h_2}{\omega^2} & 1+\frac{4\pi G\rho}{\omega^2} & 0 \end{pmatrix}\begin{pmatrix}\omega\xi\\ \eta\\ P\\ \omega S\end{pmatrix}, \qquad (1)$$

where $\xi(r)$, $\eta(r)$, and $P(r)$ define radial displacements, Eulerian pressure perturbations, and gravitational potential perturbations:

$$\delta r = \frac{h_1}{r^2}\xi(r)Y_l^m(\theta,\phi), \qquad (2)$$

$$p' = \rho h_2 \eta(r) Y_l^m(\theta, \phi), \tag{3}$$

$$\Phi' = -P(r) Y_l^m(\theta, \phi), \tag{4}$$

$$l(l+1)S(r) = r^2 \frac{dP}{dr} - 4\pi G \rho h_1 \xi, \tag{5}$$

$$h_1(r) = \exp\left(\int_0^r \frac{g}{c^2} dr\right), \tag{6}$$

$$h_2(r) = \exp\left(\int_0^r \frac{N^2}{g} dr\right), \tag{7}$$

and other notations are common. The special structure of (1) admits its reduction to the system of two first-order vector equations

$$\frac{d}{dr}\begin{pmatrix}\eta \\ P\end{pmatrix} = \omega \mathbf{A}_1(\omega, r)\begin{pmatrix}\omega\xi \\ \omega S\end{pmatrix} \tag{8}$$

and

$$\frac{d}{dr}\begin{pmatrix}\omega\xi \\ \omega S\end{pmatrix} = \omega \mathbf{A}_2(\omega, r)\begin{pmatrix}\eta \\ P\end{pmatrix}, \tag{9}$$

where the components of matrices $\mathbf{A}_1$ and $\mathbf{A}_2$ are corresponding components of matrix in (1). In solar interior (where $\mathbf{A}_1$ is regular) this system is equivalent to one second-order vector equation

$$\frac{d}{dr}\left[\mathbf{A}_1^{-1}\frac{d}{dr}\begin{pmatrix}\eta \\ P\end{pmatrix}\right] - \omega^2 \mathbf{A}_2 \begin{pmatrix}\eta \\ P\end{pmatrix} = 0 \tag{10}$$

convenient for asymptotic analysis. Studying the oscillations of different degree, we define $w \equiv \omega/L$ with $L \equiv l + 1/2$ and use w as a parameter. We are seeking the solutions of (10) in terms of Airy functions

$$\begin{pmatrix}\eta \\ P\end{pmatrix} = \left(Y_0 + \frac{1}{\omega}Y_1 + \frac{1}{\omega^2}Y_2 + \ldots\right)\begin{pmatrix}\omega^{1/6} Ai(-\omega^{2/3}\phi) \\ \omega^{-1/6} \dot{A}i(-\omega^{2/3}\phi)\end{pmatrix} \tag{11}$$

and obtain finally the second-order (uniform) asymptotic approximation for $\eta(r)$ in solar interior.

Near the surface any asymptotic solutions become invalid and must be matched with exact solutions. This is done as described in Vorontsov and Zharkov (1989) using the facts that the effects of gravity perturbation are negligible here and exact solutions depend only on frequency, at least when degree l is not too high.

The final eigenfrequency equation is

$$F(w) + \frac{1}{\omega^2}P(w) = \frac{\pi[n + \alpha(\omega)]}{\omega}, \tag{12}$$

where all the second-order effects are determined by $P(w)$:

$$F(w) = \int_{r_1}^{R} (s^2)^{1/2} dr, \tag{13}$$

$$s^2 = \frac{1}{c^2} - \frac{1}{r^2 w^2}, \tag{14}$$

$$P(w) = \frac{1}{2}\int_{r_1}^{r_m} (s^2)^{-1/2}\Big[-\frac{(\ln h)''}{2} - \frac{(\ln h)'^2}{4} - \frac{(\ln h)'}{r} - \frac{(\ln \Psi')''}{2} + \frac{(\ln \Psi')'^2}{4} - N^2 s^2 + \frac{4\pi G\rho}{c^2} + \frac{1}{4r^2}\Big] dr, \tag{15}$$

$$h = h_2/h_1, \tag{16}$$

and

$$\Psi = \mathrm{sgn}(s^2)\left|\frac{3}{2}\int_{r_1}^{r} |s^2|^{1/2} dr\right|^{2/3}. \tag{17}$$

Here r_1 is turning point corresponding to $s^2(r) = 0$. The upper limit of the integral in $P(w)$ is matching point r_m, which is somewhere in the deep adiabatic regions of the convection zone. Its exact position is insignificant because second-order terms are low there.

It can be shown that $F(w)$ and $P(w)$ satisfy following conditions:

$$\lim_{1/w \to 0} \frac{dF}{d(1/w)} = -\frac{\pi}{2} \tag{18}$$

and

$$\lim_{1/w \to 0} \frac{d^3 F}{d(1/w)^3} = 0. \tag{19}$$

These boundary conditions can be used in the inversion technique as an additional a priori information capable to improve significantly the accuracy and stability of the results near the solar center.

The study of the second-order terms show that $P(w)$ is regular at $1/w \to 0$ if (and only if) the parameter $w = \omega/L$ is defined with $L = l + 1/2$. It explains the well-known improvement of the first-order results (usually obtained with the JWKB technique) when this definition is used: this definition is needed for asymptotic expansion to be convergent.

Direct computations of p-mode frequencies were done using (12)-(17) in frequency range 1.5 mHz – 4.5 mHz for solar model 1 of Christensen-Dalsgaard. Results of comparison with exact eigenfrequencies are shown in Figs. 1 and 2 for the first- and second-order approximations, respectively. Accuracy of asymptotic description is significantly improved, especially for low-degree modes, indicating the real capabilities in sounding the structure in the solar core.

Acknowledgements

Author thanks M. A. Brodsky for valuable discussions and collaborative development of the second-order asymptotic technique in the Cowling approximation. He also thanks J. Christensen-Dalsgaard for his solar model 1 used in the computations. This work was partly supported by "Zodiac" Scientific-Methological Council.

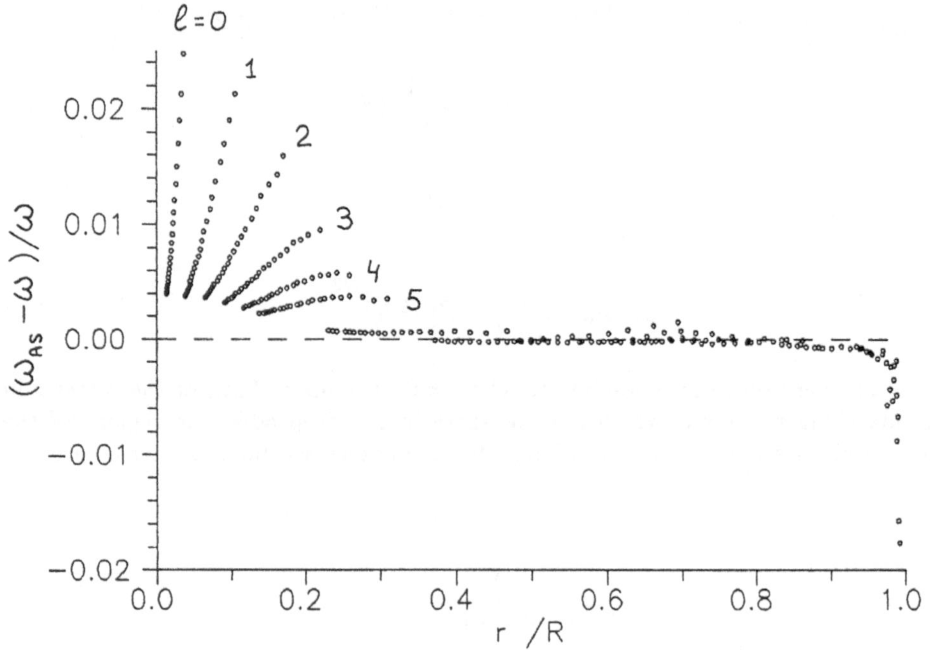

Fig. 1. Relative deviations of the first-order asymptotic eigenfrequencies from the exact eigenfrequencies versus penetration depth of p-modes. Results are shown for $l = 0$ to 5 step 1, 10 to 50 step 10, 100 to 500 step 100.

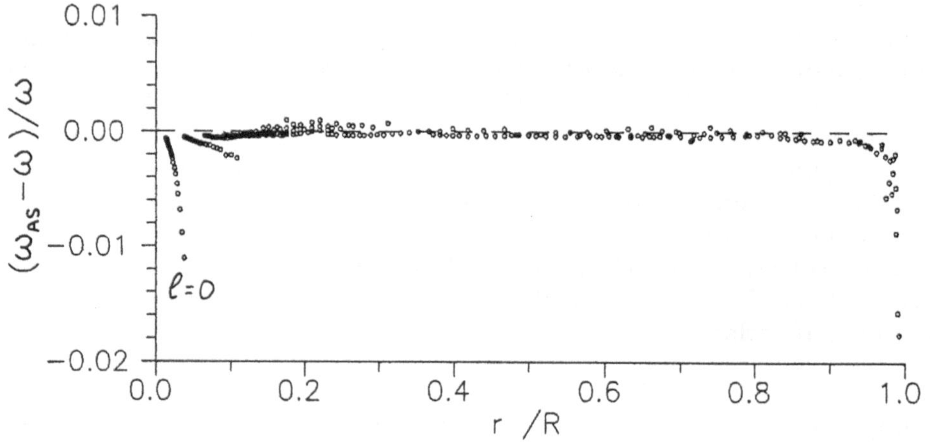

Fig. 2. The same as Fig. 1, but for the second-order asymptotic approximation.

References

Kosovichev, A.G. 1988, in *Seismology of the Sun and Sun-like Stars*, ed. E. Rolfe, ESA SP-286 (ESA Publication Division, Noordwijk), p.533.
Vorontsov, S.V. 1988, in *Seismology of the Sun and Sun-like Stars*, ed. E. Rolfe, ESA SP-286 (ESA Publication Division, Noordwijk), p.475.
Vorontsov, S.V., and Zharkov, V.N. 1989, *Soviet Sci. Rev. E, Astrophys. Space Phys. Rev.*, **7**, 1.
Woodhouse, J.H. 1978, *Geophys. J. Roy. Astron. Soc.*, **54**, 263.

III

Excitation Mechanisms of Oscillations

Excitation Mechanisms of Solar Oscillations

Yoji Osaki

Department of Astronomy, University of Tokyo, Bunkyo-ku, Tokyo 113, Japan

Abstract: Excitation mechanisms of solar oscillations are reviewed. Two excitation mechanisms have been proposed for solar p-mode oscillations: the linear overstability due to the so-called κ-mechanism and the stochastic excitation by turbulent convection. Here we discuss the stochastic excitation mechanism in a greater detail because it is more likely to be the possible explanation than the former mechanism. In this mechanism, the amplitudes of oscillations are determined by the balance of the excitation rate by turbulent convection to the damping rate. We examine these two effects separately. In particular, we discuss the acoustic energy generation by turbulent convection and the question of quadrupole versus dipole radiation.

1 Introduction

One of the most important but unsolved problems in helioseismology is the problem of the excitation and damping of solar oscillations. That is the question why the sun is oscillating in so many p-modes having the peak power at five minutes and with such observed amplitudes and line-widths. While helioseismology is very successful in proving the internal structure of the sun by using very accurate frequencies both in observation and theory, the study of the excitation mechanism of solar oscillations still remains in a very rudimentary stage. In fact, we are still struggling to explain the observed amplitudes of oscillations by order of magnitude and we still fail to explain the observed frequency dependence of amplitudes of solar oscillations. The problem of excitation and damping of solar oscillations is very important in helioseismology because the observed amplitudes and line widths of solar p-mode oscillations can also provide us another information about the interior of the sun, in particular, about the solar convection zone and its poorly understood convective motion.

As for the excitation mechanism of solar oscillations, two different mechanisms have in general been recognized: (1) the linear (or thermal) overstability of eigenmodes due to the κ-mechanism and (2) the stochastic excitation of oscillations by turbulent convection. In this review, we mainly discuss the stochastic excitation mechanism and we mention only briefly the thermal overstability mechanism because the stochastic excitation mechanism is thought to be more likely responsible for the excitation of solar oscillations.

2 Thermal Overstability

The linear (or thermal) overstability by the so-called κ mechanism is the very mechanism that is thought to be responsible for the excitation of pulsations in classical pulsating variables such as Cepheid and RR Lyrae stars. The possibility of thermal overstability of solar 5-min oscillations was discussed by Ando and Osaki (1975) among others. This mechanism or the κ mechanism works in the hydrogen ionization zone just below the solar photosphere. The opacity κ is basically due to the negative hydrogen there and it has a strong temperature dependence with $\kappa \propto T^{10}$. If we consider the radiative energy flow in the hydrogen ionization zone in a cycle of solar or stellar oscillation, the radiative flux is then blocked in the phase of compression while it is enhanced in the phase of expansion in such a circumstance. This kind of effect in the hydrogen ionization then tends to drive the oscillation, while the optically thin atmospheric layer always contributes to the damping of oscillation. By solving fully non-adiabatic equation for nonradial oscillations of a solar envelope model, Ando and Osaki (1975) found that many of solar p-modes are in fact overstable and that the highest growth rates occur for nonradial p-modes having frequencies of about 5-min in the diagnostic (k, ω)-diagram. Their results looked thus encouraging for the thermal overstability mechanism. However, numerical results of overstability are rather delicate and they may easily be changed from instability to stability because they are determined by the difference of a positive contribution (driving) to a negative contribution (damping) .

Furthermore, there exist several problems in this thermal overstability model. Firstly, the effects of coupling of oscillations with convection was completely neglected in Ando and Osaki (1975). Since the convective energy transport is the major source of energy transfer in the solar convection zone, its neglect is fatal. In fact, at least, the turbulent viscosity works to the damping of oscillations and Goldreich and Keeley (1977a) argued that thermally overstable p-modes could be damped by the turbulent viscous dissipation. The second problem of linear overstability is the problem of amplitude limitation of unstable modes. If solar p-modes are really overstable, some nonlinear mechanisms must limit the amplitudes of p-modes in finite values. However, we still do not know how to limit the amplitudes of oscillations. In fact, the observed amplitude per mode is about 10 cm s^{-1} at the solar surface. This corresponds to the radial displacement, δR, as small as about $\delta R/R \sim 10^{-8}$ where R is the solar radius. Comparing this with the corresponding value of the Cepheid variable stars with $\delta R/R \sim 0.1$, we can recognize how small the amplitudes of the solar oscillation are. Kumar and Goldreich (1989) have examined the nonlinear interactions among solar acoustic modes and they have concluded that nonlinear coupling cannot limit the growth of overstable p-modes within observed values. Unless this difficulty is overcome, the hypothesis of the stochastic excitation may be more favorable than the model of thermal overstability.

3 Stochastic Excitation Mechanism

3.1 Basic Mechanism

Goldreich and Keeley (1977b) was the first to suggest the stochastic excitation of solar oscillations by turbulent convection. The outer 30% in radius of the sun is covered by the convection zone. The turbulent convection there generates "acoustic noise" while the sun may be regarded as an acoustic cavity as far as its oscillation property is concerned. The acoustic noise generated by turbulent convection then excites resonant modes of the cavity and the observed oscillations are thought to be the manifestation of thus excited oscillation. This is the basic idea of the stochastic excitation by the turbulent convection. In fact, we observe the granulation motion that is the manifestation of the convective motion in the solar atmosphere. Its life time is about 8-min and it is comparable to the period of the solar 5-min oscillation.

In this model, the sun is regarded as an acoustic resonator having many eigenmodes with discrete frequencies. It is supposed that these eigenmode oscillations are intrinsically stable so that they are damped oscillators. The turbulent convection in the solar convection zone generates broad-band acoustic noise in frequency, which in turn excites eigenmode oscillations of the cavity at resonant frequencies. In the stochastic excitation model, the amplitudes of oscillations are determined by the balance of the excitation rate by turbulent convection to the damping rate. Major advantages of this model are that it can explain the following two points; (1) millions of modes are excited, and (2) their amplitudes are expected rather low.

Following Goldreich and Keeley (1977b), we write the energy of a particular eigenmode, its angular frequency, and damping rate by E_q, ω_q, and γ_q where the subscript, q, signifies a particular eigenmode. The energy of the mode is then given by

$$E_q = 2 \times \langle \frac{1}{2} \int_{m_\odot} \boldsymbol{v}_q^2 \, dm \rangle = \langle \int_{m_\odot} \boldsymbol{v}_q^2 \, dm \rangle, \tag{1}$$

where $\boldsymbol{v}_q$ denotes the velocity amplitude of the oscillation mode and the integration is performed over the whole mass of the sun. Here the angular bracket means time average and we note that the energy of oscillation is the sum of the kinetic energy plus the potential energy. The equation of energy of the mode is then given by

$$\frac{dE_q}{dt} = G - 2\gamma_q E_q = G - \Gamma E_q, \tag{2}$$

where G denotes the acoustic power by turbulent convection available to a particular mode in our interest, and $\Gamma = 2\gamma_q$ is the energy dissipation rate of the mode (or observationally the line width of the mode in angular frequency unit). The energy of oscillation of the mode is then given by

$$E_q = \frac{G}{\Gamma} \,. \tag{3}$$

Thus, the central problem of this model is to know theoretically how the damping rate Γ and the acoustic noise generation rate G by turbulent convection are determined, respectively, and we shall address these two problems below.

3.2. Damping Rate of Oscillations

We may divide the damping rate of oscillation into two parts:

$$\Gamma = \Gamma_{\rm dyn} + \Gamma_{\rm therm}, \tag{4}$$

where $\Gamma_{\rm dyn}$ and $\Gamma_{\rm therm}$ stand for damping rates of oscillation that are caused through the momentum equation and the thermal equation, respectively. Furthermore, each of them can be written by several contributions by different causes. In fact, the dynamical damping rate consists of those due to the turbulent viscous dissipation, due to the leakage of waves from boundaries, and due to energy transfer to other modes by nonlinear interaction. That is,

$$\Gamma_{\rm dyn} = \Gamma_{\rm visc} + \Gamma_{\rm leak} + \Gamma_{\rm coupl}, \tag{5}$$

where the turbulent viscous damping and the damping due to the leakage of waves are supposed to be always positive but the term due to the mode coupling may be either positive or negative. On the other hand, the thermal damping is due to the so-called non-adiabatic effects of oscillation and $\Gamma_{\rm therm}$ may be written into two parts:

$$\Gamma_{\rm therm} = \Gamma_{\rm rad} + \Gamma_{\rm conv}, \tag{6}$$

where $\Gamma_{\rm rad}$ and $\Gamma_{\rm conv}$ stand for the damping rates of oscillation due to radiative and convective flux variations. The thermal damping rate $\Gamma_{\rm therm}$ can be negative (i.e., "negative" dissipation) as discussed before in the thermal overstability. Generally speaking, it is very difficult to estimate theoretically the damping rate of a mode because so many different causes are involved. The most difficult among them is to estimate the term $\Gamma_{\rm conv}$ because we need to treat "time-dependent" convection, whose theory is the well-known unsolved problem in astrophysics. Under such circumstance, we think that observations rather than the theory guide us on this problem.

3.3. Acoustic Noise Generation

The generation of acoustic noise by turbulence is known as the Lighthill mechanism (Lighthill 1952). Let us consider the acoustic power generated by a turbulent eddy of size $\lambda_{\rm T}$ having velocity v_λ, density ρ, and Mach number $M = v_\lambda/c$. According to the Lighthill theory (see, e.g., Stein and Leibacher 1981), the total acoustic power generated by a turbulent eddy is given by

$$P \sim \frac{\rho v_\lambda^3}{\lambda_{\rm T}} M^{2n+1}, \tag{7}$$

where n is the multipole index. Since the turbulence is thought to be sub-sonic, the lowest order multipole is the most effective. The lowest three multipoles are;
(1) monopole radiation or a "mass source" with $n = 0$, which occurs when a parcel of gas executes periodically expansion and contraction,
(2) dipole radiation or a "momentum source" with $n = 1$, which occurs when the external force acts on the gas,
(3) quadrupole radiation with $n = 2$, which is emitted when turbulent Reynolds stresses act as the source term.

In the original Lighthill mechanism, the quadrupole radiation is the lowest order multipole generated by turbulence.

Goldreich and Keeley (1977b) presented an interesting argument that there will be an equipartition between an oscillation mode and its "resonant" eddy if the damping of the oscillation modes is due to the turbulent viscous dissipation while the excitation of oscillation is due to quadrupole radiation by the turbulent eddy. That is, the energy of an oscillation mode is given by

$$E_q \sim \mathcal{M}_\lambda v_\lambda^2, \tag{8}$$

where $\mathcal{M}_\lambda \sim \rho\lambda_{\mathrm{T}}^3$ is the mass of the resonant eddy.

3.4. Comparison with Observations

Let us now compare theory with observations. As for the theoretical estimate, we use the result of Goldreich and Keeley (1977b) which was based on the Lighthill mechanism, that is, the quadrupole radiation for the acoustic noise generation. As for the observations, Libbrecht (1988a,b) summarizes his recent observations of line widths and amplitudes of solar p-modes. Libbrecht (1988a,b) notes that damping rates and amplitudes of p-modes depend only on frequencies and they are independent of the spherical harmonic degree l of modes if it is sufficiently low, i.e., $l < 100$.

(1) Line-widths (Γ) :

Let us first examine the line-widths or damping rates of oscillations. Figure 1 compares observed line widths of solar p-mode oscillations by Libbrecht (1988a,b) with those of theoretical estimate due to the turbulent viscous damping Γ_{visc} calculated by Goldreich and Keeley (1977a). The most important feature is the observational plateau near 3 mHz in the curve of the line widths as the function of frequency of oscillation modes.

We see from Fig. 1 that the line width due to turbulent viscous damping is the same order of magnitude to the observed line width around $\nu \sim 3\,\mathrm{mHz}$, but it is definitely smaller than the observed line width for frequency ν less than 2.5 mHz.

On the other hand, Christensen-Dalsgaard, Gough, and Libbrecht (1989) compared the observed line widths with the theoretical damping rates obtained by assuming the thermal damping Γ_{therm} alone. They found that the observed line widths are more favorably compared with theoretical ones if the time-dependent convection is considered in calculating damping rates. They have shown that line widths of low frequency modes with $\nu < 2\,\mathrm{mHz}$ can only be explained by including the convective flux perturbation in the deeper part of the convective envelope while the radiative damping in the atmosphere is responsible for line widths of high frequency modes with $\nu > 4\,\mathrm{mHz}$.

In any way, the line widths of solar oscillations are not likely to be explained by turbulent damping alone but that the thermal coupling of oscillations with radiation field and convection is as important as the turbulent damping in determining damping rates of oscillations. If this is the case, the arguments such as the energy equi-partition between an eigenmode of oscillation and its resonant eddy may not be applicable for the solar oscillations because they are based on balancing turbulent stochastic excitation to turbulent viscous damping.

(2) Energy per mode (E_q) :

Libbrecht (1988a,b) has presented observed energy per mode as a function of frequency, which exhibits a peak value of 10^{28} ergs/mode near $\nu = 3\,\mathrm{mHz}$. The observations

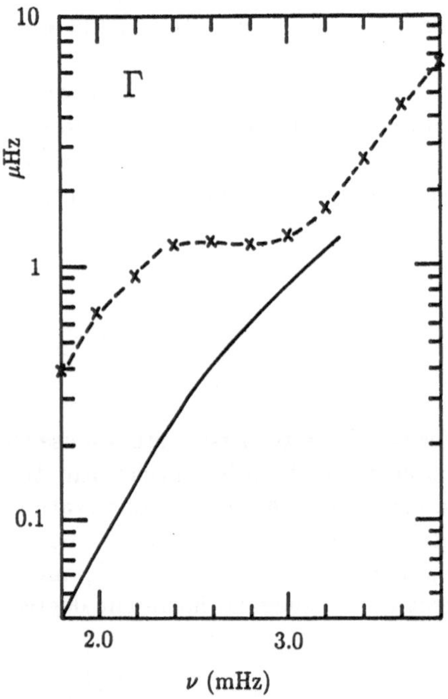

Fig. 1. Linewidths Γ as a function of frequencies. The crosses connected by dashed line show observed linewidths of modes obtained by Libbrecht (1988a) while the solid curve exhibits the calculated results of the turbulent viscous damping rate $\Gamma = \gamma_q/\pi$ by Goldreich and Keeley (1977a).

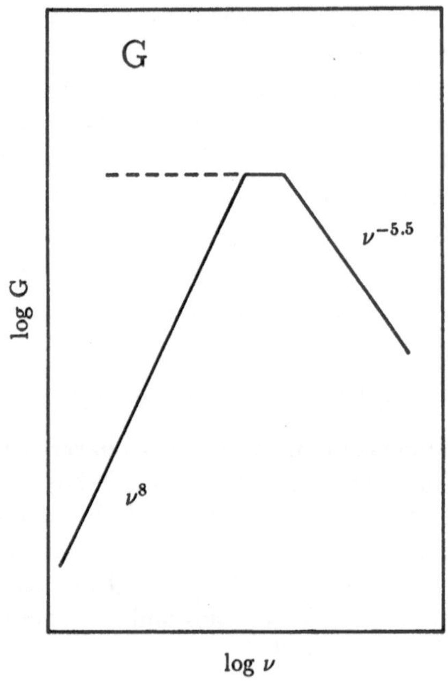

Fig. 2. Schematic drawing for the acoustic power G pumped into an individual p-mode as a function of frequency based on our model. See the text of sub-section 3.6 for a detailed explanation. Dashed line shows the low frequency acoustic power without attenuation correction. The figure is in log-log scale.

may be compared with the theoretical estimate of Goldreich and Keeley (1977b) based on the stochastic excitation model. We find that the theoretical values are smaller by a factor 10^4 around $\nu = 3\,\mathrm{mHz}$ than observed ones and its frequency dependence predicts a monotonically decreasing form: that is a disagreement with observation.

(3) Acoustic power ($G = E_q \times \Gamma$) :

Libbrecht (1988a,b) has noted that the acoustic power pumped into the mode can be estimated purely from observation by plotting $E_q \times \Gamma$. [i.e., Fig. 3 of Libbrecht (1988b)]. We can compare this observation with the theoretical estimate of acoustic power per mode, G, based on the stochastic excitation mechanism, which may be estimated from the result of Goldreich and Keeley (1977a,b). Again, the disagreement is enormous in the sense that the theoretical power is deficient by a factor 10^3 around $\nu = 3\,\mathrm{mHz}$ to explain the observed power.

3.5. Quadrupole or Dipole ?

Goldreich and Kumar (1988) have suggested under such a circumstance that the turbulent convection may emit the acoustic dipole radiation and that the acoustic emissivity is enhanced by a factor M^{-2} compared with the standard Lighthill formulation where M is the Mach number. They have argued that in the case of turbulent convection the emissivity is enhanced by a factor M^{-2} as compared to free turbulence but no enhancement of absorptivity and that the excited modes by this mechanism then attain the energies $E_q \sim \mathcal{M}_\lambda c^2$, where $\mathcal{M}_\lambda$ and c are the mass of the resonant eddy and the sound speed, respectively.

The possibility of acoustic dipole and monopole radiation from the solar convection zone was recognized long time ago by Unno (1964) who had noticed that the effects of density- and pressure-stratification in the solar convection zone can produce the dipole and quadrupole radiation. Stein (1967) examined this issue very carefully and he concluded that the main source of acoustic radiation from the solar convection zone was the quadrupole radiation although there was some contribution from the dipole and monopole radiation.

Let us now scrutinize Goldreich and Kumar's (1988) theory which claims the enhanced emission by the dipole radiation. According to the Lighthill formulation, acoustic wave generation is governed by the following equation:

$$\left(\frac{\partial^2}{\partial x_i^2} - \frac{1}{c^2}\frac{\partial^2}{\partial t^2}\right)\rho = \frac{1}{c^2}\left(\frac{\partial F_i}{\partial x_i} - \frac{\partial^2 T_{ij}}{\partial x_i \partial x_j}\right), \tag{9}$$

where the left-hand side describes the wave propagation while the right-hand side of the equation describes the source terms. The first term of the right-hand side gives the source term due to the presence of external force and this term gives rise the dipole radiation. On the other hand, the second term gives the ordinary source term due to the turbulent Reynolds stresses, producing the quadrupole radiation. In their model of the so-called turbulent pseudo-convection, Goldreich and Kumar (1988) have argued that the fluctuating buoyancy force in the turbulent convection acts like an external force of equation (9) and that it thus results in the dipole radiation.

Two questions are here raised for their formulation;

(1) A question about their formulation itself :

In the Lighthill formulation of acoustic wave generation, it is usual to put all linear terms in the left-hand side and all nonlinear terms in the right-hand side of equation. The left-hand side then describes the wave propagation and the right-hand side represents the source terms. Goldreich and Kumar (1988) have assumed the the fluctuating buoyancy force as a source term and put it in the right-hand side. However, the fluctuating buoyancy force of eddy is apparently a linear term and it should be put in the left-hand side rather than in the right-hand side.

In fact, we may write the basic equation of motion in a symbolic form:

$$\frac{\partial v}{\partial t} + \mathbf{L}(v) = \mathbf{N.L.}, \tag{10}$$

where **L** and **N.L.** denote the linear and nonlinear vector operators. Here the most important nonlinear term in the right-hand side is the Reynolds stresses of turbulent convection. We may separate this equation into two equations which describe the wave motion and the eddy motion. To do so, we write

$$\boldsymbol{v} = \boldsymbol{v}_{\mathrm{w}} + \boldsymbol{v}_{\mathrm{eddy}}, \quad \text{with} \quad |\boldsymbol{v}_{\mathrm{w}}| \ll |\boldsymbol{v}_{\mathrm{eddy}}| . \tag{11}$$

We then obtain the equation describing the wave motion:

$$\frac{\partial \boldsymbol{v}_{\mathrm{w}}}{\partial t} + \mathbf{L}_{\mathrm{w}}(\boldsymbol{v}_{\mathrm{w}}) = \mathbf{N.L.}(\boldsymbol{v}_{\mathrm{eddy}}, \boldsymbol{v}_{\mathrm{eddy}}), \tag{12}$$

and the equation describing the eddy motion:

$$\frac{\partial \boldsymbol{v}_{\mathrm{eddy}}}{\partial t} + \mathbf{L}_{\mathrm{eddy}}(\boldsymbol{v}_{\mathrm{eddy}}) = \mathbf{N.L.}(\boldsymbol{v}_{\mathrm{eddy}}, \boldsymbol{v}'_{\mathrm{eddy}}). \tag{13}$$

The fluctuating buoyancy force should then appear in the second term of the left-hand side of equation (13). That is, the fluctuating buoyancy force of the eddy must balance the eddy acceleration term and the fluctuating pressure force term in equation (13).

(2) A question about the cancellation effect :

In their formulation of the acoustic dipole radiation from the turbulent convection, Goldreich and Kumar (1988) used the so-called turbulent pseudo-convection model in which the effects of the fluctuating buoyancy force of the turbulent convection are modeled by adding and removing scalar contaminant *randomly*. The 'external' force due to the buoyancy force then produces the dipole radiation. However, in the real convection, the fluctuating buoyancy force is not random but the positive entropy perturbation produced by one side of an eddy is always accompanied with the negative entropy perturbation of the other side of the eddy, because the entropy perturbation is produced by advection due to the eddy motion. If this is the case, then the two dipoles with the opposite direction sit side by side and they are *largely cancelled*, resulting in the "quadrupole" radiation. Therefore, even if we accept the fluctuating buoyancy force as an external force, the dipole radiation may not be so effective as Goldreich and Kumar (1988) proposed.

So far we have argued that the dipole radiation may not be so effective in the deep interior of the solar convection zone. However, this does not mean that the dipole radiation is unimportant. Here we propose a possibility of the dipole radiation from the top of the solar convection zone. At the top of the solar convection zone, the rising convective elements overshoot into the radiative atmosphere. The cancellation effect of the dipole radiation discussed above may well disappear there because the symmetry between the rising and sinking motions breaks down there. In fact, Curle (1955) argued long time ago that a solid boundary of a turbulent layer can emit the dipole radiation. The radiative layer above the convective zone may act like a solid "boundary" to the convective motion and the overshooting convective elements may emit acoustic dipole radiation effectively.

3.6. A New Estimate of Acoustic Power

We now estimate the acoustic power based on the quadrupole radiation. In the previous subsection we have argued that the dipole radiation in the deep interior of the convection zone may not be so effective but the dominant source of acoustic radiation is the quadrupole radiation, although the dipole radiation may be important at the top of the convective zone. Here we shall show that the quadrupole radiation is more important than the dipole radiation even at the top of the convection zone because of the difference of frequency of emitted radiation.

In the studies so far made (i.e., Goldreich and Keeley 1977b), acoustic wave generation was thought to occur largely in the evanescent region and therefore it was rather inefficient because of an attenuation effect of the evanescent zone. We would like to point out that the frequency of emitted radiation by the quadrupole radiation is twice as high as that of the dipole radiation or an expression used for the quadrupole radiation by previous workers. If this effect is taken into account, the frequency of acoustic quadrupole radiation falls in that of propagating waves and the efficiency of acoustic radiation is much more effective than thought before.

To show this, we consider the resonant frequency of an eddy. If we write the characteristic time of the eddy (or the turn-over time) by $\tau_\lambda \sim \lambda_T/v_\lambda$, then the characteristic frequency of the eddy is given by $\omega_{\rm eddy} \sim 1/\tau_\lambda$. The typical frequency of emitted radiation by the dipole is $\omega_{\rm D} \sim \omega_{\rm eddy}$. On the other hand, the characteristic frequency emitted by the quadrupole radiation is twice as high as this frequency, i.e., $\omega_{\rm Q} \sim 2\omega_{\rm eddy}$ because the product of turbulent velocity components is involved in the Reynolds stress tensor $T_{ij} = \rho v_i v_j$. This effect was pointed out by Curle (1955) but it seems to have been overlooked for a long time in the astrophysical community.

We can now easily show that the acoustic waves generated by the quadrupole radiation occur in the propagating zone if the Mach number of the energy bearing eddy is larger than about 0.25. We first note that $\omega_{\rm eddy} \sim v_{\rm eddy}/\lambda \sim v_{\rm eddy}/H$, where H stands for the scale height. Since the acoustic cut-off frequency, below which waves can not propagate as acoustic waves, is given by $\omega_{\rm ac} \sim c/2H$, we find $\omega_{\rm Q} > \omega_{\rm ac}$, if $M = v_{\rm eddy}/c > 0.25$.

In the solar convection zone, the maximum convective velocity occurs just below the photosphere. We may estimate various quantities there as $H \sim 200\,{\rm km}$, $c \sim 10\,{\rm km\,s^{-1}}$, $v_{\rm conv} \sim 2.5\,{\rm km\,s^{-1}}$. We then find $\omega_{\rm ac} \sim c/2H \sim 0.025$, and therefore $\omega_{\rm D} \sim 0.012\,{\rm s^{-1}}$ for dipole and $\omega_{\rm Q} \sim 0.025\,{\rm s^{-1}}$ for quadrupole. We therefore find that $\omega_{\rm D} < \omega_{\rm ac}$, i.e., the dipole radiation occurs in the evanescent zone, while $\omega_{\rm Q} \sim \omega_{\rm ac}$, i.e., the quadrupole radiation is propagating.

Based on this finding, we estimate the acoustic power generated by the solar convection zone. To do so, we consider the convective layer near the top of the solar convection zone where the turbulent convective velocity attains its maximum value. Since the quadrupole radiation depends strongly on the Mach number, the most of acoustic radiation in the solar convection zone is generated there. If we approximate this zone by a layer having a thickness of H and the convective velocity $v_{\rm max}$, the total acoustic power generated is given by

$$P(\omega) \sim 4\pi R^2 H \epsilon(\omega), \tag{14}$$

where R is the solar radius and $\epsilon(\omega)$ stands for the acoustic emissivity per unit volume and per unit frequency. If we consider the quadrupole radiation and the Kolmogoroff spectrum for the turbulent convective eddy, the acoustic emissivity is given by (see, e.g., Goldreich and Kumar, 1988)

$$\epsilon(\omega) \sim \rho v_{\rm max}^2 M^5 \begin{cases} (\omega/\omega_{\rm Q})^2 & \text{for} \quad \omega \le \omega_{\rm Q} \\ (\omega/\omega_{\rm Q})^{-7/2} & \text{for} \quad \omega \ge \omega_{\rm Q}. \end{cases} \tag{15}$$

Let us now estimate the acoustic power per mode. The acoustic power per mode at the source is given by

$$G_s \sim P(\omega)\frac{\Delta k}{k}\Delta\omega, \tag{16}$$

where the subscript, s, is added to distinguish the acoustic power at the source from that in the acoustic cavity, and $\Delta k/k$ denotes the horizontal wave number band-width for a particular nonradial mode with degree ℓ and m, and $\Delta\omega$ denotes the frequency band-width. To estimate the latter two quantities, we consider a low degree ℓ mode with frequency around 3 mHz. The horizontal wave-number band-width is estimated by assuming isotropic radiation as

$$\frac{\Delta k}{k} \sim \frac{1}{2(kR)^2} = \frac{1}{\omega^2}\frac{c^2}{2R^2}, \tag{17}$$

whereas $\Delta\omega = \omega_{n+1} - \omega_n \sim 2\pi \times 135\ \mu\mathrm{Hz} \sim 0.85 \times 10^{-3}\,\mathrm{s}^{-1}$. Substituting these into equation (16), we obtain

$$G_s \sim \rho H^3 v_{\max}^2 M^3 \Delta\omega \begin{cases} (\omega/\omega_Q)^0 & \text{for} \quad \omega \leq \omega_Q \\ (\omega/\omega_Q)^{-5.5} & \text{for} \quad \omega \geq \omega_Q . \end{cases} \tag{18}$$

Since the acoustic power in Eq. (18) is very sensitive to the Mach number and since the theory of the solar convection zone is not still definite, we use Eq. (18) to estimate the maximum convective velocity necessary for the observed acoustic power rather than to estimate the acoustic power from the theoretical model of solar convection zone.

(1) Peak acoustic power per mode :

Libbrecht (1988a,b) has found that the observed peak acoustic power per mode is about $E\Gamma \sim 10^{22}$erg s^{-1}/mode for the peak frequency $\nu = 3 \sim 4$ mHz. In order to reproduce this order of power, we need $M \sim 0.3$ or $v_{\max} \sim 3\,\mathrm{km\ s}^{-1}$ in Eq. (18). This value of the maximum convective velocity is not unreasonable from the theoretical point of view.

(2) High-frequency cut-off :

The acoustic power of Eq. (18) for the higher frequency with $\nu > 4$ mHz is given by a power law with

$$G \sim \nu^{-5.5}. \tag{19}$$

This is very well compared with the observed one

$$E\Gamma \sim \nu^{-5.3}, \tag{20}$$

where the observational value for the power index in Eq. (20) is estimated from Fig. 3 of Libbrecht (1988b). We note by passing that the dipole radiation predicts the power index of -6.5.

(3) The low frequency power :

Libbrecht (1988a,b) notes that the observed low frequency power of acoustic radiation is very well represented by a power law with an index 8. On the other hand, Eq. (18) predicts the flat power in frequency. However, Eq. (18) represents the acoustic power at the source and what we need is to estimate the acoustic power transmitted into the solar acoustic cavity because the low frequency power is very much attenuated due to the evanescent nature of wave at the source. In fact, the acoustic power transmitted into the acoustic cavity is given by

$$G = G_s \exp(-\kappa d), \tag{21}$$

where $\exp(-\kappa d)$ stands for an attenuation factor due to the evanescent zone, κ the decay rate of evanescent wave with depth, and d is the depth of the evanescent zone for a given frequency wave.

Let us estimate the attenuation factor. If we write the densities of the wave generating zone and of the turning point (i.e., the bottom of the evanescent zone or the top of the acoustic cavity) by ρ_1 and ρ_2, respectively, we find approximately

$$\exp(-\kappa d) \sim \rho_1/\rho_2. \tag{22}$$

Since the temperature of the turning point of the wave is related to the frequency by

$$\omega =. \omega_{ac} \propto \left(\frac{1}{T_2}\right)^{1/2}, \tag{23}$$

we find

$$T_2 \propto \omega^2. \tag{24}$$

If we approximate the solar convective zone by a polytrope of index n, we may write $\rho \sim T^n$. In the temperature range in our interest of $10^4\,\mathrm{K} \leq T \leq 4 \times 10^4\,\mathrm{K}$, we may estimate the temperature gradient as $\nabla \sim 0.2$ or $n \sim 4$. We then find

$$G = G_s \exp(-\kappa d) \propto \frac{1}{\rho_2} \propto \nu^8. \tag{25}$$

This is the very relation which is sought after. Obviously, the various assumptions used here such as the one zone model for wave-generation zone and the polytropic stratification may not be good but it can demonstrate qualitatively how the observed power-law relation of acoustic power with a power index as large as 8 can simply be reproduced by the effect of propagation.

Figure 2 illustrates the above picture schematically.

4 Conclusion

(1) The line widths of solar p-modes :

The observed line widths of solar oscillations are not likely explained by turbulent viscous damping alone but the thermal coupling of oscillations with radiation field and convection is as important as the turbulent damping in determining damping rates of oscillations. If so, the arguments such as the energy equi-partition between an eigenmode of oscillation and its resonant eddy may not be applicable for the problem of solar oscillations because they are based on balancing turbulent excitation to turbulent damping. However, it is difficult at present to estimate reliably the damping rates of oscillation from theory.

(2) Quadrupole versus dipole radiation :

Three arguments are presented that the dipole radiation may not likely be so much important as compared to quadrupole radiation in the problem of solar oscillations. On the other hand, the quadrupole radiation may be much effective than considered by previous workers because of its propagating nature due to its twice as high frequency as considered before.

(3) Quantitative estimate of acoustic wave generation rate :

Acoustic wave generation rates have been estimated by assuming quadrupole radiation. The observed acoustic power at the peak frequency of $\nu \sim 3.5\,\mathrm{mHz}$ can be reproduced if the maximum convective velocity is about $3\,\mathrm{km\ s^{-1}}$ or if its Mach number is about 0.3. The estimated maximum convective velocity is not unreasonable from the present theory of the solar convection zone. The frequency dependence of acoustic generation rate was also discussed. Observed acoustic power per mode shows the high frequency cut-off having the power law dependence in frequency with an index of about −5.3, and this can very well be explained by the quadrupole emission with Kolmogoroff spectrum for turbulent convection, which predicts the power law with an index of −5.5. On the other hand, the dipole radiation has the frequency dependence with the −6.5 power. We have presented an argument that the observed frequency dependence of the low frequency acoustic power of $G = E\Gamma \propto \nu^8$ simply reflects the propagation nature of low-frequency wave under density stratification in the solar convection zone and not the emission mechanism.

References

Ando, H., and Osaki, Y. 1975, *Publ. Astron. Soc. Japan*, **27**, 581.
Christensen-Dalsgaard, J., Gough, D.O., and Libbrecht, K.G. 1989, *Astrophys. J. Letters*, **341**, L103.
Curle, N. 1955, *Proc. Roy. Soc. London A*, **231**, 505.
Goldreich, P., and Keeley, D.A. 1977a, *Astrophys. J.*, **211**, 934.
Goldreich, P., and Keeley, D.A. 1977b, *Astrophys. J.*, **212**, 243.
Goldreich, P., and Kumar, P. 1988, *Astrophys. J.*, **326**, 462.
Kumar, P., and Goldreich, P. 1989, *Astrophys. J.*, **342**, 558.
Libbrecht, K.G. 1988a, *Astrophys. J.*, **334**, 510.
Libbrecht, K.G. 1988b, in *Seismology of the Sun and Sun-like Stars*, ed. E. Rolfe, ESA SP-286 (ESA Publication Division, Noordwijk), p.3.
Lighthill, M.J. 1952, *Proc. Roy. Soc. London A*, **211**, 564.
Stein, R.F. 1967, *Solar Phys.*, **2**, 385.
Stein, R.F., and Leibacher, J.W. 1981, in *the Sun as a Star*, ed. S. Jordan, NASA SP-450, p.289.
Unno, W. 1964, in *Transactions I.A.U. XIIB* (Academic Press, New York), p.555.

What are the Observed High-Frequency Solar Acoustic Modes?

P. Kumar [1], T. L. Duvall, Jr. [2], J. W. Harvey [3],

S. M. Jefferies [4], M. A. Pomerantz [4], and M. J. Thompson [1]

[1]HAO/NCAR, Boulder, CO 80307, USA
[2]Laboratory for Astronomy and Solar Physics
NASA/Goddard Space Flight Center, Greenbelt, MD 20771, USA
[3]NSO/NOAO, Tucson, AZ 85726, USA
[4]Bartol Research Institute, University of Delaware, Newark, DE 19711, USA

Abstract: Jefferies *et al.* (1988) observe discrete peaks up to ~7mHz in the power spectra of their intermediate degree solar intensity oscillation data obtained at South Pole. This is perhaps surprising since waves with frequency greater than the acoustic cut-off frequency at the temperature minimum (~ 5.5mHz), unlike their lower frequency counterparts, are not trapped in the solar interior. We propose that the observed peaks are associated with what are principally progressive waves emanating from a broad-band acoustic source. The geometrical effect of projecting observations of these progressive waves onto spherical harmonics then gives rise to peaks in the power spectra. The frequencies and amplitudes of the peaks will depend on the spatial characteristics of the source. Partial reflections in the solar atmosphere modify the power spectra, but in this picture they are not the primary reason for the appearance of the peaks. We estimate the frequency and power which would be expected from this model and compare it with the observations. We argue that these high frequency "*mock-modes*" are not overstable, and that they are excited by acoustic emission from turbulent convection.

Solar intensity oscillation data, obtained by Jefferies *et al.* (1988) at South Pole in 1987, exhibit ridges in the ℓ–ν diagram extending beyond the observational Nyquist frequency of 6.7mHz. These ridges are clearly visible in Fig. 2 of that paper and in Fig. 4 of Harvey (1990); a section through that plot for degree $\ell = 50$ is shown in Fig. 1 below. Similar high-frequency ridges are also observed in the velocity data of Libbrecht (1988). It may at first sight be surprising that the ridges extend beyond the acoustic cut-off frequency of the atmosphere (approximately 5.5mHz)[1]. Below this frequency the power ridges are a manifestation of trapped acoustic modes. However, acoustic waves are not trapped below the temperature minimum at the observed high frequencies. Even when waves are only partially reflected, normal modes can exist: in that case the normal frequencies are

[1] The acoustic cut-off frequency in the atmosphere is some what uncertain, but we think it very unlikely that it could be as large as 6.7mHz.

complex. The real part of the frequency is determined by the properties of the atmosphere in much the same way as for a trapped mode, and the imaginary part arises from the loss of energy from the system due to the partial transmission of the waves (this is illustrated below for a simple one-dimensional example). Two places where such partial reflection might be important in the sun are in the photosphere and at the transition between the chromosphere and corona. The reflection at the photosphere, however, is small once the frequency exceeds the acoustic cut-off frequency. The amount of reflection at the chromosphere/corona transition is quite uncertain because the upper chromosphere may be sufficiently inhomogeneous for a large degree of scattering to occur. In any case, a model for the observed high frequency peaks which requires significant reflection at the transition layer, as has recently been proposed by Balmforth and Gough (1990), appears to be in conflict with the observations: the frequency spacing between adjacent order high frequency modes for such a model is smaller on average by roughly 10–15μHz than the observed frequency spacing. This indicates that reflecting the waves at the transition layer results in an acoustic cavity that is too large.

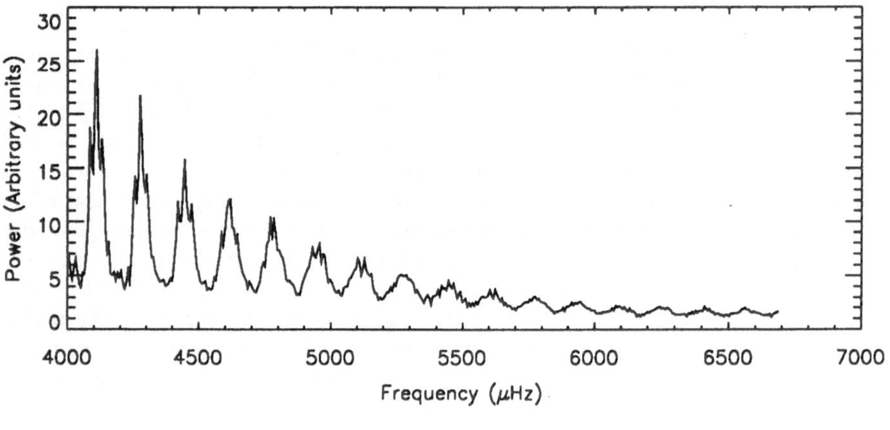

Fig. 1. Observed intensity power spectrum for $\ell = 50$ mode.

The quite different possibility which we propose here is that the observed high frequency peaks arise principally from a geometrical effect of projecting progressive waves onto spherical harmonics. We envisage that such waves are produced by broad-band acoustic noise in a layer that is located near the top of the convection zone and is roughly one pressure scale height (~200 km) thick. In this scenario, the waves are not significantly reflected in the atmosphere. But because the vertical wavelengths of the waves (~1200 km) are somewhat larger than the extent of the source layer, we argue below that in this model the frequency spectrum corresponding to a particular spherical harmonic is not smooth but has peaks similar to those observed.

To illustrate this idea, consider a point source located beneath the surface, isotropically emitting waves of frequency ω. For simplicity assume that the outward propagating waves from the source are not reflected and are observed on some exterior spherical surface. The inward propagating waves are refracted within the sun and eventually pass through the exterior surface at various distances from the source. All points on the surface will oscillate at the same frequency ω but with differing phases and amplitudes relative to the source. (For absolute simplicity one might even consider a homogeneous model: then, of course, the rays along which the waves propagate are simply straight lines.) When this global pattern of observed oscillations is analyzed in terms of spherical harmonics, some ℓ values will have large amplitude and some small depending on ω and the source position. If the source emits many frequencies then even if the spectrum of the source is flat the observed power spectrum at a given ℓ will contain peaks whose frequencies depend on the source position and the properties of the solar model.

If we consider the point source now to be distributed in the radial direction, as long as its extent is smaller than the wavelength of the emitted waves, the pattern of frequency peaks will be maintained. If we consider two sources at the same depth but at different positions, each will produce a similar spectrum of frequency peaks. Provided the sources emit incoherently, their individual spectra can be added together and the peak structure is preserved, as indeed it is in the limit of a uniform shell distribution of sources.

In the real solar atmosphere one expects even high frequency outward-propagating waves to experience some reflection. To illustrate the combined effect of source position and atmospheric properties (including partial reflection) on wave amplitude, we consider the following simple one dimensional potential well problem:

$$\frac{d^2\psi}{dx^2} + [\omega^2 - V(x)]\psi = 0, \tag{1}$$

where $V(x) = \infty$ for $x < 0$, $V(x) = 0$ for $0 < x < a$, and $V(x) = \alpha^2$ (constant) for $x > a$. The step in the potential at $x = a$ allows the possibility of total or partial reflection of the waves. This system admits both trapped waves, which are evanescent in $x > a$ and have real frequencies $\omega < \alpha$, and waves which propagate for $x > a$ and have complex frequencies with $\Re e(\omega) > \alpha$. These are analogous to solar p modes with frequencies below and above the acoustic cut-off frequency respectively. If a point source, oscillating with real frequency $\omega > \alpha$, is placed at x_0 ($< a$), the solution in $x > a$ will be a propagating wave with squared amplitude, or power, given by

$$\frac{\sin^2 \omega x_0}{\omega^2 - \alpha^2 \sin^2 \omega a}. \tag{2}$$

Thus the power has a sinusoidal frequency dependence through the numerator. When α is non-zero, this is modulated by the denominator which has local minima at those values of ω which correspond to the real part of the resonant frequency of the cavity. The successive maxima in the power spectrum, which we refer to as *mock-modes*, are determined by a combination of source position and cavity resonance. However, when $\omega \gg \alpha$, the peaks are determined principally by the source position. For the sun, the reciprocal of the separation $\Delta\omega$ between successive high-frequency *mock-modes* would be approximately twice the sound travel time along a ray path from the source to the lower turning point.

We now turn our attention to what excites these "modes." It appears very unlikely, for reasons discussed below, that they are overstable or self excited, and it is found that excitation due to acoustic emission from turbulent convection is in rough agreement with the observations.

The energy input rate for overstable modes is proportional to mode energy, E. Thus E grows exponentially with time until some nonlinear dissipation mechanism (damping rate $dE/dt \propto E^{1+|\epsilon|}$) sets in and saturates this growth. Energy leakage into the atmosphere, for high frequency *mock-modes*, is an important damping process, and the resulting dissipation rate is equal to the energy flux, F, at the surface. In terms of the surface velocity, v, sound speed, c, and the density, ρ, the energy flux is $4\pi R_{\odot}^2 \rho v^2 c$. This is proportional to the mode energy, and thus inappropriate for limiting exponential growth. Interestingly, if modes were to be overstable, the above flux puts a lower limit on their linear growth rate: $\Gamma = d\ln E/dt = F/E \sim \nu/\pi n$ (where ν and n are the mode frequency and the number of radial nodes respectively). For high frequency modes this gives an e-folding time of about an hour or less. Thus if it is assumed that these modes are self excited it is very surprising that their amplitudes do not grow to large value, since this will require a nonlinear mechanism that would quench them on a similarly short time scale. Since both radiative and turbulent damping rates are linear in mode energy, they are not suitable for quenching the overstability. Finally, nonlinear mode couplings tend to transfer energy from trapped p modes to the high frequency propagating modes (Kumar and Goldreich, 1989). Therefore, if the *mock-modes* are overstable there does not appear to be any way of limiting their amplitude at the observed small value. Thus we suspect that these waves are not self-excited.

Could mode couplings excite the *mock-modes*? We estimate the expected velocity amplitude at the photospheric level for *mock-modes* due to this process, and show that this does not appear likely. The observed p-mode energies and linewidths give the total energy dissipation rate from all trapped p modes to be approximately equal to 10^{28}erg s^{-1} [Libbrecht (1988); but using the more recent data of Kaufman (1988)]. A fraction of this energy, of order 10% (Kumar and Goldreich 1989), is carried by propagating waves. It follows from the conservation of angular momentum in 3-mode couplings that the largest degree for a *mock-mode* should be about twice the largest ℓ for trapped modes, *i.e.* 6000, and that the number of resonant triplets for a mode of degree ℓ increases approximately linearly with ℓ. If the energy lost from the trapped modes due to nonlinear couplings is distributed uniformly over the high frequency modes then the typical surface velocity for the *mock-modes*, v, can be obtained by equating the total energy flux in the *mock-modes* at the surface with the energy input rate i.e. $4\pi R_{\odot}^2 \rho v^2 c 6000^2 \sim 10^{27}\,\mathrm{erg\,s^{-1}}$, where ρ, and c are the photospheric density and sound speed respectively. Substituting the appropriate values we find that $v \sim 4 \times 10^{-2}\,\mathrm{cm\,s^{-1}}$. However, it can be shown that the energy input rate in a mode due to nonlinear couplings is proportional to the number of resonant triplets feeding energy into it, and thus is proportional to the mode degree. Therefore, the velocity amplitude of a *mock-mode* of $\ell = 50$ would be smaller that the value estimated above by a factor of $(50/6000)^{1/2}$. This yields a surface velocity $\sim 4 \times 10^{-3}\,\mathrm{cm\,s^{-1}}$, which is smaller than the observed value by about two orders of magnitude (Libbrecht 1988). Thus these modes are unlikely to be excited due to nonlinear coupling with trapped p modes.

We finally consider excitation due to acoustic emission from turbulent convection, and show that it gives power in rough agreement with observations. The inhomogeneous wave equation governing the excitation of radial modes due to Reynolds stress is

$$\frac{\partial}{\partial r}\left[\frac{c^2}{r^2}\frac{\partial}{\partial r}(r^2\rho\xi)\right]+\frac{\partial}{\partial r}\left[\frac{P}{C_v}\frac{\partial s}{\partial r}\xi\right]+\frac{g}{r^2}\frac{\partial}{\partial r}(r^2\rho\xi)-\rho\frac{\partial^2\xi}{\partial t^2}=\nabla\cdot(\rho\mathbf{VV})_r\equiv S(t,\mathbf{x}), \quad (3)$$

where ξ is the radial displacement, $\mathbf{V}$ is the turbulent velocity, s is the entropy, and other symbols have the standard meaning. The displacement amplitude, ξ_ω, for a wave of frequency ω, can be easily determined using the Green's function, $G_\omega(r,r')$, of the above equation and the source, S_ω, and is given below

$$\xi_\omega(r)=\int d^3x'\, G_\omega(r,r')S_\omega(\mathbf{x}'). \quad (4)$$

We adopt the usual, highly simplified, picture of turbulent convection as consisting of uncorrelated critically damped eddies of size roughly a pressure scale height and less. This enables us to evaluate the above integral and we obtain the following expression for displacement amplitude squared:

$$\xi_\omega^2 \sim 4\pi R_\odot^2\int dr\, \rho^2 h^3 V_h^4\left|\frac{\partial G_\omega}{\partial r}\right|^2, \quad (5)$$

where h is the size of the largest eddy at r which has characteristic turnover time less than or equal to mode period ω^{-1}. The velocity of sub-energy bearing eddies, V_h, is taken to be related to energy bearing eddy of size, H, by the Kolmogoroff spectrum, i.e. $V_h=V_H(h/H)^{1/3}$, where V_H is the velocity of energy bearing eddy. This is a reasonable approximation since Reynolds stress dominates buoyancy force for eddies smaller than one scale height.

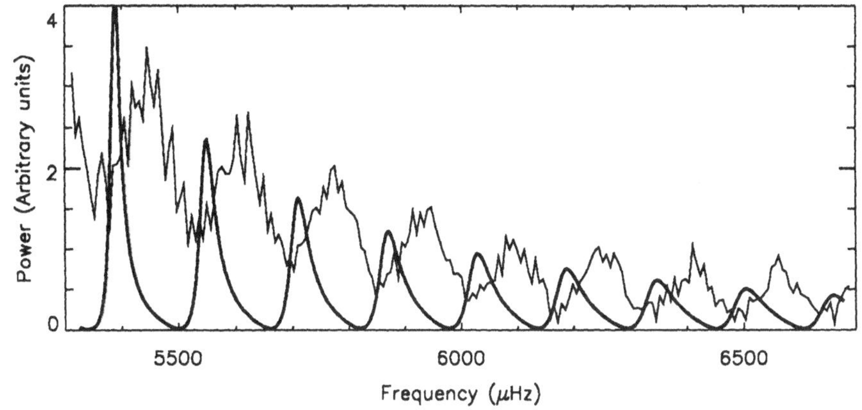

Fig. 2. Numerically computed velocity power spectrum (thick line), and the observed power spectrum for $\ell = 50$.

We have evaluated ξ_ω^2 numerically using Eq. (5) and a standard solar model (kindly provided by J. Christensen-Dalsgaard), and the result is displayed in Fig. 2, together with the observed intensity spectrum. Of course the calculated power spectrum includes any effects of partial reflection arising from the detailed structure of the solar model. Note that the observed spectrum is broadened by ℓ leakage and is also perturbed by aliases of higher frequency features that are folded about the Nyquist frequency. The observed and theoretical powers have similar frequency dependence[2], and the frequency spacings are also in rough agreement. However, the position of the peaks in the theoretical calculation are shifted by a substantial amount to lower frequencies with respect to the observed peaks. This might be due to incorrect positioning of the source in our calculation, or it could arise from other simplifications in our modeling of wave propagation in the atmosphere. We are currently investigating these issues (Kumar *et al.*, in preparation).

In conclusion, we note that our model accounts for frequency peaks above the acoustic cut-off frequency, and reproduces the observed inter-peak spacing and the variation of peak amplitude with frequency. The calculated peak positions are at variance with the observations, however, and unless this is resolved our model cannot be claimed to be on very firm ground.

Acknowledgements

We thank Tim Brown and Peter Goldreich, and the other participants at the Santa Barbara helioseismology workshop, for many useful discussions. We are grateful to Douglas Gough and Neil Balmforth for a preprint of their paper, and PK thanks Mark Rast for help with the IDL graphics package. This research was supported in part by NSF grant DPP-8715791. The National Center for Atmospheric Research is sponsored by the National Science Foundation, and the NOAO is operated by the Association of Universities for Research in Astronomy, Inc. under cooperative agreement with the National Science Foundation.

References

Balmforth, N., and Gough, D.O. 1990, *Astrophys. J.*, submitted.

Harvey, J. W. 1990, these proceedings.

Jefferies, S.M., Pomerantz, M.A., Duvall, T.L., Jr., Harvey, J.W., and Jaksha, D.B. 1988, in *Seismology of the Sun and Sun-like Stars*, ed. E. Rolfe, ESA SP-286 (ESA Publication Division, Noordwijk), p.279.

Kaufman, J.M. 1988, in *Seismology of the Sun and Sun-like Stars*, ed. E. Rolfe, ESA SP-286 (ESA Publication Division, Noordwijk), p.31.

Kumar, P., and Goldreich, P. 1989, *Astrophys. J.*, **342**, 558.

Libbrecht, K.G. 1988, *Astrophys. J.*, **334**, 510.

Libbrecht, K.G. 1990, Caltech BBSO preprint no. 0305.

[2] Comparison of intensity and velocity power is not meaningless since we expect the relative amplitudes of velocity and intensity variations to be roughly constant in the frequency range considered here, *cf.* Libbrecht (1990).

Driving and Damping of Oscillations

Robert F. Stein [1] and Åke Nordlund [2]

[1]Dept. Physics-Astronomy, Michigan State University,
East Lansing, MI 48824, USA
[2]Copenhagen University Observatory,
Oster Voldgade 3, 1350 Copenhagen K, Denmark

Abstract: We have simulated the upper 2.5 Mm of the solar convection zone using a realistic, three-dimensional, compressible, hydrodynamic computer code. P-mode oscillations are excited at the eigenfrequencies of the simulation volume. We have calculated the time averages of the work terms in the kinetic energy equation, using the internal energy equation to evaluate the fluctuations in the gas pressure. This calculation shows that the modes are excited near the surface by the divergence of the convective flux and damped by the divergence of the radiative flux. The fundamental mode is also spuriously driven at the lower boundary, by density and turbulent pressure fluctuations induced when downward plunging convective plumes pass through the lower boundary of the simulation.

1 Introduction

Among the outstanding questions about the global solar oscillation modes is what excites them and what limits their amplitude. This paper reports on our ongoing attempts to answer these questions. Several possible driving and damping processes have been explored by others: Radiative energy exchange may make the modes overstable (Ando and Osaki 1975, but see Christensen-Dalsgaard and Frandsen 1983). Interaction with turbulent convection may stochastically drive and damp the modes (Goldreich and Keeley 1977, Goldreich and Kumar 1988). Energy transfer to the chromosphere damps the modes, particularly at high frequencies. [See also the review by Libbrecht (1988).] Here we investigate these processes as they act in a numerical simulation of the upper solar convection zone.

2 Vertical (Radial) Modes

We have modeled the upper 2.5 Mm of the solar convection zone and the photosphere using a three-dimensional, compressible, hydrodynamic computer code, including a realistic equation of state and radiative transfer (Stein and Nordlund 1989). In the course of simulating four solar hours, radial and non-radial p–modes were excited. We study these modes with the aim of gaining insight into the processes at work on the sun. The modes are similar to the solar modes, but with a much sparser spectrum, because of the small size of our computational domain ($6 \times 6 \times 3$ Mm). Figure 1 shows the velocity spectrum at the solar surface, for vertical (radial) modes.

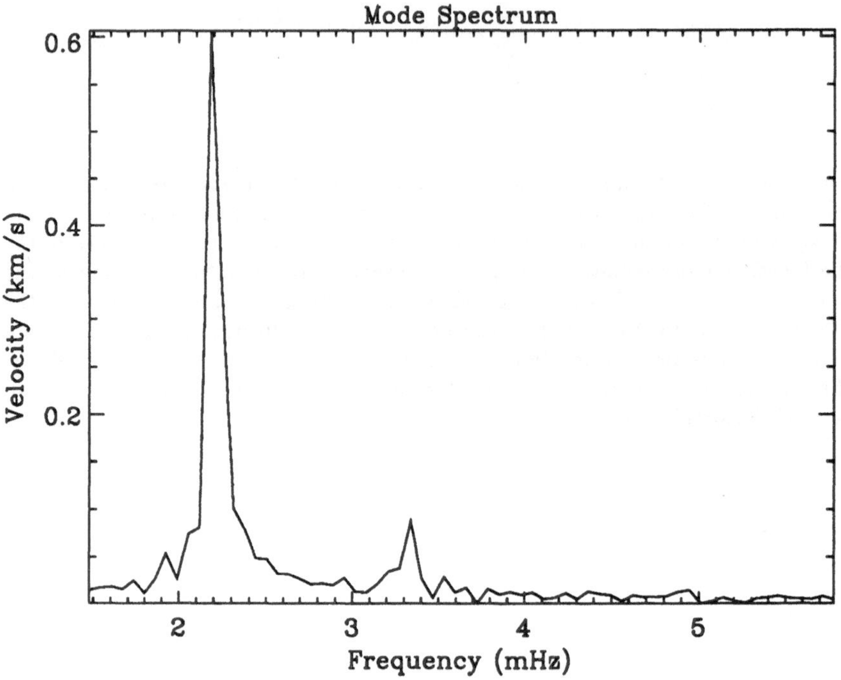

Fig. 1. Velocity spectrum at the solar surface.

Two modes, the fundamental (with $\nu = 2.18$ mHz) and first overtone (with $\nu = 3.34$ mHz), are clearly visible. Obviously, these modes do not have the amplitude spectrum of the solar modes, but new studies, just begun, simulating a much deeper portion of the convection zone, show a mode amplitude spectrum that is similar to that of the solar p–modes. In this preliminary, shallow, calculation the fundamental mode grows in amplitude continuously during the course of the four solar hours, unlike on the sun where the modes are saturated (Fig. 2).

Note the fairly smooth growth in the mode amplitude. At a depth near one megameter, where the first harmonic has a node and there is no beating, the amplitude grows extremely smoothly. This indicates that the excitation mechanism for the fundamental is not stochastic, while the slight decrease in growth rate with time indicates that the

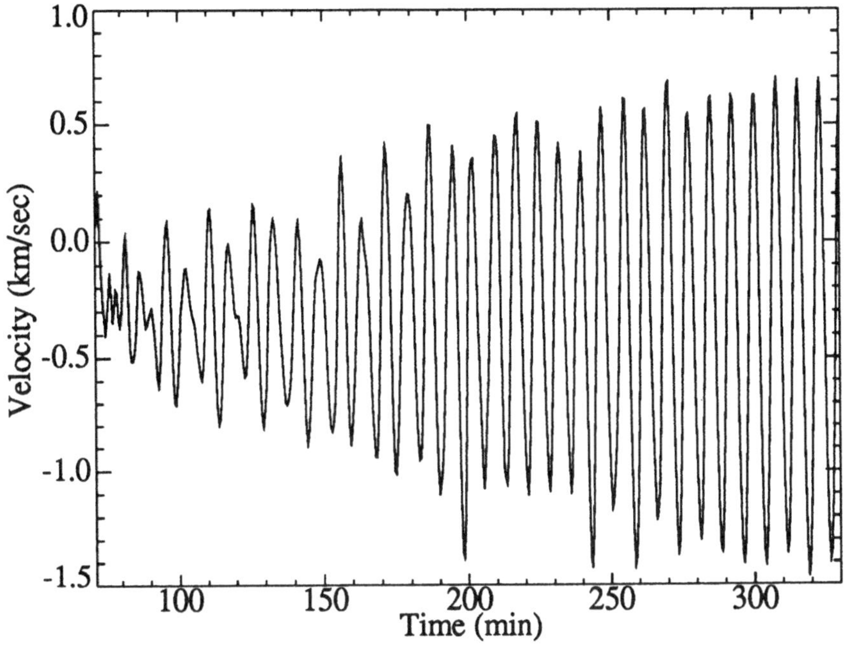

Fig. 2. Velocity of vertical modes at the solar surface as a function of time. The velocity is primarily that of the fundamental mode, with some beating due to the first harmonic.

mode is not growing exponentially. The nearly linear growth with time indicates that the excitation is locked in phase with the mode, that is, the interaction is resonant. In this figure, we cannot distinguish the time behavior of the first harmonic in the presence of the large amplitude fundamental. However, we also see non-radial modes, and these are clearly stochastically excited, for their amplitude changes randomly with time (Fig. 3). The first harmonic vertical mode behaves similarly.

3 Mode Driving

In order to determine the driving and damping mechanisms of these oscillations, we analyze the kinetic energy equation. We only consider vertical (radial) modes, because for other modes the non-linear terms would involve different horizontal spatial dependencies. We average the kinetic energy equation over the horizontal plane (using a mass weighted average for the velocity) and get the kinetic energy equation for the vertical modes,

$$\bar{\rho}\frac{D}{Dt}\left(\frac{1}{2}\bar{u}_z^2\right) = -\frac{\partial}{\partial z}[(\bar{P} + < \rho u_z^{'2} >)\bar{u}_z] - (\bar{P} + < \rho u_z^{'2} >)\frac{D \ln \bar{\rho}}{Dt} + \bar{\rho}\bar{u}_z g. \quad (1)$$

The overbars indicate horizontal averages and refer to the mode, while the primes refer to the horizontally fluctuating or convective motions. P is the gas pressure and $< \rho u_z^{'2} >$

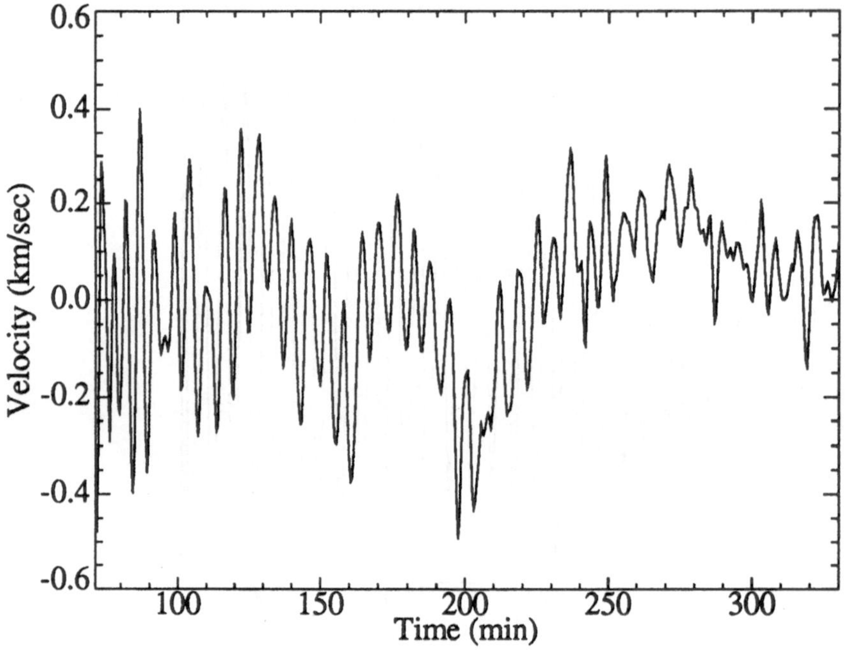

Fig. 3. Velocity of a non-radial mode at the solar surface as a function of time.

is the turbulent pressure. We can use the internal energy equation to express the gas pressure in terms of the heating contributions,

$$\frac{D\bar{P}}{Dt} - \frac{\Gamma_1 \bar{P}}{\bar{\rho}} \frac{D\bar{\rho}}{Dt} = (\Gamma_3 - 1)\rho Q_{\text{total}}, \tag{2}$$

where

$$\Gamma_3 - 1 = \frac{1}{\rho} \left(\frac{\partial P}{\partial e} \right)_{\rho}, \tag{3}$$

and

$$\rho Q_{\text{total}} = -\frac{\partial}{\partial z} < P u'_z + \rho e' u'_z > + < \rho Q_{\text{rad}} > + < \boldsymbol{u}' \cdot \nabla P > . \tag{4}$$

Here e is the internal energy, $< P u'_z + \rho e' u'_z >$ is the convective flux, and Q_{rad} is the radiative heating. Working in the frequency domain, the average rate of working is half the real part of the complex conjugate of the pressure times the derivative of the logarithm of the density, or

$$W_\omega = \frac{1}{2} \text{Re} \left[\frac{1}{\bar{\rho}} \left(\frac{\partial P}{\partial e} \right)_{\rho} < \rho Q_{\text{total}} >^*_\omega \ln \bar{\rho}_\omega \right] + \frac{1}{2} \text{Im} [P^*_{\text{turb}\,\omega} \omega \ln \bar{\rho}_\omega]. \tag{5}$$

We have explored the values of these various work terms for the fundamental and first harmonic vertical modes. The main result of our analysis, so far, is that the oscillations

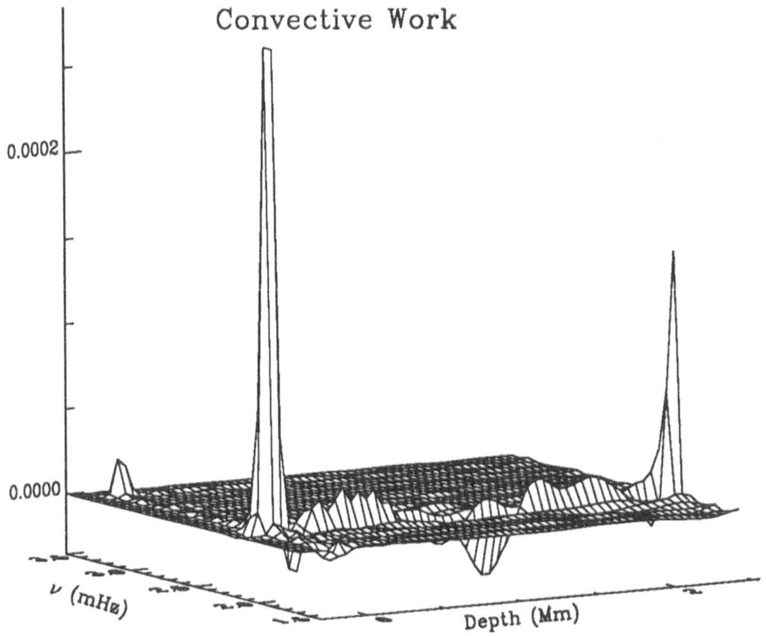

Fig. 4a. Convective work.

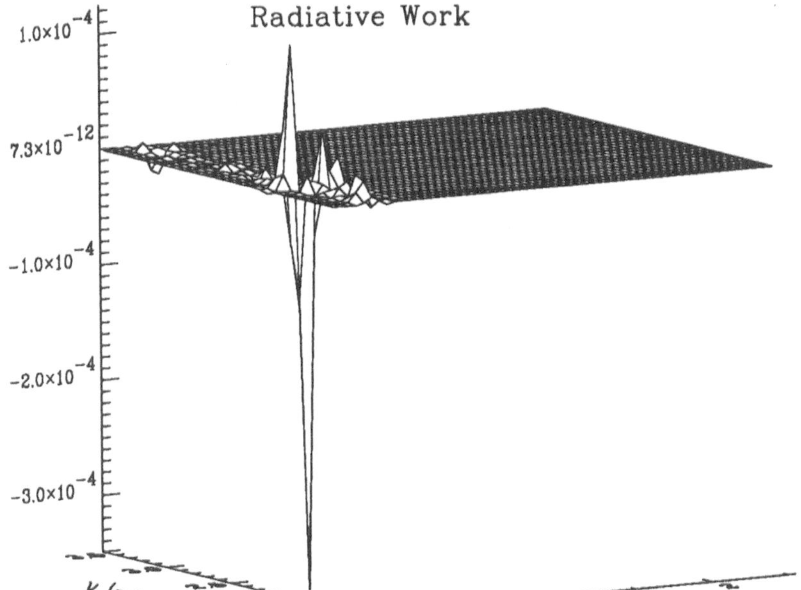

Fig. 4b. Radiative work.

are excited and damped near the solar surface, where energy transport switches from convection to radiation. The radiative flux divergence stabilizes and the convective flux divergence destabilizes the oscillations. Radiation cools these layers. It tends to remove energy at maximum compression and hence damps the oscillation modes. Since the total flux tends to remain constant, it is reasonable that convection and radiation work in opposite directions and nearly cancel one another. The average rate of work by these processes is shown in Fig. 4 as a function of frequency and depth. The turbulent pressure and convective velocity work contributions are generally an order of magnitude smaller. Excitation and damping below the surface are reduced because when hydrogen is undergoing ionization heat goes into ionization rather than thermal energy, so $\Gamma_3 - 1$ becomes small and the pressure fluctuations are reduced.

4 Fundamental and First Overtone

Figure 5 shows that the fundamental mode is destabilized by the divergence of the convective flux at the bottom boundary as well as at at the surface. It is damped by the divergence of the radiative flux at the surface.

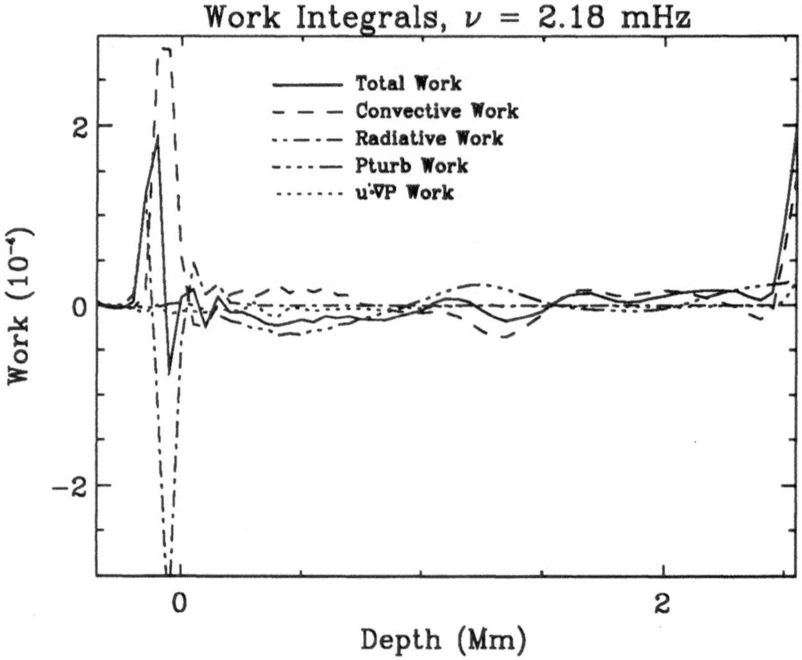

Fig. 5. Rate of energy input to the fundamental radial mode from the various processes.

The driving at the bottom boundary is probably due to fluctuations induced when convective plums descended through the boundary. In this calculation, we employed a constant pressure lower boundary condition, so that when a high density convective finger passed through the boundary a buoyancy fluctuation was induced near the boundary. Also, we find that turbulent pressure boundary work contributes significantly to the spurious driving of the fundamental mode at the lower boundary. Figure 6 shows the phase of the convective flux with respect to the mode velocity at the surface. Since the convective flux is primarily carried by downward plunging cool plumes (Stein and Nordlund 1989), the well defined decreasing phase with increasing depth indicates that these downward motions tend to be locked in phase with the oscillations. They must therefore be sensitive to the oscillation and be preferentially launched at a particular phase of the wave. Work near the bottom boundary accounts for 27% of the total energy input to the fundamental mode. In our new simulation, we have changed the lower boundary condition to make the bottom boundary a velocity node for radial motions. This eliminates spurious boundary driving.

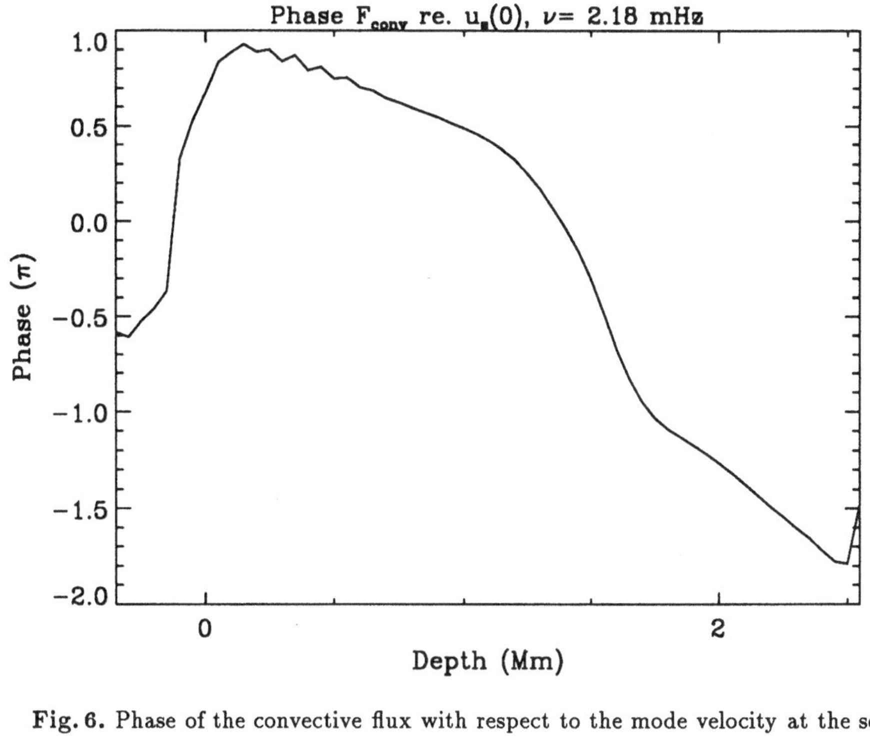

Fig. 6. Phase of the convective flux with respect to the mode velocity at the solar surface.

In contrast to the fundamental, the first harmonic is driven and damped exclusively at the solar surface. From Fig. 7 it can be seen that the divergence of the convective flux destabilizes the mode, while the radiative cooling stabilizes the mode. Again, these two terms nearly cancel each other.

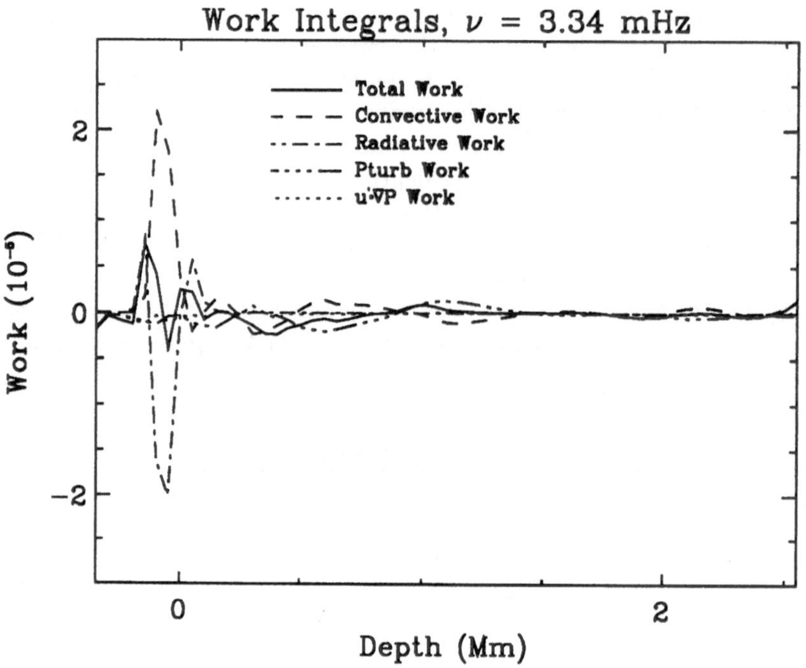

Fig. 7. Rate of energy input to the first harmonic radial mode from the various processes.

5 Concluding Remarks

We have seen that (except for the special case of the fundamental, where due to our constant pressure lower boundary condition substantial boundary work occurs) the oscillations are driven and damped at the solar surface. Thus our simulation includes the essential mode physics. If account is taken of the small mode inertia, these simulation results may be used to study the solar p-mode interaction with convection. We are in the process of performing this investigation.

This work was supported in part by grants from the National Aeronautics and Space Administration (NAGW 1695), the Danish Natural Science Research Council and the Danish Space Board. The authors wish to express their appreciation for this support.

References

Ando, H., and Osaki, Y. 1975, *Publ. Astron. Soc. Japan*, **27**, 581.
Christensen-Dalsgaard, J., and Frandsen, S. 1983, *Solar Phys.*, **82**, 165.
Goldreich, P., and Keeley, D. A. 1977, *Astrophys. J.*, **212**, 243.
Goldreich, P., and Kumar, P. 1988, *Astrophys. J.*, **326**, 462.
Libbrecht, K.G. 1988, in *Seismology of the Sun and Sun-Like Stars*, ed. E.J. Rolfe, ESA SP-286 (ESA Publication Division, Noordwijk), p.3.
Stein, R.F., and Nordlund, Å. 1989, *Astrophys. J. Letters*, **342**, L95.

Nonlinear Oscillations in the Convective Zone

Wasaburo Unno [1] and Da-run Xiong [2]

[1]Res. Inst. Sci. Tech., Kinki University, Higashi-Osaka-shi, Osaka 577, Japan
[2]Purple Mountain Observatory, Nanjing, People's Republic of China

Abstract: Radiation-hydrodynamics for stellar oscillations in convective zone (Xiong 1989) is reformulated into a form suitable for non-linear dynamical interpretation of semi-regular and irregular variability of red super-giant stars.

1 Introduction

Semi-regular and irregular variables have been rather poorly observed, because neither definite periods nor well-defined light curves that are worthy of being reported can not be determined. Recently, however, those variables have been noticed as rich manifestation of nonlinear chaotic dynamical systems (Yuasa *et al.* 1989). Saitou *et al.* (1988) showed that this is the case by proposing a simple one-zone model of semi-regular variability by introducing mathematical model of saturation of radiative excitation mechanism. However, since most of energy flux is carried by convection in irregular variables, the modulation of convective flux by oscillation should be crucial in the nonlinear dynamics of those stars (Kamijo 1963). Formulation of basic equations for this problem has been made recently by Xiong (1989). The treatment is general but somewhat involved especially for the radiative transfer and for the treatment of velocity and temperature correlations. In the present study, simplifications are made to modify Xiong's equations into more convenient form to study the problem using the gradient diffusion approximation to evaluate nonlinear fluctuating terms. One-zone modeling is not intended here but it should be straightforward in the local theory. However, displacement of convective zone with pulsation, which is the nonlocal effect, will be the problem of further consideration.

2 Separation of Averages and Fluctuations

Basic Equations of hydrodynamics determining density ρ, velocity $\boldsymbol{u}$, and pressure p (or temperature T) are given by

$$\frac{d\rho}{dt} + \rho\nabla\cdot\boldsymbol{u} = 0, \tag{1}$$

$$\rho \frac{d\boldsymbol{u}}{dt} = -\nabla p - \rho \nabla \phi + \mathrm{div} \mu \sigma(\boldsymbol{u}), \tag{2}$$

and

$$\rho \frac{dE}{dt} + p \nabla \cdot \boldsymbol{u} = \rho \frac{dh}{dt} - \frac{dp}{dt} = -\nabla \cdot \boldsymbol{F}_{\mathrm{R}} + \mu \sigma(\boldsymbol{u}) \cdot \nabla \boldsymbol{u}, \tag{3}$$

where μ denotes the molecular viscosity, σ the stress tensor, E the internal energy, $h = E + p/\rho$ the enthalpy, and $\boldsymbol{F}_{\mathrm{R}}$ the radiative flux given by

$$\boldsymbol{F}_{\mathrm{R}} = -K_{\mathrm{R}} \nabla T \tag{4}$$

and

$$K_{\mathrm{R}} = C_p \kappa_{\mathrm{R}}, \tag{5}$$

κ_{R} being the radiative thermometric conductivity,

$$\kappa_{\mathrm{R}} = 4acT^3/(3C_p \overline{\kappa} \rho), \tag{6}$$

where $\overline{\kappa}$ denotes the opacity. Molecular viscosity is unimportant in stellar envelopes, but it is introduced here for the purpose of guiding the eddy viscosity formulation.

Variables are separated into averages indicated by overbar and fluctuations indicated by prime so that

$$\rho = \overline{\rho} + \rho', \tag{7a}$$

$$p = \overline{p} + p', \tag{7b}$$

$$T = \overline{T} + T', \tag{7c}$$

$$\boldsymbol{u} = \overline{\boldsymbol{u}} + \boldsymbol{u}', \tag{7d}$$

etc. The way of defining averages are:

$$\overline{\rho'} = 0, \tag{8a}$$

$$\overline{p'} = 0, \tag{8b}$$

and

$$\overline{\boldsymbol{F}'} = 0, \tag{8c}$$

for energies (flux) per unit volume (area), and

$$\overline{\rho \boldsymbol{u}'} = 0, \tag{8d}$$

$$\overline{\rho T'} = 0, \tag{8e}$$

$$\overline{\rho E'} = 0, \tag{8f}$$

and

$$\overline{\rho h'} = 0, \tag{8g}$$

for momentum and energies per unit mass. Then, Eq. (1) is easily separated into

$$\overline{\rho} D_t (1/\overline{\rho}) = \nabla \cdot \overline{\boldsymbol{u}} \tag{9}$$

and

$$\overline{\rho} D_t (\rho'/\overline{\rho}) + \nabla \cdot (\rho \boldsymbol{u}') = 0, \tag{10}$$

where

$$D_t = \frac{\partial}{\partial t} + \overline{\boldsymbol{u}} \cdot \nabla. \tag{11}$$

Decomposing $\rho\boldsymbol{u}$ and $\rho\boldsymbol{u}\boldsymbol{u}$ as follows,

$$\rho\boldsymbol{u} = \overline{\rho\boldsymbol{u}} + (\rho\boldsymbol{u}' + \rho'\overline{\boldsymbol{u}}) \tag{12}$$

and

$$\rho\boldsymbol{u}\boldsymbol{u} = (\overline{\rho\boldsymbol{u}\boldsymbol{u}} + \overline{\rho\boldsymbol{u}'\boldsymbol{u}'}) + [\rho'\overline{\boldsymbol{u}\boldsymbol{u}} + \rho\overline{\boldsymbol{u}}\boldsymbol{u}' + \rho\boldsymbol{u}'\overline{\boldsymbol{u}} + (\rho\boldsymbol{u}'\boldsymbol{u}')'], \tag{13}$$

we transform Eq. (2) into

$$\overline{\rho} D_t \overline{\boldsymbol{u}} = -\nabla\overline{p} - \overline{\rho}\nabla\overline{\phi} - \nabla \cdot \overline{\rho\boldsymbol{u}'\boldsymbol{u}'} + \mathrm{div}\overline{\mu\sigma} \tag{14}$$

and

$$\overline{\rho} D_t(\rho\boldsymbol{u}'/\overline{\rho}) = -\nabla p' - \rho(\boldsymbol{u}' \cdot \nabla)\overline{\boldsymbol{u}} - \rho'(\nabla\overline{\phi} + D_t\overline{\boldsymbol{u}}) - \mathrm{div}[(\rho\boldsymbol{u}'\boldsymbol{u}')' - (\mu\sigma)'], \tag{15}$$

where ϕ' is neglected for simplicity. Similar treatment for Eq. (3) is much more involved. We first rewrite Eq. (3) in the following form,

$$\overline{\rho} D_t(\rho E/\overline{\rho}) + p\nabla \cdot \overline{\boldsymbol{u}} + \nabla \cdot (\rho h \boldsymbol{u}') - \boldsymbol{u}' \cdot \nabla p = -\nabla \cdot \boldsymbol{F}_{\mathrm{R}} + \mu\sigma \cdot \nabla\boldsymbol{u}, \tag{16}$$

where E denotes the internal energy ($E = h - p/\rho$) and use the following equation,

$$-\boldsymbol{u}' \cdot \nabla p = \rho\boldsymbol{u}' \cdot (\nabla\overline{\phi} + D_t\overline{\boldsymbol{u}}) + \rho u'_j u'_k \partial_k \overline{u_j} + \partial_t(\rho u'^2/2) + \nabla \cdot (\rho u'^2/2)(\overline{\boldsymbol{u}} + \boldsymbol{u}')] - \boldsymbol{u}' \cdot \mathrm{div}\, \mu\sigma(\boldsymbol{u}), \tag{17}$$

to manipulate Eq. (16) into

$$\begin{aligned} &\overline{\rho} D_t[(\rho/\overline{\rho})(E + u'^2/2)] + \nabla \cdot [\rho(h + u'^2/2)\boldsymbol{u}'] + (\rho u'_j u'_k + p\delta_{jk} - \mu\sigma_{jk})\partial_j \overline{u_k} \\ &\quad - (\rho/\overline{\rho})\boldsymbol{u}' \cdot (\nabla \cdot \overline{\rho\boldsymbol{u}'\boldsymbol{u}'} + \nabla\overline{p} - \nabla \cdot \overline{\mu\sigma}) = -\nabla \cdot [\boldsymbol{F}_{\mathrm{R}} + \boldsymbol{F}_{\mathrm{C}} - \boldsymbol{u}' \cdot \mu\sigma(\boldsymbol{u})], \end{aligned} \tag{18}$$

where

$$\boldsymbol{F}_{\mathrm{C}} = \rho(h' + u'^2/2)\boldsymbol{u}'. \tag{19}$$

The last term, $-\boldsymbol{u}' \cdot \mu\sigma(\boldsymbol{u})$, in Eq. (18) indicating the viscous stress energy flux and the effect of density variation in all the viscous terms will be neglected hereafter. At this stage, we introduce Stokes' assumption for the turbulent Reynolds stress, $\rho u'_i u'_j$, referring to the form of the third and the fourth terms of Eq. (18),

$$\rho u'_i u'_j = p_{\mathrm{t}}\delta_{ij} - \mu_{\mathrm{t}}\sigma_{ij}(\boldsymbol{u}), \tag{20}$$

$$\sigma_{ij} = \partial_i u_j + \partial_j u_i - (2/3)\partial_k u_k \delta_{ij}, \tag{21}$$

$$p_{\mathrm{t}} = (2/3)\rho E_{\mathrm{t}} = (1/3)\rho u'^2, \tag{22}$$

and

$$\mu_{\mathrm{t}} = R_{\mathrm{eff}}^{-1}\overline{\rho}(\overline{u'^2}/3)^{1/2} l_e, \tag{23}$$

where $R_{\mathrm{eff}}(\sim 10)$ denotes the effective Reynolds number and $l_e(\sim H_p$: scale height) the mixing length. Turbulence is assumed to be homogeneous and isotropic so that p_{t} is a scalar function of position. The origin of the eddy viscosity μ_{t} is the decaying correlation among velocity components brought from upstream in such a away that (see Nakano *et al.* 1979)

$$\rho Q' u'_j = -R_{\text{eff}}^{-1} \overline{\rho} (\overline{u'^2}/3)^{1/2} l_e \partial_j Q, \tag{24}$$

in accordance with the mixing-length formulation and in consistence with the Stokes' assumption (19). Equation (24) describes the gradient diffusion approximation in which Q is treated as a conserving quantity for a mass element to be mixed. Therefore, if Q is a purely oscillating quantity in timescale τ_o shorter than the characteristic time of turbulence τ_t, the resulting nonlinear effect acting in the same phase as that of the linear disturbance Q' will be much reduced so that

$$R_{\text{eff}} = 10[1 + (\tau_t/\tau_o)^2]. \tag{25}$$

Equation (17) can now be rewritten as

$$\begin{aligned} &\overline{\rho} D_t[(\rho/\overline{\rho})(E + E_t)] + \nabla \cdot [\rho(\overline{h} + \overline{E_t})\boldsymbol{u}'] + (p + p_t)\nabla \cdot \overline{\boldsymbol{u}} - (\mu + \mu_t)\sigma(\boldsymbol{u}) \cdot \nabla\overline{\boldsymbol{u}} \\ &\quad - (\rho\boldsymbol{u}'/\overline{\rho}) \cdot [\nabla(\overline{p} + \overline{p_t}) - \nabla \cdot (\mu + \mu_t)\sigma(\overline{\boldsymbol{u}})] = -[\boldsymbol{F}_R + \boldsymbol{F}_C]. \end{aligned} \tag{26}$$

The results are summarized below. The average (oscillation) velocity field is described by

$$\overline{\rho} D_t(1/\overline{\rho}) = \nabla \cdot \overline{\boldsymbol{u}}, \tag{27}$$

$$\overline{\rho} D_t \overline{\boldsymbol{u}} = -\nabla(\overline{p} + \overline{p_t}) - \overline{\rho}\nabla\overline{\phi} + \text{div}[\mu_t \sigma(\boldsymbol{u})], \tag{28}$$

and

$$\overline{\rho} D_t(\overline{E} + \overline{E_t}) + (\overline{p} + \overline{p_t})\nabla \cdot \overline{\boldsymbol{u}} = -\nabla \cdot (\overline{\boldsymbol{F}_R} + \overline{\boldsymbol{F}_C}) + \mu_t \sigma(\overline{\boldsymbol{u}}) \cdot \nabla\overline{\boldsymbol{u}}. \tag{29}$$

The fluctuating (convective) velocity field is described by

$$\overline{\rho} D_t(\rho'/\overline{\rho}) + \nabla \cdot (\rho \boldsymbol{u}') = 0, \tag{30}$$

$$\overline{\rho} D_t(\rho\boldsymbol{u}'/\overline{\rho}) = -\nabla(p' + p'_t) - (\rho\boldsymbol{u}' \cdot \nabla)\overline{\boldsymbol{u}} - \rho'(\nabla\overline{\phi} + D_t\overline{\boldsymbol{u}}) + \text{div}[\mu_t \sigma(\boldsymbol{u}')], \tag{31}$$

and

$$\begin{aligned} &\overline{\rho} D_t[\rho(h' + h'_t)/\overline{\rho}] - D_t(p' + p'_t) + \rho\boldsymbol{u}' \cdot [\nabla(\overline{h} + \overline{h_t}) - (1/\overline{\rho})\nabla(\overline{p} + \overline{p_t})] \\ &\quad = -\nabla \cdot [\boldsymbol{F}'_R + \boldsymbol{F}'_C - (\overline{p_t}/\overline{\rho})\rho\boldsymbol{u}'] - \rho' D_t(\overline{h} + \overline{h_t}), \end{aligned} \tag{32}$$

where the enthalpy of turbulence, h_t, has been defined by

$$h_t = E_t + (p_t/\rho) = (5/3)E_t. \tag{33}$$

The molecular viscosity, μ, has been neglected in comparison with the eddy viscosity, μ_t. The manipulation of Eq. (33) from Eq. (26) is somewhat involved but straightforward.

3 The Boussinesq Convection

Equations (28) to (32) are appropriate for three dimensional computation of convection-pulsation systems. Here we are not particularly interested in time-dependent pattern of turbulent convection. In that case, the simplest description of convection are given by the Boussinesq approximation, unless the stochastic excitation of oscillations by turbulence should be considered.

First, we try to eliminate the pressure term in Eq. (31), assuming cellular cell on the uniform average background. Taking divergence of Eq. (31), we obtain an approximate expression of $(p' + p'_{\rm t})$ by use of the Boussinesq version of Eq. (30)

$$k^2(p' + p'_{\rm t}) = \rho u'_k \partial_k (ik_j \overline{u}_j) - ik_z \rho' \overline{g}, \tag{34}$$

where $k_i(= \pi/l_e)$ denotes the i-component of wave vector and $\overline{g}$ the gravity,

$$\overline{g} = (\nabla \overline{\phi} + D_t \overline{u}). \tag{35}$$

Then, omitting cells of different phase pattern, we rewrite Eq. (31) in the following form,

$$[\overline{\rho} D_t + \mu_{\rm t} k^2](\rho u'_i / \overline{\rho}) = -b_i^2 \rho u'_k \partial_k \overline{u}_i - b_z^2 \rho' \overline{g}_i, \tag{36}$$

where

$$\rho' = -\alpha T', \tag{37}$$

$$\alpha = \rho_{\rm t}(\overline{p}/\overline{T}), \tag{38}$$

and

$$b_i^2 = 1 - (k_i/k)^2 = b^2(\text{isotropic}) = 2/3. \tag{39}$$

The Boussinesq version of Eq. (32) is given by

$$[\overline{\rho} D_t + (\kappa_{\rm R} + \kappa_{\rm t}) k^2](C_p \rho T' / \overline{\rho}) = -\rho u'_j [\partial_j (\overline{h} + \overline{h_{\rm t}}) - \overline{\rho}^{-1} \partial_j (\overline{p} + \overline{p_{\rm t}})], \tag{40}$$

where $p'_{\rm t}$ and $h'_{\rm t}$ as well as the terms that are proportional to p' and ρ' have been neglected, since their main contribution arises in higher harmonics and not in the representative eddy pattern. The fluctuating convective flux, $\boldsymbol{F}'_{\rm C}$, has been replaced by $-\kappa_{\rm t} \overline{T} \nabla S'$ (S : entropy; $\kappa_{\rm t}$: turbulent conductivity $\sim 2.5\mu_{\rm t}$) and its divergence by $C_p \kappa_{\rm t} k^2 T'$, using the gradient diffusion approximation and assuming short convective scale length k^{-1}. The latter assumption is not valid, but is sufficient for semi-quantitative study. Equations (36) and (40) describe the Boussinesq convection in oscillating envelope.

4 Turbulent Pressure and Convective Flux

Oscillation is influenced by convection through $\overline{p_{\rm t}}$, $\mu_{\rm t}$ and $\overline{\boldsymbol{F}_{\rm C}}$. To complete formulation, equations determining these quantities are necessary in the form as given below.

Multiplying u'_j and (36) together, adding the result with the same equation but with subscripts i and j exchanged: $i \rightleftharpoons j$, neglecting ρ' in consistence with the Boussinesq approximation, and taking the average, we obtain ,

$$(\overline{\rho} D_t + 2\mu_{\rm t} k^2)(X_{ij}/\overline{\rho}) = -b^2 (X_{jk} \partial_k \overline{u}_i + X_{ik} \partial_k \overline{u}_j) + b^2 \alpha (\overline{g}_i H_j + \overline{g}_j H_i), \tag{41}$$

where

$$X_{ij} = \overline{\rho u'_i u'_j}, \tag{42}$$

and

$$H_i = \overline{C_p \rho T' u'_i} \approx \overline{F_{{\rm C},i}}. \tag{43}$$

Putting $X_{ij} = \overline{p_{\rm t}} \delta_{ij}$ (isotropic), we have by contraction

$$(D_t + 2\mu_t k^2)(\overline{p_t}/\overline{p}) = (2b^2/3)[-\overline{p_t}(\nabla \cdot \overline{\boldsymbol{u}}) + \alpha \overline{\boldsymbol{g}} \cdot \boldsymbol{H}]. \tag{44}$$

Similarly, using Eqs. (36) and (40), we obtain

$$[\overline{\rho} D_t + (\mu_t + \kappa_t + \kappa_R)k^2](H_i/\overline{\rho}) = -b^2(\boldsymbol{H} \cdot \nabla)\overline{u}_i + b^2 \rho_t G \overline{g}_i - \overline{p_t}[\partial_i(\overline{h} + \overline{h_t}) - \overline{\rho}^{-1}\partial_i(\overline{p} + \overline{p_t})] \tag{45}$$

and

$$\begin{aligned}[\overline{\rho} D_t + 2(\kappa_t + \kappa_R)k^2 + (C_{p,T} + 1)\overline{\rho}(D_t\overline{T}/\overline{T})](G/\overline{\rho}) \\ = -2(C_p\overline{T})^{-1}\boldsymbol{H} \cdot [\nabla(\overline{h} + \overline{h_t}) - \overline{\rho}^{-1}\nabla(\overline{p} + \overline{p_t})],\end{aligned} \tag{46}$$

where

$$G = \overline{C_p \rho T'^2}/\overline{T}, \tag{47}$$

$$\rho_t = -(\partial \log \rho / \partial \log T)_p, \tag{48}$$

and

$$C_{p,T} = (\partial \log C_p / \partial \log T)_p. \tag{49}$$

Equations (44), (45), and (46) describe $p_t, H_i \, (i = x, y, z)$, and G. The eddy viscosity, μ_t, and conductivity, κ_t, can be evaluated in terms of $\overline{p_t}$, $\overline{\rho}$, and l_e. The kinetic energy transport, $\overline{\boldsymbol{F}_K} = (1/2)\overline{u'^2 \rho \boldsymbol{u}}$ in $\overline{\boldsymbol{F}_C}(= \boldsymbol{H} + \overline{\boldsymbol{F}_K})$ is estimated in the gradient diffusion approximation,

$$\overline{\boldsymbol{F}_K} = -(1/2)\kappa_t \nabla(\overline{p_t}/\overline{\rho}). \tag{50}$$

5 Summary

Nonlinear oscillations coupled with the nonlinear Boussinesq (local) convection applicable to semi-regular and irregular variability are formulated as given by Eqs. (26), (27), (28), (44), (45), and (46). One zone modeling which is useful for the interpretation of chaotic behavior of these stars will be the next step of our study. whether the Lagrangian displacement of convective zone with pulsation can reasonably be taken into account in the one-zone modeling requires further consideration.

Acknowledgements

The present study is supported partly by the Exchange of Researchers Program between Japan Society for Promotion of Sciences (JSPS) and Academia Sinica.

References

Kamijo, F. 1963, *Publ. Astron. Soc. Japan*, **15**, 1.
Nakano, T., Fukushima, T., Unno, W., and Kondo, M. 1979, *Publ. Astron. Soc. Japan*, **31**, 713.
Saitou, M., Takeuti, M., and Tanaka, Y. 1989, *Publ. Astron. Soc. Japan*, **41**, 297.
Yuasa, M., Unno, W., and Chikawa, M. 1989, *Science and Technology* (Kinki University), 1, 35.
Xiong, D.-R. 1989, *Astron. Astrophys.*, **209**, 126.

Asymptotic Analysis of Inertial Waves in the Convective Envelope of the Sun

Umin Lee and Hideyuki Saio

Department of Astronomy, University of Tokyo, Bunkyo-ku, Tokyo 113, Japan

Abstract: Inertial waves are propagative in convective regions of rotating stars. Half of the inertial waves have positive energy of oscillations and the other half negative energy. The inertial waves with negative energy become overstable when they are in resonance with waves having positive energy such as internal gravity waves or when they dissipate energy of oscillations through nonadiabatic effects. We calculate a frequency spectrum of inertial (oscillatory convective) modes with negative energy propagating in the surface convective zone of the sun by using an asymptotic method of nonradial oscillations of rotating stars. It is shown that the inertial modes have large amplitudes only at high latitudes. The inertial modes with negative energy have very low frequencies seen in the corotating frame and hence if they are observed in an inertial frame their frequencies are approximately equal to $-m\Omega_{\odot}$.

Not only does rotation modify low frequency nonradial oscillations of rotating stars, but rotation induces new kinds of waves in the stars. Inertial waves originate from rotation as do Rossby waves. The restoring force for the inertial waves is the Coriolis force. Their frequencies are low and less than the twice of the rotation frequency. Inertial waves propagating in the convective region of rotating stars may be regarded as oscillatory convective modes (Lee and Saio 1986). It may be possible to observe the velocity fields caused by the inertial oscillations in the convective envelopes of late type stars. In this paper we study the properties of the inertial waves in the sun by employing an asymptotic method. Somewhat different approach was taken to investigate these waves by Dziembowski *et al.* (1987).

Nonradial oscillations of rotating stars are significantly modified by rotation when they have frequencies (in the corotating frame) comparable to or less than the frequency of rotation. Their θ and ϕ dependence is in general represented by a polynomial expansion in terms of the spherical harmonic functions $Y_l^m(\theta,\phi)$ for a given m. Oscillations of rotating stars are separated into two groups, which are respectively called *even* modes and *odd* modes (e.g. Berthomieu *et al.* 1978). For example, the Euler perturbation of the pressure $p'(r,\theta,\phi,t)$ is given as (see e.g. Zahn 1966)

$$p'(r,\theta,\phi,t) = \sum_{j=1}^{j=\infty} p'_{l_j}(r) Y_{l_j}^m(\theta,\phi) e^{i\omega t}, \tag{1}$$

where ($'$) denotes an Euler perturbation, and ω is a frequency observed in the corotating frame, and for $j = 1,\ 2,\ 3,\cdots$, l_j's are given as $l_j = |m| + 2(j-1)$ for even modes and as $l_j = |m| + 2(j-1) + 1$ for odd modes. When we use the polynomial expansion of the perturbed quantities in terms of Y_l^m, the governing equations of nonradial oscillations of rotating stars are converted to an infinite system of the first order linear ordinary differential equations (e.g. Lee and Saio 1986). It is very difficult to solve them exactly. Two approximations which we resort to in the following discussions are the traditional approximation, in which the horizontal component of the angular velocity vector of rotation is neglected, and the Cowling approximation, in which the Euler perturbation of the gravitational potential is neglected. Under these approximations, we introduce a new set of basis functions for the polynomial expansion of the perturbed quantities (Berthomieu *et al.* 1978; Lee and Saio 1987a, 1989). The new basis functions $\tilde{Y}_l^m$ are defined by using Y_l^m as

$$\tilde{Y}_{l_j}^m(\theta,\phi) \equiv \sum_{k=1}^{\infty} B_{kj} Y_{l_k}^m(\theta,\phi), \tag{2}$$

which means, when we write $Y_l^m \equiv \bar{P}_l^m e^{im\phi}/\sqrt{2\pi}$ and $\tilde{Y}_l^m \equiv \tilde{P}_l^m e^{im\phi}/\sqrt{2\pi}$,

$$\tilde{P}_{l_j}^m(x) \equiv \sum_{k=1}^{\infty} B_{kj} \bar{P}_{l_k}^m(x) = \sum_{k=1}^{\infty} B_{kj}\,(-1)^{\frac{m+|m|}{2}} \left[\frac{2l_k+1}{2}\frac{(l_k-|m|)!}{(l_k+|m|)!}\right]^{1/2} P_{l_k}^m(x), \tag{3}$$

where B_{kj} are elements of a matrix $\mathbf{B}$ which diagonalizes the matrix $\mathbf{W}$, and $P_l^m(x \equiv \cos\theta)$ is the associated Legendre polynomial (see e.g. Lee and Saio 1990a). The matrix $\mathbf{W}$ is the operator of the Laplace's Tidal Equation (see Chapman and Lindzen 1970) and we denote its eigenvalues as $\lambda_{l_j m}$, which are functions of $\nu \equiv 2\Omega/\omega$ for given values of j and m. The governing equations given under the two approximations are then reduced to

$$\frac{dz_1^j}{d\ln r} = \left(\frac{V}{\Gamma_1} - 3\right) z_1^j + \left(\frac{\lambda_{l_j m}}{c_1\bar{\omega}^2} - \frac{V}{\Gamma_1}\right) z_2^j, \tag{4}$$

$$\frac{dz_2^j}{d\ln r} = (c_1\bar{\omega}^2 + rA) z_1^j + (1 - U - rA) z_2^j \tag{5}$$

for $j = 1, 2, 3, \cdots$, where $z_2^j \equiv \sum_{k=1}^{\infty} B_{kj} p'_{l_k}(r)/\rho g r$ and $z_1^j \equiv \sum_{k=1}^{\infty} B_{kj} S_{l_k}(r)$, and $rS_{l_k}(r)$ is the coefficient of $Y_{l_k}^m$ when we expand the radial component of the displacement vector $\boldsymbol{\xi}$ in terms of the spherical harmonic functions. Here and hereafter, the frequencies with a bar on them mean dimensionless frequencies normalized by $\sqrt{GM_\odot/R_\odot^3}$, and other symbols have their usual meanings. The eigenvalues $\lambda_{l_j m}$ tend to $l_j(l_j+1)$ as $\nu \to 0$. Note that equations (4) and (5) are the same as those of nonradial oscillations of non-rotating stars given in the Cowling approximation if $\lambda_{l_j m}$ is replaced by $l_j(l_j+1)$.

Using an asymptotic method of nonradial oscillations of rotating stars (Berthomieu *et al.* 1978; Lee and Saio 1987a, 1989), we calculate the eigenfrequencies of inertial oscillations propagative in the convective envelope of the sun. The square of the radial wave number of low frequency nonradial oscillations of a rotating star is given by

$$k_r^2 \simeq -\frac{\lambda_{l_j m}}{r^2}\frac{rA}{c_1\bar{\omega}^2} = \frac{\lambda_{l_j m}}{r^2}\frac{\bar{N}^2}{\bar{\omega}^2}, \tag{6}$$

where N is the Brunt-Väisälä frequency. Nonradial waves are propagative in the regions where k_r^2 is positive. In the radiative region waves associated with positive $\lambda_{l_j m}$ are propagative, while in the convective region propagative are waves associated with negative

$\lambda_{l_j m}$. The former is internal gravity waves and the latter inertial waves. There are an infinite number of $\lambda_{l_j m}$'s with $l_j \geq |m|$ for a given value of m. In the frequency range of $\nu \equiv 2\Omega/\omega \geq 1$, some of $\lambda_{l_j m}$'s are negative. For a negative $\lambda_{l_j m}$, the eigenfrequencies of inertial oscillations are obtained from the dispersion relation given by

$$(n+\alpha)\pi = \int_{R_c}^{R_\odot} k_r dr = \int_{R_c}^{R_\odot} \sqrt{\lambda_{l_j m}\frac{\bar{N}^2}{\bar{\omega}^2}}\frac{dr}{r}, \tag{7}$$

where α is a numerical constant and R_c is the inner edge of the convective envelope.

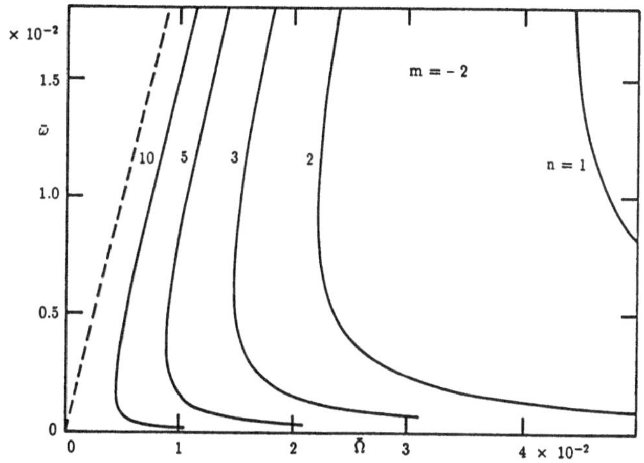

Fig. 1. Loci of inertial (oscillatory convective) modes in the $\bar{\omega} - \bar{\Omega}$ plane for $m = -2$.

Figure 1 gives the eigenfrequencies $\bar{\omega}$ of inertial oscillations (even modes) as functions of $\bar{\Omega}$ for $m = -2$. The dashed line in this figure is the line of $2\bar{\Omega}/\bar{\omega} = 1$. Integers n denote the radial order of the inertial modes and are equal to the number of nodes of the eigenfunctions. Since our full numerical calculations show $\alpha \ll 1$ for low values of m, we have used $\alpha = 0$ for simplicity. We have obtained the inertial modes associated with the largest of the negative λ_{lm}'s (i.e. having the smallest absolute value). These modes have the smallest horizontal wave numbers. For a negative m prograde waves have positive ω and retrograde waves negative ω, and hence all the inertial waves depicted in Fig. 1 are prograde waves. For a given value of Ω greater than some critical value, there are two inertial modes ω_H and ω_L ($\omega_H \geq \omega_L$) with a radial order n. Lee and Saio (1990b) showed that the inertial modes ω_L have negative energy while the inertial modes ω_H positive energy of oscillations. The inertial waves with negative energy cause overstability when they are in resonance with other nonradial waves with positive energy (Lee and Saio 1989; see also Cairns 1979). They also become overstable when energy is dissipated by nonadiabatic effects (Lee and Saio 1987b). Thus, it is reasonable to assume that the inertial oscillations with negative energy are selectively excited and the velocity fields caused by them become observable.

In Fig. 2 given are the frequencies $\bar{\omega}_L$ at $\bar{\Omega} = \bar{\Omega}_\odot$ of the five overtones (for each m) of the inertial oscillations with negative energy. We have given the inertial modes associated with the largest of the negative λ_{lm}'s for given values of m and Ω, though there should be inertial modes associated with other negative λ_{lm}'s. Since the inertial waves have very low frequencies ω_L, their horizontal motions are dominant over the radial ones, and the toroidal components of the velocity fields become significant (Lee and Saio 1990a). The θ dependence of the eigenfunctions of inertial oscillations is given by $\tilde{P}_l^m(\cos\theta)$. Figure 3 is an example, in which $\tilde{P}_l^m$ associated with the largest of the negative λ_{lm}'s is depicted as a function of $\cos\theta$ for $m = -2$ and $\nu = 5$. As shown in this figure the amplitudes of the inertial oscillations are well confined in the regions at high latitudes, which is different from the case of low frequency internal gravity waves of rotating stars. The amplitudes of the low frequency *gravity* waves are large near the equator (Lee and Saio 1990a). The inertial waves can be used to probe the structure of the surface convective zone of the sun because their frequency spectrum directly reflects it.

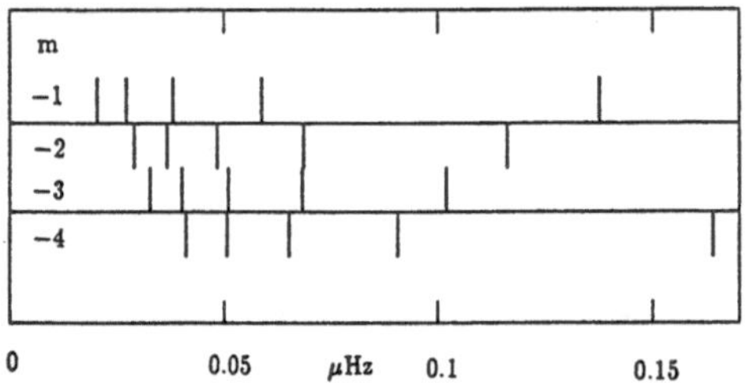

Fig. 2. Frequencies of inertial (even) modes $\bar{\omega}_L$ for $m = -1, -2, -3$ and -4. Only five overtones of the inertial modes are given for each m.

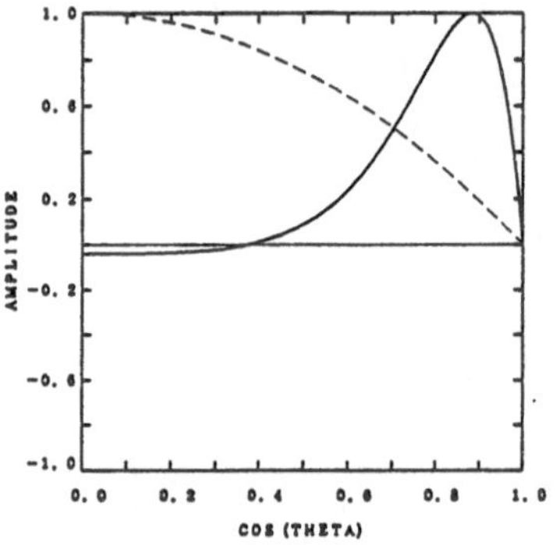

Fig. 3. The θ dependence of the eigenfunction of inertial modes for $m = -2$ and for $\nu = 5$. The dashed curve is for $\nu = 0$. The amplitudes are normalized unity at the maximum.

Reference

Berthomieu, G., Gonczi, G., Graff. Ph., Provost, J., and Rocca, A. 1978, *Astron. Astrophys.*, **70**, 597.

Cairns, R.A. 1979, *J. Fluid Mech.*, **92**, 1.

Chapman, S., and Lindzen, R.S. 1970, *Atmospheric Tides* (Reidel, Dordrecht).

Dziembowski, W., Kosovichev, A., and Kozlowski, M. 1987, *Acta Astron.*, **37**, 331.

Lee, U., and Saio, H. 1986, *Monthly Notices Roy. Astron. Soc.*, **221**, 365.

Lee, U., and Saio, H. 1987a, *Monthly Notices Roy. Astron. Soc.*, **224**, 513.

Lee, U., and Saio, H. 1987b, *Monthly Notices Roy. Astron. Soc.*, **225**, 643.

Lee, U., and Saio, H. 1989, *Monthly Notices Roy. Astron. Soc.*, **237**, 875.

Lee, U., and Saio, H. 1990a, *Astrophys. J.*, **349**, 570.

Lee, U., and Saio, H. 1990b, submitted to *Astrophys. J.*

Zahn, J.-P. 1966, *Ann. d'Astrophys.*, **29**, 313.

IV

Observations of Solar Oscillations

Trends in Helioseismology Observation and Data Reduction

J. W. Harvey

National Solar Observatory, National Optical Astronomy Observatories[1]
P. O. Box 26732, Tucson AZ 85726, U.S.A.

Abstract: Rapid growth of observational helioseismology continues unabated. Several major trends are obvious in instrumentation for the observation of p-mode oscillations. These include great improvements in angular resolution and fidelity, temporal coverage, signal quality and the planning of long-term and cooperative measurements. While progress in the observation of g modes is slow, p-mode seismic imaging is advancing rapidly. Major trends are also affecting the reduction of p-mode observations. Among these are the development of techniques for handling huge amounts of data and reduction methods which either suppress or allow for imperfect data. These imperfections include leakages in temporal and spatial domains as well as random noise and various systematic biases. Analysis of oscillation spectra is moving to fitting of individual spectral features rather than the use of dangerous cross-correlation methods. Special reduction methods are being developed for seismic imaging and localized oscillation observations. Recent observational results have mainly dealt with the form of internal solar rotation, changes in the frequencies of p modes perhaps related to the solar activity cycle and seismic imaging. In the next decade a flood of high quality data, reduced and analyzed with improved techniques and with cooperation of solar modelers and theoreticians, will lead to a high fidelity picture of the structure and dynamics of the solar interior.

1 Introduction

Observation and data reduction are essential parts of helioseismology. These aspects, as well as the rest of the science of helioseismology, are growing rapidly. This paper is not a complete review of helioseismology observation and data reduction for several reasons. First, the field has become too large to cover comprehensively in a short space. Second, the growth rate is so rapid that it is nearly impossible to obtain knowledge of the latest results before a meeting such as this one. The papers presented at this seminar provide the best access to the latest results. Third, several recent reviews of helioseismology with extensive discussions of observations and data reduction have been presented by Duvall (1990), F. Hill *et al.* (1990), H. Hill *et al.* (1990), Libbrecht and Morrow (1990), Unno *et*

[1] Operated by the Association of Universities for Research in Astronomy, Inc., under contract with the National Science Foundation.

al. (1989), Vorontsov and Zharkov (1989), Brown (1988), Harvey (1988), Libbrecht (1988) and van der Raay (1988). Finally, thirteen months ago the author attempted to review an aspect of helioseismology and the effort became outdated within hours. Therefore, this paper will concentrate on trends rather than specific results. The emphasis here is on p modes rather than the more elusive g modes.

The doubling time for helioseismology publications is currently about four years which is shorter than nearly all astronomical disciplines. Surprisingly, an inevitable slowing of the growth rate is only just detectable. This indicates that helioseismology is still in a phase of youthful growth. We will examine some directions of this growth but, as in the case of many youths, it is not entirely clear what is in the future and we are assured of many surprises, frustrations and delights. One trend is easy to discern: much of the easy work has been done and the field of helioseismology has entered a phase where spectacular results will be harder to achieve.

Table 1 is a brief history of observational helioseismology. Prior to 1974 might be termed the dark ages when the true nature of the solar oscillations had not been demonstrated. The period from 1974 to roughly 1985 might be called the age of exploration and discovery. This period continues today but, except for low frequencies and high degrees, much of the spectral range of p modes has been explored and its gross characteristics defined. We are now in an age of developing a detailed description of the p-mode spectrum. This is an exciting time since much of the promise of helioseismology will be realized in practice and our knowledge of the solar interior will improve greatly. Much remains yet to be done in the exploration of g modes and local seismic imaging.

Table 1. A partial chronology of observational helioseismology

date	event
1960	5 minute oscillation discovered
1965	2 d observations of velocity; first $k-\omega$ diagram
1974	clear demonstration of modal nature of oscillations
1975	evidence for long period oscillations of diameter
1976	0 d observations with resonance cells
1977	1.5 d spectrograph observations of high degrees
1979	1.5 d spectrograph observations of intermediate degrees
1980	0 d observations of low degrees from South Pole
1980	clear demonstration of global nature of oscillations
1980	first 0 d observations with a spacecraft
1981	0 d network observations of low degrees
1981	2 d observations of full disk from South Pole
1983	2 d observations with magneto-optical filter
1984	2 d full disk observations with Fourier tachometer
1985	2 d full disk observations with Lyot filter
1985	1 d observations with heterodyne spectrometer
1986	2 d observations with tunable Fabry-Perot etalon
1986	discovery of local acoustic absorption of p modes

2 Trends in Observations

Observational helioseismologists have a clear goal to provide as nearly perfect observations of solar oscillations as possible. Meeting this goal is limited by imperfect technology and finite budgets. Faced with these conditions, the easy observations have been made first. Now p-mode observations are pushing to low and high frequencies and to detailed examinations of high degrees and the fine structure of line shapes. Larger groups of people are becoming involved as the sophistication of the observations and data reduction increases. Second and third generation instruments are being developed.

2.1 Improving Spatial Quality

For the case of p modes, it is now relatively easy to define an ideal observation. This is summarized in Table 2. Observers have not achieved this ideal observation but the improvements since 1974 have been spectacular.

Table 2. Characteristics of an ideal helioseismology observation

parameter	value	spectral result
angular sample	1 arc second	$\ell_{max} = 3000$
angular range	full disk	$\ell_{min} = 0$
time sample	60 s	$\nu_{max} = 8.3$ mHz
time duration	10^8 s	$\delta\nu = 10$ nHz
signal sensitivity	$\ll$ solar noise	negligible instrumental noise
signal range	linear over full range	no harmonic distortion

In the early days of observational helioseismology, spatial coverage was typically restricted to a few hundred fairly large pixels covering a limited part of the solar disk. This was changed by the introduction of charged-coupled-device (ccd) arrays of detectors and computer hardware to record and process large numbers of measurements. Figure 1 shows a snapshot of the 5-min intensity oscillations obtained with a 4-arc-second-pixel ccd array operated at South Pole in 1988 (Jefferies *et al.* 1989). The current state of the art is a million, 2-arc-second pixels covering the full disk of the sun both for Doppler shift and intensity oscillation observations. It is now possible to obtain 2048 × 2048 ccds from at least two vendors and 4096 × 4096 ccds have been manufactured (Janesick *et al.* 1989). Thus, achieving essentially ideal spatial coverage required for helioseismology (at least from one vantage point of the sun) is no longer limited by detector technology.

Other factors will limit spatial resolution in the near future. From the ground, distortion and blurring of the solar image caused by Earth's atmosphere (seeing) seriously limit our ability to measure high-degree oscillations (F. Hill 1984). To reduce this problem, more attention will be paid to observing solar oscillations from sites with excellent seeing. From space, where seeing is not a problem, the main difficulties are inadequate telemetry bandwidth and/or continuity. One can observe essentially perfect images but cannot transmit all the data to the ground. A geosynchronous satellite would offer a good

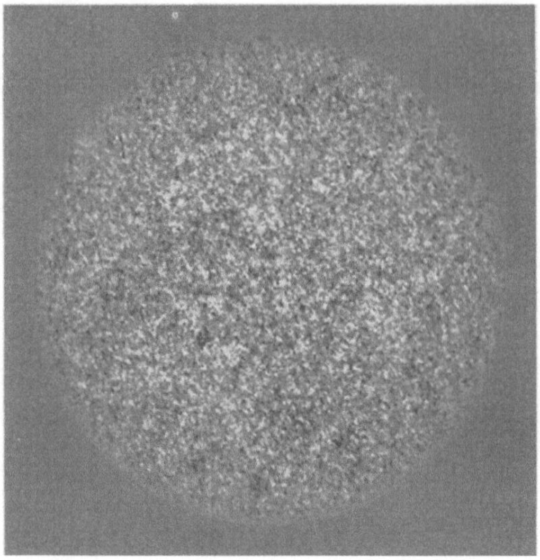

Fig. 1. Solar intensity oscillations revealed as the difference between two one-minute integrations taken two minutes apart. Note the visibility of oscillations even at the solar limb. The pixel size is 4 arc seconds in these 1988 South Pole observations (Jefferies *et al.* 1989). The current state of the art provides twice this resolution over the entire disk.

platform for nearly continuous oscillation observations at a distance close enough for a large telemetry bandwidth.

Good angular resolution is not enough. Geometric distortion of disk images must be either small or well calibrated. In early imaging p-mode observations this factor was not well controlled and, as a result, the wavenumber scale was uncertain. More subtle effects were image drift and distortion caused by the optical systems used. Observers are now able to monitor image scale and position variations by the simple technique of observing the entire disk and watching the limb position. Distortion is controlled by using high quality optical systems and calibrating residual distortion in the focal plane.

There is potential for future improvement in coverage of the disk. If we could observe the entire sphere of the sun for a long duration then identification of individual oscillation modes would be simple. At present, with only the visible disk available, there is confusion between oscillation modes with similar values of degree and angular order. These similar modes overlap in spectral analyses and make it difficult to determine the fine details of the oscillation spectrum. Even worse, the confusion may lead to systematic errors of frequency measurements. Doppler observations do not show p modes well near the limb so the effective sampling is less than the full disk, in this case making the confusion problem worse. Intensity observations do show oscillations right to the limb but the signal to background ratio of intensity oscillations is smaller than the case for Doppler observations. In addition, the foreshortening near the limb is so extreme that high-degree intensity oscillations are difficult to observe there.

The only evident observational solution to this problem is to place helioseismology instruments above widely different solar longitudes thus creating the ultimate helioseismology network. It is not likely that this solution will be implemented soon. However, the IPHIR instrument on board one of the PHOBOS probes did provide the first p-mode helioseismology data from a different vantage point than Earth (Fröhlich *et al.* 1988) and a comparison of simultaneous ground and space data will be interesting. Observations from the Pioneer Venus Orbiter have been used to deduce the rotation of the solar interior based on low frequency oscillations (Wolff and Hoegy 1989).

2.2 Improving Temporal Quality

A great deal of effort has been devoted by observational helioseismologists to improving temporal coverage. The main reasons are to minimize temporal sideband structure in the spectra of solar oscillations and to obtain good frequency resolution. Sideband structure is reduced by increasing the observational duty cycle and resolution is increased by longer duration observations. Figure 2 illustrates improvements over the years.

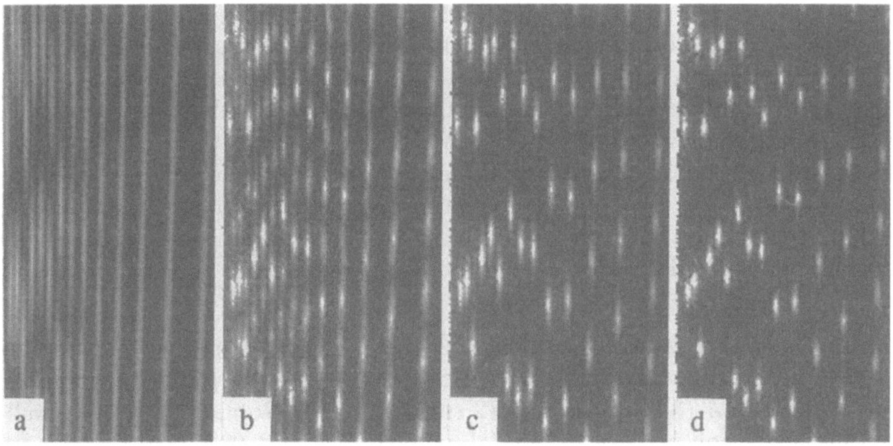

Fig. 2. A small portion of the spectrum of p-mode oscillations observed from South Pole for 20 days in 1987 (Jefferies *et al.* 1988). Abscissa is ℓ value from 0 to 99 and ordinate is frequency covering a range of about 80 μHz. The intrinsic frequency resolution is about 0.6 μHz. a) Convolved with a 12 hour window function. This simulates the resolution typical of a decade ago. b) Convolved with a day-night, 50% duty cycle. This simulates the spectrum derived from a single, low-latitude observatory. c) Actual observed spectrum with weather interruptions. Note the absence of daily sidelobe structure made possible by observing from South Pole. d) Deconvolved by the actual window function. This restores the spectrum to what would be observed if there were no interruptions (at the expense of increased noise).

The duration of non-imaging oscillation observations has increased from campaigns of one or a few days to seasons to nearly continuous at present. Observing duty cycle, or continuity, has improved by setting up global networks (Aindow *et al.* 1988; Fossat

1988; Jiménez *et al.* 1988), by operating from South Pole (Fossat *et al.* 1987) and from space (Fröhlich *et al.* 1988). While good duty cycles can be obtained using networks (F.Hill 1990) the merger of observations from different instruments is challenging. This problem is avoided at South Pole but weather degrades the duty cycle (Harvey 1989). A compromise was suggested by Fossat *et al.* (1989) who have proposed an Antarctic network of two non-imaging instruments to obtain a higher duty cycle than weather permits at a single Antarctic site. Lynch (1989) has proposed an international Antarctic station at a location likely to be substantially clearer than South Pole. Such a site would be valuable for helioseismology. The non-imaging ACRIM and IPHIR instruments flown in space suffered from relatively low duty cycles. The non-imaging helioseismology instruments under development for the SOHO mission should provide nearly perfect temporal sampling of low-degree oscillations.

The history of temporal improvements in imaged helioseismology is much the same as for non-imaging instruments. Improvements have been relatively delayed because of the greater complexity and larger data production of imaging equipment. Single sites are now typically operated for a season such as local summer. Durations are of the order of 100 days to provide a frequency element of about 100 nHz. This good frequency resolution is a considerable help in isolating spectral features from diurnal sidebands. Nevertheless, these sidebands are a serious source of noise and duty cycle improvements are in progress. The longest duration successful imaging observations from South Pole were obtained in 1988 for about 480 hours at a duty cycle of about 75% and 123 hours at 92% (Jefferies *et al.* 1989). Even with a duty cycle considerably less than 100%, South Pole observations offer an advantage which has only recently been realized. Because there are no strong periodicities in South Pole data gaps it is possible and practical to correct spectra for an imperfect duty cycle. This is done by simply deconvolving the observed spectrum with the temporal spectral window function as shown in Fig. 2. Such a procedure should work well for any non-periodically interrupted helioseismology data set. In particular, network observations should benefit from this technique.

At present, at least four ground-based networks are being developed for imaging helioseismology to improve durations and duty cycles among other goals. These include the GONG (6 sites), SCLERA (3 sites), Hawaii (2 sites) and JPL/Rome (2 sites). Six well-selected sites provide a non-periodic, annual duty cycle of about 92% while fewer sites lead to lower duty cycles (F. Hill 1990). Our goal of 100% duty cycle, long duration observations of oscillations is nearly in reach. New observing programs are being developed to enable imaging observations for as long as a sunspot cycle. One example is the Solar Cycle Telescopes project of the National Astronomical Observatory of Japan (Hiei, private communication). Another example is a high-degree helioseismograph collaborative project of Bartol, NASA and NSO.

Imaging helioseismology has not yet been done from space. Such observations will be possible with the photospheric imager on the Solar-A satellite but its orbit will not permit continuous coverage and it is likely that telemetry and primary observing constraints will limit the amount of helioseismology performed with Solar-A. Two instruments on the SOHO will provide imaged observations with nearly perfect temporal coverage of low degrees but temporal coverage of high degrees will be compromised by limited telemetry. Excellent temporal coverage from space remains for some future mission. It is likely that successful g-mode observations will require excellent temporal characteristics.

2.3 Improving Signal Quality

Solar p-mode oscillations are manifest in several measurable quantities. The principal ones in current use are Doppler shift, intensity variation and limb darkening variation. The latter method suffers from poor resolution in degree and angular order and so has not been adopted by most observers. Results from the first two methods are in excellent agreement. In order to produce a high-quality Doppler oscillation signal, a sensitivity to patterns over the solar disk with amplitudes of the order of 1 mm s^{-1} is required together with a reasonably linear response over a range of about $\pm 3 \times 10^6$ mm s^{-1}. For intensity measurements, a sensitivity to patterns with amplitudes of the order of 1 part in 10^6 with a dynamic range of about 1 (to accommodate limb darkening and sunspots). Achieving these degrees of performance requires attention to instrumental details. Observers have gradually improved their instrumental performance so that these levels are being reached. One important aid in this effort has been analytical study of instrument requirements and performance (e.g. Appourchaux 1989; Hoyng 1989). A trouble area remains for the lowest degree oscillations. These oscillations do not average to zero across the solar disk and therefore require instrumental stability as well as sensitivity at the high performance levels given above.

In Doppler measurements, only alkali vapor resonance cells have demonstrated stability performance at near the desired level. However, other Doppler detectors are being improved to nearly the same performance. An example is shown in Fig. 3. Resonance cells have suffered from a limited range of linear response in the past but this problem has been reduced by the development of tunable and multiple pass band cells (e.g. Lin and Kuhn 1989). Whether these new cell designs retain the excellent stability of simpler designs remains to be seen.

For intensity measurements, stability depends critically on the repeatability of shuttering and integration intervals as well as on the stability of the detector. CCD detectors in particular must have excellent stability of signal gain and bias. It is now possible to detect $\ell = 0$ p modes in imaging intensity observations (e.g. Jefferies *et al.* 1988).

Variations of Earth's atmosphere introduce noise. For high degrees, the main problem is image motion. This can be reduced by high-speed image stabilization techniques and by modulation of the signal at frequencies in excess of image motion frequencies. Both techniques are now in use at several ground-based instruments. Space based observations should not suffer from these problems. At low degrees, the main problem is varying atmospheric transmission or extinction. Aside from observing from excellent sites, one instrumental approach to reducing this problem has been described by McLeod and Isaak (1988). Because of their differential nature, imaging Doppler measurements should be relatively immune to the low-degree atmospheric noise problem.

2.4 Other Observational Trends

One promising development is the emergence of cooperative observing campaigns. Two or more groups agree to observe at the same time but in different ways. Such joint observations will lead to an improved understanding of the oscillation physics but, just as importantly, this is the best way to assess the magnitude of systematic errors in various data sets.

As the quality and duration of observations improves, there is increasing interest in the extremes of the $\ell - \nu$ range of p modes. At high frequencies, the phenomenon of

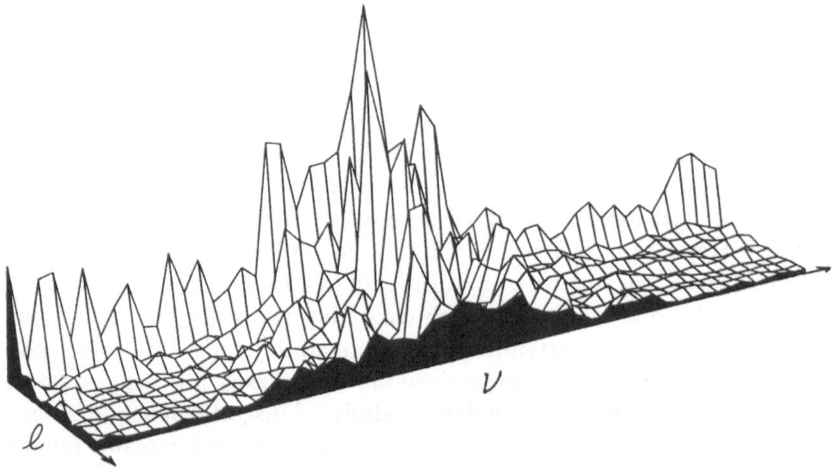

Fig. 3. A portion of an $\ell - \nu$ power spectrum of $m = 0$ p modes obtained with the breadboard GONG instrument for 760 minutes on June 8, 1989. Frequency runs from left to right over the range 1.6 to 4.9 mHz. The degree ranges from 0 in the back of the plot to 10 in the front. Note that, although degree zero is noisier than higher degrees, the p modes are readily detected at all degrees.

pseudomodes (Kumar and Goldreich 1989) has attracted both observational and theoretical interest. While the value of such pseudomodes for interior helioseismology appears limited, aliasing of these oscillations into the frequency range of global p modes is an undesired source of noise. Figure 4 illustrates this phenomenon.

Interest in low frequencies has always been strong. It is now clear that solar rather than instrumental noise sources are the main limitations at low frequencies. The long coherence time of low frequency oscillations offers the possibility of using long duration observations to try to detect weak oscillations (e.g. Anguera Gubau *et al.* 1990). Further progress is likely to depend on development of new observing techniques. One idea being explored is to use differential observational methods. For example, it is well known that p-mode oscillations of Doppler shift are about a quarter cycle out of phase with intensity oscillations. On the other hand, the intensity and Doppler patterns of granulation appear to be in phase. Under these conditions, noise from granulation can be suppressed in a suitable observation. Unfortunately, the main sources of noise at low frequencies are active regions and the supergranulation network. It is not yet clear what strategy will best suppress these sources of noise. Progress toward detecting the elusive g modes will probably require additional work on differential observational techniques to extract the signals from noise.

Fig. 4. A portion of an m-averaged spectrum of p modes observed from South Pole (Jefferies *et al.* 1988). The degree range runs from 0 (bottom) to 150 (top). The frequency ranges from 3.3 mHz (left) to 6.7 mHz (center) and the spectrum is reflected about the Nyquist frequency. Pseudomodes and their high frequency aliases are seen at frequencies above 5.5 mHz.

3 Trends in Data Reduction

In parallel with improved observations, better data reduction methods have been developed. The goal is to produce high fidelity oscillation spectra with a minimum of distortion and to accurately characterize the individual spectral features. A major trend is the development of efficient methods for handling increasing amounts of helioseismology data. Fortunately, improvements of computing equipment are rapid enough to keep pace with the data flow. An example is high-density, helical-scan, magnetic tape recorders which became available in 1988.

Roughly speaking, each pixel in an oscillation image represents an opportunity to investigate a single oscillation mode. With million-pixel images now available, we are obliged to pick and choose which modes to reduce even with the best available equipment. It simply is not practical to reduce all $\sim 10^7$ modes in the p-mode oscillation spectrum. The current trend is to do all the possible modes up to some degree of order a hundred or so and then apply limits at higher degrees. Various limit strategies can be imagined and have been tried. No clear trend has emerged but an argument can be made for reducing all the possible modes in small degree ranges. The reason for this is to allow one to detect and correct spectral leakage effects and to fully exploit the noise reduction of m-averaging.

3.1 The Traditional Global P-Mode Reduction

There is a commonly used procedure for the reduction of both imaged Doppler and intensity p-mode observations (cf. Brown 1988). The result is the desired spectrum in units of ν, ℓ and m:

1. Correct the raw observations for known static instrumental signatures.
2. Correct for known time variations in the data.
3. Spatially remap the observations to a grid in longitude and sine latitude.
4. Apply a longitudinal Fourier transform to each sine latitude row.
5. For each longitudinal wavenumber step do a transformation of the set of sine latitude components to a series of associated Legendre function coefficients. The result is a set of coefficients of spherical harmonic functions for the entire image at that time.
6. Produce an edited time series of the spherical harmonic coefficients.
7. Do Fourier transforms of the time series to produce the final spectrum.

3.2 Trends in Editing

A number of improvements to the basic reduction scheme have been developed. One important matter is editing of imperfect images. Clouds and other temporally varying defects must be detected and substandard images rejected. Every observing group has developed its own strategy for dealing with this need. A powerful technique now entering service is to make a movie of the data. Seriously defective images are usually eliminated from a time series. Slightly degraded images can often be restored to adequate quality, depending on the nature of the degradation. An obvious criterion for rejecting an image is whether including it would add more noise than leaving it out. But the decision is not so simple because some defects mainly affect quality at high degrees and others mainly at low degrees. There is still considerable art in editing imperfect images.

3.3 Trends in Temporal Aspects

An additional reduction step has recently been added by most observing groups between steps 2 and 3. This consists of normalizing either the cleaned up images or the spatially remapped images to a temporal running mean image. This is easy and has the nice benefit of keeping strong low frequency signals and noise from leaking to higher frequencies. The technique has been particularly valuable for low degrees observed with imaging instruments. One could use sophisticated temporal combinations of images in making a weighted running mean image but in practice a simple uniform weighting over 20 to 30 minutes works well. Attenuation of various temporal frequencies caused by this technique is easily computed and corrected in the final spectra.

It is generally the case that solar oscillation observations are taken at well controlled, even time steps so that a simple fast Fourier transform of time series of spherical harmonic coefficients is all that is needed to produce good frequency spectra. As observational precision increases, temporal corrections will be required for changes in the earth-sun distance. Here, the recent development of a fast algorithm for transforming unequally spaced data (Press and Rybicki 1989) should be very helpful. Similar methods for correlation analyses of unequally spaced data have also become available (Scargle 1989). The success of deconvolving spectra derived from non-periodically gapped observations should become a routine reduction tool in temporal reduction.

3.4 Trends in Spatial Aspects

A major part of the reduction of helioseismology observations is the spatial transformation from an observed image to a set of spherical harmonic coefficients. This is computationally demanding and discovery of a fast way of doing the transformation would be a real advance. Brown (private communication) has proposed that a spatial remapping of observed data with respect to the disk center would permit an efficient two-dimensional Fourier transform to produce a useful approximation of spherical harmonic coefficients at high degrees. As more helioseismology data of higher resolution becomes available the search for efficient transform algorithms will become more urgent.

A number of geometric problems which were previously neglected are now being addressed during reduction of imaging observations. Seeing blurs images and reduces the observed amplitude of oscillation patterns with increasing degree. This can be monitored and corrected (e.g. Kaufman 1988). Atmospheric refraction distorts the solar image. This can be calculated and appropriate corrections made. Errors in the image scale translate to errors in the ℓ and m scales. Careful use of the solar limb keeps these errors to less than 0.1%. Interpolation of images from one coordinate system to another introduces noise. This can be minimized by using the best interpolation algorithms but there is a tradeoff with computing effort. Variations in atmospheric and instrumental scattered light can be a serious source of noise. The amount of scatter can be measured and corrections made to the individual images or to the spherical harmonic coefficients.

There is no clear trend in the handling of spatial apodization of images. The question is what to do as one approaches the limb. If data from too close to the limb are accepted then the noise in final spectra increases. On the other hand rejecting data from near the limb decreases the ℓ and m resolution. If the transition from rejection to acceptance is abrupt then spatial sidelobes are introduced into the final spectra. The tradeoff is the familiar one of good ℓ and m resolution against low spatial leakage of unwanted features. Unfortunately, the solutions that worked well in the temporal domain are not applicable in the spatial domain. An optimum solution to the problem has not yet emerged.

Woodard (1989) has pointed out that spherical harmonic patterns of oscillation modes will be spatially distorted by differential rotation. The result is a systematic confusion of ℓ and m values at high degrees unless appropriately distorted functions are used for the spatial transformations. This is an area for further work and it is too early to see a trend toward coping with the problem.

3.5 Trends in Spectral Analysis

When a good spectrum of oscillations has been produced, the next task is to analyze the characteristics of its spectral features. The first issue is what to do with spectral leakage which remains in even the best spectra and which corrupts individual features. One approach is to try to invert the observed spectrum to be free of spurious responses. This has not been very successful because of the intrinsically low signal-to-noise ratio of oscillation power spectra. The widely used approach is to create a parametric model of the spectrum and fit the model to the observations with the parameters varied in a way to minimize some function of the residuals. The current trend is to use better models in this process. Models now allow for asymmetric spectral leakage, varying background noise and the correct statistics of noise (Duvall 1990). A dozen or more parameters may be used in the fitting of a single spectral feature. As much prior knowledge as possible is built into the models but there is always a tradeoff between assuming too much to reduce the number of parameters to be fit at the expense of introducing biases in the results. This fitting procedure is laborious and attempts are continuing to automate it and to improve its efficiency.

The large number of modes that can be measured obliges observers to average their results down to a manageable size. Many different averaging schemes have been used. One of these has recently proven to be very dangerous. In the study of frequency dependence on m, a widely used reduction is to first average frequency spectra at a given value of ℓ over the range of m values with assumed frequency shifts applied to remove the effects of rotation and asphericity. This spectrum is then cross correlated with the individual spectra at each value of m and the frequency offset of the peak of the cross correlation function is used to estimate the m-dependence of frequency. The process is iterated to convergence. A wide range of frequencies is usually processed at one time and the results therefore are a weighted average over a number of n values. The problem which has recently been recognized is noise in the spectra. The m-averaged spectrum has a component of noise in it from each of the individual m spectra. At each m value, this noise will cause a significant correlation at a frequency offset corresponding exactly to that assumed in doing the m averaging. The result is that the process returns essentially the same model which was assumed in the first place (depending on how the peak of the cross correlation function is determined). The problem can be reduced either by restricting the cross correlation to just those frequencies where mode power is observed or by deleting the target spectrum from the m-averaged spectrum. The problem can be eliminated by using a noise-free model of the m-averaged spectrum. In view of this serious problem, all published results on m dependence of frequency shifts based on cross correlation must be used with caution. It is likely that this explains some of the systematic discrepancies between results from different observers.

Another questionable averaging procedure used by observers is to group results by ℓ or frequency without attention to the possibility of meaningful variations depending on both of these quantities. A trend is now emerging to analyze and average results with attention to multivariate dependence possibilities. As observers become more involved with doing inversions, there will be less need for arbitrary averaging to present results.

3.6 Other Reduction Trends

Some specialized reduction techniques are emerging that are specifically tailored for localized seismology. The basic observations are the same as for the more traditional reduction methods. One of these techniques involves fitting ellipses to frequency cuts of oscillation spectra in full two-dimensional wavenumber space (F. Hill 1988). This is a substantial improvement over previous analyses of high degree spectra because all wavenumbers are analyzed rather than a single set. The rapidly evolving technique of local acoustic imaging has led to development of a method to produce surface maps of acoustic absorptivity (Braun *et al.* 1990). A first exploration has been made of the intriguing possibility of imaging the back side of the sun by taking advantage of its acoustic lensing properties (Lindsey and Braun 1990).

4 Conclusion

It is always dangerous to extrapolate present trends to the future, especially in a rapidly developing field such as helioseismology. The major trend is to acquire much more and higher quality helioseismology data than have been available in the past. This trend should continue during the next decade. Reduction methods are becoming more sophisticated in order to keep up with the improvements of data quality. Collaboration between observers and theoreticians is becoming even closer than in the past with major benefits to the entire field. Evidence of this can be found in the increasing numbers of authors on some helioseismology papers. Current research on p modes centers on the form of internal solar rotation and time variability of oscillation parameters such as frequency, frequency splitting, amplitude and coherence time. It is quite possible that a new trend will emerge as a result of this meeting.

Acknowledgements

I am indebted to my colleagues Tom Duvall, Stuart Jefferies and Martin Pomerantz for sharing many adventures in observational and data reduction aspects of helioseismology and for agreeing to let me present unpublished results from our collaboration. I am also grateful to colleagues who have provided results in advance of publication useful in the preparation of this paper. Partial support for my participation in this meeting was kindly provided by the organizers.

References

Aindow, A., Elsworth, Y. P., Isaak, G. R., McLeod, C. P., New, R., and van der Raay, H. B. 1988, in *Seismology of the Sun and Sun-Like Stars*, ed. E. J. Rolfe, ESA SP-286 (ESA Publication Division, Noordwijk), p.157.

Anguera Gubau M., Pallé, P. L., Pérez Hernández, F., and Roca Cortés, T. 1990, *Solar Phys.*, submitted.

Appourchaux, T. 1989, *Astron. Astrophys.*, **222**, 361.

Braun, D. C., LaBonte, B. J., and Duvall, T. L., Jr. 1990, *Astrophys. J.*, in press.

Brown, T. M. 1988, in *Advances in Helio- and Asteroseismology*, eds. J. Christensen-Dalsgaard and S. Frandsen, IAU Symp. No. 123, (Reidel, Dordrecht), p.453.

Duvall, T. L., Jr. 1990, in *Inside the Sun*, eds. G. Berthomieu and M. Cribier, IAU Colloq. No. 121 (Kluwer, Dordrecht), in press.

Fossat, E. 1988, in *Seismology of the Sun and Sun-Like Stars*, ed. E. J. Rolfe, ESA SP-286 (ESA Publication Division, Noordwijk), p.161.

Fossat, E., Gelly, B., Grec, G., and Pomerantz, M. 1987, *Astron. Astrophys.*, **177**, L47.

Fossat, E., Gelly, B., Grec, G., and Schmider, F.-X. 1989, in *Astrophysics in Antarctica*, eds. D. J. Mullen, M. A. Pomerantz, and T. Stanev, AIP Conference Proc. 198 (AIP, New York), p.231.

Fröhlich, C., Bonnet, R. M., Bruns, A. V., Delaboudiniere, J. P., Domingo, V., Kotov, V. A., Kollath, Z., Rashkovsky, D. N., Toutain, T., Vial, J. C., and Werhli, C. 1988, in *Seismology of the Sun and Sun-Like Stars*, ed. E. J. Rolfe, ESA SP-286 (ESA Publication Division, Noordwijk), p.359.

Harvey, J. 1988, in *Seismology of the Sun and Sun-Like Stars*, ed. E. J. Rolfe, ESA SP-286 (ESA Publication Division, Noordwijk), p.55.

Harvey, J. 1989, in *Astrophysics in Antarctica*, eds. D. J. Mullen, M. A. Pomerantz, and T. Stanev, AIP Conference Proc. 198 (AIP, New York), p.227.

Hill, F. 1984, in *Solar Seismology from Space*, NASA JPL 84-84, p.255.

Hill, F. 1988, *Astrophys. J.*, **333**, 996.

Hill, F. 1990, in *Inside the Sun*, eds. G. Berthomieu and M. Cribier, IAU Colloq. No. 121 (Kluwer, Dordrecht), in press.

Hill, F., Deubner, F.-L., and Isaak, G. 1990, in *The Solar Interior and Atmosphere* , eds. A. N. Cox and W. C. Livingston (Univ. of Arizona Press, Tucson), Ch. 10, submitted.

Hill, H., Fröhlich, C., Gabriel, M., and Kotov, V. A. 1990, in *The Solar Interior and Atmosphere*, eds. A. N. Cox and W. C. Livingston (Univ. of Arizona Press, Tucson), Ch. 14, submitted.

Hoyng, P. 1989, *Astrophys. J.*, **345**, 1088.

Janesick, J., Elliott, T., Bredthauer, R., Cover, J., Schaefer, R., and Varian, R. 1989, *Proc. SPIE*, **1071**, 115.

Jefferies, S. M., Pomerantz, M. A., Duvall, T. L., Jr., and Harvey, J. W. 1989, *Antarctic J. U. S.*, **24**, in press.

Jefferies, S. M., Pomerantz, M. A., Duvall, T. L., Jr., Harvey, J. W., and Jaksha, D. B. 1988, in *Seismology of the Sun and Sun-Like Stars*, ed. E. J. Rolfe, ESA SP-286 (ESA Publication Division, Noordwijk), p.279.

Jiménez, A., Pallé, P. L., Roca Cortés, T., Andersen, B. N., Domingo, V., Jones, A., Alvarez, M., and Ledezma, E. 1988, in *Seismology of the Sun and Sun-Like Stars*, ed. E. J. Rolfe, ESA SP-286 (ESA Publication Division, Noordwijk), p.163.

Kaufman, J. M. 1988, in *Seismology of the Sun and Sun-Like Stars*, ed. E. J. Rolfe, ESA SP-286 (ESA Publication Division, Noordwijk), p.31.

Kumar, P., and Goldreich, P. 1989, *Astrophys. J.*, **342**, 558.

Libbrecht, K. G. 1988, *Space Sci. Rev.*, **47**, 275.

Libbrecht, K. G., and Morrow, C. A. 1990, in *The Solar Interior and Atmosphere*, eds. A. N. Cox and W. C. Livingston (Univ. of Arizona Press, Tucson), Ch. 11, submitted.

Lin, H., and Kuhn, J. R. 1989, *Solar Phys.*, **122**, 365.

Lindsey, C., and Braun, D. 1990, *Solar Phys.*, in press.

Lynch, J. 1989, in *Astrophysics in Antarctica*, eds. D. J. Mullen, M. A. Pomerantz, and T. Stanev, AIP Conference Proc. 198 (AIP, New York), p.249.

McLeod, D. B., and Isaak, G. R. 1988, in *Seismology of the Sun and Sun-Like Stars*, ed. E. J. Rolfe, ESA SP-286 (ESA Publication Division, Noordwijk), p.223.

Press, W. H., and Rybicki, G. B. 1989, *Astrophys. J.*, **338**, 277.

van der Raay, H. B. 1988, in *Seismology of the Sun and Sun-Like Stars*, ed. E. J. Rolfe, ESA SP-286 (ESA Publication Division, Noordwijk), p.339.

Scargle, J. D. 1989, *Astrophys. J.*, **343**, 874.

Unno, W., Osaki, Y., Ando, H., Saio, H., and Shibahashi, H. 1989, *Nonradial Oscillations of Stars*, 2nd ed. (Univ. of Tokyo Press, Tokyo).

Wolff, C. L., and Hoegy, W. R. 1989, *Solar Phys.*, **123**, 7.

Woodard, M. F. 1989, *Astrophys. J.*, **347**, 1176.

Vorontsov, S. V., and Zharkov, V. N. 1989, *Sov. Sci. Rev. E*, **7**, 1.

The Spectrum of Solar p-Modes and the Solar Activity Cycle

P. L. Pallé, C. Régulo, and T. Roca Cortés

Instituto de Astrofísica de Canarias, 38200 La Laguna, Tenerife, Spain

Abstract: Solar cycle variations on the power and frequencies of the low l p–mode solar acoustic oscillations are investigated using an extensive set of observations obtained at Observatorio del Teide (Izaña, Tenerife). The radial velocity of integrated sunlight has been monitored, by means of a resonant scattering spectrophotometer, at several epochs (basically each year) from 1977 to 1989. The latest data (1988 and 1989) confirm previously found results (Pallé *et al.*, 1988, 1989a,b):

a) A variation of nearly 40% peak to peak in the power of the low l solar p–modes, being higher when the solar activity is at its minimum.
b) A null variation in frequency for $l = 0$ and a decrease of $\sim 0.5\ \mu$Hz for $l = 1$ when solar activity goes from maximum to minimum (similar results for $l = 2$ and $l = 3$) by using a cross-correlation technique.

1 Power of Acoustic Modes

Integrated sunlight data from 1977 to 1989 obtained at the Observatorio del Teide (Izaña, Tenerife), using a resonant scattering spectrophotometer (Brookes *et al.* 1978), is analyzed to look for changes in the power of low degree acoustic modes along the solar cycle (Pallé *et al.*, 1989b). Data taken each day are individually reduced (Pallé, 1986) and joined together in sets of time strings of 60 days span each. Power spectra of each set is calculated using an interactive sine wave fitting procedure between 2 and 3.8 mHz (Pallé *et al.*, 1986).

In 1979 and 1983 no useful observations are available; from 1982 to 1989, two or more independent sets of 60 continuous days are available every year. The series used have duty cycles higher than 30%. The duty cycle is calculated as the percentage of hours of observations including nights; therefore the best possible duty cycle is $\sim 50\%$ in summer days.

The quality of the series used is very important when we are looking for changes in the power of low degree acoustic p–modes because in these 60 days series, non negligible first order sidebands appear (at $\pm 11.57\,\mu$Hz) for every peak present in the spectra, due to the observing window function. Moreover, the worse the duty cycle, the more power goes into the sidebands. This is a problem because the frequency separation between modes of degree l=0 and 2 is $\sim$ 9 μHz and its relative power is roughly 1:1; furthermore, due to the rotational splitting for l=2 modes, some of the sidebands power can contribute to the calculated power per mode. As far as modes l=1 and 3 are concerned, this problem is somewhat less severe because their separation is $\sim$ 15 μHz and their relative power is roughly 10:1. Therefore, the most reliable power measurement is probably for l=1 where the effect of mixing peak power with power of the sidebands is very small.

It is very convenient, first of all, to get rid of the background noise in each spectrum. In order to do so, every spectrum has been divided into 9 intervals of 0.2 mHz each. In each interval, a linear fit to the power estimates is made ignoring the signal present and their first order sidebands; then, these fits, representing the noise level, are subtracted from the corresponding intervals in the spectrum.

The power in each mode $0 \leq l \leq 3$ is calculated for each spectra of the 60 day time series as follows:

$$P_l = \sum_{n=13}^{25} \sum_{i=-k}^{k} (A_{i,n}^2 - A_{\text{noise}}^2) \tag{1}$$

where: $A_{i,n}$ is the amplitude at frequency ν_i , around a peak representing the mode of degree l and order n, as found by Jiménez *et al.* (1988); A_{noise} is the mean amplitude of the noise level calculated as described above; k has been set at $\delta\nu$, where $\delta\nu$ is the line width as measured in Elsworth *et al.* (1990) and n stands for the order of the modes.

If we have more than one series for anyone year, a mean value is used. The values found for P_l (l=0, 1 ,2 and 3) are shown in Fig. 1. A peak to peak variation of 40% for l=0 and l=1 calculated between 1980 and 1986-87 can be deduced, this change is correlated with the solar activity cycle, the power being higher when the solar activity is at its minimum. For l=2 and 3 the change in not so clear, these modes are (very) broad due to rotational splitting, then, the power of the sidebands of the neighbour peaks have some contribution to the power calculated and the modes are not as well defined as l=0, 1.

If the result found for the different p–modes are combined, the Fig. 2 is obtained, where the power has been calculated as follows:

$$P = \sum_{l=0}^{2} \frac{P_l}{S_l} = \sum_{l=0}^{2} \frac{1}{S_l} \sum_{n=13}^{25} \sum_{i=-k}^{k} (A_{i,l,n}^2 - A_{\text{noise}}^2) \tag{2}$$

where: S_l is the sensitivity of the integrated sunlight velocity measurements as defined in Pallé *et al.* (1989c)

The interpretation of this effect could be attributed to the absorption of mode power by magnetic structures (sunspots, active regions, etc...) already found for higher l modes by Braun *et al.* (1988), but it seems unlikely. The reason being that, if magnetic structures absorb the same amount of p-mode power for $l \leq 3$ as for higher l modes, which is $\sim$ 50%, then, since at the maximum of solar activity, the surface covered by active regions can be 10% at most hence the power absorbed by them would be less than 5%, which is clearly not enough to explain the observed effect.

Therefore, it is more likely that changes in the efficiency of the excitation mechanism of such modes could be the cause. Assuming turbulent convection to be the responsible for exciting these modes with energies $E \sim c^2 \rho L^2 H$ (Libbrecht, 1988), where ρ is the density, c the sound speed, H the scale height and L the characteristic size of the granules,

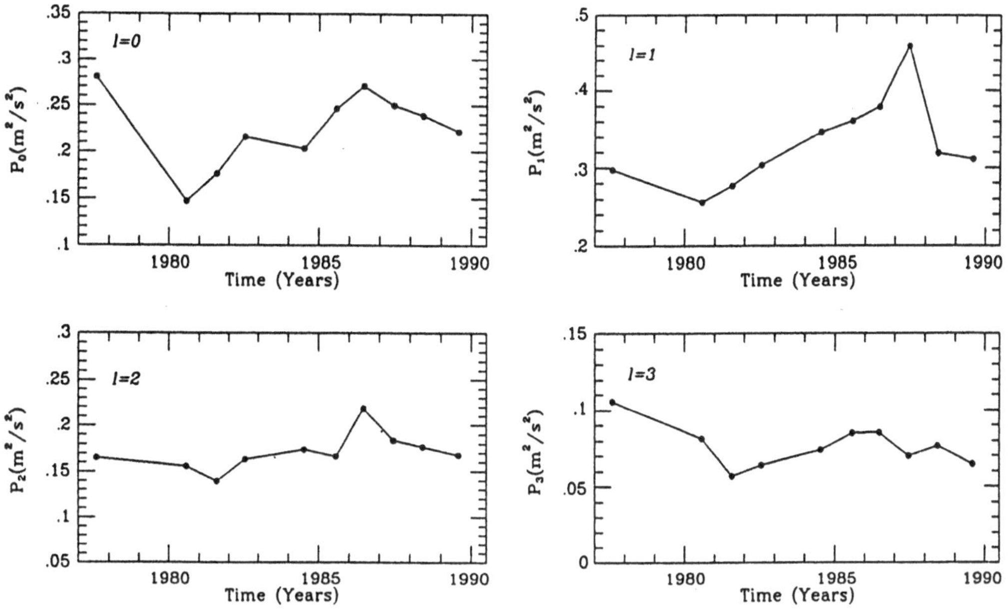

Fig. 1. Yearly variation of the solar p–modes power of degree l=0, 1, 2 and 3.

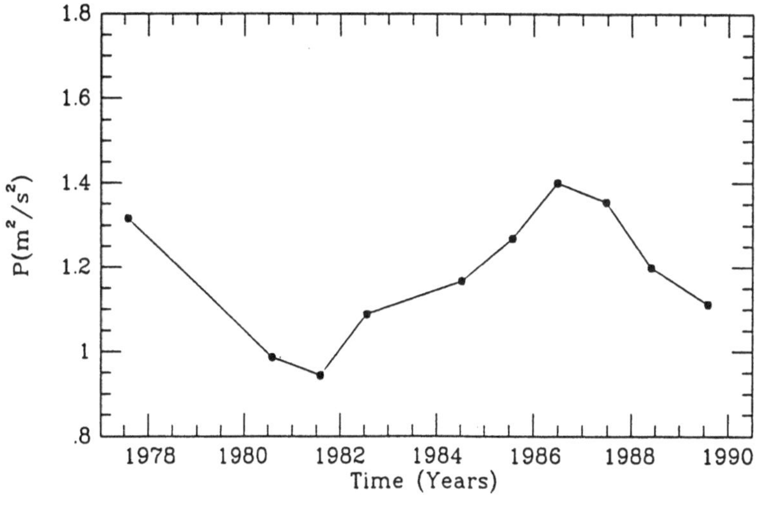

Fig. 2. Power per mode summed over all modes with $l \leq 2$.

then a maximum variation of $\sim$ 20% in L along the solar cycle would account for the observed effect. Observations of such variations have already been made by Muller and Roudier (1984) with similar numerical results (in fact their observations would imply a change of only 10% on L).

2 The Frequencies of Acoustic Modes

To look for variations in the frequencies of the acoustic p–modes, all the 60 continuous days sets available from 1977 through 1989 are used.

To compare the different spectra, a cross-correlation technique is used. The power spectrum of 1981 data series is used as the reference against which all the others are cross-correlated, because it is very close to solar maximum and has one of the best duty cycle (42%). To measure the position of the cross-correlation peak, the centroid is calculated (Pallé *et al.* 1988; 1989a,b). In Fig. 3 the centroids of all series used are shown; in it the centroids of the cross-correlation functions of the summer months are joined by a full line. The variation of $-0.37 \pm 0.04\mu$Hz already found (Pallé *et al.* 1988; 1989a,b) from maximum to minimum to the solar cycle is confirmed by the latest data. The different result obtained for different l values is confirmed too. To look for this effect, four spectra were obtained from each calculated power spectrum, each one containing only the information around the "isolated" first order sideband of modes l=0, 1, 2, and 3 respectively. The sidebands are then used instead of the real peaks due to the proximity of the neighbour peak sideband for each mode. All modes have at least one sideband that can be isolated without problems, since it is far away from any other feature.

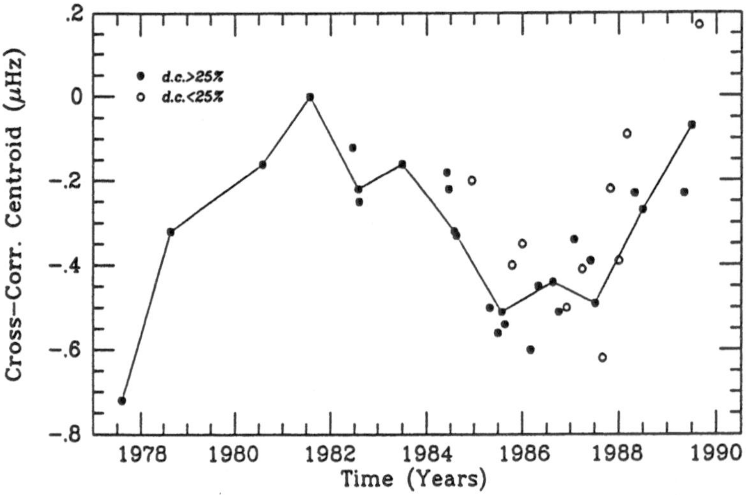

Fig. 3. Centroids calculated from the cross-correlation functions of power spectra of all available data. Unfilled circles stand for those series whose duty cycle is less than 25%.

The centroids of the cross-correlation function for each p-mode are shown in Fig. 4. From this picture we can calculate the relative shift between maximum and minimum, the result being: -0.04 ± 0.1, -0.42 ± 0.06, -0.38 ± 0.14 and -0.47 ± 0.10 μHz for l=0, 1, 2 and 3 respectively. The dependence of this variation on the l value of the modes (the variation being absent for $l = 0$) suggest that other interpretations than merely a shift of frequencies in the acoustic mode spectrum across the solar cycle are plausible: an amplitude modulation between modes of the same multiplet and/or an asymmetric change of the splitting (probably due to magnetic fields) through the solar cycle. Therefore, the results found might come partially from an actual dynamic effect of an internal field (which would also shift the $l = 0$ mode), amplified due to the solar cycle amplitude modulation, as unresolved magnetically split lines weight differently to the rotationally split ones. A straightforward numerical simulation of such an explanation has been performed; it is found that a 25% amplitude variation of the modes within a multiplet across the solar cycle would yield the observed results.

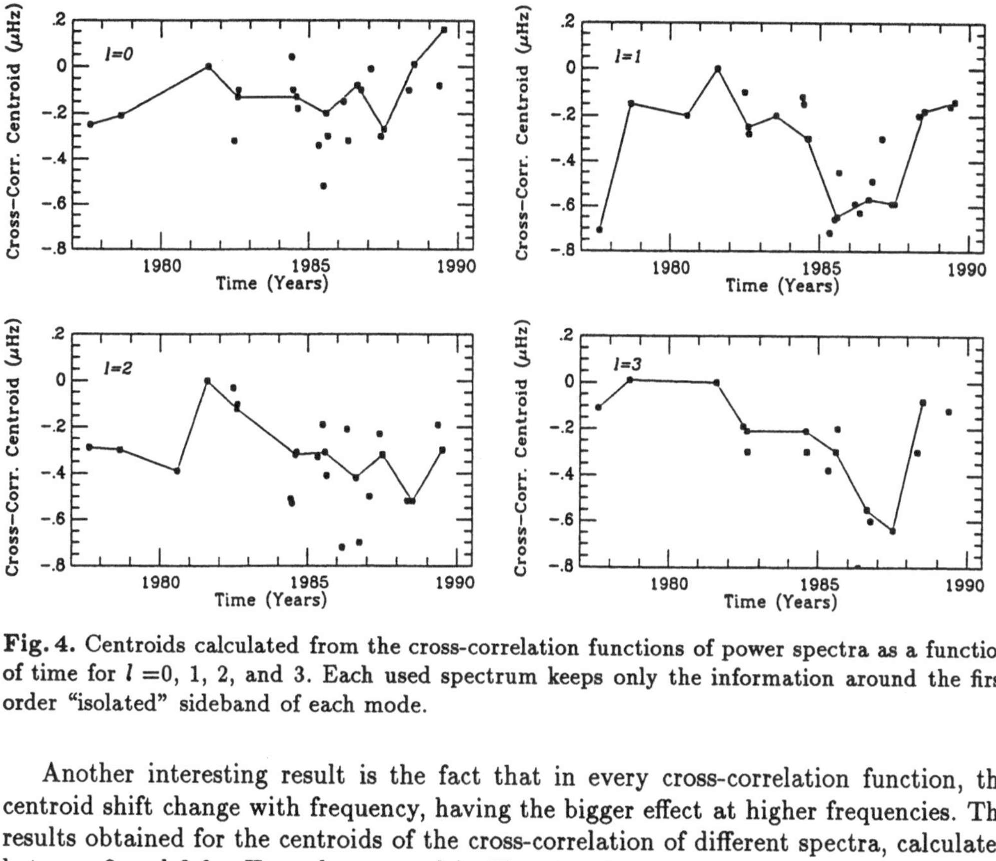

Fig. 4. Centroids calculated from the cross-correlation functions of power spectra as a function of time for l =0, 1, 2, and 3. Each used spectrum keeps only the information around the first order "isolated" sideband of each mode.

Another interesting result is the fact that in every cross-correlation function, the centroid shift change with frequency, having the bigger effect at higher frequencies. The results obtained for the centroids of the cross-correlation of different spectra, calculated between 2 and 3.8 mHz and presented in Figs. 3 and 4, are mean value. If this interval is divided in three parts: [2→2.6], [2.6→3.2] and [3.2→3.8], the relative change between the third and the first part is ~ 2:1. This result would agree with the one already found by Libbrecht and Woodard (1990) for higher l modes ($5 \leq l \leq 60$).

References

Braun D.C., Duvall T.L., Jr., and La Bonte B.J. 1988, *Astrophys. J.*, **335**, 1015.

Brookes J.R., Isaak G.R., and van der Raay H.B. 1978, *Monthly Notices Roy. Astron. Soc.*, **185**, 1.

Elsworth Y., Isaak G.R., Jefferies S.M., McLeod C.P., New R., Pallé P.L., Régulo C., and Roca Cortés T. 1990, *Monthly Notices Roy. Astron. Soc.*, in press.

Jiménez J.A., Pallé P.L., Pérez J.C., Régulo C., Roca Cortés T., Isaak G.R., McLeod C.P., and van der Raay H.B. 1988, in *Proc. IAU Symp. No. 123, Advances in Helio- and Asteroseismology*, eds. J. Christensen-Dalsgaard and S. Frandsen (Reidel, Dordrecht), p.205.

Libbrecht K.G. 1988, in *Seismology of the Sun and Sun-like Stars*, ed. E. Rolfe, ESA SP-286 (ESA Publication Division, Noordwijk), p.3.

Libbrecht K.G., and Woodard M.F. 1990, These Proceedings.

Muller R., and Rodier T. 1984, in *The Hydromagnetics of the Sun*, eds. T.D. Guyene and J.J. Hunt, ESA SP-220 (ESA Publication Division, Noordwijk), p.51.

Pallé P.L. 1986, Ph. D. Thesis, Univ. de La Laguna, Tenerife.

Pallé P.L., Pérez J.C., Régulo C., Roca Cortés T., Isaak G.R., McLeod C.P., and van der Raay H.B. 1986, *Astron. Astrophys.*, **170**, 114.

Pallé P.L., Régulo C., and Roca Cortés, T. 1988, in *Proc. IAU Symp. No. 123, Advances in Helio and Asteroseismology*, eds. J. Christensen-Dalsgaard and S. Frandsen (Reidel, Dordrecht), p.285.

Pallé P.L., Régulo C., and Roca Cortés T. 1989a, *Astron. Astrophys.*, **224**, 253.

Pallé P.L., Régulo C., and Roca Cortés T. 1989b, in *Proc. IAU Colloq. No. 121, Inside the Sun*, in press.

Pallé P.L., Pérez Hernández F., Roca Cortés T., and Isaak G.R. 1989c, *Astron. Astrophys.*, **216**, 258.

Helioseismology from the South Pole: Results from the 1987 Campaign

S. M. Jefferies [1], T. L. Duvall, Jr. [2], J. W. Harvey [3],

and M. A. Pomerantz [1]

[1]Bartol Research Institute, University of Delaware, Newark, DE 19711, USA
[2]Laboratory for Astronomy and Solar Physics,
NASA/Goddard Space Flight Center, Greenbelt, MD 20771, USA
[3]National Solar Observatory, National Optical Astronomy Observatories,[1]
Tucson, AZ 85726, USA

Abstract: This paper presents some results on the frequencies and line widths of features in solar p-mode spectra obtained from 460 hours of observations made at South Pole in 1987. To investigate the possibility of temporal variations in these quantities, a comparison is made with measurements obtained from data taken in 1981. The differences between the frequencies measured from the 1981 and 1987 data sets appear to be independent of both frequency ($2.4 \leq \nu \leq 4.8$ mHz) and degree ($3 \leq \ell \leq 98$). The mean difference ($\nu_{1981} - \nu_{1987}$) averaged over ν and ℓ is found to be 224 ± 19 nHz. The line width measurements display the same variation with ν as that previously reported (Libbrecht 1988a), an increase with ℓ (Duvall *et al.* 1988) and with solar activity. Measurement of the rotational splittings of sectoral modes ($m = \pm\ell$) in the range ($3 \leq \ell \leq 15$), shows no indication of a dependence on the depth of the lower turning points of these modes.

1 Introduction

The accurate and precise measurement of the frequencies and line widths of modes in a solar oscillation power spectrum permits an insight into the properties of the solar interior. The measurement of mode frequencies allows a determination of the sound speed and angular velocity profiles of the solar interior, and measurement of the line widths gives information on the damping and excitation mechanisms of the oscillations.

Information on the rotation profile in the outer $\sim 2/3$ of the Sun has come from spatially resolved observations. These data have suggested (Duvall *et al.* 1986, Libbrecht 1988b, Brown *et al.* 1989) that the differential rotation observed at the solar surface is conserved throughout the convection zone and then disappears in the radiative region.

[1] Operated by the Association of Universities for Research in Astronomy, Inc., under contract with the National Science Foundation.

Information on the rotation profile near the solar core however, has come primarily from zero angular resolution experiments. Some reports (Claverie *et al.* 1981, van der Raay *et al.* 1986, Jefferies *et al.* 1988b) claim that the core is rotating more rapidly than the surface rate while others (Woodard 1984a, Henning and Scherrer 1986, Fröhlich *et al.* 1989) claim that it is equal to the surface rate. A major problem with the unimaged measurements is that all the m components at a given ℓ value (such that $\ell + m$ =even), are present in the same spectrum. This makes difficult a determination of the frequency difference between adjacent m components (which is related to the rotation rate). The only results from spatially resolved experiments (Duvall and Harvey 1984) (which allow spectra to be computed for each ℓ and m value), have associated uncertainties that preclude a resolution of the contradictory results from the unimaged experiments.

Information on the sound speed profile of the solar interior is obtained from the inversion of measured mode frequencies (Christensen-Dalsgaard *et al.* 1985). Comparisons of mode frequencies measured at different epochs of solar cycle 21, for both low and intermediate degree modes, have shown some interesting discrepancies (Woodard and Noyes 1985, Duvall *et al.* 1988, Gelly *et al.* 1988, Pallé *et al.* 1990) and hint at the possibility of a solar cycle variation in the frequencies. On the other hand, some reports (Jefferies *et al.* 1988a, Rhodes *et al.* 1988) suggest no temporal changes in the frequencies.

Two objectives for our South Pole observations in 1987 were: a) to address the issue of temporal variations in the mode frequencies (by observing at a time of low solar activity and comparing the results with those obtained from a similar experiment executed at a time of high solar activity) and b) to measure rotational splittings of modes with ℓ values ranging from as low an ℓ value as possible up to ℓ=150, thus enabling a more accurate determination of the rotational profile.

2 Observations and Reductions

The 1987.9 experimental set-up (Jefferies *et al.* 1988c) closely resembled that used in the 1981.9 South Pole observations (Duvall *et al.* 1986). Observations were obtained over a period of 460 hours with a duty cycle of 54%, and consisted of four sequences of ~60 hours each. The last of the observing sequences was taken during an ice crystal storm that only affected the measurement of modes with ℓ=0 and 1.

The reduction of the image data to power spectra in ℓ, m and ν is described in detail in another publication (Jefferies *et al.* 1988c). Briefly, the image reduction process consisted of eight stages:

1) Corrections to images, including dark exposure subtraction, flat-field normalization, removal of cross-talk effects and deconvolution of secondary images owing to multiple reflections in the instrument.
2) Remapping images onto a standard grid with uniform steps in the sine of the latitude and longitude difference from the central meridian. This involved finding the coordinates of the limb for each image and accurately determining the position angle of the rotation axis.
3) Rejection of bad data.
4) Removal of limb-darkening effects from the time series of maps using a running mean produced from 15 maps centered on the map to be corrected.

5) Computing the time series of spherical harmonic coefficients for all m values for $\ell \leq 20$ and for even m values for $21 \leq \ell \leq 150$. (Note that before the spatial filters were applied, each map had its mean value subtracted from it and placed in the ℓ=0 time series data.)
6) Correcting the spherical harmonic coefficients for the effects of varying scattered light.
7) Fourier transforming the time series of spherical harmonic coefficients to produce power spectra.
8) Corrections to the power spectra for clock rate errors and the frequency response variation of the running mean filter.

After step (6), the time series of spherical harmonic coefficients were examined for outlying data points. It was found that spikes would occur for some ℓ and m values but not enough to warrant the rejection of the whole map. Here, data identified as being 'bad' in the time series for a given ℓ and m value, led to the corresponding data points being removed from the time series for all m values computed at that ℓ value. This resulted in different window functions for each ℓ value.

Data from the 1981 South Pole experiment was also reduced using the same reduction techniques that were employed with the 1987 data. This re-reduction was done to minimize any systematic effects owing to differences in the reduction techniques.

3 Analysis and Results

For both the 1981 and 1987 data sets, m-averaged spectra were produced by first removing from each (ℓ, m) spectrum a dependence of frequency on m using (Duvall *et al.* 1986):

$$\Delta\nu = L\sum_{i=1}^{i=5} a_i P_i\left(-\frac{m}{L}\right), \tag{1}$$

where $\Delta\nu$ is the frequency shift due to the m dependence, $L = [\ell(\ell+1)]^{1/2}$, P_i are the Legendre polynomials of degree i, and then averaging the spectra in m (using equal weight) for each ℓ value. The coefficients a_i used for 1981 are averages of those given in (Duvall *et al.* 1986) and for 1987 averages of the values in (Jefferies *et al.* 1988c).

Observation of only one hemisphere of the Sun results in the spatial filters employed in the data reduction being imperfect (owing to the orthogonality properties of the spherical harmonic functions used to describe the oscillations). This leads to leakage of power into the target (ℓ, m) spectrum from modes with the same ℓ value but different m values (m-leakage), and from modes with different ℓ and m values (ℓ-leakage). A consequence of the imperfect spatial response function is that in an m-averaged spectrum, the spectral line profile at any (ℓ, n) value is broadened by m-leakage. This broadening should be symmetric about the line center. Also, use of Eq. (1) to remove the frequency dependence on m is only correct for modes of degree equal to that of the target spectrum and results in the line profiles of modes from ℓ-leakage, being "smeared out" in frequency by the m-averaging process.

The temporal coverage for the 1981 data is good (89% duty cycle), consequently there is negligible leakage of power from one frequency to another in the power spectra. The same is not true of the window function for the 1987 data set. However, the forward

Fourier transform of the power spectrum of the 1987 window function does not contain any zero crossings and it is possible to deconvolve the effect of the imperfect window function from each of the power spectra. Because the deconvolution process tends to increase the amount of noise in a spectrum, the deconvolution of the window function was only performed on the m-averaged spectra.

The spectral features in each m-averaged power spectrum were fit by least-squares using the model given by:

$$M_i = \sum_{j=1}^{j=N} \frac{A_j \Gamma_j}{(\nu_i - \nu_{0j})^2 + \Gamma_j^2} + \sum_{k=1}^{k=n} c_k \nu_i^{k-1}, \tag{2}$$

where A_j is the maximum signal power, Γ_j is the half-width at half maximum, ν_{0j} is the mode frequency, ν_i is the frequency in channel i, N is the number of modes in the fitting interval and c_k are coefficients describing the background power. The fitting interval around each target mode was $(\nu_{\ell,n-1} + \nu_{\ell,n})/2$ to $(\nu_{\ell,n} + \nu_{\ell,n+1})/2$. The background was assumed to be linear ($n = 2$) over this interval and the weight assigned to each point given by $(M_i/n_{\rm spec})^{-1}$, where $n_{\rm spec}$ is the number of spectra used in forming the m-averaged spectrum. The justification for using this weighting is that solar oscillation power spectra are governed by statistics that follow an exponential distribution (Woodard 1984b, Duvall and Harvey 1986, Duvall 1990):

$$P_i = \frac{1}{M_i} \exp\left(\frac{-O_i}{M_i}\right), \tag{3}$$

where P_i is the probability density function in channel i, O_i is the observed spectrum and M_i is the limit spectrum or the expectation value (estimated by the model). When $n_{\rm spec}$ spectra are averaged together, this probability density function changes and tends towards a Gaussian distribution as $n_{\rm spec}$ increases (Anderson *et al.* 1990).

Although the use of Lorentzian profiles to model the spectral features is appropriate when dealing with an oscillation spectrum at a given ℓ and m, as pointed out above, the m-averaging process will modify the spectral line profiles. From calculations of spatial response functions, we find that the effect of m-leakage is to inflate the line widths by a small fraction of their natural values. The extent to which the line profile of a spectral feature is distorted from Lorentzian due to ℓ leakage and the effect of such a distortion on the values of fitted parameters is under investigation. For the results presented in this paper, we assumed that the distortion from the Lorentzian profile is negligible and that this effect will only inflate the line widths by a small fraction of their natural values. Both the frequency and line width measurements discussed below, were obtained from fits to the m-averaged data.

3.1 Frequency Measurements

Mode frequencies from the 1981 and 1987 data sets were determined over the ranges ($6 \leq n \leq 29$) and ($2 \leq n \leq 29$) respectively. The 1987 frequencies were subtracted from the 1981 frequencies and the variation of the frequency difference with ℓ and ν is shown in Figs. 1 and 2 respectively. The differences deviate substantially from preliminary results we presented earlier (Jefferies *et al.* 1988c). This is because a clock rate error correction was originally applied with the wrong sign to the 1981 frequencies. The effect of this error is to increase 1981 frequencies by about 350 nHz in the results presented in Duvall *et al.* (1988) and to strengthen the suggestion made there of a solar cycle dependence of frequencies.

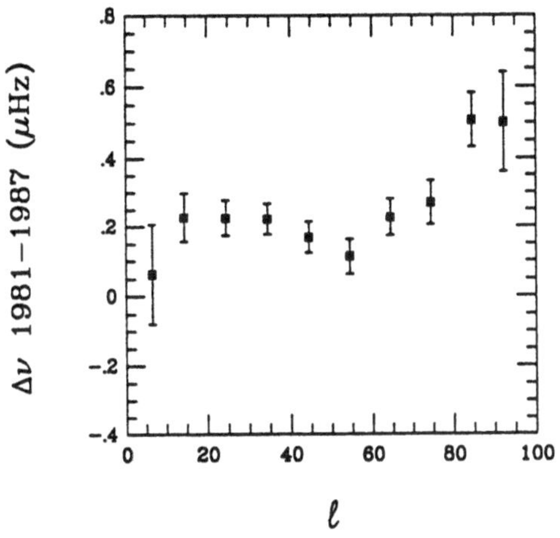

Fig. 1. The weighted mean difference (and one standard deviation errors) of p-mode frequencies between 1981 and 1987 grouped into 10-degree bins.

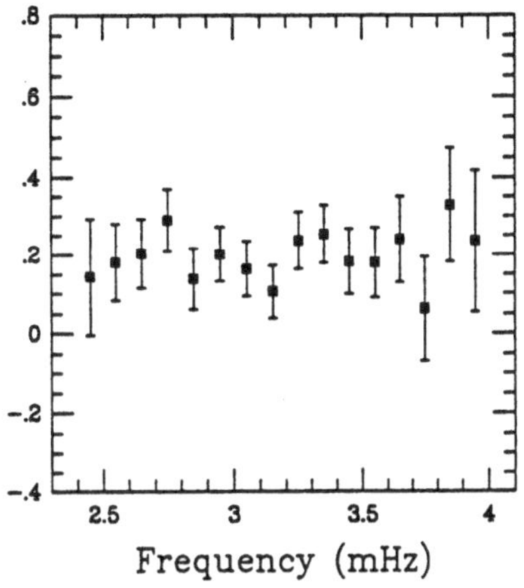

Fig. 2. The same as Fig. 1 except the averaging is over 100 μHz regions and only values in the ℓ range 20 to 70 were used.

Aside from an average difference of about 200 nHz which we ascribe to a solar cycle dependence, Fig. 1 appears to show a gradually increasing difference with increasing ℓ. However, because of the decline in quality of the 1981 data set in the regions ($\ell < 20$) and ($\ell > 80$), we caution against over interpretation of fine structure in Fig. 1. Figure 2 suggests that there may be a slight increase in the frequency difference with increasing ν although the errors admit a frequency difference independent of ν.

The weighted mean difference of the frequencies, averaged over the ℓ range ($3 \leq \ell \leq 98$) and frequency range ($2.4 \leq \nu \leq 4.8$ mHz) was found to be 224 $\pm$19 nHz. Our frequency change results are consistent with reports of increasing frequency with increasing solar activity, and relatively larger increases at higher frequencies and greater ℓ values (Duvall *et al.* 1988, Libbrecht and Woodard 1990). This issue will be better addressed with the analysis of data obtained at South Pole in November 1988 (Jefferies *et al.* 1989), a time of increasing solar activity.

3.2 Line Width Measurements

The variation of line width with frequency for modes with ℓ values in the range ($80 \leq \ell \leq 100$) for both the 1981 and 1987 data sets, is shown in Fig. 3.

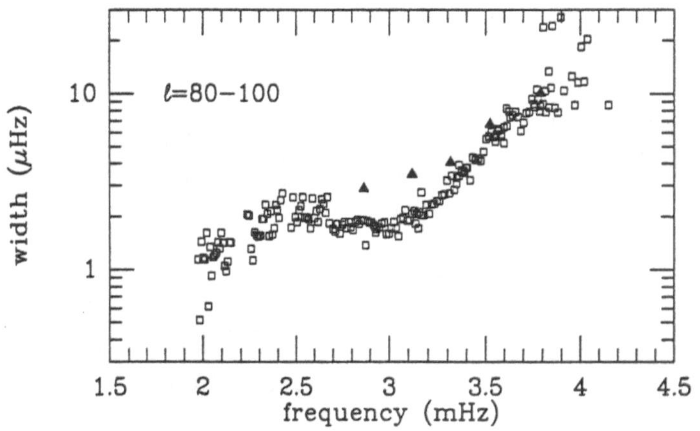

Fig. 3. Measurement of line width for 1981 (solid triangles) and 1987 (open boxes). There are 70 measurements for the 1981 data which have been binned in frequency with an equal number of measurements in each bin.

The line widths as measured from the power spectra contain a component due to the frequency resolution ($\Delta\nu_{\rm res} \simeq 1/T$) inherent in observations made over a finite length of time T. This extra width component can be calculated and removed as it arises from the convolution of a sine function of width $\Delta\nu_{\rm res}$, with the natural line profile of the mode (*i.e.* a Lorentzian). The line widths shown in Fig. 3 have been corrected for this effect. In addition, the line widths can also be inflated from the use of incorrect a_i values in (1) when forming the m-averaged spectra. However, for typical uncertainties in the a_i coefficients encountered over the degree range ($80 \leq \ell \leq 100$), the change in the line width from this process is a small fraction of the observed line width.

The functional form of the variation with frequency of the 1987 data points is consistent with other observations (Libbrecht 1988a) of line widths for modes in the range ($19 \leq \ell \leq 24$) but is systematically larger. This shift suggests a dependence of the line width on ℓ. The increase in line width as ℓ increases is consistent with other reports (Duvall *et al.* 1988, Elsworth *et al.* 1988).

The line widths of modes in 1981 are significantly larger than those in 1987. It would thus appear that solar p modes (at least over the range $80 \leq \ell \leq 100$), have line widths that are greater during periods of high solar activity and that may be changing with the solar cycle. If the modes are stochastically excited by convection and intrinsically damped (Goldreich and Kumar 1988), then the line widths are determined by the damping rate. This would imply that the damping rate of the modes increases with solar activity. An obvious candidate for increased damping at larger values of ℓ is increased absorption of p-mode power in active regions at times of high solar activity.

3.3 Rotational Splittings between Degrees 3 and 15

A maximum likelihood fitting routine that employs the probability density function described by Eq. (3) (Anderson *et al.* 1990), was used to determine the frequencies of sectoral modes ($m = \pm\ell$) in the range $3 \leq \ell \leq 15$ for the 1987 data. The algorithm allows for the effect of an imperfect window function at each iteration, by convolving the model given by Eq. (2) (with a constant background), with the power spectrum of the window function. The fitting region around each mode is selected to minimize the effects of ℓ-leakage. The rotational splitting is estimated using:

$$\Delta\nu_{\rm rot} = \frac{\nu_{n,\ell,m=-\ell} - \nu_{n,\ell,m=+\ell}}{2\ell}, \tag{4}$$

and is plotted in Fig. 4 [along with splittings obtained from velocity data taken at Kitt Peak in 1983 (Duvall and Harvey 1984)] as a function of L/ω which is related to the lower turning point of a mode.

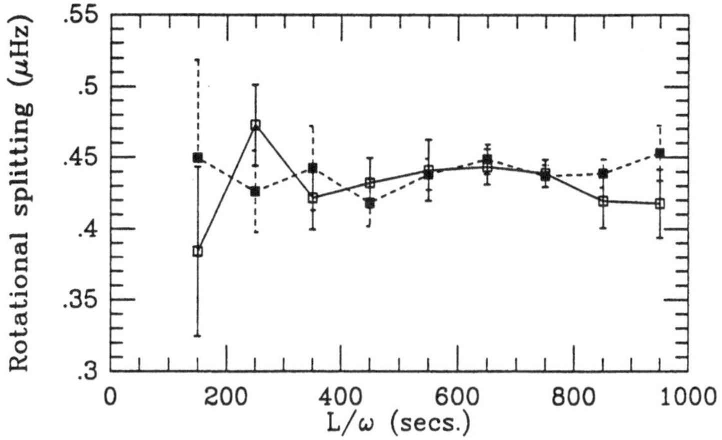

Fig. 4. Rotational splitting of the sectoral modes over the degree range ($3 \leq \ell \leq 15$). The error flags for both the 1983 data (filled boxes) and the 1987 data (open boxes) represent one standard deviation of the weighted means.

Although modes with $\ell = 1$ and 2 are observed in the 1987 data, m-leakage problems cause the $m = -\ell$ mode to leak into the $m = +\ell$ spectrum (and vice versa) at a level of 47% for $\ell = 1$ and 8% for $\ell = 2$. Since the frequency separation between the sectoral modes at these low ℓ values is of the same order as the intrinsic frequency resolution of the data set, it is difficult to obtain reliable estimates of the splittings. This m-leakage problem is $< 1\%$ for $\ell = 3$ and decreases with increasing ℓ. The data in Fig. 4 have been corrected for systematic shifts owing to ℓ-leakage problems (primarily from modes with ℓ values of ($\ell_{\rm target}\pm 4$).

Figure 4 does not show any indication of the rotational splitting increasing as the lower turning point of the modes approaches the core. However, what is really required to settle the dispute of the rotation rate near the core, are good measurements of the ℓ=1 and 2 splittings.

The estimates presented here for the rotational splitting of modes with $\ell < 15$, can be improved by measuring the frequencies of all modes at each ℓ value, in all the spectra where they can be detected (*i.e.* including when they appear as "leaks"). We have found that in some of the spectra where the $\ell = 1$ and 2 sectoral modes appear as "leaks," the m-leakage problem discussed above is greatly reduced and should thus allow for an improved estimate of the splitting of these modes.

References

Anderson, E., Duvall, T.L., Jr., and Jefferies, S.M. 1990, submitted to *Astrophys. J.*

Brown, T.M., Christensen-Dalsgaard, J., Dziembowski, W.A., Goode, P., Gough, D.O., and Morrow, C.A. 1989, *Astrophys. J.*, **343**, 526.

Christensen-Dalsgaard, J., Duvall, T.L., Jr., Gough, D.O., Harvey, J.W., and Rhodes, E.J., Jr. 1985, *Nature*, **315**, 378.

Claverie, A., Isaak, G.R., McLeod, C.P., van der Raay, H.B., and Roca Cortés, T. 1981, *Nature*, **293**, 443.

Duvall, T.L., Jr. 1990, in *Inside the Sun, IAU Colloq. No. 121*, eds. G. Berthomieu and M. Cribier (Kluwer, Dordrecht), in press.

Duvall, T.L., Jr., and Harvey, J.W. 1984, *Nature*, **310**, 19.

Duvall, T.L., Jr., and Harvey, J.W. 1986, in *Seismology of the Sun and the Distant Stars*, ed. D.O. Gough (Reidel, Dordrecht) p.105.

Duvall, T.L., Jr., Harvey, J.W., Libbrecht, K.G., Popp, B.D., and Pomerantz, M.A. 1988, *Astrophys. J.*, **324**, 1158.

Duvall, T.L., Jr., Harvey, J.W., and Pomerantz, M.A. 1986, *Nature*, **321**, 500.

Duvall, T.L., Jr., Harvey, J.W., and Pomerantz, M.A. 1987, in *Advances in Helio- and Asteroseismology, IAU Symp. No. 123*, eds. J. Christensen-Dalsgaard and S. Frandsen (Reidel, Dordrecht), p.37.

Elsworth, Y., Isaak, G., Jefferies, S.M., McLeod, C.P., New, R., Pallé, P.L., Régullo, C., and Roca Cortés, T. 1988, in *Seismology of the Sun and Sun-Like Stars*, ed. E. Rolfe, ESA SP-286 (ESA Publication Division, Noordwijk), p.27.

Fröhlich, C., Toutain, T., Bonnet, R.M., Bruns, A.V., Delaboudiniere, J.P., Domingo, V., Kotov, V.A., Kollath, Z., Rashkovsky, D.N., Vial, J.C., and Wehrli, Ch. 1989, submitted to *Nature.*

Gelly, B., Fossat, E., and Grec, G. 1988, in *Seismology of the Sun and Sun-Like Stars*, ed. E. Rolfe, ESA SP-286 (ESA Publication Division, Noordwijk), p.275.

Goldreich, P., and Kumar, P. 1988, *Astrophys. J.*, **326**, 462.

Henning, H.M., and Scherrer, P.H. 1986, in *Seismology of the Sun and the Distant Stars*, ed. D.O. Gough (Reidel, Dordrecht) p.55.

Jefferies, S.M., McLeod, C.P., van der Raay, H.B., Pallé, P.L., and Roca Cortés, T. 1988b, in *Advances in Helio- and Asteroseismology, IAU Symp. No. 123*, eds. J. Christensen-Dalsgaard and S. Frandsen (Reidel, Dordrecht), p.25.

Jefferies, S.M., Pallé, P.L., van der Raay, H.B., Régulo, C., and Roca Cortés, T. 1988a, *Nature*, **333**, 646.

Jefferies, S.M., Pomerantz, M.A., Duvall, T.L., Jr., and Harvey, J.W. 1989, *Antarctic J. U. S.*, **24**, in press.

Jefferies, S.M., Pomerantz, M.A., Duvall, T.L., Jr., Harvey, J.W., and Jaksha, D.B. 1988c, in *Seismology of the Sun and Sun-Like Stars*, ed. E. Rolfe, ESA SP-286 (ESA Publication Division, Noordwijk), p.279.

Libbrecht, K.G. 1988a, *Astrophys. J.*, **334**, 510.

Libbrecht, K.G. 1988b, in *Seismology of the Sun and Sun-Like Stars*, ed. E. Rolfe, ESA SP-286 (ESA Publication Division, Noordwijk), p.131.

Libbrecht, K.G., and Woodard, M.F. 1990, these proceedings.

Pallé, P. L., Régulo, C., and Roca Cortés, T. 1990, these proceedings.

Rhodes, E.J., Jr., Woodard, M.F., Cacciani, A., Tomczyk, S., Korzennik, S.G., and Ulrich, R.K. 1988, *Astrophys. J*, **326**, 479.

van der Raay, H.B., Pallé, P.L., and Roca Cortés, T. 1986, in *Seismology of the Sun and the Distant Stars*, ed. D.O. Gough (Reidel, Dordrecht) p.215.

Woodard, M.F. 1984a, *Nature*, **309**, 530.

Woodard, M.F. 1984b, Ph.D. Thesis, University of California.

Woodard, M.F., and Noyes, R.W. 1985, *Nature*, **318**, 449.

Observations of Solar Cycle Variations in Solar p-Mode Frequencies and Splittings

K. G. Libbrecht and M. F. Woodard

Big Bear Solar Observatory, California Institute of Technology, Pasadena, CA 91125, USA

Abstract: We discuss here two sets of helioseismology data acquired at Big Bear Solar Observatory during the summers of 1986 and 1988. Each data set consists of roughly 60,000 full-disk Doppler images of the sun, accumulated over a four-month time span. These data clearly show that solar p-mode frequencies change with time, and that the measured frequency shifts $\Delta\nu = \nu_{88} - \nu_{86}$ depend strongly on frequency and only weakly on ℓ for $5 \leq \ell \leq 60$. The frequency dependence is well described by $\Delta\nu \propto M^{-1}(\nu)$, where $M(\nu)$ is the mode mass for low-ℓ modes. Such a frequency dependence is expected if the effective sound speed perturbation is located predominantly near the solar surface. It should be possible to invert the frequency shift measurements to determine some aspects of the structure of solar activity as a function of depth. The data also show that the even-index splitting coefficients depend strongly on frequency, again being well described by $\alpha_{2j}(\nu) \propto M^{-1}(\nu)$. This functional form is expected if the sound speed perturbation responsible for $\Delta\nu$ is localized in solar latitude. Latitude inversions of the time-dependent splitting and $\Delta\nu$ measurements show that the perturbation is strongest in the active latitudes, but includes a weak polar component.

1 Introduction

Ever since global p-mode oscillations were discovered in the sun, it has been hoped that through measurements of mode frequency changes we would be able to learn something about the interior structure of the solar dynamo. Early attempts to accurately measure frequencies of low-ℓ modes ($\ell \lesssim 3$) showed that frequency changes with time are quite small—$\Delta\nu < 1$ μHz between solar minimum and maximum for modes with frequencies $\nu \approx 3$ mHz—so a detection of solar dynamo effects requires frequency determinations with an accuracy of at least a part in 10,000. While there are indications in more recent low-ℓ data that the mode frequencies at solar maximum are perhaps ~0.4 μHz greater than at solar minimum, the signal-to-noise ratio in the published observations is not very high, and there is some disagreement between different observations (Gelly *et al.* 1988, Pallé *et al.* 1988). Measurements of intermediate-ℓ mode frequencies ($5 \lesssim \ell \lesssim 100$) have also been used to search for frequency variations with time, but again with the recent measurements the signal-to-noise has been fairly low, and different observations have yielded different results (Jefferies *et al.* 1988, Duvall *et al.* 1988, Rhodes *et al.* 1988).

One area where many different observers have been in relatively good agreement is the time variation of the even-index p-mode splitting coefficients. If we expand the p-mode frequencies of a single ($n\ell$) multiplet as a sum of Legendre polynomials

$$\nu_{n\ell m} = \nu_{n\ell} + \sum_{i=1} \alpha_i(n,\ell) P_i(m/L), \tag{1}$$

with $L = \sqrt{\ell(\ell+1)}$, then the α_i with i even measure a latitude-dependent variation in the propagation speed for acoustic waves, while the larger odd-i terms measure a corresponding east-west propagation asymmetry (*i.e.* the solar rotation). Kuhn (1988) first suggested that the observed $\bar{\alpha}_2$ and $\bar{\alpha}_4$ [here averaging $\alpha_i(n,\ell)$ over observed n and ℓ] varied systematically with solar cycle, and all the measurements to date, made by several different observers, support this idea (see Fig. 5 below).

We present here new observations of intermediate-ℓ p-mode frequencies, made in 1986 and 1988, the former very near solar minimum. Comparing measurements from the two years, we are able to clearly see mode frequency shifts with high signal-to-noise. The accuracy of our $\Delta\nu$ measurements are over an order of magnitude greater than any previous observations. Furthermore the 1986 and 1988 data sets are nearly identical, being taken at the same site with the same equipment, so the measured frequency differences appear to be largely untroubled by systematic errors.

2 Data and Analysis

The data presented here were obtained at Big Bear Solar Observatory using our dedicated helioseismology telescope (Libbrecht and Zirin 1986), the observations spanning approximately four months in the summer of 1986 and again in 1988. The system uses a Zeiss 0.25 Å birefringent filter in combination with a KD*P electro-optical crystal to produce full-disk solar images in the red and blue wings of the 6439 Å Ca line. By digitizing and differencing image pairs from the two wings of the line, Doppler velocity images of the sun are formed. Each minute 752 image pairs are averaged together into a single Doppler image, which is then stored on tape for later processing. Approximately 60,000 images were obtained during each of the 1986 and 1988 observing seasons. Almost no changes to the observing hardware were made between 1986 and 1988: the Zeiss filter was not retuned, and only cosmetic changes were made to the data acquisition software. In addition, 1986 was at solar minimum, while by 1988 the new cycle had reached considerable amplitude. These two facts mean that these observations are ideally suited to look for solar cycle frequency changes.

The data were processed by fitting each image to all projected spherical harmonics with $\ell \leq 140$, and the fit coefficients were subsequently Fourier transformed to produce power spectra $S_{\ell m}(\nu)$ for each ℓ and m. The compute-intensive parts of the analysis were performed at the San Diego Supercomputer Center using a CRAY X-MP. The 1986 data set with $\ell \leq 60$ has already been used to measure p-mode amplitudes and linewidths (Libbrecht 1988a, 1988b), and also to measure p-mode splittings (Libbrecht 1988c, 1989).

3 Mode Frequency Shifts

In order to accurately measure p-mode multiplet frequencies the individual $S_{\ell m}(\nu)$ power spectra at fixed ℓ were first corrected for the known rotational splitting and combined to form averaged power spectra $S_\ell(\nu)$. The individual multiplet features in the $S_\ell(\nu)$ spectra were then fit to sums of Lorentzians to determine the multiplet frequencies $\nu_{n\ell}$ and corresponding uncertainties $\sigma_{n\ell}$ (Libbrecht *et al.* 1990). A plot of the derived 1986 $\nu_{n\ell}$ with uncertainties is shown in Fig. 1.

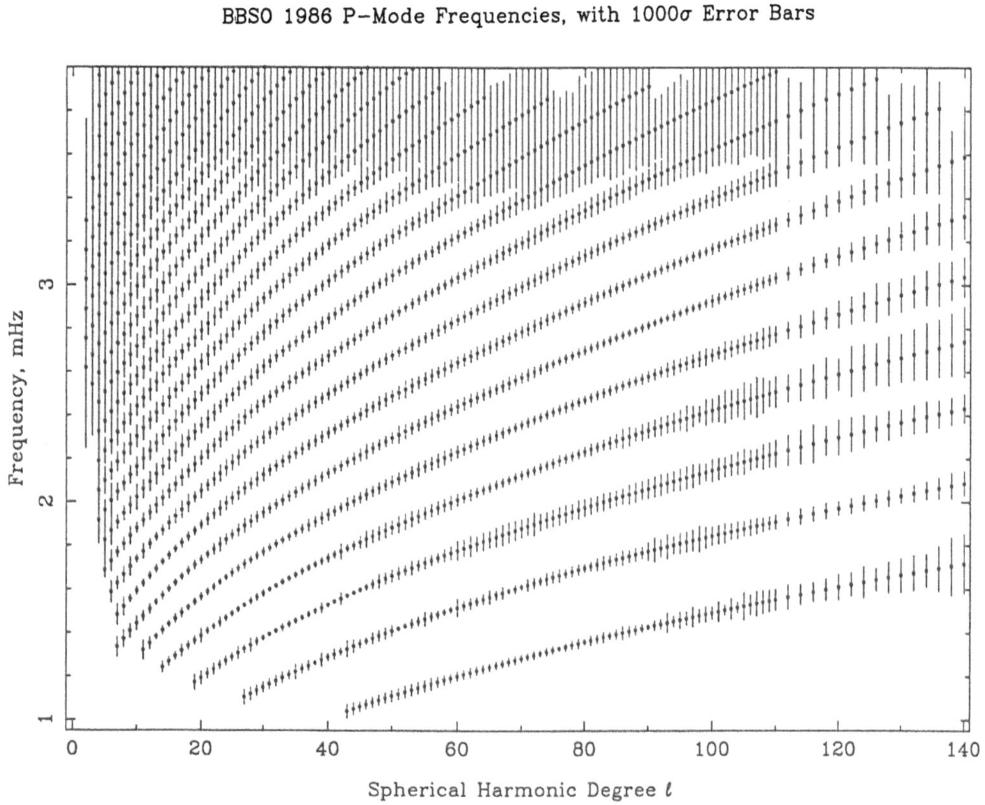

Fig. 1. Plot showing the observed 1986 p-mode frequencies from BBSO, along with error bars which have been magnified 1000 times. Modes with periods up to 16 minutes are seen, and the 1σ frequency uncertainties are as low as 10 nHz for some modes. The lowest frequency ridge shown has $n = 1$.

The 1988 data were in an identical format, and they were analyzed using essentially the same software to produce a second table of 1988 $\nu_{n\ell}$ and $\sigma_{n\ell}$ (at present the 1988 table is only complete up to $\ell = 63$). Taking the difference between the two frequency tables yields frequency shifts $\Delta\nu_{n\ell} = \nu_{n\ell}(1988) - \nu_{n\ell}(1986)$.

A quick look at the data revealed that the $\Delta\nu_{n\ell}$ are strongly dependent on mode frequency $\nu_{n\ell}$, and only weakly dependent on ℓ. Averaging $\Delta\nu_{n\ell}$ over ℓ for $5 \leq \ell \leq 60$

gives the plot shown in Fig. 2. For comparison we have plotted the inverse mode mass at $\ell = 20$, $M_{20}^{-1}(\nu)$ (the mode mass is defined as the ratio of the energy in a mode to the square of its surface velocity), scaled to best fit the data. The residual ℓ dependence is shown in Fig. 3, by fitting the data over small ranges in ℓ to the same function of frequency, $M_{20}^{-1}(\nu)$, and plotting the fit coefficient, appropriately scaled to give $\Delta\nu$ at 3 mHz.

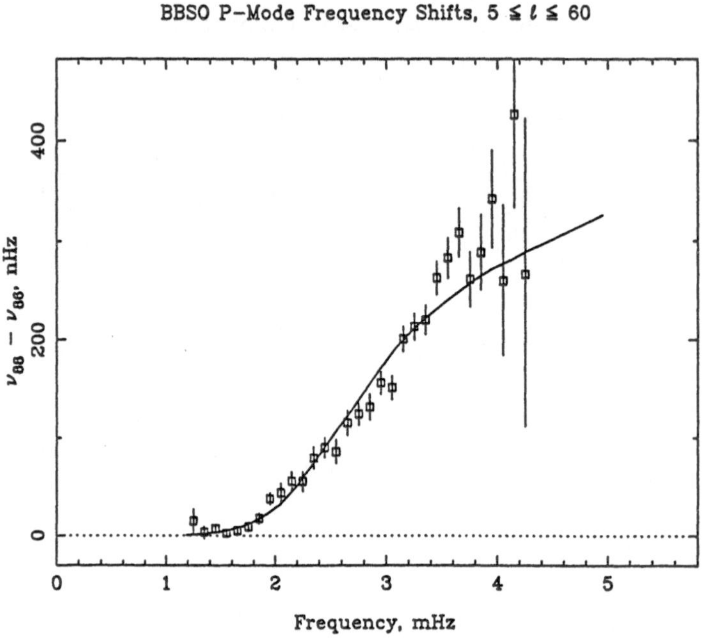

Fig. 2. P-mode frequency differences between 1988 and 1986, as a function of frequency, after averaging over $5 \leq \ell \leq 60$. Also plotted is the inverse mode mass for $\ell = 20$, $M_{20}^{-1}(\nu)$, using mode masses at $\tau_{5000} = 0.05$, which were calculated by P. Kumar (private communication) using a solar model by Christensen-Dalsgaard (1982). The curve was scaled to best fit the data.

The direct cause of the observed frequency shifts is probably changes in the thermal and magnetic structure of the solar interior. We argue that the most significant changes occur near the solar surface as follows. On the basis of asymptotic theory, if the perturbation were to extend over a significant fraction of the solar radius, the (fractional) mode frequency shift should depend mainly on the ratio $\nu_{n\ell}/\ell$ which labels different acoustic rays. Since the observed frequency shift depends almost entirely on mode frequency we conclude that either the relevant changes occur mainly in a thin layer or in one of the evanescent regions of the modes. The effect of perturbing a thin layer in the propagating regions of the modes was studied by Thompson (1988), who found a frequency dependence in $\Delta\nu$ which is much too weak. Thus the dominant effect in the frequency shift data is not the direct result of, say, magnetic field changes at the base of the convection zone, although the effect of a hypothetical perturbed layer could conceivably show up in a more careful analysis. A perturbation confined mainly to the evanescent regions of the modes near the center of the Sun can also immediately be ruled out because the lower turning point of a mode depends mainly on $\nu_{n\ell}/\ell$, which implies that a strong increase

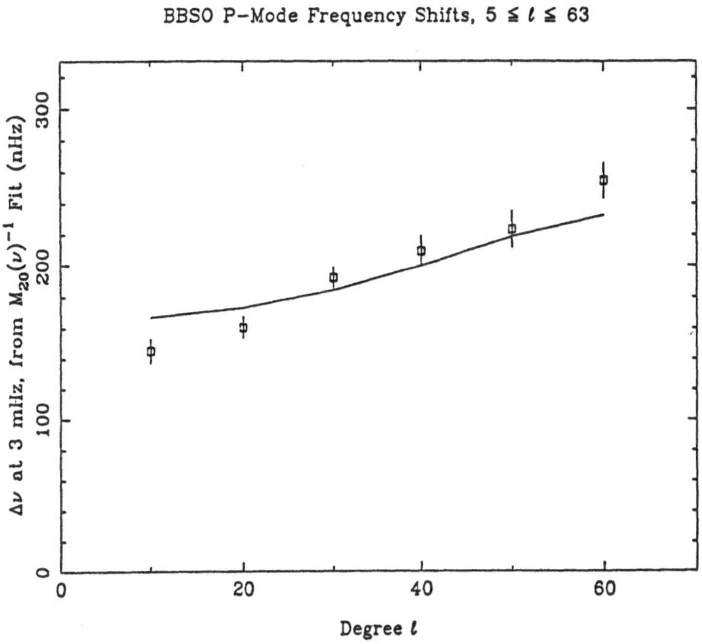

Fig. 3. P-mode frequency differences between 1988 and 1986, at $\nu = 3$ mHz, as a function of ℓ. These points were obtained by fitting the $\Delta\nu(\nu)$ data at different ranges in ℓ to the function $M_{20}^{-1}(\nu)$ shown in Fig. 2. The resulting fit coefficients were scaled to give $\Delta\nu(\ell)$ at 3 mHz. The line plotted is the inverse mode mass, scaled to fit the data; this curve was generated by fitting mode masses to $M_{20}^{-1}(\nu)$, a procedure analogous to that used for the $\Delta\nu$ data.

in frequency shift with frequency be accompanied by a strong decrease in frequency shift with ℓ, contrary to our measurements. The remaining possibility is that the changes occur mainly in the evanescent layers near the surface. Qualitatively the observed frequency dependence can be understood from the fact that the upper reflection point of the modes is deeper in the Sun for lower-frequency modes than for higher-frequency modes, and consequently higher-frequency modes are more sensitive to changes near the surface. However, in our view, it is quite possible that the thermal or magnetic restructuring near the solar surface is the indirect effect of a more deep seated dynamo.

The surface evanescent layer of a typical observed mode extends to a depth of less than $\sim 1\%$ of the solar radius, so the perturbation cannot be significant much deeper than this. Since the surface motion of these modes is mostly radial, it is useful consider the effect of a radially symmetric perturbation of the thermal stratification on the frequencies of $\ell = 0$ p-modes. Starting from the Lagrangian form of the linear adiabatic wave equation (e.g., Cox 1980), it is straightforward to show that the angular eigenfrequency of a mode, ω, is given by

$$\omega^2 = \int_0^{M_\odot} dm \left\{ f \left(\frac{dV}{dm} \right)^2 - gV^2 \right\} \Big/ \int_0^{M_\odot} dm \; V^2, \tag{2}$$

where the various quantities in the integrands are regarded as functions of the interior mass variable, m, and $M_\odot$ is the mass of the Sun. The function V is the velocity eigenfunction of the mode, while f and g are given by

$$f \equiv [4\pi r^2 \rho c]^2 \tag{3}$$

and

$$g \equiv \frac{4Gm}{r^3} + 4\pi r^2 \frac{d}{dm}\left(\frac{2\rho c^2}{r}\right), \tag{4}$$

where r, ρ, and c, are the radius, mass density, and adiabatic sound speed, and G is the gravitational constant. Although expression (2) is rigorously valid only in the absence of magnetic fields, it should be possible to include magnetic effects, in a crude way, by regarding c as an effective sound speed.

The shift in mode frequency resulting from a sufficiently small change in the stratification of the solar model can be calculated by replacing the functions f and g in Eq. (2) by their Lagrangian variations Δf and Δg. The contribution of the term in Eq. (2) involving g, which is basically a buoyancy term, can be ignored for the presently observed p-modes. The denominator in expression (2) is simply the mode energy, equal by definition to $M(\nu)V_s^2$, where V_s is the surface velocity amplitude of the mode and $M(\nu)$ is again the mode mass. Thus to a good approximation the perturbed ω^2 can be written

$$\Delta\omega^2 = \int_0^{M_\odot} dm\, \Delta f \left(\frac{dV}{dm}\right)^2 \Big/ M(\nu)\, V_s^2. \tag{5}$$

To gain further insight we consider the effect of a perturbation confined to the photosphere. A simple calculation based on Eq. (5) and the form of the mode eigenfunctions for an isothermal photosphere (e.g., Cox 1980) gives the rough frequency dependence of the frequency shift:

$$\Delta\nu \propto \frac{\nu^3}{M(\nu)}. \tag{6}$$

(This expression breaks down for modes which are close to or above the photospheric acoustic cutoff frequency). Since the observations are reasonably well fit by a simple $M(\nu)^{-1}$ dependence, we conclude that more of the solar envelope is involved in producing the measured frequency shifts than just the photosphere. P. Goldreich *et al.* (work in progress) have found that the observed $\propto M(\nu)^{-1}$ dependence of the frequency shift can follow from quite reasonable assumptions about the how the entropy in the convection zone is perturbed as the result of solar-cycle changes.

We note also that the ℓ-dependence of the frequency shift is somewhat stronger than the simple inverse mode mass dependence, as seen in Fig. 3. However, it is perhaps premature to speculate on this aspect of the data until we better understand the basic $M(\nu)^{-1}$ dependence. Furthermore we will be able to address this point better in the near future when our data extend up to $\ell = 140$.

4 Frequency Splittings

The above frequency shifts indicate that between 1986 and 1988 there was a change in the structure of surface layers of the sun, a change that affected the propagation of acoustic waves in that region. If this perturbation is also localized in latitude, as one might expect for a solar cycle effect, then it should also be observable in the even index splitting coefficients $\alpha_{2j}(n,\ell)$. In particular, the α_{2j} should show the same frequency dependence as the $\Delta\nu$. To investigate this, we fit sections of power spectra $S_{\ell m}(\nu)$ around each $\nu_{n\ell}$ to the expansion in $\alpha_i(n,\ell)$ in Eq. (1) up to $i = 6$ [see Libbrecht (1989) for details of the fitting algorithm]. Averaging over ℓ for $5 \le \ell \le 60$ we obtain the results shown in Fig. 4 for the 1986 and 1988 data.

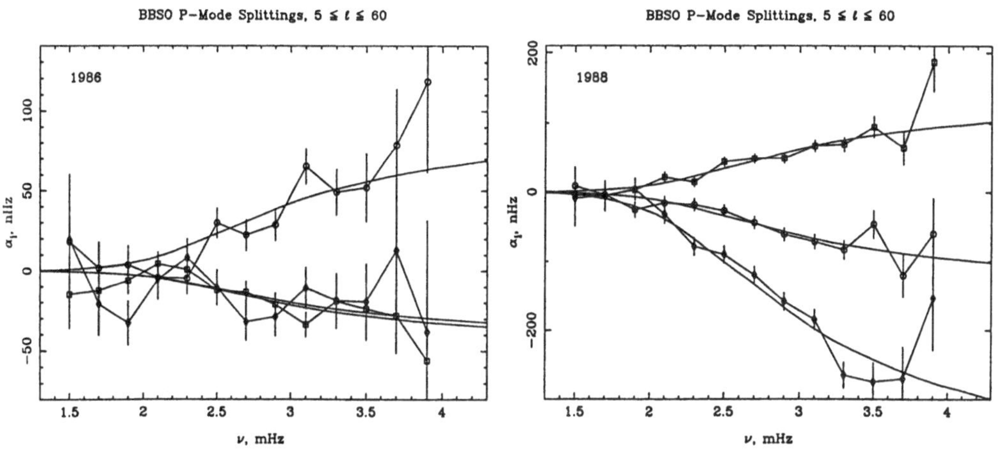

Fig. 4. Plot of the even-index splitting coefficients, α_i, $i = 2, 4, 6$, as a function of frequency for 1986 and 1988, averaging over $5 \le \ell \le 60$. α_2, α_4, and α_6 are represented by boxes, circles, and diamonds, respectively. Lines connect the data points, and fits to $M_{20}^{-1}(\nu)$ are also drawn for each of the α_i. Note the obvious frequency dependence, particularly in 1988, which is well represented by $M_{20}^{-1}(\nu)$, just like the $\Delta\nu$ measurements in Fig. 2.

Clearly all the α_{2j} show a frequency dependence that is consistent with the inverse mass function, as expected. These data also show that the α_{2j} depend only weakly on ℓ for $5 \le \ell \le 60$, consistent with the $\Delta\nu$ in Fig. 3. Averaging the $\Delta\nu$ and α_{2j} over $5 \le \ell \le 60$, and fitting each to $M_{20}^{-1}(\nu)$, we obtain the table of fit coefficients given in Table 1.

As mentioned above, plots of α_2 and α_4 vs. time, averaged over ℓ and n, have shown a clear solar cycle dependence, using data from many observers. However we now see that the α_{2j} depend strongly on frequency, so a simple average over n will give different results depending on how the modes are weighted (for example, brightness data tend to show less power at low frequencies than Doppler data). Nevertheless, we can compare

Table 1. Splitting coefficients at 3 mHz, from $M_{20}^{-1}(\nu)$ fits, in nHz.

	1986	1988	1988 − 1986
$\Delta\nu$	-	-	197.7±4.0
α_2	−23.6±3.8	69.3±3.9	92.9±6.7
α_4	46.8±5.4	−69.1±5.0	−115.9±7.4
α_6	−21.8±6.4	−202.7±7.1	−180.9±9.6

the present data with earlier data by examining the α_{2j} evaluated at a frequency of 3 mHz. This is shown in Fig. 5.

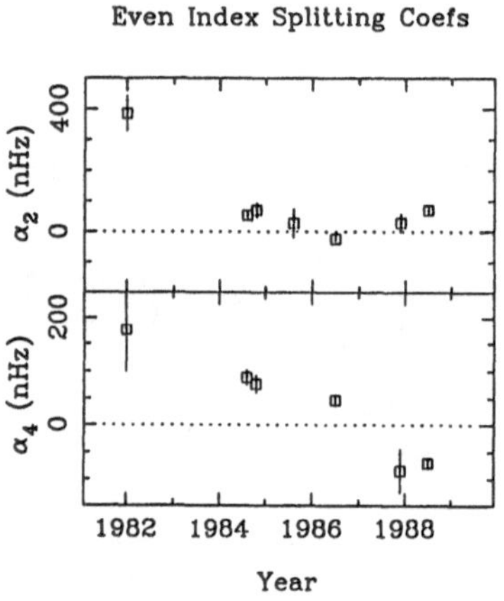

Fig. 5. Even index splitting coefficients as a function of time. The sources for points earlier than 1986 are given in Libbrecht (1989), the 1987.9 point is from Jefferies *et al.* (1988), and the 1986.5 and 1988.5 points are from Table 1.

5 Latitude Inversions

It is straightforward and instructive to invert the α_{2j} and $\Delta\nu$ to obtain a measure of the strength of the perturbation as a function of solar latitude. Following Kuhn (1988), Gough (1988), and Goode and Kuhn (1990), we assume an effective sound speed perturbation

$$\frac{\Delta c}{c} = \sum_j \beta_{2j} f(r) P_{2j}(\cos\theta), \tag{7}$$

where θ is the colatitude. With this, the observed splitting coefficients will be

$$\alpha_{2j} = C(\nu)(-1)^j \frac{(2j-1)!!}{(2j)!!}\beta_{2j}, \tag{8}$$

where $C(\nu)$ is a function which depends on the depth dependence of the perturbation. If the perturbation changes with time then we find that $\Delta\nu(\nu) = C(\nu)\Delta\beta_0$, which relates the frequency shifts to the splittings. The $C(\nu)$ gives the observed frequency dependence in the $\Delta\nu$ and α_{2j}, and if we use the fit coefficients in Table 1 then the depth dependence of the perturbation can be absorbed into a single arbitrary scale factor.

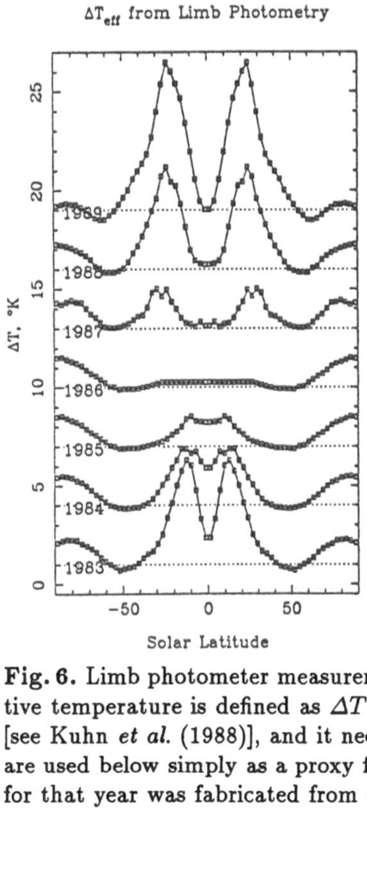

Fig. 6. Limb photometer measurements for 1983-89, from Kuhn *et al.* (1988). The relative effective temperature is defined as $\Delta T = (\Delta F/4F)5770$ K, where F is the the solar limb brightness [see Kuhn *et al.* (1988)], and it need not signify a physical temperature on the sun. These data are used below simply as a proxy for solar activity. No data were available in 1986, so the curve for that year was fabricated from an "interpolation" between 1985 and 1987.

We expect such an inversion of the fit coefficients in Table 1 will show that the effective perturbation will be primarily confined to the active latitudes on the sun. To investigate this, we have used the limb photometer measurements of Kuhn *et al.* (1988) as proxy data, to outline the active latitudes. Figure 6 shows the relative effective temperature from the limb measurements for 1983-89. This is defined as $\Delta T = (\Delta F/4F)5770$ K, where F is the limb brightness. The physical significance of ΔT need not be considered at present ... for now we will use these data only as proxy data for solar activity. These are not ideal proxy data, since these are only *relative* effective temperature measurements ... there could be offsets in ΔT from year to year from instrumental drifts. The data sets from different years were shifted relative to one another such that the effective temperature outside the active latitudes remained roughly constant. Also, no data existed for 1986, so that year's curve was fabricated from a best-guess "interpolation" between 1985 and 1987 measurements.

The results of the inversions are shown in Fig. 7, again for 1986 and 1988, as well as 1988 − 1986. Since β_0 cannot be determined absolutely, the 1986 and 1988 inversions were arbitrarily displaced to match the proxy data. Comparing 1988 and 1986 data we can determine $\Delta\beta_0$, however, along with $\Delta\alpha_{2j}$, so the 1988 − 1986 inversion contains no arbitrary displacement. All the inversions were scaled using a single scale factor.

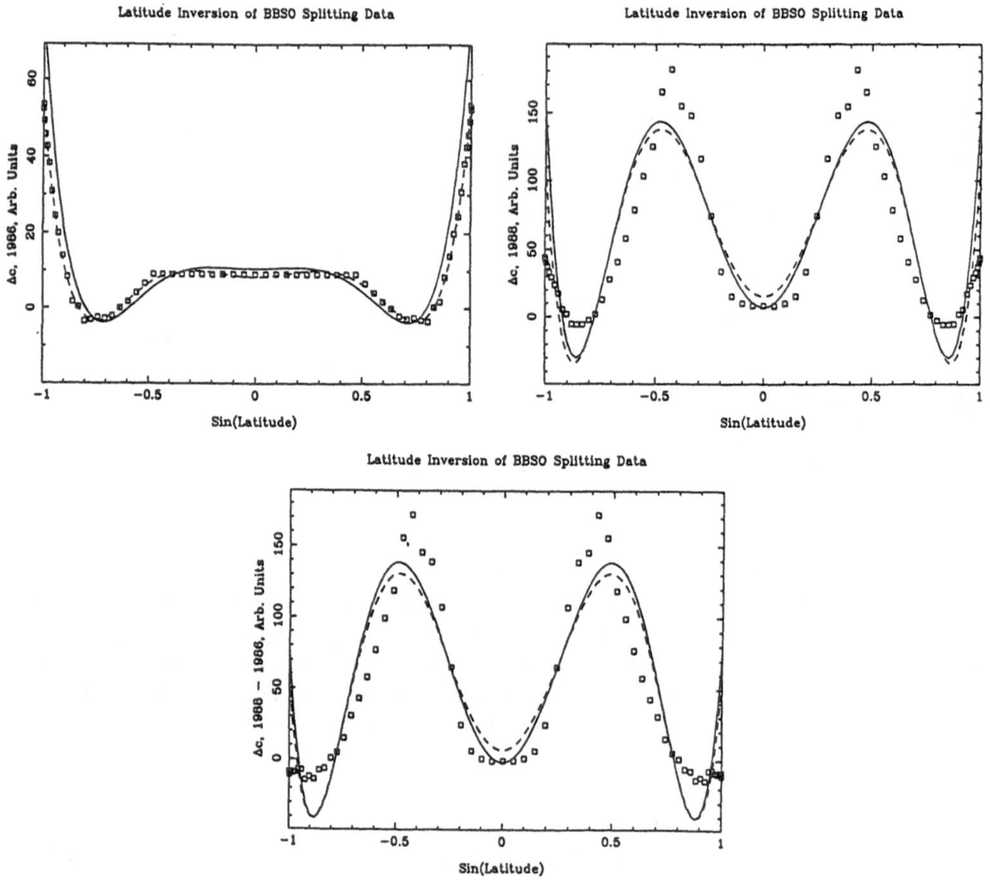

Fig. 7. Latitude inversions of the BBSO splitting data, compared with the limb brightness measurements. The solid lines are the inversions, the points are the limb measurements, and the dotted lines are fits to the limb measurements as in Eq. (5), keeping only terms up to β_6.

While the proxy data used here are far from perfect, the fit to the splitting inversions is remarkably good. The splittings clearly give the expected result in that the effective sound speed perturbation is confined mostly to the active latitudes when the activity is strong. Note, however, that the 1986 splittings show a substantial perturbation coming from the polar region, which again matches the limb brightness measurements. The presence of the polar perturbation results from the fairly large α_4 seen in 1986. This detail provides an interesting piece of information: the source of the perturbation cannot be due entirely to the very strong atmosphere perturbation in active regions, since there are no active regions at the solar poles. Weaker magnetic fields, or large-scale temperature variations, must contribute significantly to the perturbation.

6 Discussion

In summary, with these new precise observations of p-mode frequencies and frequency splittings we have done the following: 1) measured changes in p-mode frequencies with time, and found them to be strongly frequency dependent, and only weakly ℓ-dependant for $5 \leq \ell \leq 60$; 2) shown that $\Delta\nu$ is well represented by an inverse mass function, which suggests that the effective sound speed perturbation responsible for the frequency shifts is localized near the solar surface; 3) shown that the even-index splitting coefficients are also frequency dependent like $\Delta\nu$, showing that the perturbation is localized in latitude, being strongest in the active latitudes; 4) shown that a significant polar perturbation is visible at solar minimum, so the perturbation is not caused entirely by strong fields in the active latitudes.

The frequency shifts and splittings thus form a consistent picture of the solar cycle affecting the p-modes. The latitude dependence of the perturbation is mostly confined to the active latitudes, as expected. These data open up the possibility of inverting $\Delta\nu(\nu,\ell)$ and $a_{2j}(\nu,\ell)$ to determine the depth dependence of the perturbation, and hopefully construct a model which describes physically the structure of the solar cycle with depth.

Acknowledgements

We are grateful to the cast of observers at Big Bear Solar Observatory who diligently carried out these observations, particularly Bill Marquette and Randy Fear. We also thank Peter Goldreich and Pawan Kumar for many informative discussions. These data were analyzed in part using the facilities of the San Diego Supercomputer Center, and the work was supported in part by NSF ATM-8604632 and NSF PYI award AST-8657393.

References

Christensen-Dalsgaard, J. 1982, *Monthly Notices Roy. Astron. Soc.*, **199**, 735.

Cox, J.P. 1980, *Theory of Stellar Pulsation* (Princeton Univ. Press, Princeton).

Duvall, T.L., Jr., Harvey, J.W., Libbrecht, K.G., Popp, B.D., and Pomerantz, M.A. 1988, *Astrophys. J.*, **324**, 1158.

Gelly, B., Fossat, E., and Grec, G. 1988, *Seismology of the Sun and Sun-like Stars*, ed. E. Rolfe, ESA SP-286 (ESA Publication Division, Noordwijk), p.275.

Goode, P.R., and Kuhn, J.R. 1990, preprint.

Gough, D.O. 1988, *Seismology of the Sun and Sun-like Stars*, ed. E. Rolfe, ESA SP-286 (ESA Publication Division, Noordwijk), p.679.

Jefferies, S.M., Pomerantz, M.A., Duvall, T.L., Jr., Harvey, J.W., and Jaksha, D.B. 1988, *Seismology of the Sun and Sun-like Stars*, ed. E. Rolfe, ESA SP-286 (ESA Publication Division, Noordwijk), p.279.

Kuhn, J.R. 1988, *Astrophys. J. Letters*, **331**, L131.

Kuhn, J.R., Libbrecht, K.G., and Dicke, R.H. 1988, *Science*, **242**, 908.

Libbrecht, K.G. 1988a, *Astrophys. J.*, **334**, 510.

Libbrecht, K.G. 1988b, *Seismology of the Sun and Sun-like Stars*, ed. E. Rolfe, ESA SP-286 (ESA Publication Division, Noordwijk), p.3.

Libbrecht, K.G. 1988c, *Seismology of the Sun and Sun-like Stars*, ed. E. Rolfe, ESA SP-286 (ESA Publication Division, Noordwijk), p.131.

Libbrecht, K.G. 1989, *Astrophys. J.*, **336**, 1092.

Libbrecht, K.G., Woodard, M.F., and Kaufman, J.M. 1990, submitted to *Astrophys. J. Suppl.*

Libbrecht, K.G., and Zirin, H. 1986, *Astrophys. J.*, **308**, 413.

Pallé, P.L., Régulo, C., and Roca Cortés, T. 1988, *Seismology of the Sun and Sun-like Stars*, ed. E. Rolfe, ESA SP-286 (ESA Publication Division, Noordwijk), p.285.

Rhodes, E.J., Jr., Woodard, M.F., Cacciani, A., Tomczyk, S., Korzennik, S.G., and Ulrich, R.K. 1988, *Astrophys. J.*, **326**, 479.

Thompson, M.J. 1988, *Seismology of the Sun and Sun-like Stars*, ed. E. Rolfe, ESA SP-286 (ESA Publication Division, Noordwijk), p.321.

Measuring Solar Structure Variations from Helioseismic and Photometric Observations

J. R. Kuhn

Dept. Physics-Astronomy, Michigan State University,
E. Lansing, MI 48824, USA

Abstract: There has been some discussion at this meeting of small changes in frequencies and frequency splittings observed during the solar cycle. Previous observations of frequency changes at the level of 10^{-4} have been convincingly confirmed by several observations presented at this meeting (Jefferies *et al.* 1990; Libbrecht and Woodard 1990; Pallé *et al.* 1990). It has been noted that solar photometric (differential and absolute) observations are of comparable accuracy and show similar solar cycle variations. I will summarize here how the latest photometric and helioseismic data may be accounted for by changes in the the convection zone during a solar cycle. Some of these data and this model have been further described elsewhere (Kuhn 1989; Goode and Kuhn 1990).

1 Do shape or internal structure changes account for the photometric and helioseismic data?

We may consider the solar luminosity to be a function of the sun's radius and effective temperature. Similarly we can express changes in the luminosity in terms of changes in the effective area and brightness of the photosphere. In the case of axially symmetric deviations from spherical symmetry, a measurement of the solar limb shape and the latitudinal surface brightness distribution determine the solar luminosity. Kuhn *et al.* (1988) (henceforth KLD) described measurements of the solar limb effective temperature between 1983 and 1987. While those observations could not distinguish a latitudinally constant effective temperature variation from one year to the next, they did show that limb brightness varied with the solar cycle. By extrapolating the limb data across the full solar disk KLD reproduced the solar irradiance variations found by ACRIM (Willson and Hudson 1988). The limb photometry has now been extended through the summer of 1989.

Data from Figure implies that the quadrupolar component of the limb brightness profile varies by about 2 degrees between 1983 and 1987. On the same timescale variations in the observed solar limb shape (the solar oblateness) are no larger than about 10^{-2} arcseconds. Thus we find that the fractional radius (shape) variations are much smaller

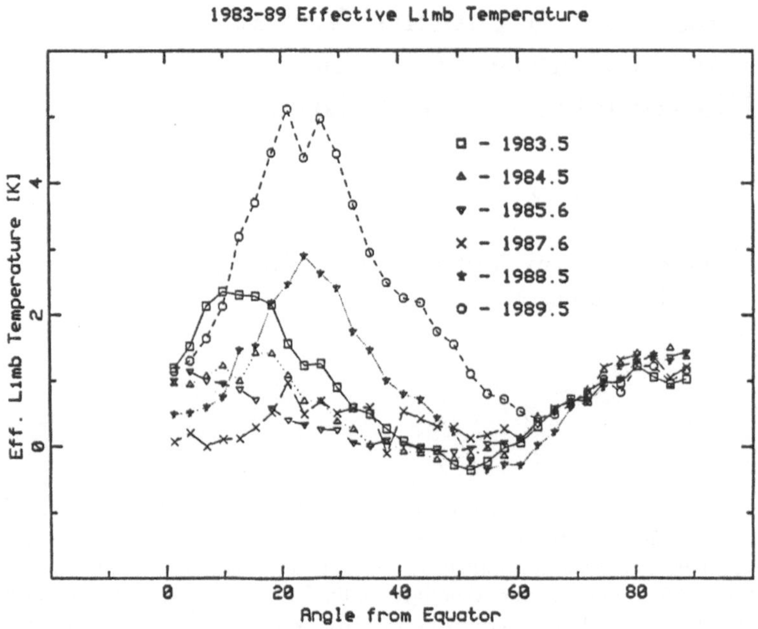

Fig. 1. Non-facular Effective Limb Temperature Variation

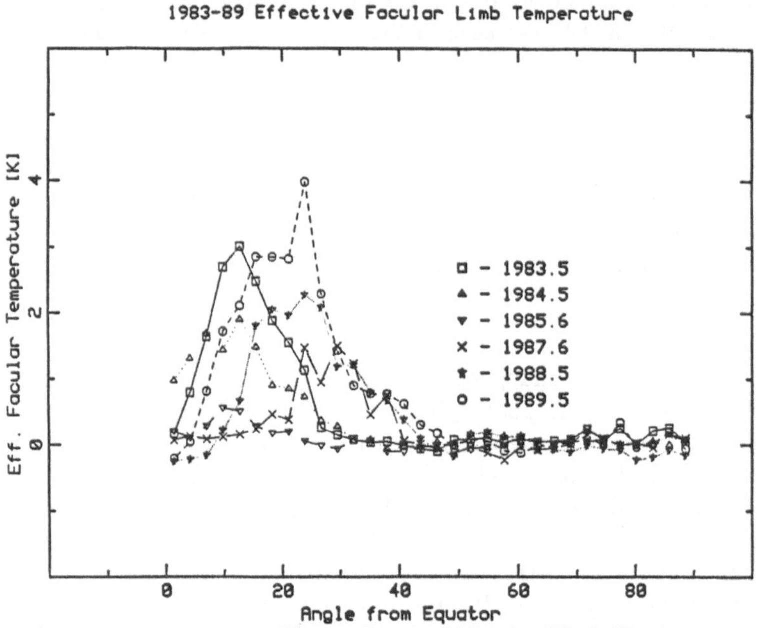

Fig. 2. Facular Effective Limb Temperature

than the fractional luminosity or temperature changes. Our measurements imply that $\frac{\delta r/r}{\delta L/L} \lesssim 0.03$.

The helioseismic splitting observations are also reproduced by a solar model which has a negligible variation in the outer acoustic boundary position (Kuhn 1988a,b, 1989; Goode and Kuhn 1990). A constant solar shape, but spatially and temporally variable thermal (sound speed) structure near the photosphere accounts for both the photometric and helioseismic variability.

2 Does the solar luminosity vary with the solar cycle?

The ACRIM observations imply that either the solar luminosity is constant but redistributed into the equatorial plane, or that the net luminosity varies during the sun's magnetic cycle. In fact both sunspots and faculae tend to redistribute energy from the normal to the photosphere, to transverse directions. Thus the appearance of faculae near the pole might account for the irradiance maximum near solar maximum. As we see in Figs. 1 and 2 the polar regions show no significant brightness variation (facular or isotropic) during the cycle – in contrast to the ACRIM observations. Thus, the differential photometry, described here, and the full disk absolute ACRIM photometry suggest that it is the solar luminosity that evolves during the magnetic cycle.

It is notable that the brightness (or effective temperature) outside of the active latitude bands shows the same latitudinal form throughout the solar cycle. If we also assume that the mean brightness in these latitudes does not vary then we can easily calculate the solar luminosity variation (as in KLD). Figure 3 shows the luminosity variation implied by different assumptions for the contrast function that describes the limb brightening. As I have previously argued, this limb brightening is not facular – and as Fig. 3 shows, we do reproduce the ACRIM irradiance variation if we treat the limb brightness as isotropic (subject to a limb darkening function). Consistent with the previous discussion we have assumed that the solar radius and shape is constant to compute the luminosity variations. This comparison supports two important conclusions: 1) that geometric limb shape change is negligible compared with changes in the brightness, and 2) that the net solar luminosity varies with the magnetic cycle.

3 Where does the solar cycle flux perturbation originate from in the sun?

The net flux excess from the active latitudes is apparently not balanced by an energy flux deficit from other latitudes. The solar luminosity also appears to vary during the 11 year sunspot cycle. The mechanism that regulates this oscillation must operate on a characteristic timescale of several years. Two possibilities stand out: 1) the energy is regulated by, for example, magnetic fields near the photosphere, or 2) a deeper (perhaps the dynamo field) regulates the emergent energy flux at the photosphere. The former possibility encounters a timescale problem. If we imagine that magnetic fields near the photosphere gate the solar luminosity then we must contrive to bottle up this energy over timescales much longer than the characteristic thermal timescale of this part of the convection zone (cf. Spruit 1987). It seems that a more natural explanation is to let the dynamo magnetic field gate the energy from the base of the convection zone. There we do not have a problem storing energy over timescales of years.

A thin (of small radial extent) toroidal magnetic field trapped near the boundary between the radiative and convective regions of the sun could yield a flux deficit or excess

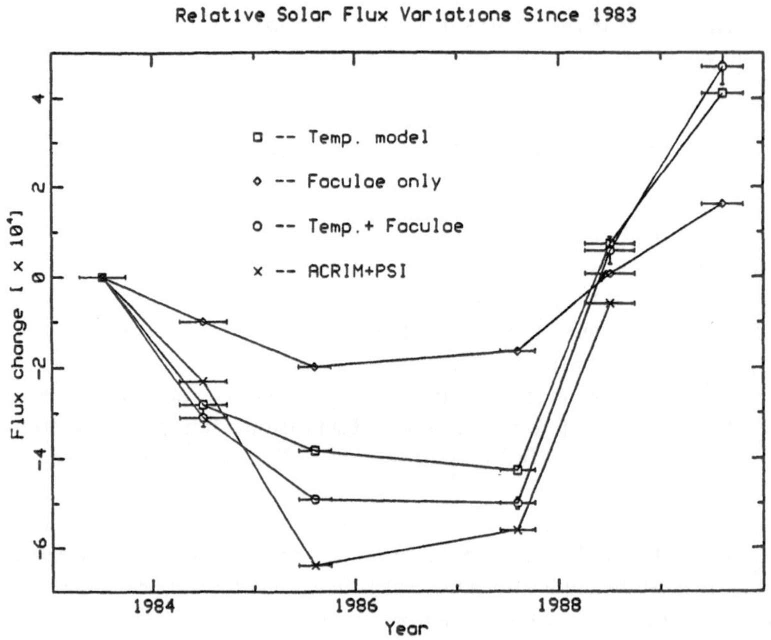

Fig. 3. Solar irradiance variations implied by limb photometry and ACRIM results

in the overlying convection zone compared to a non-field region. If the field extends into the convection zone then, since energy is transported by fluid motion, the net flux above the field is reduced. On the other hand if the field is confined to the radiative zone then the density and photon free path length is increased. Thus the energy flux above the field region is now increased, since the underlying opacity is decreased. As the toroidal field evolves latitudinally and radially during the solar cycle the flux incident on the base of the convection zone would then also vary. The resulting positive or negative thermal shadow of the field at the base of the convection zone could be seen in the photospheric brightness distribution.

4 How does a deep latitude dependent flux perturbation affect the mean radial stratification in the convection zone?

One approach to this problem is to compare two convection zones with different incident fluxes from below. I calculate these models using the mixing length formalism in (Spruit 1974). These are obviously idealized one-dimensional models. If we imagine that the two models, placed side by side, reach pressure equilibrium through some, as yet, unspecified Reynolds stress from induced circulation currents then we can compute the fractional temperature variation between models as a function of depth. Figure 4 shows $\delta T/T$ between two models that differ in incident flux by 1%. The details of this curve may not be physically interesting, but the qualitative result, that the fractional temperature perturbation is strongly peaked in the superadiabatic zone just below the photosphere, is notable. The excess flux is easily carried by convection until the convective efficiency drops as the solar medium becomes transparent to radiation (within a few hundred kilometers of the photosphere). As the convective efficiency falls the fractional temperature perturbation increases in response to the flux excess from below. The curve in Fig. 4 shows just this response. We note that $\delta T/T$ then falls to the radiative value corresponding to approximately a 1% flux perturbation at the surface.

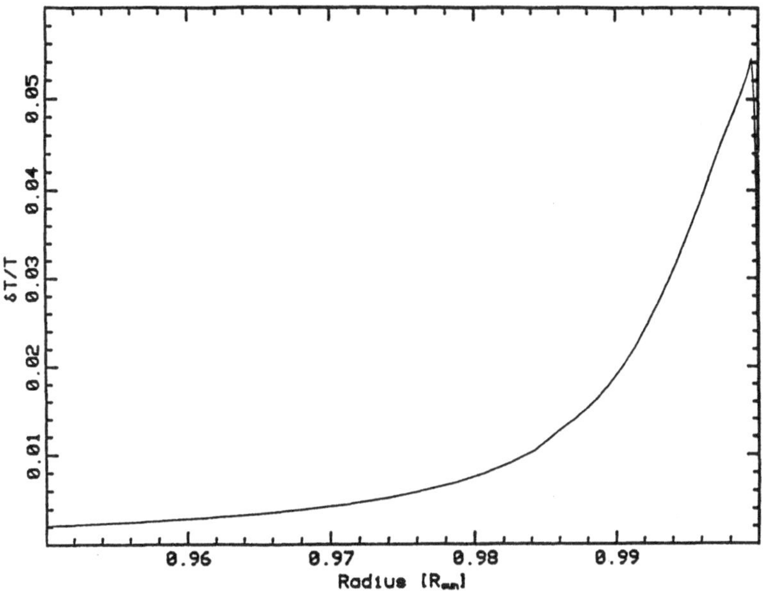

Fig. 4. Fractional temperature perturbation from a deep flux perturbation

5 How do the helioseismic data contribute to this interpretation?

We have previously shown that the mean splittings averaged over radial order n and angular order l vary with the solar cycle in a regular pattern. In fact we can accurately predict the mean splittings just from the differential photometry (Kuhn 1988a, 1989; Goode and Kuhn 1990). Earlier inversion of the l dependent splittings (Kuhn 1988b; Goode and Kuhn 1990) were noisy and hinted that the asphericity is peaked near the surface. Improvements in the splitting data, described at this meeting, now allow a higher resolution inversion for the radial temperature perturbation. A complete inversion has not been completed but a simple solution for the depth dependence of the fractional temperature asphericity has been computed for some of the frequency dependent splittings reported at this meeting. Figure 5 shows the expected frequency dependence of the splittings for several sets of perturbations of varying depth plotted against the data from (Libbrecht and Woodard 1990). In these calculations a constant asphericity extends from the surface to the depth labeled next to each of the curves in this figure. Below this depth the asphericity is zero. It is quite apparent that the asphericity is best described by a perturbation near the surface. In this model the data is best fit by a shallow perturbation, extending downward less than a few hundred kilometers below the photosphere. These data and fits are encouraging support for the model described above.

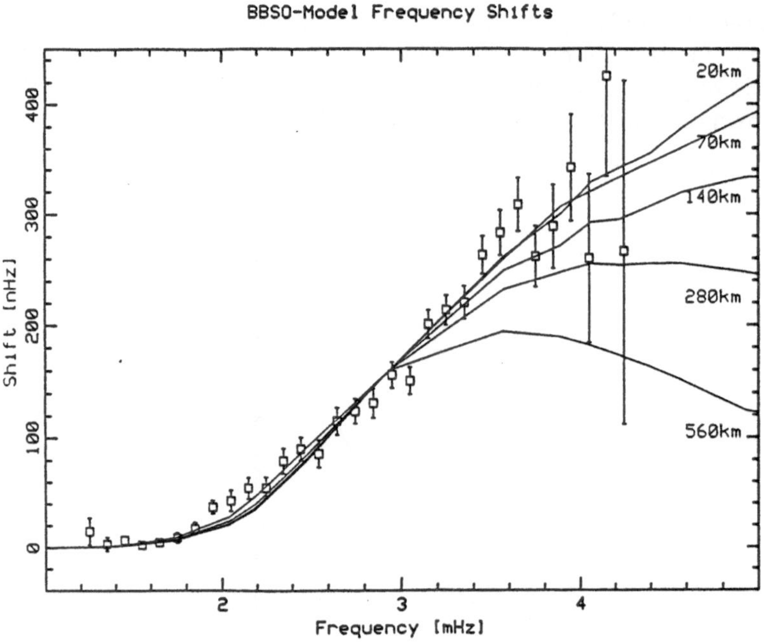

Fig. 5. Frequency dependence of frequency shifts as a function of depth due to a constant fractional temperature perturbation

Acknowledgements

I'm grateful to K. Libbrecht for providing the splitting observations used in Fig. 5. This research was supported by NSF grant ATM-8901689 and by a Sloan foundation grant.

References

Goode, P., and Kuhn, J.R. 1990, *Astrophys. J.*, in press.

Jefferies, S.M., Pomerantz, M., Duvall, T.L., Jr., and Harvey, J.H. 1990, these proceedings.

Kuhn, J.R. 1988a, *Astrophys. J. Letters*, **331**, L131.

Kuhn, J.R. 1988b, in *Seismology of the Sun and Sun-like Stars*, ed. E. Rolfe, ESA SP-286 (ESA Publication Division, Noordwijk), p.87.

Kuhn, J.R. 1989, *Astrophys. J. Letters*, **339**, L45.

Kuhn, J.R., Libbrecht, K.G., and Dicke, R.H. 1988, *Science*, **242**, 908.

Libbrecht, K.G., and Woodard, M.F. 1990, these proceedings.

Pallé, P.L., Régulo, C., and Roca Cortés, T. 1990, these proceedings.

Spruit, H.C. 1974, *Solar Phys.*, **34**, 277.

Spruit, H.C. 1987, in *Solar Radiative Output Variations*, ed. P. Foukal (NCAR, Boulder), p.254.

Willson, R.C., and Hudson, H.S. 1988, *Nature*, **332**, 810.

Evidence for Radial Variations in the Equatorial Profile of the Solar Internal Angular Velocity

Edward J. Rhodes, Jr. [1], Alessandro Cacciani [2], and

Sylvain G. Korzennik [3]

[1]Department of Astronomy, University of Southern California,
Los Angeles, CA 90089, USA, and
Jet Propulsion Laboratory, Calif. Institute of Technology,
Pasadena, CA 91109, USA
[2]Department of Physics, University "La Sapienza" of Rome, Rome, Italy
[3]Department of Astronomy, University of California at Los Angeles,
Los Angeles, CA 90024, USA

Abstract: We present evidence that the solar internal angular velocity, at least as measured in the equatorial plane, shows systematic radial variations in the outer half (by radius) of the solar interior. Specifically, we employ the rotationally-induced frequency splittings of both high- and intermediate-degree sectoral p-mode oscillations to demonstrate that the internal angular velocity rises inwardly from the observed spectroscopic rotation rate of the photospheric gas to a higher value that is at least equal to the observed rotation rate of sunspots, if not higher, in the outer third of the convection zone before decreasing inward of the convection zone to a value which is at least two percent below the photospheric gas rotation rate. By making the assumption that the observed splittings are sensitive to solar rotation at the mid-points of the p-mode eigenfunctions we obtain an angular velocity profile which rises from 452 nHz at the photosphere to 462 nHz at a depth of about five percent of the solar radius below the photosphere. A comparison of this inferred angular velocity profile with that obtained from a formal inversion of these splittings (which is reported elsewhere in these proceedings by Korzennik *et al.*) suggests that the angular velocity might actually exceed the magnetic rotation rate over much of the convection zone before decreasing inwardly toward the center of the sun.

1 Introduction

One of the most active branches of helioseismology is the use of p-mode oscillation frequency splittings in the study of the internal rotation of the sun. The first attempt at determining the solar internal angular velocity from the observed frequency splittings employed high-degree p-modes (Duebner, Ulrich, and Rhodes 1979), while most of the more recent studies have been restricted to low- and intermediate-degrees ($\ell \leq 120$). About one year ago two new attempts at measuring high-degree ($120 \leq \ell \leq 600$) frequency splittings were presented at the Tenerife Symposium on the Seismology of the Sun and Sun-Like Stars (Rhodes *et al.* 1988; Woodard and Libbrecht 1988). However, both of these studies were progress reports which were based upon the analysis of only a single day of observational data and hence both of these studies were only able to present observed high-degree splittings which were roughly consistent with observed photospheric rotation rates. Neither study incorporated enough observational data to allow for any possible discrimination between the photospheric rotation rate of the gas (Snodgrass 1984) and of magnetic features (Snodgrass 1983). As a result of these two papers it became clear at the Tenerife meeting that many more consecutive days of Dopplergrams would have to be analyzed in a single time series in order for a substantial improvement to be made in this situation.

During the fourteen months which have elapsed since the Tenerife meeting, we have concentrated our efforts on the reduction of 19 additional days of observations from our 1988 observing campaign. In this paper we will present both intermediate- and high-degree ($\ell \leq 500$) p-mode frequency splittings that are based upon a time series comprised of 20 consecutive days of full-disk Dopplergram observations. The splitting results which we will present will be limited to the solar equatorial plane and, where there is overlap in degree with earlier intermediate-degree studies, we will compare our recent equatorial results with the results of those earlier studies.

2 The Observations

The summer of 1988 marked the second summer in which we combined a high resolution (1024×1024 pixel) CCD camera with the Na version of the Cacciani magneto-optical filter (MOF) at the Mount Wilson Observatory's 60-Foot Solar Tower Telescope. By combining this CCD camera, which was developed and built at the Jet Propulstion Laboratory, with the MOF, we were able to obtain time series of filtergrams and Dopplergrams which had a high enough spatial resolution (~2.4 arcseconds per pixel) to allow us to extend our previous 60-Foot Tower intermediate-degree p-mode analyses (Rhodes *et al.* 1987, 1988 and 1990) to higher degrees.

Our 1988 observing campaign began on July 1 and continued throughout the remainder of the year. A partial record of the daily number of hours of clear observations which we obtained in July, August, and September, 1988, is shown here as Fig. 1. In this figure we show the actual hours of available observations on each day as the vertical rectangles which are plotted as functions of Pacific Standard Time. The figure begins on July 1 (day number 183) and ends on September 10 (day number 254). The 20 consecutive days which were reduced for this current study are indicated as the filled-in rectangles beginning on July 1, 1988. The open rectangles indicate subsequent observations which are now being reduced.

On each of the 20 observing days employed in this study one pair of full-disk filtergrams was obtained each minute for up to 11 hours per day. One image of each pair was obtained in the blue wing of the two Na D lines, while the second filtergram was obtained in the red wing. Five seconds elapsed between the acquisition of the two filtergrams in each pair and the two images were stored on a magnetic disk before the start of the subsequent minute. The images were then archived on 9-track, 6250-bpi magnetic tapes.

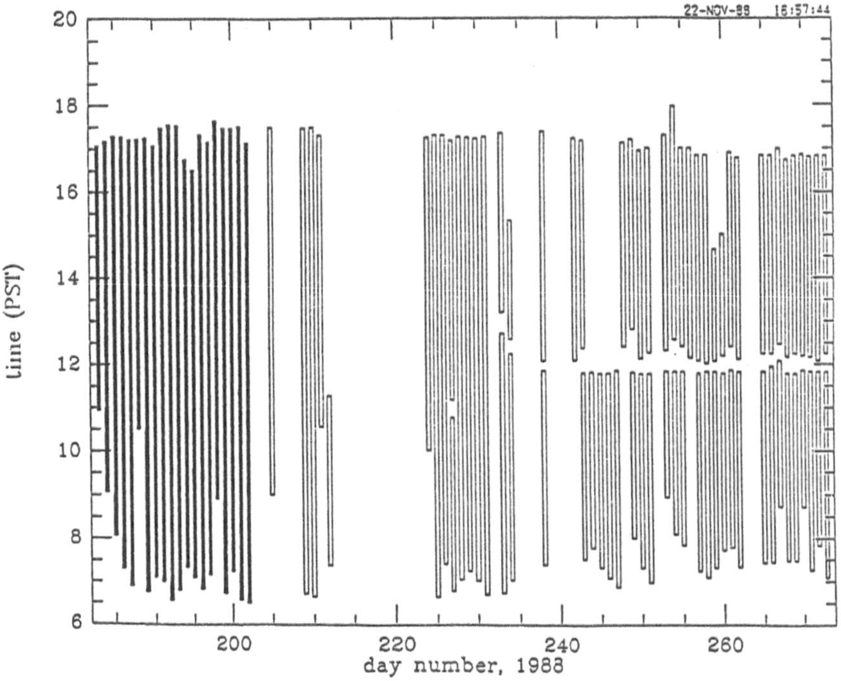

Fig. 1. Log of hours of observation (plotted vertically in Pacific Standard Time) for the first 70 days (plotted horizontally) of our 1988 observing campaign. The 20 consecutive days of observations which were employed in this paper are shown as the 20 filled-in rectangles at the left side.

During the 27,799 consecutive minutes which elapsed between 8:40 PST on July 1, 1988, and 16:00 PST on July 20, we obtained 11,879 pairs of filtergrams. After eliminating those observations which were obtained when the sun was close to either horizon we generated a time series of 10,775 Dopplergrams having an effective duty cycle of 39%.

Next,we computed the integrated Doppler signal for each uncalibrated Dopplergram and employed an ephemeris program which computed the Mt. Wilson- sun center velocity to convert those raw Dopplergrams into series of calibrated Dopplergrams in which the pixel values were expressed in m/sec. This calibration operation was performed separately upon each of the 20 different observing days.

From 19 of these 20 daily sets of Dopplergrams, we computed 90,601 spherical harmonic coefficients, or one for each even azimuthal order, m, for each degree, ℓ, ranging between zero and 600. (For one of the 20 days we inadvertently carried out the spherical harmonic decomposition up to $\ell = 500$, rather than up to 600. Consequently, in this paper we will restrict our results to $\ell = 500$.)

Once we had generated the time series of spherical harmonic coefficients for each day separately, we interspersed zeroes during each of the nighttime data gaps and then we appended additional zeroes at each end of the resulting 27,661-point time series until we had formed 65,536-point time series. For each of the 63,001 different (ℓ, m) combinations between (0,0) and (500,500) we then computed a 65,538-point power spectrum.

3 The Rotational Frequency Splittings

From the 250 pairs of even-ℓ sectoral power spectra (i.e. those for which $m = \pm\ell$), we obtained estimates of the rotational frequency splittings by employing a cross-correlation analysis in which we cross-correlated the corresponding retrograde ($m = \ell$) and prograde ($m = -\ell$) portions of each power spectrum for frequencies between 1800 and 4800 μHz.

The frequency splittings which resulted when we binned the raw sectoral splittings from our cross correlation analysis are shown in Fig. 2. Also shown in this figure is the rotation rate of the photospheric gas as determined spectroscopically (Snodgrass 1984, the lower line) and the photospheric rotation rate of magnetic features such as sunspots (Snograss 1983, the upper line). The 1988 sectoral splittings appear to be consistent with the photospheric gas rotation rate (i.e. 452 nHz) for degrees between 250 and 500. Between $\ell = 250$ and $\ell = 140$ the splittings increase with decreasing degree until they lie close to the magnetic feature rotation rate of 462 nHz. Between $\ell = 40$ and $\ell = 140$ the splittings are roughly constant at 462 nHz. Below $\ell = 40$ the splittings decrease systematically with decreasing degree.

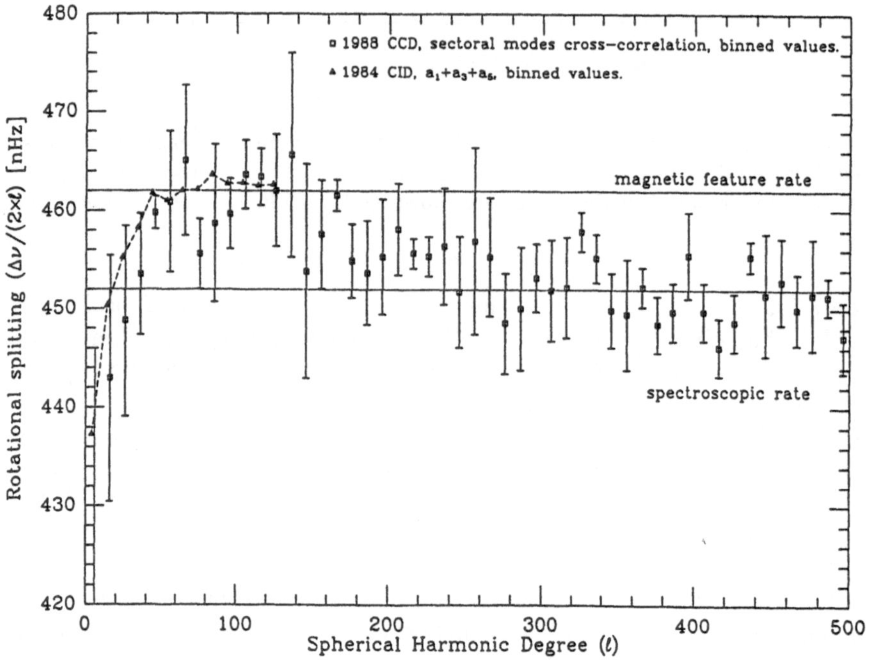

Fig. 2. Binned values of the 1988 sectoral mode frequency splittings are shown as the open squares. Raw splittings were binned in 10-degree wide bins. Below $\ell = 120$ corresponding splittings which were obtained from the 1984 Mt. Wilson CID observations by summing the odd Legendre expansion coefficients are shown as the open triangles at the left.

Figure 2 also includes estimates of the frequency splittings which were obtained at the 60-Foot Tower in 1984 using a CID cameras (Tomczyk *et al.* 1988) for comparison. These frequency splittings were obtained by employing a full Legendre polynomial expansion of the $m-\nu$ tesseral power spectra after which the odd Legendre expansion coefficients were summed (i.e. $\Delta\nu/\sqrt{\ell(\ell+1)} = a_1 + a_3 + a_5$). This comparison shows that our 1988 intermediate-degree splittings show the same degree-dependent variations to which we recently drew attention in our 1984 results (Rhodes *et al.* 1990).

Because the differences between our 1984 and 1988 results for $2 \leq \ell \leq 120$ might appear to be due to temporal variations in those splittings during the four years which separated the two sets of observations, we also carried out the full Legendre polynomial expansion analysis on our more recent 1988 tesseral power spectra for all $\ell \leq 225$. The comparison of the frequency splittings obtained by applying both of these two different methods to the same set of 1988 power spectra is shown in Fig. 3.

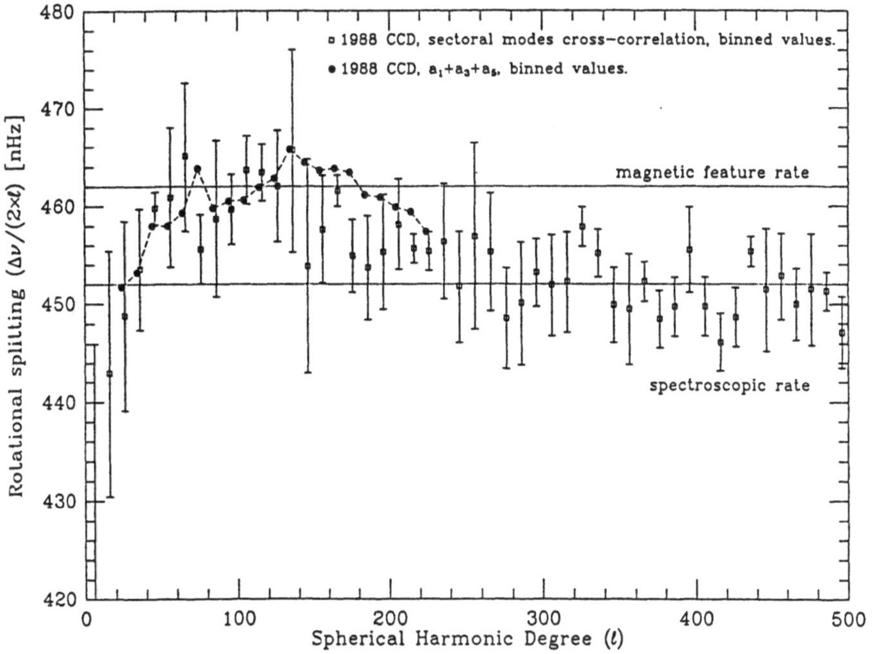

Fig. 3. Comparison of 1988 Mt. Wilson binned frequency splittings obtained from the cross-correlation of sectoral power spectra (the open squares) and from the full Legendre polynomial expansion method for $\ell \leq 225$ (the solid dots). Both sets of splittings show the same peak near $\ell = 120$ and decrease toward both higher and lower degrees.

The differences shown in Fig. 3 between $\ell = 120$ and 200, suggest that the intrinsically noisier sectoral splittings cannot be used reliably to search for temporal variations in the frequency splittings.

4 Search for Possible Temporal Variations

An alternative comparison in which we used the more accurate Legendre coefficient methods in analyzing both the 1984 and the 1988 observations is shown in Fig. 4.

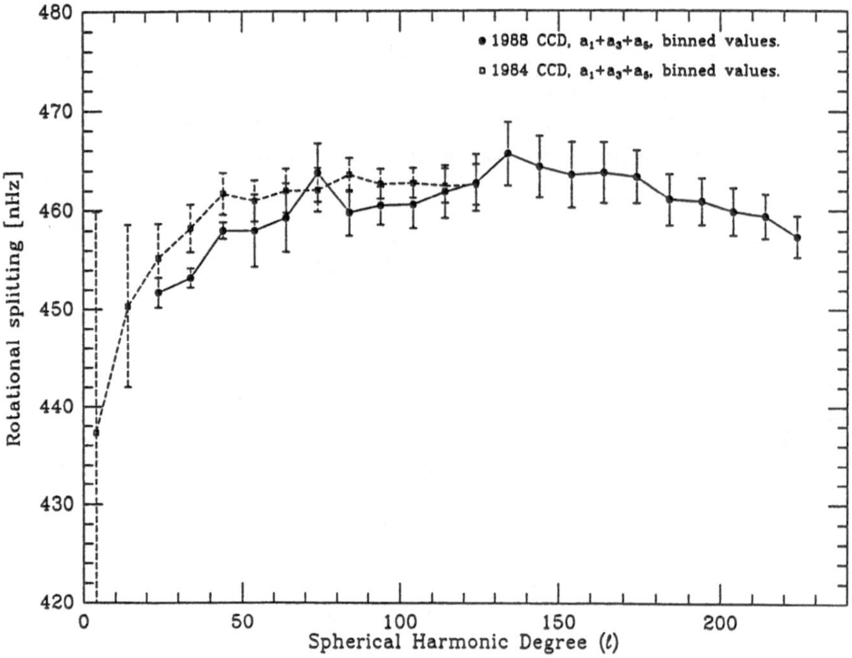

Fig. 4. Comparison of binned 1984 CID and 1988 CCD splittings obtained using Legendre polynomial expansion coefficients in both cases. With the exception of the bin for $\ell = 70 - 80$, the 1984 and 1988 splittings differ at the one sigma level.

Since the two sets of splittings shown in Fig. 4 were computed using the same method and roughly similar amounts of data (35 days in 1984 for $\ell \leq 46$, 16 days in 1984 for $46 < \ell \leq 120$ and 20 days in 1988 for $20 \leq \ell \leq 120$) any differences should be more significant than those shown in Fig. 3. However, the differences between the 1984 and 1988 splittings shown in Fig. 4 differ at only the one sigma level and so it is still premature to claim that there are temporal shifts in the intermediate-degree frequency splittings based on this comparison.

In order to compare our 1988 Mt. Wilson intermediate-degree splittings with those of an independent observer, we also compared our binned, Legendre expansion-derived splittings with those of Libbrecht (1989) for $10 \leq \ell \leq 60$. The result of this comparison is shown in Fig. 5.

In Fig. 5 the Big Bear results fall between those of Mt. Wilson 1984 and 1988 and the uncertainties in the 1988 Mt. Wilson and 1986 Big Bear splittings are comparable for $\ell \leq 50$. Again, the differences between the two sets of splittings are marginal; however, combining Figs. 4 and 5, there is a possible systematic decrease in the intermediate-

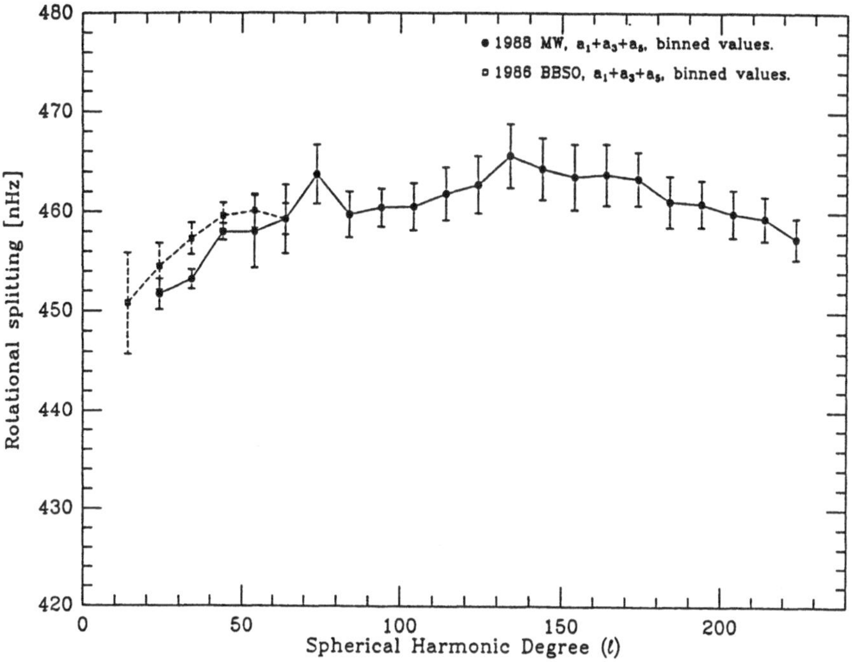

Fig. 5. Comparison between binned 1988 Mt. Wilson frequency splittings with 1986 Big Bear splittings (Libbrecht 1989). In both cases the splittings were obtained from the Legendre polynomial expansion method. Overlap in degree occurred for $10 \leq \ell \leq 60$.

degree frequency splittings between 1984 and 1988. However, more data will have to be analyzed in 1988 and also in 1985 and 1987 before any concrete statements can be made about any such temporal changes.

5 Internal Angular Velocity Estimates in the Equatorial Plane

In order to convert the frequency splittings shown in Fig. 2 into estimates of the internal angular velocity along the solar equatorial plane, we first made the same simple mapping approximation which we recently employed in the analysis of some of our 1984 Mt. Wilson observations (Rhodes *et al.* 1990) and second we carried out formal inversions of a combination of our 1984 and 1988 frequency splitting datasets (Korzennik *et al.* 1990, elsewhere, these proceedings). In our first approach we simply assumed that p-modes of varying degrees are most sensitive to the angular velocity at a radius within the sun that is located half-way between the inner and outer reflecting boundaries for that mode. We then simply computed the mid-points of those p-modes having frequencies closest to 3 mHz which had degrees corresponding to the center of each of our 10-degree-wide bins. We then simply assumed that the binned observational frequency splitting within a given bin was equal to the solar angular velocity at the mid-point of the corresponding p-mode. The angular velocity profiles which resulted from applying this simple assumption to both our 1988 CCD sectoral mode splittings (the squares) and to the 1984 CID Legendre coefficient sums (the triangles) are shown here in Fig. 6.

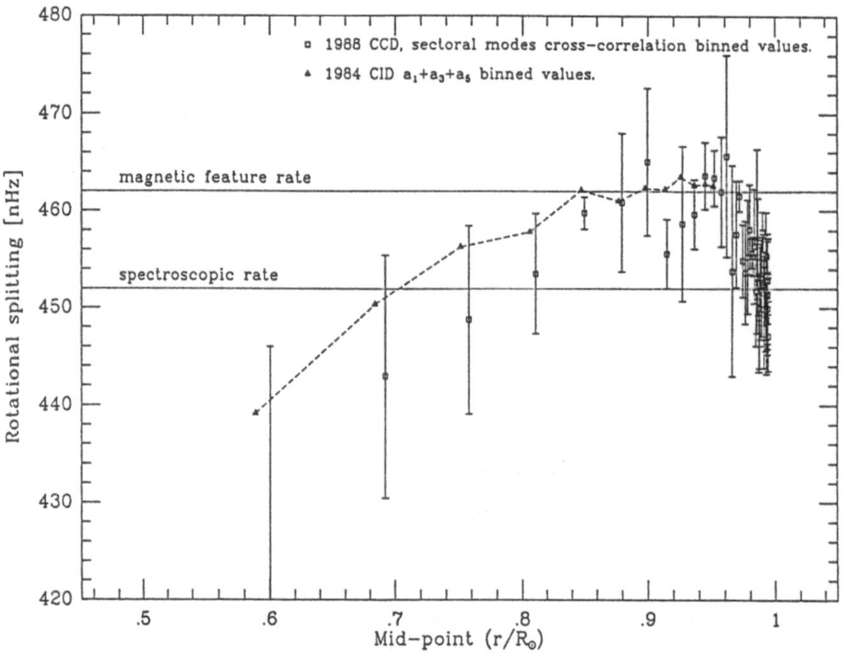

Fig. 6. Inferred equatorial angular velocity profiles obtained from the assumption that the observed frequency splittings are equal to the internal angular velocity at a radius half-way between the surface and the innermost turning point of the p-modes located within each bin.

The 1988 sectoral splittings (the squares) shown in Fig. 6 show evidence of a rise in the angular velocity below the photosphere until it is equal to the magnetic feature rotation rate at a depth of four or five percent of the solar radius below the photosphere. The equatorial angular velocity then appears to retain this elevated value until a depth of roughly 15% of the solar radius. Inward of this point the angular velocity profile appears to drop systematically along the equator toward the point at which the radius is roughly equal to one-half of the solar radius.

An expanded view of the outer portion of Fig. 6 is shown in Fig. 7. Here the two inferred angular velocity profiles shown in Fig. 6 are repeated as the solid dots and the dashed profile, while the results of one of the inversions computed by Korzennik *et al.* (1990) is shown as the solid profile.

The comparison of the two inferred profiles with the profile computed by the formal inversion suggests that the sun's angular velocity may actually exceed the rotation rate of sunspots in the outer portion of the convection zone before returning to that value in the inner portion of the convection zone. Both the simple mapping assumption and the formal inversion suggest that the angular velocity is equal to the observed spectroscopic rotation rate in the photosphere before rising to equal or exceed the sunspot rotation rate inward of the photosphere.

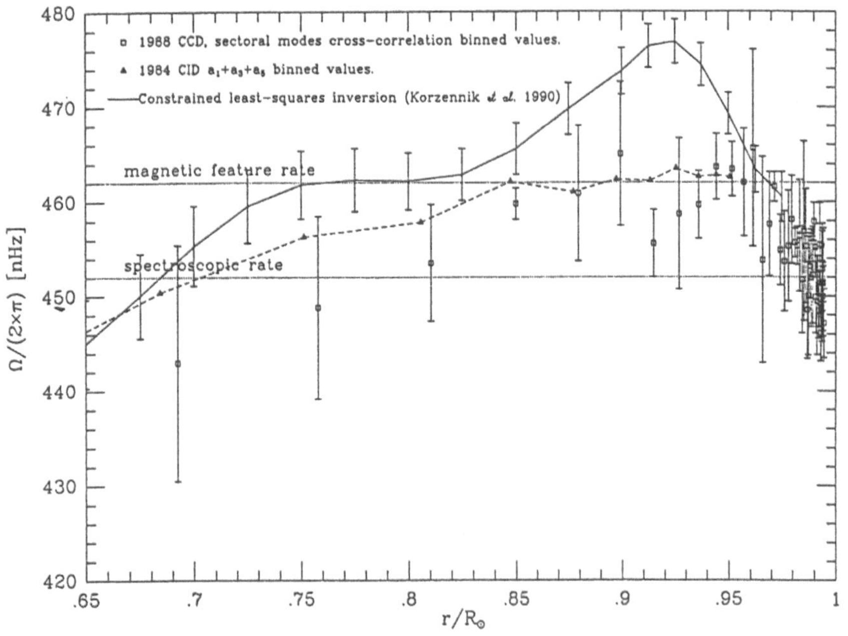

Fig. 7. Comparison of inferred internal angular velocity profiles from 1988 binned sectoral splittings (solid dots) and 1984 binned Legendre sum splittings (dashed profile) with inversion results computed by Korzennik *et al.* (1990) from combination of both 1984 and 1988 observational splittings.

6 Conclusions

Our 1988 observations of sectoral p-mode frequency splittings up to $\ell = 500$ have allowed us to extend helioseismic estimates of the solar equatorial angular velocity closer to the solar surface than has been possible in any previous study. These new splittings have shown that, over the outer two or three percent of the solar interior, the equatorial angular velocity is equal to the spectroscopically-determined rotation rate of the photospheric gas. Another unique aspect of these new splitting results is our suggestion of the existence of a radial gradient in the angular velocity close to the photosphere which has a magnitude such that the angular velocity rises from the spectroscopic rotation rate to a value that is equal to, or possibly up to two percent greater than, the observed rotation rate of sunspots as observed in the photosphere at a sub-photospheric depth of from four to eight percent of the solar radius. Such a gradient, if confirmed by the analysis of additional high-degree splittings that is now in progress, will reconcile many previously disparate measurements of solar rotation and will also suggest that sunspot magnetic fields originate at depths that are at least four percent of the solar radius below the photosphere if not from the inner half of the solar convection zone. Lastly, our 1988 splittings are also consistent with numerous earlier studies in showing that the solar angular velocity decreases inwardly below the base of the convection zone. However, these 1988 results are not yet precise enough to allow us to unambiguously claim the existence of temporal variations in the angular velocity profile between 1984 and 1988. The analysis of additional 1988 CCD observations will be necessary before we will be able to make definitive statements about the presence or absence of such temporal changes.

Acknowledgements

We wish to thank the following individuals for their assistance: Maynard Clark, Martin Iedema and Tim Purdy of the U.S.C. Department of Astronomy, and Dennis Smith and Robert Tooper of the U.S.C. University Computing Services. We also express our appreciation to the Mount Wilson and Las Campanas Observatories of the Carnegie Institution of Washington for providing us with continued access to the Mount Wilson 60 Foot Solar Tower. The research described in this paper was supported in part by NASA grant NAGW-13 and NSF grant INT-8400213 to the University of Southern California and by NASA grant NAGW-472 to the University of California at Los Angeles. A portion of the research was also performed by the Jet Propulsion Laboratory, California Institute of Technology, under contract with the National Aeronautics and Space Administration. Significant portions of the computations necessary for this research were carried out on the JPL Cray X-MP/18 supercomputer, on the USC Alliant FX/80 mini-supercomputer, and on the USC IBM 3090 Model 160E. A. Cacciani was supported by the National Research Council as a Resident Research Associate at the Jet Propulsion Laboratory during a portion of this work. A. Cacciani also acknowledges support from the University of Rome, from the Italian Consigli Nazionale delle Richerche (CNR) and from the Ministero della Publica Instruzione.

References

Deubner, F.L., Ulrich, R.K., and Rhodes, E.J., Jr. 1979, *Astron. Astrophys.*, **72**, 177.

Korzennik, S.G., Cacciani, A., Rhodes, E.J., Jr., and Ulrich, R.K. 1990, these proceedings.

Libbrecht, K.G. 1989, *Astrophys. J.*, **336**, 1092.

Rhodes, E.J., Jr., Cacciani, A., Korzennik, S., Tomczyk, S., Ulrich, R.K., and Woodard, M.F. 1988, in *Seismology of the Sun and Sun-Like Stars*, ed. E. Rolfe, ESA SP-286 (ESA Publication Division, Noordwijk), p.73.

Rhodes, E.J., Jr., Cacciani, A., Korzennik, S., Tomczyk, S., Ulrich, R.K., and Woodard, M.F. 1990, *Astrophys. J.*, **351**, in press.

Rhodes, E.J., Jr., Cacciani, A., Woodard, M., Tomczyk, S., Korzennik, S., and Ulrich, R.K. 1987, in *The Internal Solar Angular Velocity*, eds. B. Durney and S. Sofia (Reidel, Dordrecht), p.75.

Snodgrass, H.B. 1983, *Astrophys. J.*, **270**, 288.

Snodgrass, H.B. 1984, *Solar Phys.*, **94**, 13.

Tomczyk, S., Cacciani, A., Korzennik, S. G., Rhodes, E. J., Jr., and Ulrich, R. K. 1988, in *Seismology of the Sun and Sun-Like Stars*, ed. E. Rolfe, ESA SP-286 (ESA Publication Division, Noordwijk), p.141.

Woodard, M.F., and Libbrecht, K. G. 1988, in *Seismology of the Sun and Sun-Like Stars*, ed. E. Rolfe, ESA SP-286 (ESA Publication Division, Noordwijk), p.67.

The Effect of Large-Scale Flows on Oscillation Ring Diagrams

Frank Hill

National Solar Observatory, P.O. Box 26732, Tucson, AZ 85726, USA

Abstract: The effect of different flow scenarios on the solar oscillation ring diagrams is examined. Several combinations of longitudinal and latitudinal flow components as a function of depth are used to predict possible distributions of ring positions via the forward problem. The predicted positions are combined with plausible unperturbed dispersion relation parameters to produce sets of artificial rings. These rings are then fitted with ellipses, and the measured ellipse parameters are compared with observational results. The results indicate that the data are more compatible with the presence of local maxima in the depth dependence of both horizontal flow components, rather than constant slope solutions.

1 Introduction

Three-dimensional (k_x, k_y, ω) power spectra of solar Doppler data show sets of rings of enhanced power that are the signature of the five-minute p-mode oscillations. The equation describing the rings at a given temporal frequency ω_d is

$$\omega_d = ck^p + < U_x > k_x + < U_y > k_y, \tag{1}$$

where k is the total horizontal wavenumber, k_x and k_y are the components of k, $< U_x >$ and $< U_y >$ are weighted averages over depth of U_x and U_y (the horizontal components of the flow field inside the Sun), and c and p are parameters in the power law representation of the dispersion relation of the oscillations unperturbed by subsurface flows. The rings can be approximated by ellipses, whose central coordinates x_0 and y_0 are proportional to $< U_x >$ and $< U_y >$. The semi-major and minor axes, a and b, of the ellipses provide information on the values of c and p (Hill 1988a).

Observed rings have been used to infer the horizontal components of the flow field as a function of both depth and heliographic position (Hill 1989). The results show the presence of a local maximum in the longitudinal (east-west or rotational) component of some $400\,\mathrm{m\,s^{-1}}$ at a depth of 5000 km in the solar convection zone. A similar local maximum is seen in the latitudinal (north-south or meridional) component, but the maximum is broader, has a lower peak value, and is centered at a shallower depth than the feature in the longitudinal component. In order to further assess the reality of this feature, forward calculations have been carried out to test the validity of the inverse procedures that were used to infer the velocity curves. This paper reports the results of a series of forward calculations.

2 Method

The forward calculations begin by creating possible realizations of U_x and U_y, and forming the integral of the product of these velocity curves with the weighting functions, or kernels, that correspond to the modes contained in the rings. This provides $< U_x >$ and $< U_y >$, which are used in the ring equation with choices of c and p to generate artificial rings. Ellipses are then fitted to the artificial rings, and the measured artificial ellipse parameters, x_0, y_0, a, and b, are compared with the observed values.

The choice of c and p proved to be crucial for the success of the forward calculation. To find appropriate values of these parameters, the forward calculation was performed using many trial combinations of c and p along with the U_x and U_y curves inferred from the data. The resulting (c, p) pair that minimized the rms deviation between the observed and calculated values of a and b for each set of rings was then adopted for the remainder of the calculations. This is not an entirely satisfactory procedure, for it ignores any variation of c and p with ω. It would be best to develop a reliable method of directly and accurately measuring $c(\omega)$ and $p(\omega)$ from the data.

Three different classes of curves for U_x and U_y were considered. The first class consisted simply of the curves actually inferred from the data. The second class comprised curves that had the same qualitative local maximum feature as the curves inferred from the data. Three parameters, the depth at which the maximum occurred, the width of the peak, and the height of the peak, were varied in order to assess the sensitivity of the observations to these quantities. The third class of curves were simple constant-slope straight lines, parameterized by the value of the velocity at a depth where the natural logarithm of the pressure is 28, or a physical depth of about 50,000 km.

3 Results

The results of the forward calculation for the curves that were inferred from the data are shown in Figs. 1 and 2. Figure 1 compares the calculated and observed values of x_0. The comparison is quite good for the p_3 and p_4 rings, and not so good for the other three rings. The discrepancy is most likely caused by the neglect of the frequency dependence of c and p in the forward calculation.

Figure 2 shows the results for y_0. The values calculated from the inferred curve are in excellent agreement with the observed values.

The results for the constant slope curves are shown in Figs. 3 and 4. In Fig. 3, U_y has been set to zero while U_x has been varied, thus the computed values of y_0 are identically zero and the variations are limited to x_0. Similarly in Fig. 4, U_x has been set to the surface value of $1860\,\mathrm{m\,s^{-1}}$ while U_y has been varied; again the computed values of x_0 do not depend upon the U_y curves.

The results for the local maximum curves are shown in Figs. 5, 6, and 7. These curves are parameterized by three quantities: H, the height of the local maximum over the

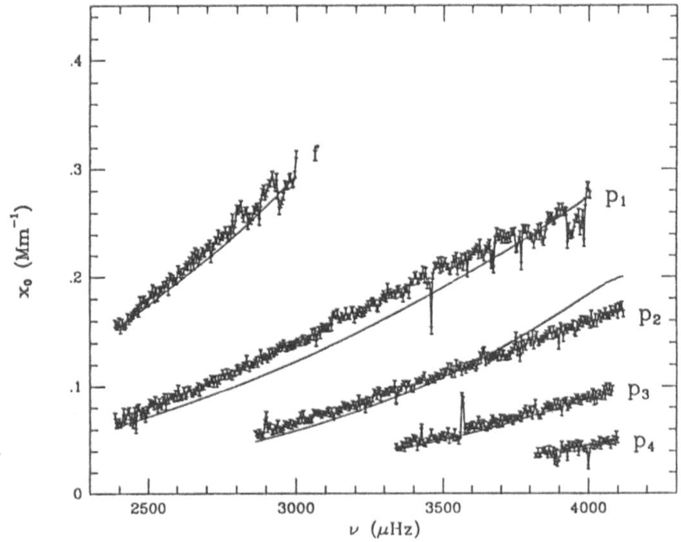

Fig. 1. The calculated (smooth line) and observed values of x_0 for the curves inferred from the data.

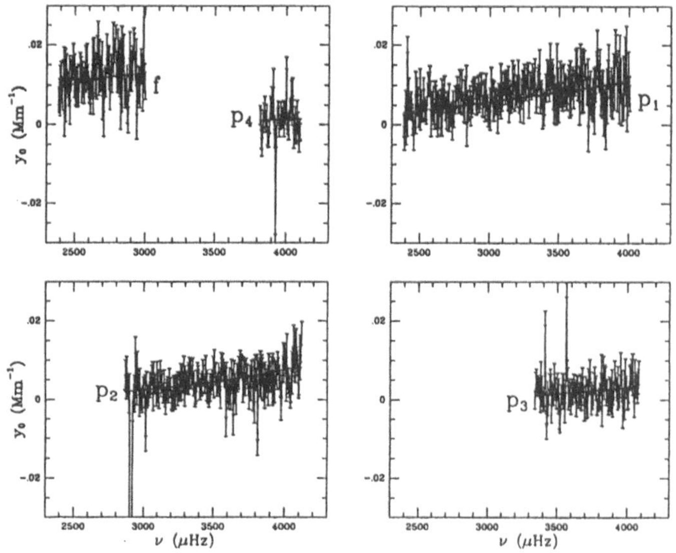

Fig. 2. The calculated (smooth line) and observed values of y_0 for the curves inferred from the data.

surface value (taken to be $1860\,\mathrm{m\,s^{-1}}$), D, the depth of the maximum velocity (ranging from $\ln P = 18$ to 22, or physical depths of about 2,500 to 10,000 km), and W, the HWHM of the maximum (ranging $\ln P = 2$ to 6). In each of the figures, two of the three parameters are held constant while the third is varied. In all of these cases, U_y was set to zero, independent of depth.

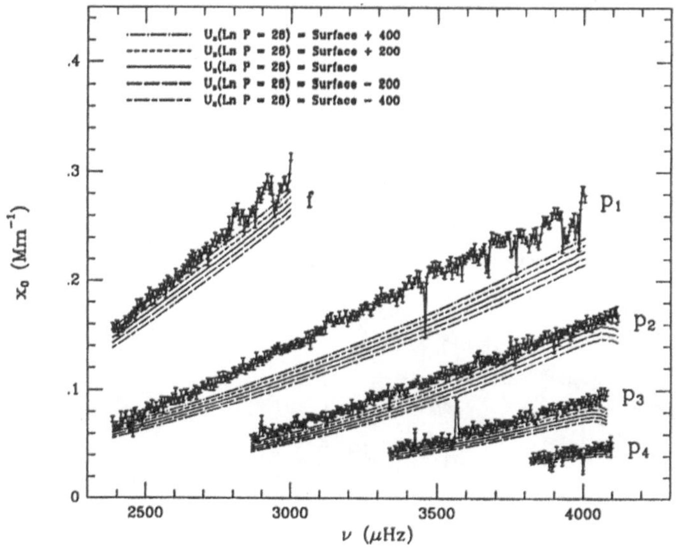

Fig. 3. The calculated values of x_0 for various constant slope models of U_x, with $U_y = 0$, independent of depth.

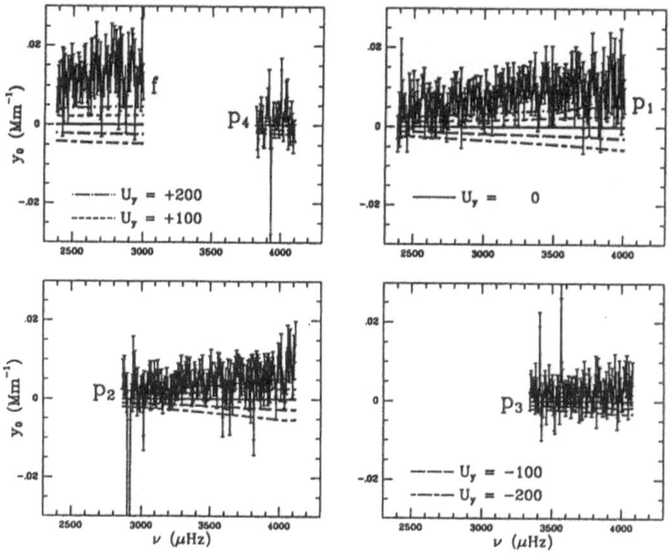

Fig. 4. The calculated values of y_0 for various constant slope models of U_y, with $U_x = 1860\,\mathrm{m\,s^{-1}}$, independent of depth.

4 Discussion

The measurement of c and p remains as an impediment to a fully successful forward calculation of the rings. The method used in this paper is not satisfactory because it ignores the frequency variation of c and p, and thus results in a fairly poor fit to x_0 for the f, p_1, and p_2 rings. A previously developed method for measuring c and p (Hill 1988b), resulted in frequency-dependent values of c and p. The use of these values in the forward calculation using the inferred curve resulted in much better fits to the observed x_0 for the p_1 and p_2 rings, at the expense of worse fits to the f and p_4 rings. Clearly work remains to be done in this area. Nonetheless, by using identical values of c and p for these calculations, some feel for the effect of different classes of velocity curves on the rings can be gained.

Comparison of the computed curves in Figs. 3 and 5 through 7 shows that the values of x_0 measured from the observed rings are in general closer to the calculated values for the local maximum velocity curves than for the constant slope curves. This suggests that a local maximum does exist in the east-west, or rotation rate in the shallow layers below the photosphere. This suggestion is further strengthened by the curve inferred directly from the rings (Hill 1989), by the rotation curve inferred from a more traditional two-dimensional power spectral analysis of the same data (Hill *et al.* 1988), and by recent results from whole-disk measurements (Brown, private communication 1989; Rhodes *et al.* 1990). The calculated values of y_0 for the inferred U_y curve are in excellent agreement with the observed values, and are in better agreement than the values predicted by curves with constant slopes.

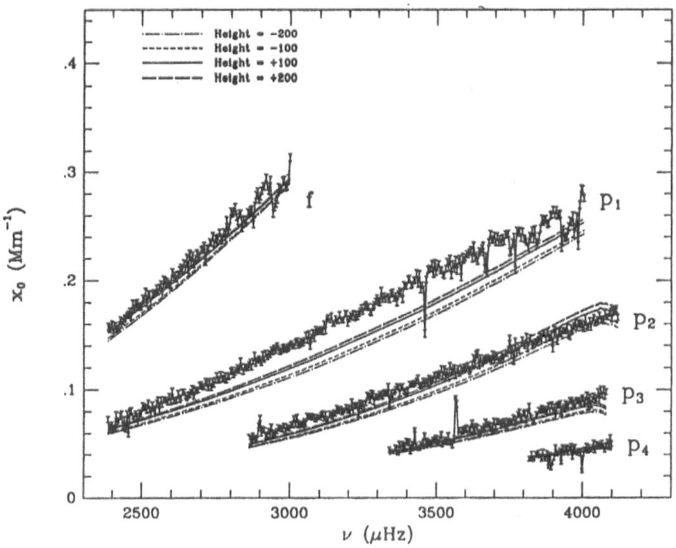

Fig. 5. The calculated values of x_0 for various local maximum models of U_x. In these curves, W=4, D=20, and H ranges from -200 to $+200\,\mathrm{m\,s^{-1}}$.

There is thus additional evidence for both meridional and rotational flows that have local maxima in the layers immediately below the photosphere. The maxima occur in the region between the hydrogen and first helium ionization zones, and may be due either to the deflection of large scale vertical convective motions into horizontal flows (Latour *et al.* 1983), or to the conservation of angular momentum in the outer convection zone (Hathaway 1982). Future application of the ring diagram analysis to full-disk data should

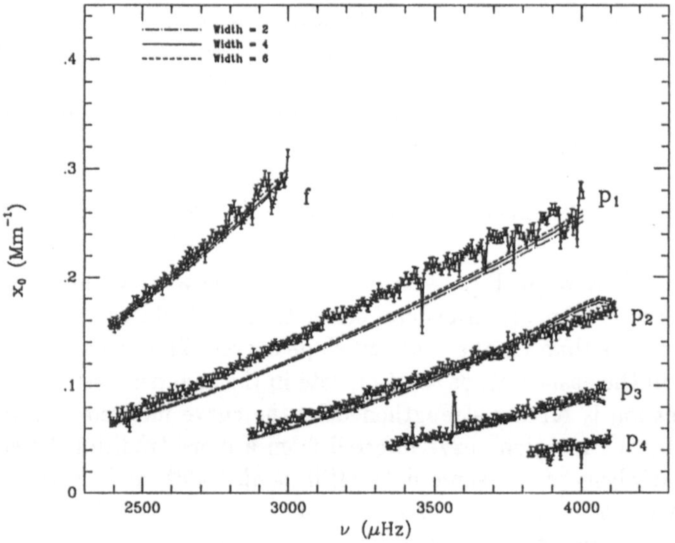

Fig. 6. As Fig. 5, except that $D = 20$, $H = +200$, and W ranges from 2 to 6.

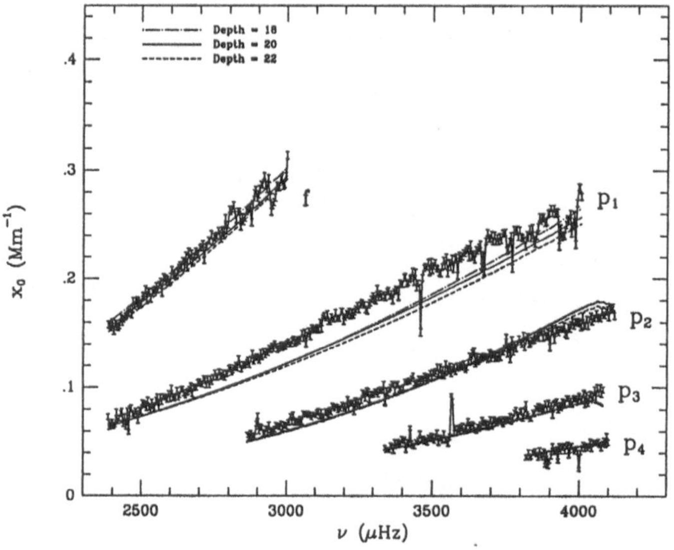

Fig. 7. As Fig. 5, except that W=4, H=+200, and D ranges from 18 to 22.

lead to a substantial increase in our knowledge of the flows within the solar convection zone.

Acknowledgements

I am indebted in all my work to discussions with Jack Harvey, Tom Duvall, John Leibacher, Larry November, Deborah Haber, Douglas Gough, Juri Toomre, and Bernard

Durney. The National Solar Observatory is a division of the National Optical Astronomy Observatories, which is operated by the Association of Universities for Research in Astronomy, Inc. under contract with the National Science Foundation.

References

Hathaway, D.H. 1982, *Solar Phys.*, **77**, 341.

Hill, F. 1988a, *Astrophys. J.*, **333**, 996.

Hill, F. 1988b, in *Seismology of the Sun and Sun-like Stars*, ed. E.J. Rolfe, ESA SP-286 (ESA Publication Division, Noordwijk), p.103.

Hill, F. 1989, *Solar Phys.*, in press.

Hill, F., Gough, D., Toomre, J., and Haber, D.A. 1988, in *Advances in Helio- and Asteroseismology, IAU Symp. No. 123*, eds. J. Christensen-Dalsgaard and S. Frandsen (Reidel, Dordrecht), p.45.

Latour, J., Toomre, J., and Zahn, J.-P. 1983, *Solar Phys.*, **82**, 387.

Rhodes, E.J., Jr., Cacciani, A., Korzennik, S., Tomczyk, S., Ulrich, R.K., and Woodard, M.F. 1990, *Astrophys. J.*, in press.

Observations of p-Mode Absorption in Active Regions

D. C. Braun [1 2], T. L. Duvall, Jr. [1], and S. M. Jefferies [3]

[1]Laboratory for Astronomy and Solar Physics,
NASA/Goddard Space Flight Center, Greenbelt, MD 20771, U.S.A.
[2]NAS-NRC Resident Research Associate
[3]Bartol Research Institute, University of Delaware, Newark, DE 19711, U.S.A.

Abstract: We present here a summary of results on the interaction of p-modes with solar active regions based on observations made at the Kitt Peak Solar Vacuum Telescope and the geographic South Pole. A travelling wave decomposition of p-modes is performed in a cylindrical coordinate system centered on the active regions. Significant absorption of p-mode wave power is observed to occur in all of the regions and is a function of horizontal wavenumber (k) — increasing linearly with k up to some maximum value and remaining constant for higher wavenumbers. The maximum fractional absorption of incident power is about 0.2 for small pores and 0.4 for typical isolated sunspots (radius = 15 Mm). A maximum of 70% absorption is seen in the large sunspot group of March 1989 (radius = 60 Mm). No convincing variation of the absorption with temporal frequency (i.e. radial order) is seen, although not entirely ruled out considering the relative errors involved with the power measurements. No significant difference in the amount of p-mode absorption is detected between equal 3-hour time intervals before and after a class X4 flare in the March 1989 region. No excess of outgoing waves following the time of the flare is detected. These observations do not support the suggestion that large flares may excite observable acoustic waves in the photosphere.

1 Introduction

The interaction of p-modes with solar active regions has recently been investigated and it has been observed that solar magnetic fields act as strong sinks of p-mode wave energy (Braun *et al.* 1987, 1988). Attempts to explain the underlying physical processes involved have been made with varying degrees of success (Hollweg 1988; Knölker and Bogdan 1988; Campbell and Roberts 1989; Chitre and Davila 1989; Lou 1988, 1989). Further observations and theoretical investigations are warranted since the possibility may exist to use p-modes as a seismic probe of magnetic fields beneath the photosphere. For the purpose of stimulating further investigations, a summary of recent results based on observations made at the Kitt Peak Solar Vacuum Telescope and the geographic South Pole is presented here.

2 Vacuum Telescope Observations

2.1 p-Mode Absorption in NOAA #5395

The giant sunspot group which appeared in March of 1989 was the largest to date of the current activity cycle and was the site of several noteworthy flares. Observations of the velocity in a $512'' \times 570''$ field centered on this region were made with the 512 channel magnetograph at the Vacuum Telescope on Kitt Peak every 95 s for a 6.76 hour interval beginning 10 March 1989 17:16:00 U.T.

Initial reduction of the images proceeded as described by Braun and Duvall (1990). The de-streaked and registered velocity images were collapsed by pixel averaging to produce a final set of images with $3'' \times 3''$ spatial resolution.

The velocity oscillations were decomposed into their radially propagating Hankel wave-components in the following manner. First, the velocities were interpolated onto a cylindrical grid centered on the sunspot group and Fourier transformed in azimuth. The output was then fit to an orthogonal set of Hankel functions of the first and second kinds computed over a radial range $r_{\rm min} = 96''$ (69 Mm) to $r_{\rm max} = 345''$ (249 Mm). A Fourier transform in time was performed on the Hankel-coefficients to produce the power of inward and outward propagating waves as a function of radial wavenumber k, polar azimuthal order m, and temporal frequency ν. The power spectra were then summed over azimuthal orders $-5 \leq m \leq 5$. In order to estimate the background contribution, measurements from the power spectra at 2 and 5 mHz were made and fit with cubic spline functions in wavenumber. The noise at intermediate frequencies was estimated by linear interpolation and subtracted from the spectra. The inward and outward propagating wave power was then summed over frequency and the absorption coefficient $\alpha \equiv (P_{\rm in} - P_{\rm out})/P_{\rm in}$ was determined as a function of radial wavenumber (Fig. 1).

The most striking result demonstrated by Fig. 1 is the magnitude of p-mode absorption at high wavenumbers. Between $0.4 \leq k \leq 0.8\,{\rm Mm}^{-1}$ ($280 \leq l \leq 560$) the mean value of α is 0.66 ± 0.02. Between $0 \leq k \leq 0.4\,{\rm Mm}^{-1}$ the absorption coefficient rises linearly from a near zero level to the aforementioned maximum value. Qualitatively, this behavior is similar to that observed in much smaller sunspots although the wavenumber at which maximum absorption is reached is somewhat lower for this spot. A summary of the properties of several sunspots is shown in Table 1, which lists the sunspot group radius (or penumbral radius in the case of isolated spots), the maximum value of p-mode absorption, and the wavenumber at which the maximum value is reached. Data for the 1983 and 1986 active regions is taken from Braun *et al.* (1988).

A legitimate question one might ask from looking at the table is why the 1989 March 10 active region does not show a maximum absorption of 100% – this region is four times the size of the 1983 spots, yet absorbs only 25% more high-degree p-mode power. It is possible that some fraction of the incident p-mode energy is reflected at the boundary between the sunspot magnetic field and quiet-Sun region.

With one exception, the 1983 and 1986 sunspots had the maximum absorption occurring at p-mode wavelengths roughly equal to (or slightly smaller than) the sunspot radius. (The exception to this was the magnetic pore observed on 1986 November 20. This pore, however, was surrounded by a large plage region which accounted for the majority of the observed p-mode absorption). The peak absorption in the large March 10 region does occur at a lower wavenumber than the other sunspots, but nevertheless corresponds to a wavelength significantly smaller than the linear size of the region. This may suggest

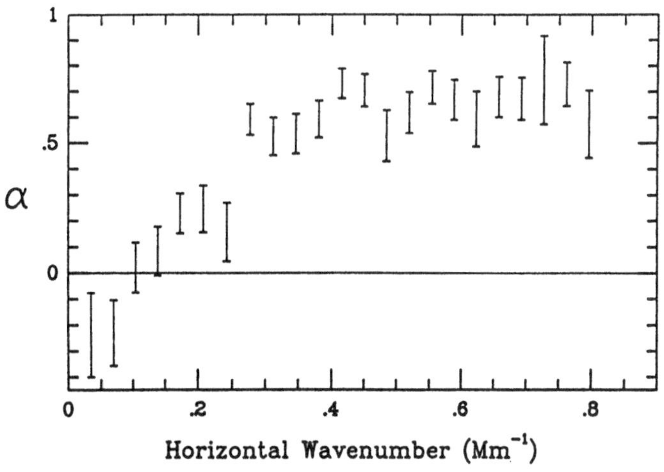

Fig. 1. p-Mode absorption vs. wavenumber in NOAA #5395

that the p-modes are effectively coupling with smaller features inside the region, or the absorption may be occurring over a range of depths where there is a significant decrease in the horizontal size of the active region.

Table 1. p-Mode absorption in active regions

Date	Spot or group radius(Mm)	Maximum absorption α	k (Mm^{-1})	λ (Mm)
1986 Nov 20	4*	0.3	0.6	10
1983 Feb 2	12	0.4	0.7	9
1983 Jan 1	15	0.45	0.6	10
1986 Oct 25	16*	0.45	0.5	13
1989 Mar 10	60	0.7	0.4	16

* Some absorption occurs within surrounding plage.

2.2 A Search for Flare-excited Acoustic Waves

A major flare (of class X4.5/3B) began in NOAA #5395 approximately two hours after the onset of our observing sequence, reaching maximum phase on 10 March 19:08 U.T.

The chance occurrence of this flare during our observing sequence provided us with an opportunity to look for effects of the flare upon high-degree p-mode oscillations. Specifically, it has been suggested that flares may excite observable acoustic waves in the photosphere (Wolff 1972; Haber *et al.* 1988a, b).

To test this possibility, the entire 6.76 hour time series was divided in half and the analysis outlined in section 2.1 was performed separately for each interval. Considering the group velocities of acoustic waves with $\nu \simeq 3\,\mathrm{mHz}$ and our value of r_{min}, any flare induced waves with $k \geq 0.4\,\mathrm{Mm}^{-1}$ should contribute to an excess of outward propagating power exclusively in the latter ("post-flare") interval.

Since P_{in} is constant with time, the difference in the absorption coefficient for the two intervals directly yields the fractional excess of any flare-induced outgoing wave power. Using α in this way also cancels any spurious changes in wave power due to atmospheric seeing fluctuations. Between horizontal wavenumbers $0.4 \leq k \leq 0.8\,\mathrm{Mm}^{-1}$ we determine a mean value of $\Delta\alpha$ (pre-flare − post-flare) of 0.01 ± 0.05, which is consistent with no change in the amount of p-mode absorption. This analysis was repeated for multiple positions of the coordinate system origin in and around the visible active region and flare site with similar (negative) results.

A search for acoustic "pulses" emanating from the flare was performed by spatially and temporally filtering the amplitudes and phases in the outgoing k-ν quadrant for the entire observing run and then performing the appropriate inverse Hankel and temporal transforms. No obvious enhancements of outgoing power at any wavenumber ($0.1 \leq k \leq 0.8\,\mathrm{Mm}^{-1}$) were seen along the trajectories expected by the group velocity of these wavemodes. Any enhancement on the order of 20% above the background p-mode signal should have been readily apparent.

The pressure excess due to energetic electrons during the chromospheric evaporation phase of a flare may be as high as 10^3 dynes (Fisher *et al.* 1985). Assuming that this pressure pulse is attenuated by the factor $\exp(-z/2H)$ ($H \equiv$ pressure scale height) from the base of the chromosphere to the photosphere yields a maximum pressure perturbation of about 200 dynes at photospheric levels. We estimate this perturbation would give rise to a maximum excess of about $\sim 5\%$ in observed p-mode power for the duration of the impulse. This value is within the limits imposed by our observations, especially considering the likelihood that the duration of any impulse is significantly smaller than our 203 minute interval.

2.3 Spatial Mapping of p-Mode Absorption

The spatial distribution of p-mode absorption in active regions can be determined by repeating the Fourier-Hankel transform at multiple positions of the coordinate system origin. A technique for optimal construction of absorption maps has been developed and applied to several of the Vacuum Telescope observing sequences (Braun *et al.* 1990). It is directly observed that the absorption is not confined to the sunspot umbrae, but is prevalent within surrounding plage. The absorption efficiency appears to scale roughly with the surface magnetic flux density.

3 South Pole Observations

In November 1987 full disk Ca II K line images of the Sun were obtained over a period of several weeks at the geographic South Pole (Jefferies *et al.* 1988, 1990). Although the observations took place during a time of low solar activity, the passage of an isolated, moderately sized sunspot (NOAA #4890) was recorded during part of the observing sequence. A 26.9 hour (95% duty cycle) segment of the sequence was selected for the purpose of studying p-mode absorption in this sunspot of modes with spherical harmonic degrees $l \leq 380$ ($k \leq 0.55\,\mathrm{Mm}^{-1}$). The data used in this analysis spanned the interval 15 Nov. 18:12 – 16 Nov. 21:05 U.T. Preliminary data reduction preceded as described by Jefferies *et al.* (1990).

As it was the intent to include in the analysis intensity data out to heliocentric angles of 90° from the sunspot, it was necessary to construct a travelling wave description of p-modes in a spherical geometry. A spherical coordinate system was defined which placed one "pole" at the position of the visible sunspot. The "colatitude" (θ) and "longitude" (ϕ) coordinates in this system are therefore analogous to the radial and azimuthal coordinates respectively in the cylindrical case. Waves traveling away from and toward the pole can be described by functions of the form

$$\Psi \sim e^{i(\omega t + m\phi)}\{N_l^m\, P_l^m(\cos\theta) \pm i\, M_l^m\, Q_l^m(\cos\theta)\} \tag{1}$$

where N_l^m and M_l^m are the normalization factors for the associated Legendre functions of the first and second kinds (P_l^m, Q_l^m), and $\omega \equiv 2\pi\nu$.

The intensity data was interpolated onto a (θ, ϕ) grid over a colatitude range $10° \leq \theta \leq 90°$. Azimuthal transforms were computed for orders $-5 \leq m \leq 5$. Legendre transforms (of the first and second kinds) were then performed for spherical harmonic degrees $50 \leq l \leq 380$ (the Nyquist limit) and the appropriate combinations of coefficients were Fourier transformed in time to produce power of inward and outward propagating waves as a function of l, m, and ω. Power spectra were averaged over azimuthal order such that at each value of l, only the spectra for $|m| < l\,(r_{\mathrm{spot}}/R_{\mathrm{sun}})$ were included. This procedure isolates only those wave components which pass through the sunspot. The background noise contribution to the spectra was estimated between the p-mode ridges. Spline-interpolated values of the background were then subtracted from the inward and outward p-mode power levels.

Figure 2 shows the absorption coefficient as a function of degree l where the p-mode ridge power has been summed over all observed radial orders $n > 0$ (the f-mode ridge was not detected above the background). A linear increase in absorption is seen, which rises from 0% at $l = 50$ to over 20% at $l = 380$. This is in good agreement with Vacuum Telescope observations of similarly sunspots observed in January 1983 (solid triangles) and February 1983 (solid circles).

Figure 3 shows the p-mode absorption as a function of radial order, where the power of in- and outgoing waves has been summed over a range in degree. A variation with the absorption coefficient with radial order might be expected if there is a depth dependence to the absorption mechanism. For p-modes with $100 \leq l \leq 200$ there is a suggestion of a slight decrease of absorption with radial order, although within the measured errors the data is consistent with a constant absorption coefficient of about 10%. At higher degrees, it appears that most of the absorption occurs in the p_3 and p_4 ridges. Since, in this

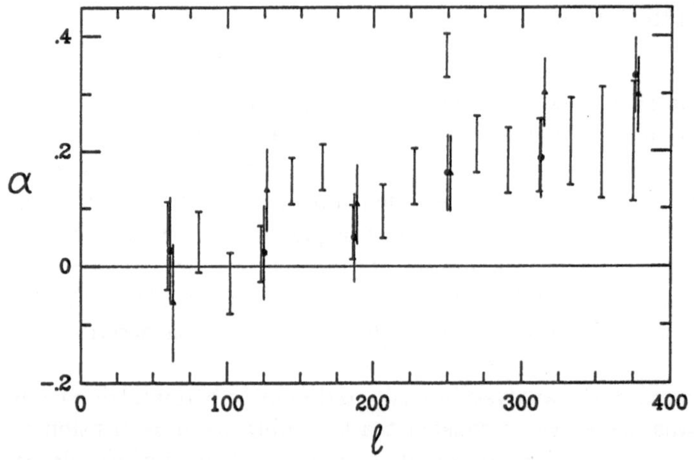

Fig. 2. p-Mode absorption vs. degree [1987 S. Pole (w/crossbars) & Vacuum Tel. (1/83 = triangles, 2/83 = circles)]

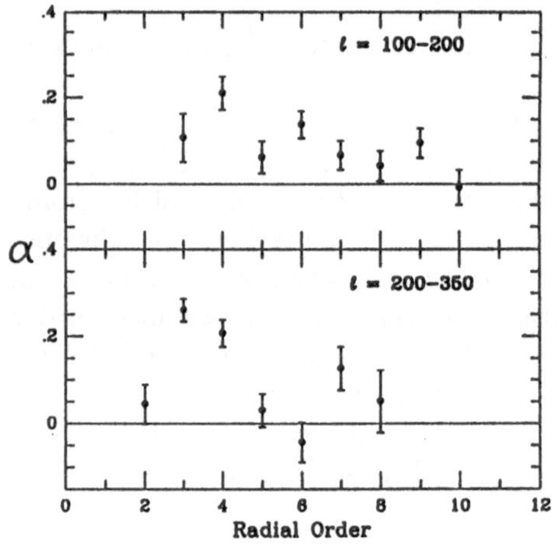

Fig. 3. p-Mode absorption vs. radial order (1987 South Pole observations).

l range, p-mode power is largely concentrated in these two ridges the possibility exists that the observed variation is an artifact, although uncertainties in the background power subtraction do not appear sufficient to cause this behavior. The frequency resolution of this data set provides a good separation of the p-mode ridges at high l and allows a determination of the background levels to within a few percent of the p-mode power. On the other hand, it seems difficult (if not impossible) to attribute this unusual variation to a real depth effect, and an absorption mechanism which scales with the amount of incoming energy has not been formulated. Observations spanning longer time periods (such as the GONG experiment) or which combine data for several sunspots may be required to unambiguously determine the variation of absorption with both degree and radial order necessary to probe the dependence of the phenomenon with depth.

Acknowledgements

Some of the concepts forming the basis of our analysis were developed in collaboration with Dr. B. LaBonte. In addition to T.D. and S.J., members of the 1987 South-Pole expedition included Drs. M. Pomerantz, J. Harvey, and Mr. D. Jaksha. Support from the National Solar Observatory in Tucson is appreciated. This work was done while D. Braun held a National Research Council-NASA/Goddard Space Flight Center Research Associateship.

References

Braun, D.C., and Duvall, T.L., Jr. 1990, *Solar Phys.*, in press.

Braun, D.C., Duvall, T.L., Jr., and LaBonte, B.J. 1987, *Astrophys. J. Letters*, **319**, L27.

Braun, D.C., Duvall, T.L., Jr., and LaBonte, B.J. 1988, *Astrophys. J.*, **335**, 1015.

Braun, D.C., LaBonte, B.J., and Duvall, T.L., Jr. 1990, *Astrophys. J.*, in press.

Campbell, W.R., and Roberts, B. 1989, *Astrophys. J.*, **338**, 538.

Chitre, S.M., and Davila, J.M. 1989, *Proc. IAU Symp. No. 142*, in press.

Fisher, G.H., Canfield, R.C., and McClymont, A.N. 1985, *Astrophys. J.*, **289**, 425.

Haber, D.A., Toomre, J., and Hill, F. 1988a, in *Advances in Helio- and Asteroseismology, IAU Symp. No. 123*, eds. J. Christensen-Dalsgaard and S. Frandsen (Reidel, Dordrecht), p.59.

Haber, D.A., Toomre, J., Hill, F., and Gough, D.O. 1988b, in *Seismology of the Sun and Sun-like Stars*, ed. E.J. Rolfe, ESA SP-286 (ESA Publication Division. Noordwijk), p.301.

Hollweg, J.V. 1988, *Astrophys. J.*, **335**, 1005.

Jefferies, S.M., Pomerantz, M.A., Duvall, T.L. Jr., and Harvey, J.W. 1990, these proceedings.

Jefferies, S.M., Pomerantz, M.A., Duvall, T.L., Jr., Harvey, J.W., and Jaksha, D.B. 1988, in *Seismology of the Sun and Sun-like Stars*, ed. E.J. Rolfe, ESA SP-286 (ESA Publication Division. Noordwijk), p.279.

Knölker, M., and Bogdan, T.J. 1988, in *Seismology of the Sun and Sun-like Stars*, ed. E.J. Rolfe, ESA SP-286 (ESA Publication Division. Noordwijk), p.265.

Lou, Y.-Q. 1988, in *Seismology of the Sun and Sun-like Stars*, ed. E.J. Rolfe, ESA SP-286 (ESA Publication Division. Noordwijk), p.305.

Lou, Y.-Q. 1989, preprint.

Wolff, C.L. 1972, *Astrophys. J.*, **176**, 88.

Frequencies, Linewidths, and Splittings of Low-Degree Solar p-Modes

P. L. Pallé, C. Régulo, and T. Roca Cortés

Instituto de Astrofísica de Canarias, 38200 La Laguna, Tenerife, Spain

Abstract: The measurement of frequencies, linewidths, and splittings of low degree solar p-modes is a difficult task due, firstly, to time scales involved and, secondly, to the spurious presence of side lobes when data is not collected continuously. However, a long set of observations have been obtained at Observatorio del Teide: a total of 52 months of data, spread over 12 years (1977 to 1989), from which the best ones have been selected. These data allows averaging of power spectra, therefore giving statistical significant profiles for the p-modes, which can be fitted to appropriate functions (i.e. Lorentzian). Such analysis is applied to the data mentioned before and preliminary results of the aforementioned parameters are obtained.

1 Introduction

Line profile determinations in the solar oscillations acoustic spectrum, are of great interest since it provides a wealth of information about the physics of the modes. The uncertainties in such a measurement are both due to the background sources of power (solar, atmospheric, instrumental, etc.) and to the modes itself. As a consequence, a particular mode of oscillation does not appear in the Fourier spectrum as one single peak but as a cluster of many of them. Increasing the length of the observations (the intrinsic frequency resolution), just increases the number of spikes and the diurnal interruptions cause the appearence of ghost peaks called sidebands (at $\pm 11.57 \mu$Hz).

Besides these difficulties, the problem of line profile parameters estimation has received previous and extensive attention but mainly applied to high degree ($l > 10$) solar oscillations (Libbrecht 1986, Duvall and Harvey 1986, Brown 1988). In this paper, such a determination is applied to integral light Doppler measurements of solar oscillations, which are sensitive to modes of $l \leq 3$. Data have been extensively collected since 1976 with a resonant scattering spectrometer (Brookes *et al.* 1978) installed at Observatorio del Teide (Izaña, Tenerife) and its high quality and length (mainly from 1980 on), provide a good base to start with. The final goal is to improve our own estimations of mode frequencies, amplitudes, and linewidth, made with more crude techniques, by fitting some appropriate functions deduced from a given model of the oscillations. Moreover, having data extended over more than one solar activity cycle, the comparison of the obtained results in opposite phases of activity, will be rather independent of the model itself and/or the techniques used, therefore providing indications of any cycle dependence of the parameters obtained.

2 Method

The method applied consists on fitting appropriate functions to the features present in the frequency spectrum of the solar oscillations. The length of time series used to obtain the spectrum, the use of single power spectrum or average of many of them and, the type of functions to fit will depend on the assumptions to be made on the physics of the solar oscillations. First of all, we assume a stochastic nature for the solar p-modes, in such a way that their profile is similar to the one corresponding to a freely decaying dampd oscillator. For weak damping and for frequencies near to the resonance one, ν_0, the input power (or dissipated power) of this oscillator is proportional to $L(\nu)$, being

$$L(\nu) = \frac{(\Gamma/2)^2}{(\nu - \nu_0)^2 + (\Gamma/2)^2} \tag{1}$$

which is a Lorentzian profile, with Γ being the full width at half maximum (FWHM) and ν_0 the mode frequency. A more realistic profile should have a contribution from the background power due to the various sources of noise. In the vicinity of each mode, this can be assumed to be a constant and therefore the observed profile can be described as:

$$P(\nu) = P_0 L(\nu) + \eta \tag{2}$$

with P_0 the maximum signal power and η the background power. Equation (2) can be fitted to real data using non-linear fitting methods and solved for the coefficients $(a_i, i = 1, \ldots, 4)$ involved: P_0, ν_0, Γ and η. In this work, a standard non-linear least squares method (Marquart Method) has been used, which requires a initial guess for the parameters and minimizes the merit function $\chi^2(a_i)$.

Once the type of functions to fit is defined, the next item is select the length of the data to use for power spectrum computation. For the lower degree solar p-modes in the range 2 to 4 mHz, it has been shown (Jefferies *et al.*,1988; Elsworth *et al.*, 1990) that the lifetime (1/width of the mode) varies between 1 month and some days. Therefore, monthly power spectra has been computed as maximum length, leaving open the possibility to average many of them. Now, the strategy for the fitting procedure should be selected. In our case we proceed as follows: select the central part of the p-mode spectrum (2 to 4 mHz) and split it in 80 μHz portions each one containing a pair l=0, 2 or a pair l=1, 3 together with their first order side bands (high and low); in this way 13 subsets for the pair l=0,2 and 10 for the l=1,3 are obtained. For a given (0,2) or (1,3) subset, the rotational splitting in $l+1$ components is assumed for the modes and also for their side bands. In total, we expect 12 peaks in a (0,2) region $(3 * [l_0, l_{2,m=-2}, l_{2,m=0}, l_{2,m=+2}])$ and 14 peaks in a (1,3) region; in the latter case, the low frequency side bands of the l=3 mode are not been considered.

The reason for this procedure is related with the adopted strategy; that is, fit in each group the sum of 12 or 14 Lorentzian at once, with initial values for P_0 , ν_0 (from Pallé *et al.*, 1986), Γ (a constant value of 0.5 μHz) and η (calculated as explained in Pallé et al, these proceedings). The interference of the sidebands and peaks will then be taken into account since the side bands structure is also fitted. In Fig. 1, examples of the (0,2) and (1,3) groups are shown in solid line.

Therefore, for a given group, we will finally obtain the value of the parameters for the solar p–modes and for their side bands.

First tests of the method on real data gave us satisfactory results in terms of fast convergence of the fit (typically less than 15 iterations) and estimation of the parameters; the only exception is for the $l=3$ modes, probably due to their poorer signal to noise ratio (less than 2).

3 Application to Real Data

As explained above, this method was applied to full disk solar measurements of oscillations made at Tenerife since 1979 up to 1989. To avoid the variations in the solar oscillations spectrum throughout the solar activity cycle (Pallé *et al.* 1989; Pallé *et al.*, these proceedings), only data obtained during the years of maximum (1980, 81, 82, and 88) or minimum (1985, 86, and 87) solar activity phases, have been considered in the present analysis. Then, 14 power spectra are obtained for maximum activity phase and 16 for the minimum; this data is the one that will be used to perform the line profile measurements. In principle, because of the statistical nature of the excitation mechanism for solar p-modes, one expects smooth profiles when average of many power spectra is made. This will greatly help the goodness of the fit, but also can induce over-estimation in the FWHM and more uncertainties in P_0 and ν_0 determinations.

To quantify this effect, single monthly spectra and also average of them, were made. When comparing, in each case, the result of the fit to a given frequency interval, we find:

a) For $l=0$ modes, an over-estimation of 0.2 μHz in Γ, as we go from one month power spectra to average of all of them. This over-estimation is less than the intrinsic resolution of the spectrum (0.4 μHz). Then we conclude that for $l=0$, is a good approach to fit the averaged power spectrum of the monthly power spectra.

b) For $l > 0$, the over-estimation is of the order or larger than 0.4 μHz, mainly due to the rotational split components of the modes, away from the central one some 0.8 μHz that, when averaged will produce a much wider peak. Uncertainties in ν_0 and P_0 are also bigger than before. As a conclusion, for $l > 0$ monthly power spectra should ideally be used.

Even knowing this, in this work we have only used the average of 14 monthly power spectra obtained during maximum solar activity and 16 corresponding to the minimum, to determine line profile parameters.

4 Results

With the method and the data shown before, different groups of modes $(l, l+2)$ were selected in the spectral range of 2 to 4 mHz: 14 groups (0,2) and 10 groups (1,3) obtained both at maximum and minimum were fitted. Figure 1-a shows an (0,2) group corresponding to $n=20$ and 19, respectively, and in Fig. 1-b one (1,3) corresponding to $n=19$ and 18, respectively, (solid lines), together with the result of the fit (dashed lines).

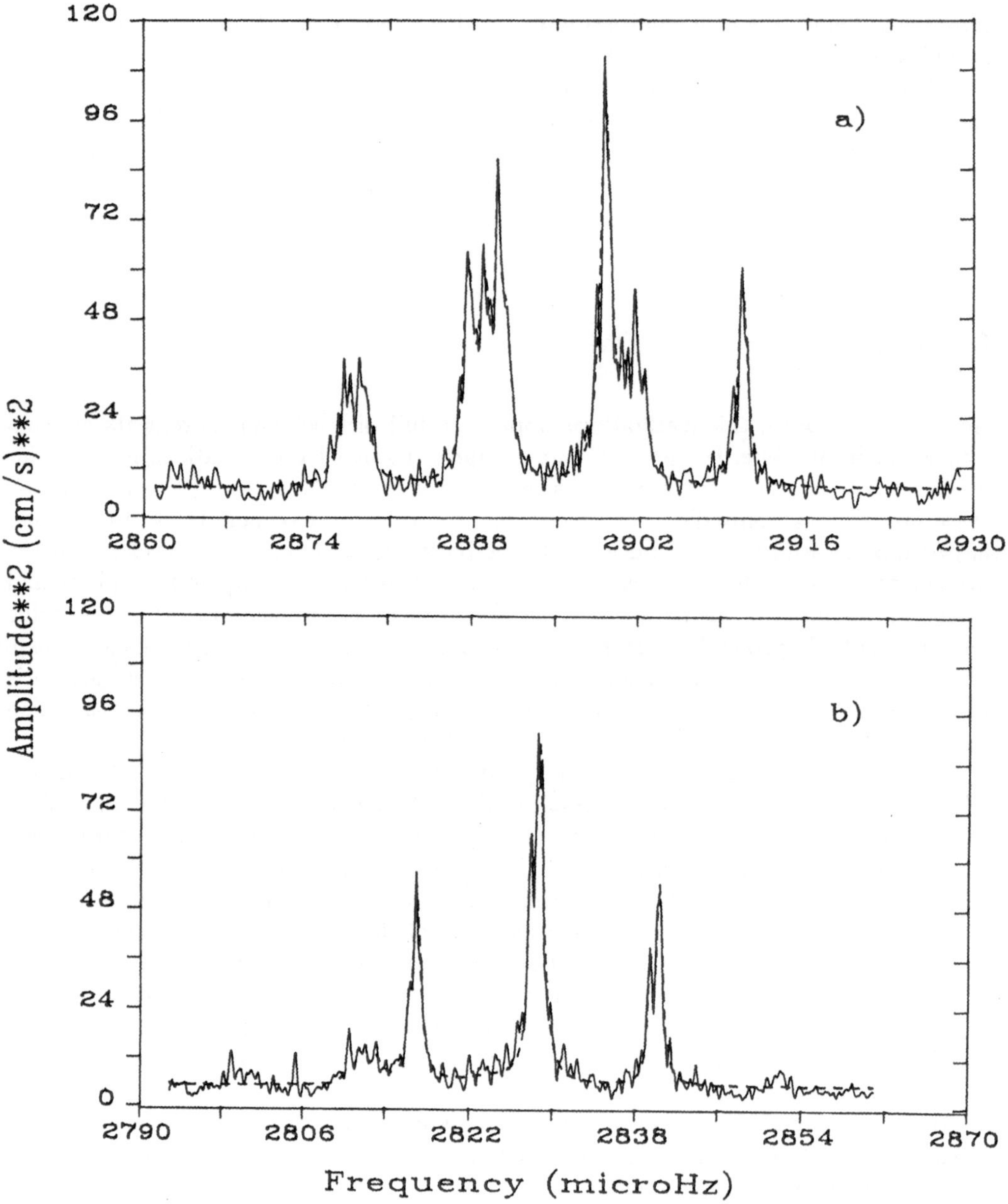

Fig. 1. Samples of the power spectra used to determine the solar p–modes line profiles. In a) a pair l=0 and l=2, together with their side bands, are shown. In b) a pair l=1 and l=3. In solid line the actual power spectrum and in dashed line the result of the fit to it.

As one can see, the number of Lorentzians simultaneously fitted in each group (12 and 14 respectively) fits rather well the averaged spectrum; also the estimation of the noise power (considered constant within each interval) is satisfactory.

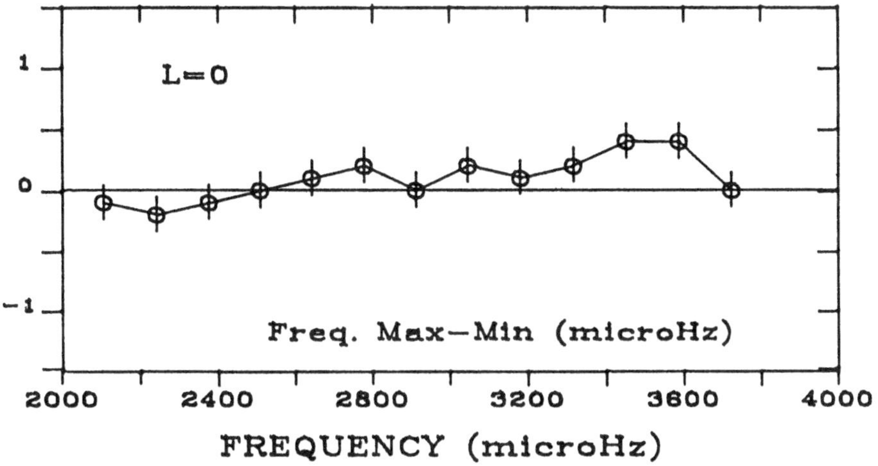

Fig. 2. Frequency differences of l=0 p–modes between maximum and minimum of the activity cycle.

In Fig. 2, the values of ν_0 obtained for the fitted modes of l=0 (from n=14 to n=26) in the maximum and minimum of activity, were subtracted and plotted. The differences, as a function of the frequency, show a slightly increasing trend towards high frequency. The mean value of all differences in less than 0.1 μHz, which is in agreement with previous results of Pallé *et al.* (1989) obtained using cross-correlation techniques.

Concerning the FWHM of the l=0 modes, and bearing in mind the over-estimation explained before, the results obtained with the present technique are also compatible with those of Elsworth *et al.* (1990), which uses a subset of the data used here and a different technique. As shown in Fig. 3, the FWHM is lower than 1 μHz for frequencies between 2 and 3.3 mHz and increases up to 2.5 μHz at 3.8 mHz. This figure also suggests a variation of the linewidth with the activity cycle; it seems to be some 30% higher in the maximum (at least in the frequency range from 2 to 3 mHz).

The results obtained for P_0 are in perfect agreement with the one obtained by Pallé *et al.* (1989), which shows an increase of some 30% in P_0 as the activity cycle goes from maximum to minimum.

In spite of the uncertainties introduced when averaging power spectra for l=1 modes, the fit to the mean power spectrum has been performed. The frequencies of the $m = \pm 1$, l=1 modes have been obtained both, at maximum and minimum. For each mode the difference between the frequencies of the m components is calculated and this is some measure of the rotational splitting; in fact, in this case, is twice the rotation rate. In Fig. 4, these values are plotted for both activity phases. The absolute values (0.8 to 1.2 μHz) are not discussed here because of limitations of the method explained before, but the differences are rather independent of the method and, if real, they would suggest a

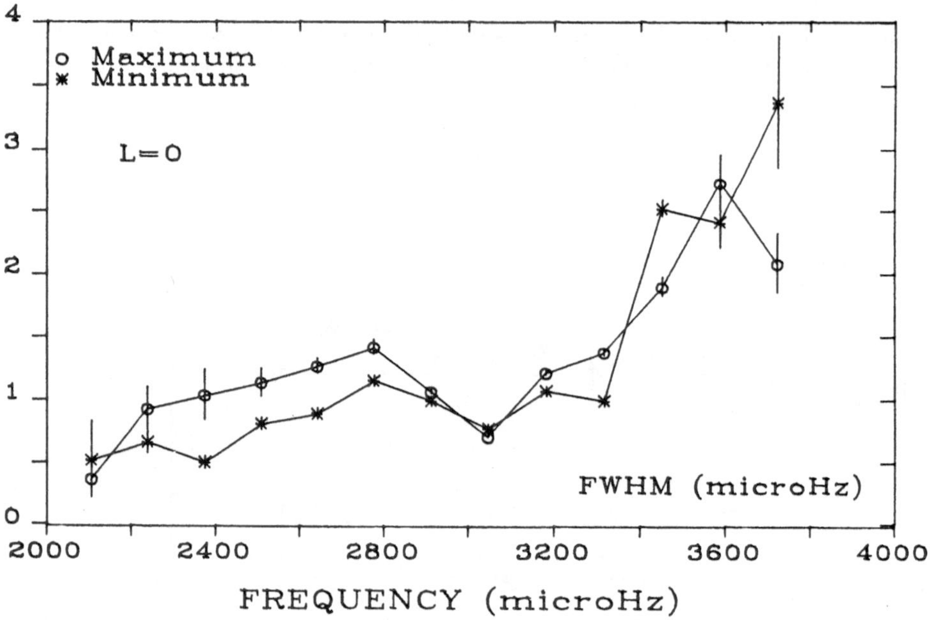

Fig. 3. Linewidths of the l=0 solar p–modes in the maximum and minimum of the activity cycle.

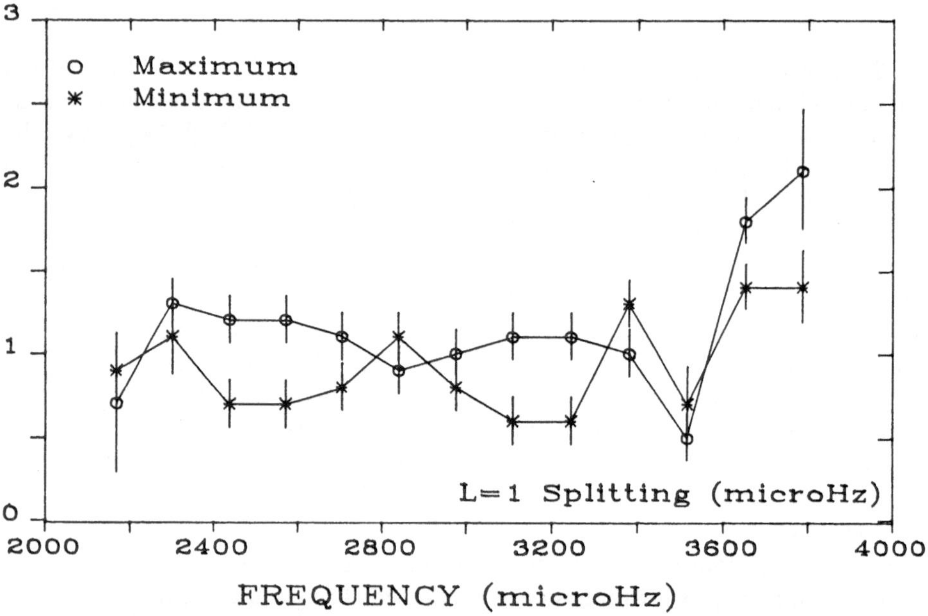

Fig. 4. Rotational splitting of the l=1 solar p–modes, measured as the difference between the $m = \pm 1$ components, for maximum and minimum of the activity cycle.

change in the rotation rate in the interior of the Sun along the activity cycle being faster in the maximum phase of activity.

The overall results obtained by this method suggest us that, at least for the $l=0$ modes it is an improvement of what was made before for line profile measurements. The method and the strategy of the fit still need some important improvements to cover all the $l < 4$ range. The work that is going on now proceeds in this direction as well as the similar kind of treatment has been used and studied by Duvall and Harvey (1986) and Anderson *et al.* (1990).

References

Anderson, E.R., Duvall, T.L., Jr., and Jefferies, S.M. 1990, submitted to *Astrophys. J.*

Brookes, J.R., Isaak, G.R., and van der Raay, H.B. 1978, *Monthly Notices Roy. Astron. Soc.*, **185**, 1.

Brown, T.M. 1988, in *Proc. IAU Symp. No. 123, Advances in Helio- and Asteroseismology*, eds. J. Christensen-Dalsgaard and S. Frandsen (Reidel, Dordrecht), p.491.

Duvall, T.L, Jr., and Harvey, J.W. 1986, in *Seismology of the Sun and distant Stars*, ed. D.O. Gough (Reidel, Dordrecht), p.105.

Elsworth, Y., Isaak, G.R., Jefferies, S.M., McLeod, C.P., New, R., Pallé, P.L., Régulo, C., Roca Cortés, T. 1990, *Monthly Notices Roy. Astron. Soc.*, **242**, 135.

Jefferies, S.M., Pallé, P.L., van der Raay, H.B., Régulo, C., and Roca Cortés, T. 1988, *Nature*, **333**, 646.

Libbrecht, K.G. 1986, *Nature*, **319**, 753.

Pallé, P.L., Pérez, J.C., Régulo, C., Roca Cortés, T., Isaak, G.R., McLeod, C.P., and van der Raay, H.B. 1986, *Astron. Astrophys.*, **170**, 114.

Pallé, P.L., Régulo, C., and Roca Cortés, T. 1989, *Astron. Astrophys.*, **224**, 253.

Pallé, P.L., Régulo, C., and Roca Cortés, T. 1990, These proceedings.

An Experiment to Measure the Solar $\ell = 1$ Rotational Frequency Splitting

A. Cacciani [1], E. Paverani [1], D. Ricci [1], P. Rosati [1],

R. M. Marquedant [2], E. J. Smith [2], and S. Tomczyk [3]

[1]Physics Department, University of Rome "LA SAPIENZA",
2 Aldo Moro, 00185 Rome, Italy,
[2]Jet Propulsion Laboratory, Pasadena, CA 91109, USA
[3]HAO/NCAR, Boulder, CO 80307, USA

Abstract: To date, only integrated light experiments have attained the high signal-to-noise ratio and frequency resolution necessary to measure the rotational frequency splitting of low degree solar p-modes. These experiments, however, are limited by the finite mode linewidths coupled with the inability of non-imaging experiments to unambiguously separate prograde and retrograde modes. In particular, the separation of the prograde and retrograde mode frequencies of the very important $\ell = 1$ spherical harmonic, dictates that the experiment have the capability to coarsely resolve the eastern from the western hemisphere of the solar disk. Initial attempts to attain the desired image resolution by masking the solar image at the focal plane of the telescope and chopping the two hemispheres on the detector have been unsuccessful due to the high velocity noise introduced by the solar rotation through image motions and guiding instabilities. In this paper we present the concept of what we call "spectroscopic masking," which provides the ability to filter oscillation modes spectroscopically, and without the need to image the Sun. This results in an optical configuration which is insensitive to image motions and guiding errors while still providing adequate spatial resolution to separate prograde and retrograde $\ell = 1$ modes. A conceptual study will be presented along with a test observing run showing the quality of the achievable data.

1 Introduction

One important application of helioseismology as a probe of the solar interior is to measure its internal rotation as a function of depth and latitude. High and intermediate degree modes provide information about the outer half of the sun while very low degree modes are candidate to bring information about the rotation of deeper layers. In particular the mode $\ell = 1$ is the lowest one that can provide rotational splitting data to probe the very core of the sun.

Various technologies have been employed to obtain high quality measurements of the solar rotational frequency splitting at high and intermediate degrees. These include spectrographs (Duvall and Harvey, 1984), interferometers (Brown and Morrow, 1987), magneto-optical filters (Tomczyk *et al.*, 1988), bi-refringent filters (Libbrecht, 1988), and aereal photometers (Duvall *et al.*, 1986). For reasons not obvious, all of these technologies have failed to provide conclusive measurements of the frequency splitting at low degree ($\ell < 5$).

To date, only integrated light experiments, primarily based on resonant scattering spectrometers (Grec *et al.*, 1980; Claverie *et al.*, 1981), have attained data of sufficient quality and frequency resolution to measure the solar rotational frequency splitting at these low degrees. These experiments, however, are limited by the finite mode linewidths coupled with the inability of non-imaging experiments to unambiguously separate prograde and retrograde modes. This has prevented these experiments from providing a conclusive measurement of low degree frequency splittings.

The ability to separate prograde and retrograde modes requires some imaging capability. In particular, the separation for mode frequencies of the $\ell = 1$ spherical harmonic dictates that the experiment have the capability to coarsely resolve the eastern from the western hemisphere of the solar disk.

Attempts have been made to attain image resolution with integrated light experiments (Cacciani *et al.*, 1981; Henning and Scherrer, 1986; Pallé *et al.*, 1989) by spatially filtering the solar image at the focal plane of the telescope. The goal of these studies was not to measure frequency splittings, however, but to shift the sensitivity of the experiments to higher ℓ values. The principal difficulty connected with the spatial filtering technique as it pertains to an $\ell = 1$ frequency splitting measurement, arises from the high velocity noise introduced by the solar rotation through image motion and guiding instabilities.

2 Our Approach

2.1 Imaging Capability with a Non-imaging Detector

Here we present a non-imaging experiment which can provide the information necessary to separate prograde and retrograde $\ell = 1$ modes. Its main advantages are simplicity and an insensitivity to image motion and guiding errors. Also, this technique can be applied to stars other than the sun.

The spectroscopic device employed for this experiment is a Magneto-Optical Filter (MOF) in a velocity measurement mode (Cacciani and Fofi, 1978). The MOF provides two narrow and tunable transmission bands that can be chopped alternately on the Red and Blue wings of a solar line. The MOF is a very stable instrument and has a high optical efficiency. Although the MOF has been used extensively as a doppler imaging device, in our $\ell = 1$ experiment we are using a single detector followed by a lock-in amplifier.

2.2 The 'Spectroscopic Masking': Concept and Simulations

The concept for this experiment was conceived while operating the MOF in an imaging mode. Under suitable MOF conditions and for each one of its two bandpasses, the doppler signals due to supergranulation, oscillations, and rotation are visible clearly only in one of the two solar hemispheres (western or eastern), while the usual algorithm to compute doppler maps, that is the difference of the the Red and Blue wings divided by their sum (Δ/Σ), continues to show fair linearity across the solar equator.

The reason for this behaviour (that we call 'spectroscopic masking') is due to the large doppler shifts caused by the solar rotation which tune different parts of the solar line into each of the two narrow spectral transmission windows of the MOF. In particular, the MOF can be adjusted in such a way that for each hemisphere, one transmission window transmits the core of the solar line (insensitive to doppler shifts) while the other window transmits a wing, and vice versa.

The idea is to take advantage of this 'spectroscopic masking' in order to obtain the desired image resolution so that separate power spectra for prograde and retrograde $\ell = 1$ modes can be computed. The important fact now is that $\ell = 1$, m=$\pm$1 modes show up through Σ even when the sun is observed from position A, as well as through Δ when the sun, 1/4 rotation earlier or later, is observed from position B (see Fig. 1). In the last case both $\ell = 1$ and $\ell = 0$ are measured through Δ and are distinguishable only from their frequencies. In general the observing direction is intermediate between A and B so that both Δ and Σ bring their contribution to the measurement.

Initial numerical simulations of the MOF response to $\ell = 0$ and $\ell = 1$, m=$\pm$1 spherical harmonics confirm the feasibility of such an experiment. A detailed report of these simulations will be given in a later paper.

2.3 Correct Interpretation of the Experimental Signal

Δ/Σ is what we get directly as an output from our existing set up. It consists of a two cell version of the MOF which alternates on a PMT the Red and the Blue wings of the solar line. The PMT is followed by a logarithmic amplifier before going to a lock-in amplifier, so that the final output is the fractional variation of the transmitted average intensity (Δ/Σ). Consider also that

$$\Delta \propto [R + S_0 + \Re(S_1)] \tag{1}$$

and

$$\Sigma \propto [N + \Im(S_1)] \tag{2}$$

where

R is the earth rotation,

S_0 and S_1 are the perturbing signals due to $\ell = 0$ and $\ell = 1$ spherical harmonics, respectively,

$\Re$ and $\Im$ are the Real and Imaginary part of S_1 at $\pi/2$ on the sun (B and A in Fig. 1), respectively,

N is the unperturbed normalization factor, that is the DC level of Σ,

so that developing to the first order to linearize the ratio Δ/Σ we have:

$$\Delta/\Sigma = R/N + S_0/N + \Re(S_1/N) - R/N \cdot \Im(S_1/N). \tag{3}$$

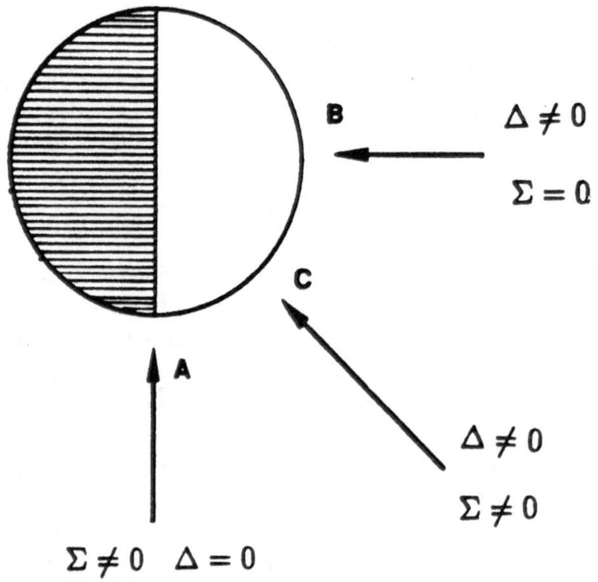

Fig. 1. Response of Σ = Blue + Red and Δ = Blue−Red to $\ell = 1$, m=±1.

The oscillatory signals S_0, S_1, and the earth rotation trend R, come out correctly normalized, and the last term in Eq. (3) is due to the spectroscopic masking technique; it also assures us that when the sun shows side A we still can have a signal from $\ell = 1$ modes.

However, in order to be able to compute individual power spectra for prograde and retrograde modes, we need to separate the two functions $\Re(S_1/N)$ and $\Im(S_1/N)$ as inputs for the FFT algorithm. In other words we need a second independent and simultaneous determination of Δ/Σ. We can also measure separately Δ and Σ, or Red and Blue, with reference to the continuum intensity.

2.4 Practical Accomplishment

As we will show at the end of this paper, the quality of our data is satisfactorily good, so we may want to use an identical experimental set up for this purpose. It has the advantage of giving directly in real time the clean ratio Δ/Σ as output from a lock-in amplifier.

With reference to Fig. 2, which is the two channel version of our existing equipment, consider the polarizing beam splitter PB before the PMT tube. Since at that position the light beam is circularly polarized, this gives two channels that are identical unless a retardation plate is put in front of it. If this retardation is 1/4 wave at 45 degrees, the signal will be split in such a way that the Red part of the line will go always to one channel while the Blue part goes to the other. When the electro-optical modulator is turned on, each PMT receives an on-off signal: the first one follows the sequence Red...Zero...Red...Zero...etc. while the second one follows the sequence Zero...Blue...Zero...Blue...etc.

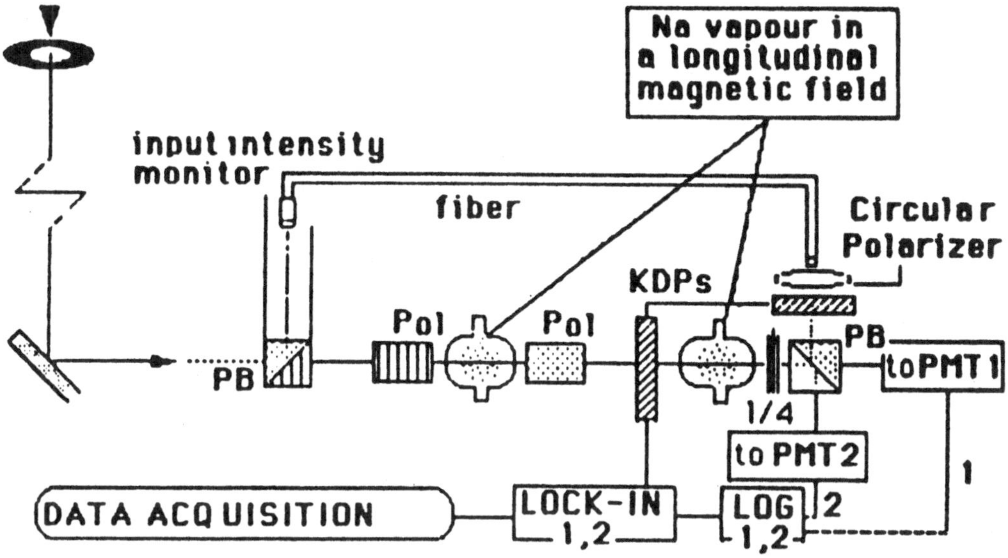

Fig. 2. Optical layout of the two channel system.

At this point, a good way to introduce the reference signal is to feed the zero phases of the sequences with light from the continuum, of a suitable amount so that the sequence will be an alternation Red...Cont...Red...etc. in one channel and Cont...Blue...Cont...etc. in the other. A logarithmic output from the PMTs can give the wanted normalized quantities Red/Cont and Blue/Cont as final output from the lock-in amplifiers.

Feeding the system with the continuum is not easy in practice. In our case we enter the MOF with a polarizer. This can have another polarizing beam-splitter PB in front of it so that one side is available to pick up the continuum signal using a fiber optic. The other end of the fiber can go straight to one side of the polarizing beam splitter in front of the PMTs. An additional circular polarizer and a second KDP will alternate the continuum signal in the desired way.

Other optical schemes are possible that can avoid the additional KDP and the fiber. One possibility is to sample the modulated signal without using a lock-in amplifier and measure the continuum signal separately. This alternative method is being tested presently at HAO.

3 Solar Oscillation Data from a Single Channel: Conclusions

Figure 3a shows a power spectrum obtained with an integrated light MOF with no spectroscopic masking. The data were taken on 67 days during the summer of 1989 at JPL in Pasadena. The spectral resolution is high enough to say that $\ell = 1$ lines are definitely more structured than $\ell = 0$ ones, suggesting the presence of a large rotational splitting (see the comparison in Fig. 3b between the 2496.5, $\ell = 0$, and the 2560.5 μHz, $\ell = 1$, lines).

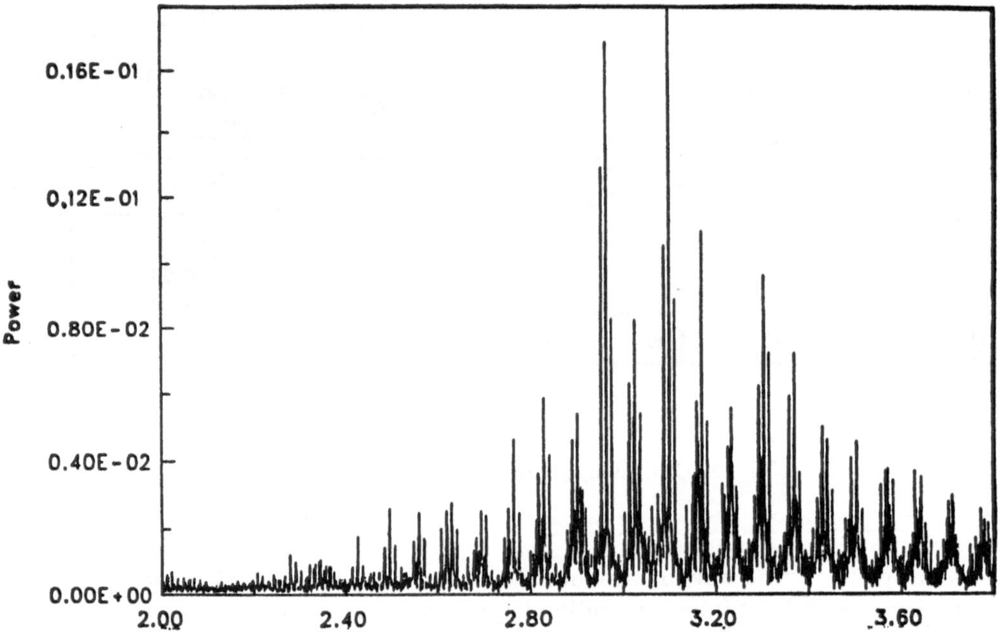

Fig. 3a. 67 days single channel power spectrum.

A quantitative determination of this splitting will be greatly facilitated by separating the prograde and retrograde power spectra and observing long enough to attain the necessary signal-to-noise ratio. A careful analysis and data reduction will be performed in the future when, if funds are available, the spectroscopic masking experiment is completed.

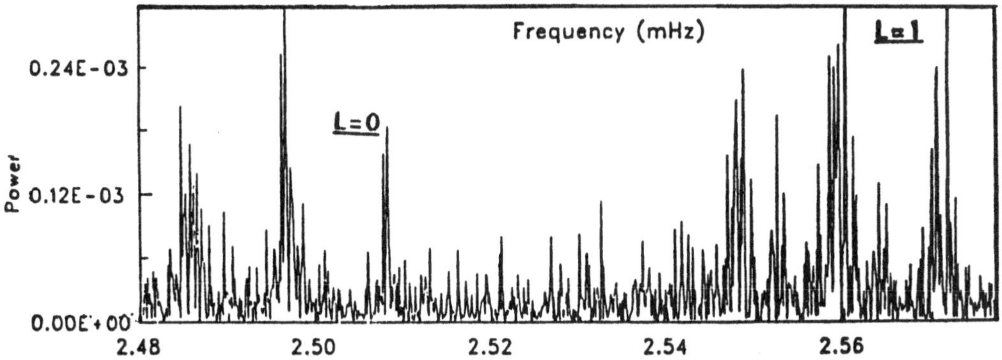

Fig. 3b. Expanded portion of the power spectrum.

References

Brown, T.M., and Morrow, C.A. 1987, in *The Internal Solar Angular Velocity*, eds. B.R. Durney and S. Sophia (Reidel, Dordrecht), p.7.

Cacciani, A., Croce, V., Fortini, T., and Torelli, M. 1981, *Solar Phys.*, **74**, 543.

Cacciani, A., and Fofi, M. 1978, *Solar Phys.*, **59**, 179.

Claverie, A., Isaak, G.R., McLeod, C.P., van der Raay, H.B., and Roca Cortés, T. 1981, *Nature*, **293**, 443.

Duvall, T.L., Jr., and Harvey, J.W. 1984, *Nature*, **310**, 19.

Duvall, T.L., Jr., Harvey, J.W., and Pomerantz, M. 1986, *Nature*, **321**, 500.

Grec, G., Fossat, E., and Pomerantz, M. 1980, *Nature*, **288**, 541.

Henning, H.M., and Scherrer, P.H. 1986, in *Seismology of the Sun and the Distant Stars*, ed. D.O. Gough (Reidel, Dordrecht), p.55.

Libbrecht, K.G., 1988, in *Seismology of the Sun and Sun-like Stars*, ed. E. Rolfe, ESA SP-286 (ESA Publication Division, Noordwijk), p.131.

Pallé, P.L., Pérez Hernández, F., Roca Cortés, T., and Isaak, G.R. 1989, *Astron. Astrophys.*, **216**, 253.

Tomczyk, S., Cacciani, A., Korzennik, S.G., Rhodes, E.J., Jr., and Ulrich, R.K. 1988, in *Seismology of the Sun and Sun-like Stars*, ed. E. Rolfe, ESA SP-286 (ESA Publication Division, Noordwijk), p.141.

Phase Relation between Velocities and Temperature Fluctuations of the Solar 5-Minute Oscillation

K. Ichimoto, S. Hamana, K. Kumagai, T. Sakurai, and E. Hiei

National Astronomical Observatory, Mitaka, Tokyo 181, Japan

Abstract: Phase relations between the velocities and temperature oscillations in the solar photosphere are investigated on the k-ω diagram. Distributions of the phase differences on the k-ω plane are roughly reproduced by a simple analytical model, but the detailed fitting is not satisfactory. In the 5-minute band, temperature reaches its peak when the atmosphere is moving downward. The amount of the phase difference between temperature and velocity suggests the radiative damping time of 1-40 s. Identification of the g-mode oscillation is not clear.

1 Introduction

Although 5-minute oscillation of the sun is most conspicuous in the velocity field, it is known that other observational quantities like the intensity or the width of absorption lines also show a 5-minute variation. These fluctuations are caused by the change of temperature and density in the solar atmosphere. The phase relations between velocities and temperature or between velocities at different depths are expected to give us important clues concerning (1) the energy transport by the waves in the solar atmosphere, (2) the radiative damping of the waves, (3) the existence of the g-mode oscillation, etc. So far a number of authors have studied the phase relations between the velocities and the intensity fluctuations.

In this paper, by using a time series of spectra taken on the solar disk, we derived a temperature fluctuation in the lower photosphere. The phase differences between the velocities and the temperature oscillations are investigated on the k-ω diagram. Interpretations will then be made with an analytical model calculation.

2 Observation and Data Reduction

Time series of the spectra including lines of FeI 5434 Å and MnI 5432 Å was obtained with the CCD camera installed on the 25-cm coude-type coronagraph at the Norikura Solar Observatory. The entrance slit of the spectrograph was placed along the solar equator, where the field of view in the east-west direction was 9.3′. Solar images were averaged over 5′ in the north-south direction by rotating a prism in front of the slit. Thus 512 frames of spectra were recorded during 256 minutes.

Typical equivalent widths of MnI 5432 Å and FeI 5434 Å lines are 46 mÅ and 184 mÅ respectively, and their line cores are formed in the lower and upper photosphere. From the line profile of MnI 5432 Å, we determined the velocity, equivalent width and continuum intensity. We also determined the velocity from the core of FeI 5434 Å line. It is noted that the equivalent width of MnI 5432 Å line is not sensitive to the adopted value of microturbulence velocity because its line absorption coefficient is intrinsically broadened by the hyperfine splitting. Hence this line is suitable for the investigation of temperature and density variations (Elste 1986). An LTE calculation of line profile shows that the continuum intensity and the equivalent width have different dependences on the temperature and the density fluctuations in the solar atmosphere; continuum intensity is relatively sensitive to the change of the density, while the equivalent width is relatively sensitive to the change of the temperature. After examining these dependences, we can derive the temperature fluctuation from the observed two quantities. In this way, we obtained the time variations of the velocity in the upper photosphere, and of the velocity, temperature and density in the lower photosphere.

3 Power Spectra

Figure 1 shows the k-ω diagram obtained from the velocity and the temperature variations. The solid lines show the theoretical boundaries between propagating and evanescent modes and between evanescent and gravity modes for $T = 5500$ K, respectively. The dashed curve indicates the relation $\omega^2 = gk$, which nearly traces the f-mode of oscillation. Both diagrams show the well known ridges of power corresponding to different n values. It is noted, however, that $n = 1$ ridge, which is clear in the velocity power, is almost absent in the power of the temperature variation. As a whole the ratio between the temperature and the velocity amplitudes increases with the increasing frequency in the 5-minute band. This tendency may be understood qualitatively by the fact that the wave compressibility is reduced when the frequency approaches that of the f-mode. The k-dependence of the temperature-velocity power ratio is not clearly seen in out data.

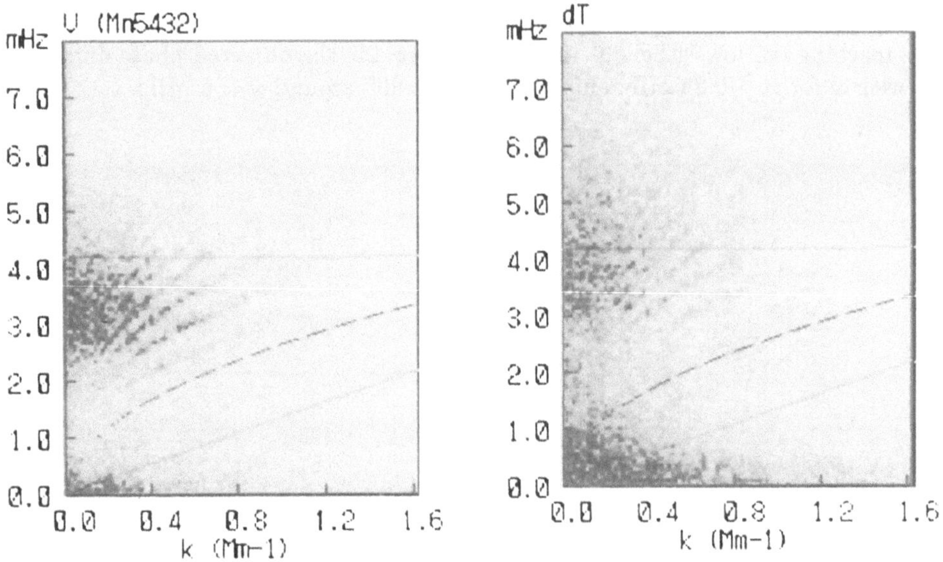

Fig. 1. Power spectra of the velocity derived from MnI 5432 Å (left) and of the temperature variation (right). Temperature variation is obtained from the equivalent width of MnI 5432 Å and the continuum intensity. The solid lines show the theoretical boundaries between propagating and evanescent modes (around 4mHz) and between evanescent and gravity modes for $T = 5500$ K, respectively. The dashed curve shows the relation $\omega^2 = gk$.

4 Phase Relations

The observed phase difference between the two velocities (derived from Mn 5432 Å and FeI 5434 Å) are shown on the k-ω plane in Fig. 2a. In the range where the k-ω diagram shows a significant power of the 5-minute oscillation, the phase difference increases gradually with an increasing ω. This confirms the theoretical prediction that the wave changes its characteristic from the evanescent to the propagating mode as the frequency increases. Existence of g-mode oscillation, which would be identified by the negative phase difference at the large k and the small ω region (Deubner and Fleck 1989), is not clear in our observation.

Figure 2b shows the velocity-temperature phase difference spectrum. The downward velocity is taken as positive, and the phase difference of 90° means that the maximum temperature occurs when the displacement of the atmosphere reaches the lowest height. The distribution of the phase difference can be characterized by (1) a gradual increase with increasing ω for $\omega > 2.5$ mHz and $k < 0.8$ Mm^{-1}, (2) a rather quick change from $-80°$ to $30°$ at $\omega = 2 - 3$ mHz, (3) weak k-dependence of the frequency at which this change occurs. The data for $k > 0.8$ Mm^{-1} may not be reliable due to poor S/N ratio.

Figure 3 shows the vertical cross section of Fig. 2b for $0.083 < k < 0.40$ Mm^{-1}. In the 5-minute band, the velocity variation proceeds that of the temperature about 35°, i.e. the temperature reaches the peak value when the atmosphere is moving downward; 55° before reaching the lowest height. As shown in Fig. 2b, the observed phase difference increases with ω for $\omega > 0.25$ mHz and then exceeds 90° around $\omega = 6$ mHz.

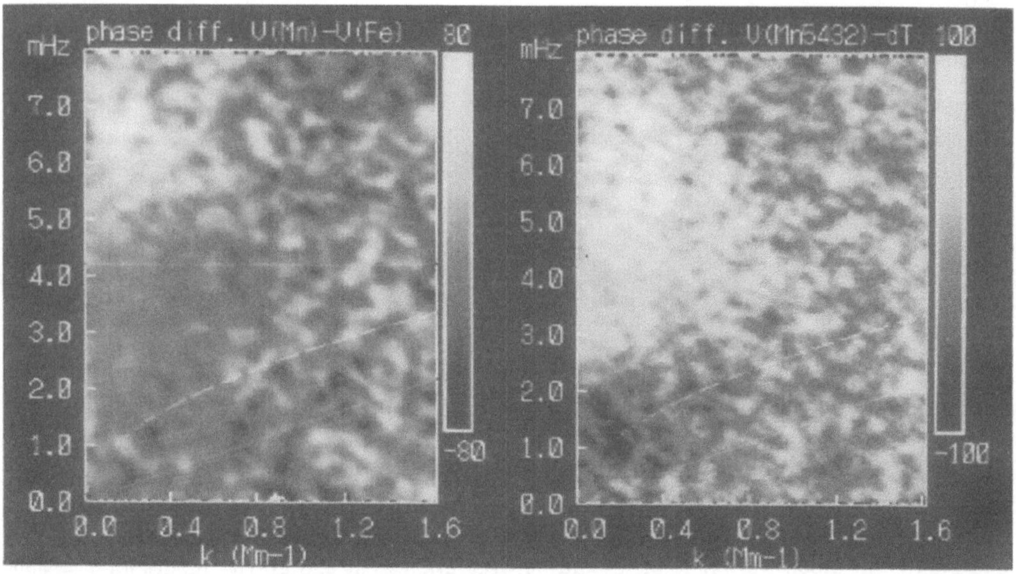

Fig. 2a. Phase difference between the velocities derived from MnI 5432 Å and FeI 5434 Å. Solid and dashed curves are the same as those in Fig. 1.

Fig. 2b. Phase difference between the velocity and temperature variations.

To interpret the observed results, we made a theoretical calculation of the phase differences according to Souffrin's (1972) formula. Isothermal atmosphere with $T = 5500$ K is assumed. The thin curves in Fig. 3 show the calculated phase differences for $k = 0.4$ Mm^{-1}, where each curve corresponds to different radiative damping time scale. A discrete change occurs at the f-mode frequency (1.66mHz) where the amplitude of the temperature variation becomes zero. Above this frequency, the phase difference increases gradually with increasing ω. It means the change of the wave characteristics from the evanescent to the propagating mode. The observed phase difference of 35° in the 5-minute band suggests the radiative damping time scale of 1–40 s at the lower photosphere. As a whole the observed distribution of the phase difference seems to be reproduced by the theoretical calculation, but the detailed fitting is not satisfactory. Probably both the noise in the observation and the idealization in the model cause the discrepancy.

It is noted that, because the temperature reaches the maximum when the atmosphere is moving downward, the wave must carry the energy downward. It may cause a "mechanical cooling" of the temperature minimum layer.

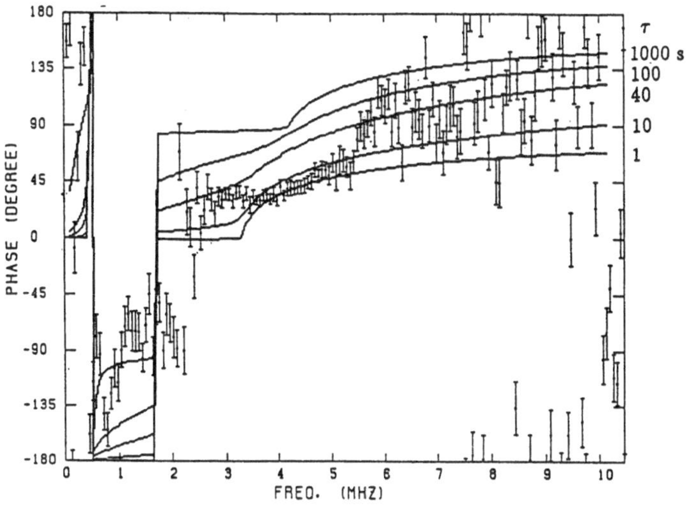

Fig. 3. Velocity-temperature phase difference against ω. The observed results for $0.08 < k < 0.40\,\mathrm{Mm}^{-1}$ are averaged. Solid curves show theoretical phase differences for $k = 0.4\ \mathrm{Mm}^{-1}$ and $T = 5500\,\mathrm{K}$ with different relaxation time.

References

Deubner, F.-L., and Fleck, B. 1989, *Astron. Astrophys.*, **213**, 423.

Elste, G. 1986, *Solar Phys.*, **107**, 47.

Souffrin, P. 1972, *Astron. Astrophys.*, **17**, 458.

Influence of Photospheric 5-Minute Oscillations on the Formation of Chromospheric Fine Structures

Yoshinori Suematsu

National Astronomical Observatory, Mitaka, Tokyo 181, Japan

Abstract: We present a basic idea how 5-minute oscillations can disturb the solar chromosphere significantly and show some results of numerical simulations in which the idea was examined.

It has been well known that 5-min oscillations are the surface manifestation of acoustic modes which are trapped in the convection zone. The upper side of two reflection layers for this trap is located just below the photosphere where the pressure scale height becomes much less than the wavelength of the acoustic waves of about 5 min period. In other words, the acoustic cutoff frequency there is smaller than the frequencies of these waves. The waves whose frequencies are smaller than the cutoff frequency become evanescent; the photosphere moves up and down as a whole.

The acoustic cutoff frequency ω_c is given by

$$\omega_c = \frac{c}{2\Lambda}, \tag{1}$$

where c is the sound velocity and Λ is the pressure scale height. In the photosphere, this frequency is about $0.03\ \mathrm{s}^{-1}$ which corresponds to a period of about 200 s.

The situation, however, should be modified when strong magnetic fields exist in the atmosphere (Michalitsanos 1973). When the plasma β (the ratio of gas pressure to magnetic pressure) is less than unity, the acoustic waves are considered to be MHD slow-mode waves which propagate almost parallel to the magnetic field lines. When the field lines are inclined from the vertical, the pressure scale length along the field lines can become larger than the scale height of non-magnetic atmosphere and the wavelength of the waves of 5 min period.

For a rough approximation, we will assume that magnetic flux tubes are rigid; the plasma β is much less than unity, and that the slow-modes are regarded as acoustic waves. In this situation, the acoustic cutoff frequency ω_{tc} in inclined flux tubes is modified to

$$\omega_{tc} = f\omega_c \cos\theta, \tag{2}$$

where θ is the inclination angle from the vertical. For an isothermal atmosphere, the factor f is given by

$$f^2 = \left(\frac{H_p}{H_A}\right)^2 + \left(2 - \frac{4}{\gamma}\right)\left(\frac{H_p}{H_A}\right) + 1, \tag{3}$$

where H_p and H_A are the pressure scale length and the scale length of cross section along the flux tube, respectively, and γ is the ratio of specific heats. We used a relation $H_p = \Lambda/\cos\theta$ in deriving (2). Roberts and Webb (1978) have derived the cutoff frequency for the vertical flux tubes. The factor f is very close to unity, because the ratio H_p/H_A is estimated to be much less than unity in the solar atmosphere. Therefore the cutoff frequency is determined solely from the inclination angle. In order that the acoustic wave of 5 min period propagates along the flux tubes in the photosphere, the inclination angle must exceed 50° because of $\cos\theta = \omega_{tc}/\omega_c < 200/300$.

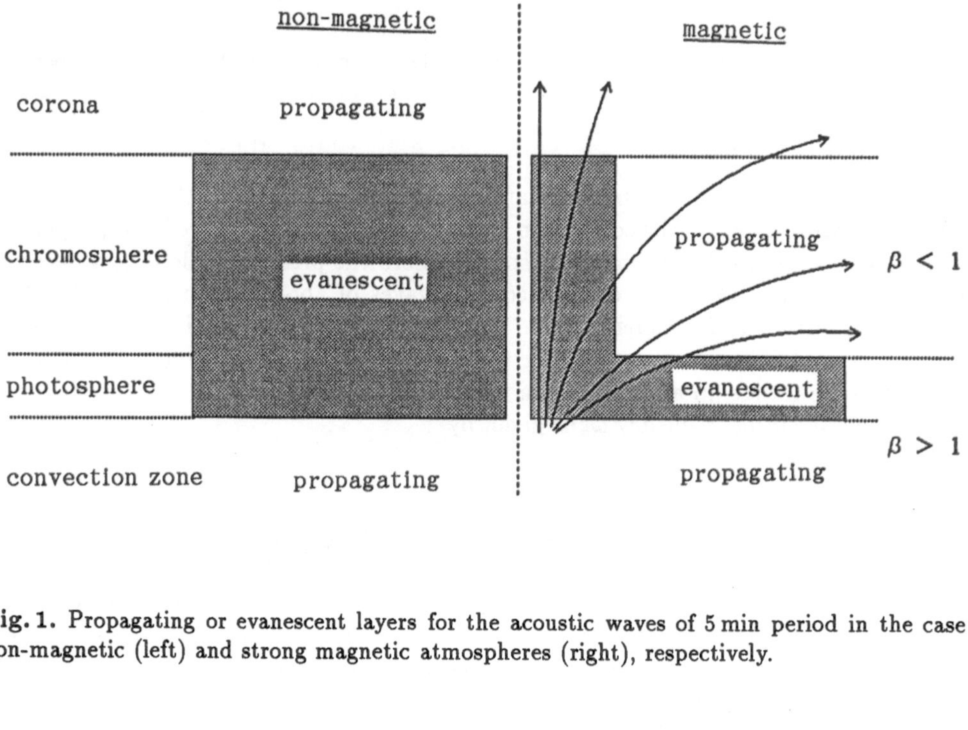

Fig. 1. Propagating or evanescent layers for the acoustic waves of 5 min period in the case of non-magnetic (left) and strong magnetic atmospheres (right), respectively.

Observations seem to suggest that the plasma β is nearly unity and the field lines are nearly vertical in the lower photosphere. This implies that the acoustic wave of 5 min period is still evanescent there. Above the upper photosphere, however, magnetic fields spread out to hold the pressure equilibrium. Then it is possible that some field lines take the plasma β less than unity and are inclined more than 50° from the vertical. Hence the width of the evanescent layer is very thin for these field lines (Fig. 1). At the upper photosphere, a possible acoustic energy flux $\rho v^2 c$ would be about 1.7×10^7 erg cm^{-2} s^{-1}, if we take $\rho = 10^{-8}$ g cm^{-3}, $v = 0.5$ km s^{-1} and $c = 7$ km s^{-1}. Hence the chromosphere and corona would be disturbed significantly, if the evanescent waves could drive propagating waves efficiently along the filed lines of plasma β less than unity and of the inclination angle larger than 50°.

It is very interesting to investigate the response of the upper atmosphere embedded in strong magnetic fields to the 5-min oscillations, because this might be related to the formation of chromospheric fine structures such as spicules. Recently some numerical simulations showed that slow-mode shocks could be responsible for such fine structures (Hollweg 1982; Hollweg *et al.* 1982; Suematsu *et al.* 1982; Suematsu 1985; Sterling and Hollweg 1989). These simulations, however, used photospheric perturbations whose periods are smaller than the acoustic cutoff period of the photosphere and the origin of the perturbations were not so obvious.

Here we studied non-linear responses of the upper atmosphere embedded in magnetic fields to 5-min oscillations using a one-dimensional hydrodynamic code. In the calculations, we considered the atmospheric regions from the convection zone to the corona. It was assumed that the magnetic field is vertical in the convection zone and is potential above the photosphere, having an open field configuration; the field lines spread out rapidly in the chromosphere and become nearly vertical in the corona. The magnetic field strength in the corona is as small as about 1% of the photospheric field. The calculations were made along each magnetic flux tube which consists of a small bundle of field lines. We neglected non-adiabatic effects.

The 5-min oscillations were generated in consequence of pressure perturbations at a depth of $-1500\,\mathrm{km}$ in the convection zone, in order to simulate the observed photospheric 5-min oscillations. Maximum velocity amplitudes in the photosphere were around $0.5\,\mathrm{km\,s^{-1}}$.

Some results of the calculations are shown in Fig. 2. We found that no appreciable chromospheric motions are generated along nearly vertical flux tubes, because the acoustic waves of 5 min period cannot propagate through the photosphere as expected from the above theory. The acoustic waves become standing waves in the convection zone and 3 min oscillations appear in the chromosphere. The same results have been obtained by Leibacher *et al.* (1982) who studied 5-min oscillations in a purely vertical geometry from the convection zone to the chromosphere.

On the other hand, along the flux tubes of inclination angle of about 50° at a chromospheric level, the waves of about 5 min period were generated and grew to a train of shock waves. The shocks were so strong that the chromosphere was forced to oscillate in the period of about 5 min, increasing its average height. After 15 min, the chromosphere-corona transition region reached up to a height of about 8000 km in the corona.

The waves of about 5 min period were also generated along the flux tubes nearly horizontal in the chromosphere, but they could not grow to shocks strong enough to lift the chromospheric material high up into the corona. The reason is that the shocks must propagate for a long distance before they arrive at the corona. Sterling and Hollweg (1989) have obtained similar results. They investigated atmospheric motions due to short period photospheric perturbations in the flux tube which is vertical in the photosphere, horizontal in the chromosphere, and vertical again in the corona.

It seems that the field inclination is very crucial to driving appreciable chromospheric motions due to the 5-min oscillations; its angle must be just around 50° in the chromosphere. It should be noted, however, that the inclination angles in the corona are not so crucial, although they would affect an ultimate height of the transition region.

In conclusion, 5-min oscillations will yield thin elongated structures, showing up-and-down motions in the corona, only along the magnetic field lines whose inclination angles are around 50° in the chromosphere and only when the plasma β there is much less than

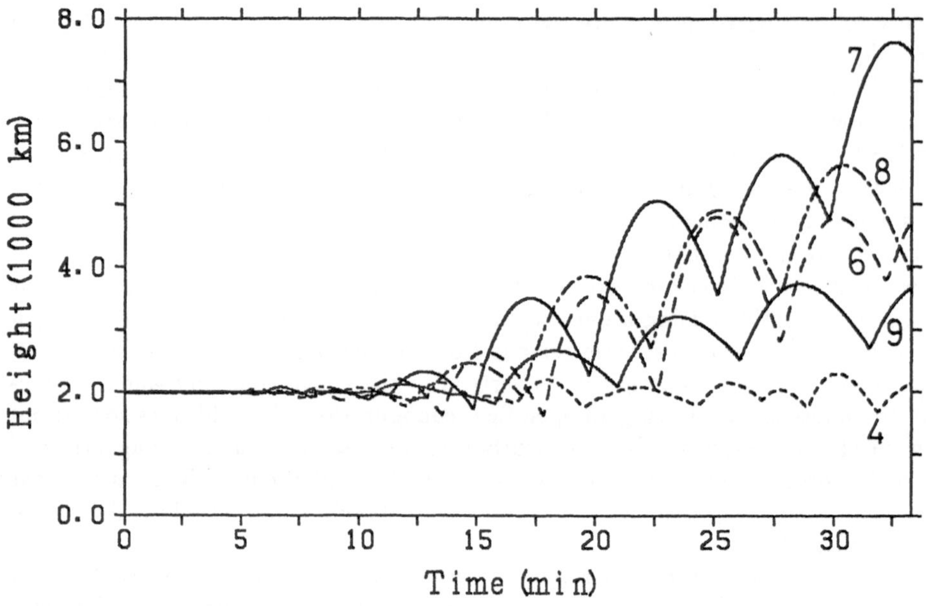

Fig. 2. Lagrangian motions of the fluid parcels initially located near the chromosphere-corona transition region. The ordinate indicates the vertical height from the level of $\tau_{5000} = 1$. The numerals denote the flux tubes; the inclination angles at the height of 1000 km are 27°, 45°, 53°, 62°, and 70°, for the tubes 4, 6, 7, 8, and 9, in this order.

unity. Such field configurations would possibly be realized at supergranulation boundaries. Furthermore physical conditions of the thin elongated structures should be quite similar to those of the chromosphere. Hence the 5-min oscillations are likely to contribute to the formation of the chromospheric fine structures such as mottles or spicules observed at the supergranulation boundaries.

References

Hollweg, J.V. 1982, *Astrophys. J.*, **257**, 345.
Hollweg, J.V., Jackson, S., and Galloway, D. 1982, *Solar Phys.*, **75**, 35.
Michalitsanos, A.G. 1973, *Solar Phys.*, **30**, 47.
Leibacher J., Gouttebroze, P., and Stein, R.F. 1982, *Astrophys. J.*, **258**, 393.
Roberts, B., and Webb, A.R. 1978, *Solar Phys.*, **56**, 5.
Sterling, A.C., and Hollweg, J.V. 1989, *Astrophys. J.*, **343**, 985.
Suematsu, Y. 1985, *Solar Phys.*, **98**, 67.
Suematsu, Y., Shibata, K., Nishikawa, T., and Kitai, R. 1982, *Solar Phys.*, **75**, 99.

p-Mode Analysis of the IPHIR Data

Claus Fröhlich [1] and Thierry Toutain [2]

[1]Physikalisch-Meteorologisches Observatorium Davos, World Radiation Center, CH-7260 Davos-Dorf, Switzerland
[2]Institut d'Astrophysique Spatiale, F-91370 Verrières-le-Buisson, France

Abstract: The results of the IPHIR experiment on the USSR planetary mission to Phobos presented here are from data gathered during 160 days of the cruise phase of PHOBOS II to Mars, launched on 12 July 1988. The long uninterrupted observation produces a spectrum of the solar p-mode oscillations in the 5-minute range with a very high signal-to-noise ratio. Frequency and line shape determination is limited by the lifetime of the modes and the 'noise' from stochastic excitation. The temporal variation of the amplitudes of $l = 0 \ldots 2$ is discussed.

1 Introduction

The experiment IPHIR (*I*nter*P*lanetary *H*elioseismology by *IR*radiance measurements) uses a very simple sunphotometer (SPM) to observe the global Sun in three wavelength channels (at 335, 500, and 865 nm). It was realized by an international consortium lead by PMOD/WRC and consisting of the Laboratoire de Physique Stellaire et Planétaire (now Institut d'Astrophysique Spatiale, France), the Space Science Department (SSD of ESA), the Crimean Astrophysical Observatory (KrAO, USSR) and the Central Research Institute for Physics (CIRP, Hungary). A description of IPHIR can be found in Fröhlich *et al.* (1988) and the first results of the p-mode analysis have been presented in Fröhlich *et al.* (1990).

2 Data Evaluation

The data coverage is excellent with virtually no gaps due to transmission or operations during the 161 days of cruise phase from 14 July to 22 December 1988. The evaluation procedure has been given in Fröhlich *et al.* (1990) where also the method for the treatment of the unexpected pointing effect is described in detail. By that method the time series can be smoothed to such a degree, that the remaining noise in the power spectrum from 1.5 to 6 cmHz is well below the ppm range and most probably of solar origin as shown in Fig. 1.

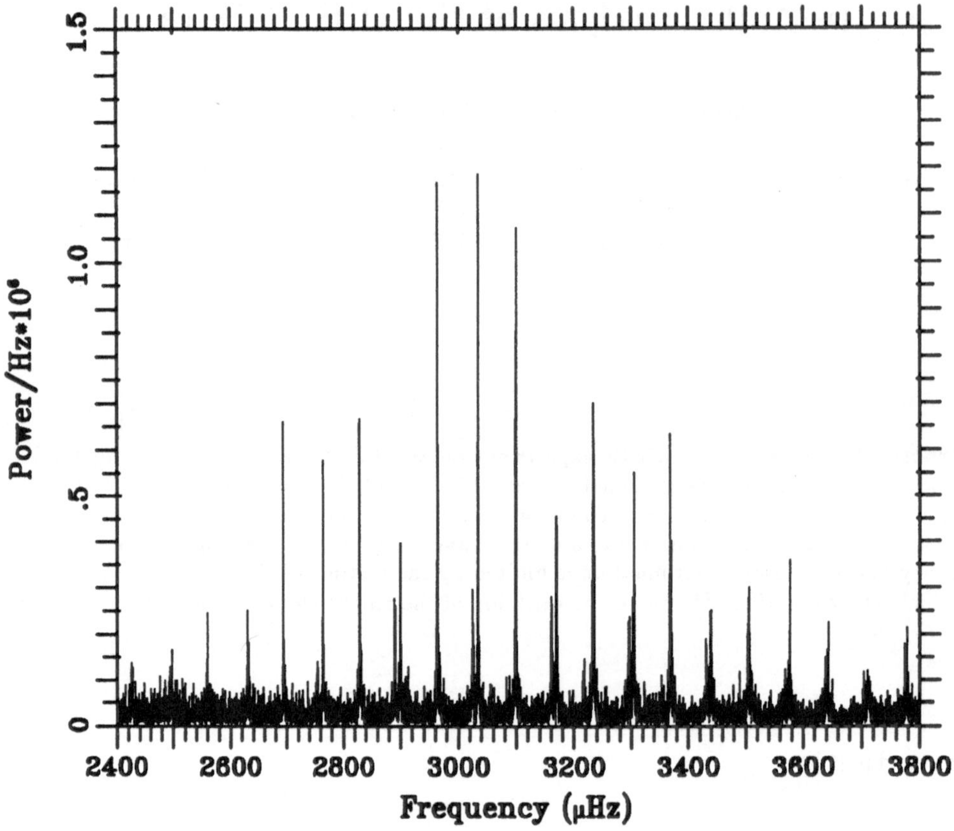

Fig. 1. Power spectrum of the low degree p-modes from 160 days of the IPHIR experiment.

3 Frequency and Line-Shape Analysis

From the expanded plots of the individual lines (top plots of Fig. 2) it becomes clear that even with the statistics of 160 days the lines are not well defined and the least-squares fitting of Lorentzians may yield ambiguous results. The difference between these results and formerly published ones (e.g. Woodard and Hudson 1983, Elsworth *et al.* 1988) may be that our spectra are cleaner due to the long uninterrupted time series of IPHIR whereas the others are influenced by observational gaps and hence redistribution of line power into aliases. The appearance of the lines, however, is what might be expected from stochastic excitation where peaks are randomly distributed around the centroid frequency and the highest peak is not necessarily located at the center (Lazrek *et al.* 1988).

For the superposition of the $l = 0, 1$, $n = 18, ..., 23$ lines (bottom panels of Fig. 2) quadratic fits of the form

$$\nu_{l,n} = a \cdot (n + l/2 - 21.5)^2 + b \cdot (n + l/2 - 21.5) + c \tag{1}$$

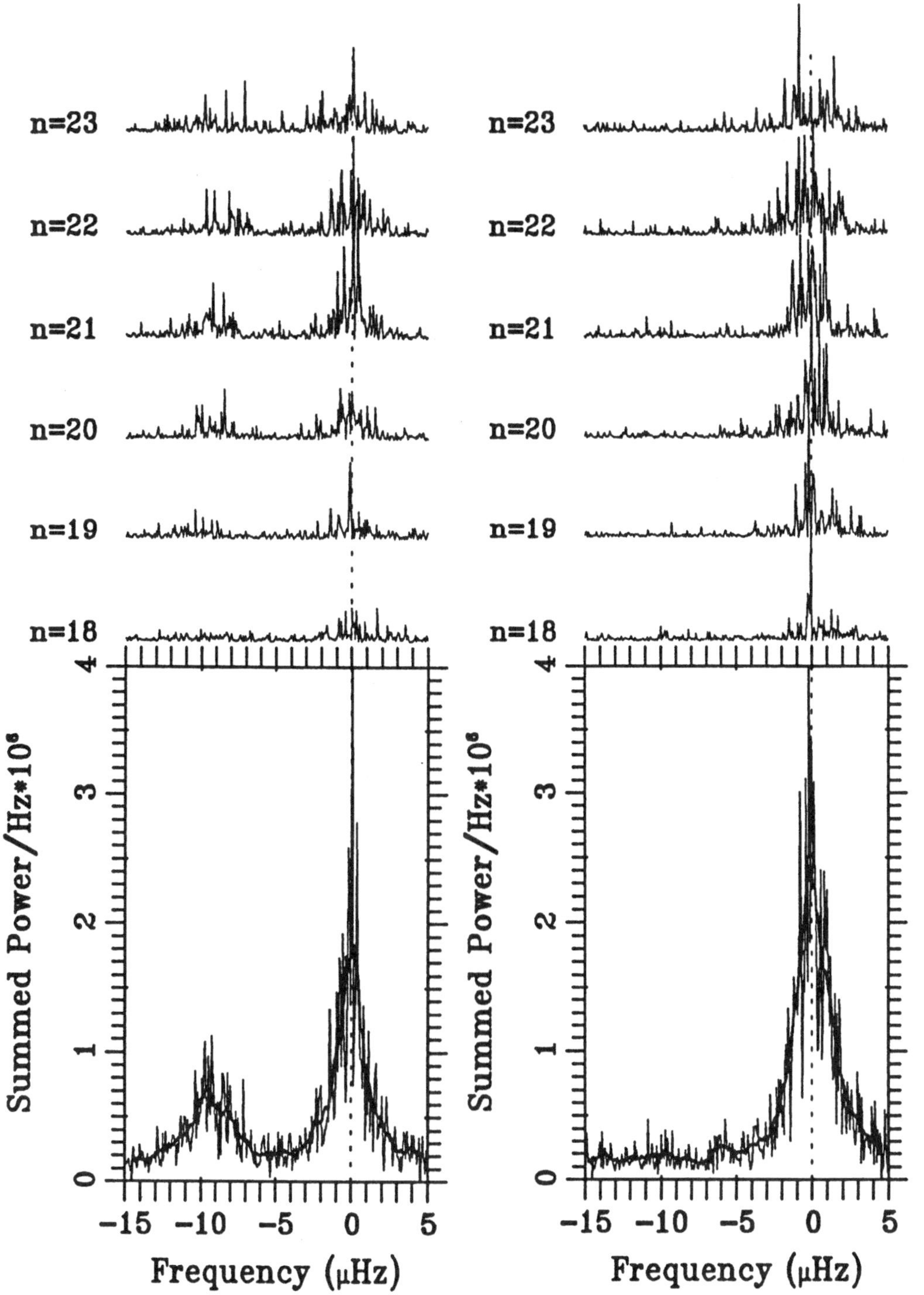

Fig. 2. Superposition of the $l = 0, 1$ p-modes from 160 days of the IPHIR experiment. Top panels show the individual lines for $n = 18, ..., 23$ and bottom panels the superposition. The zero corresponds for each line to the frequency value determined by the parabolic fit.

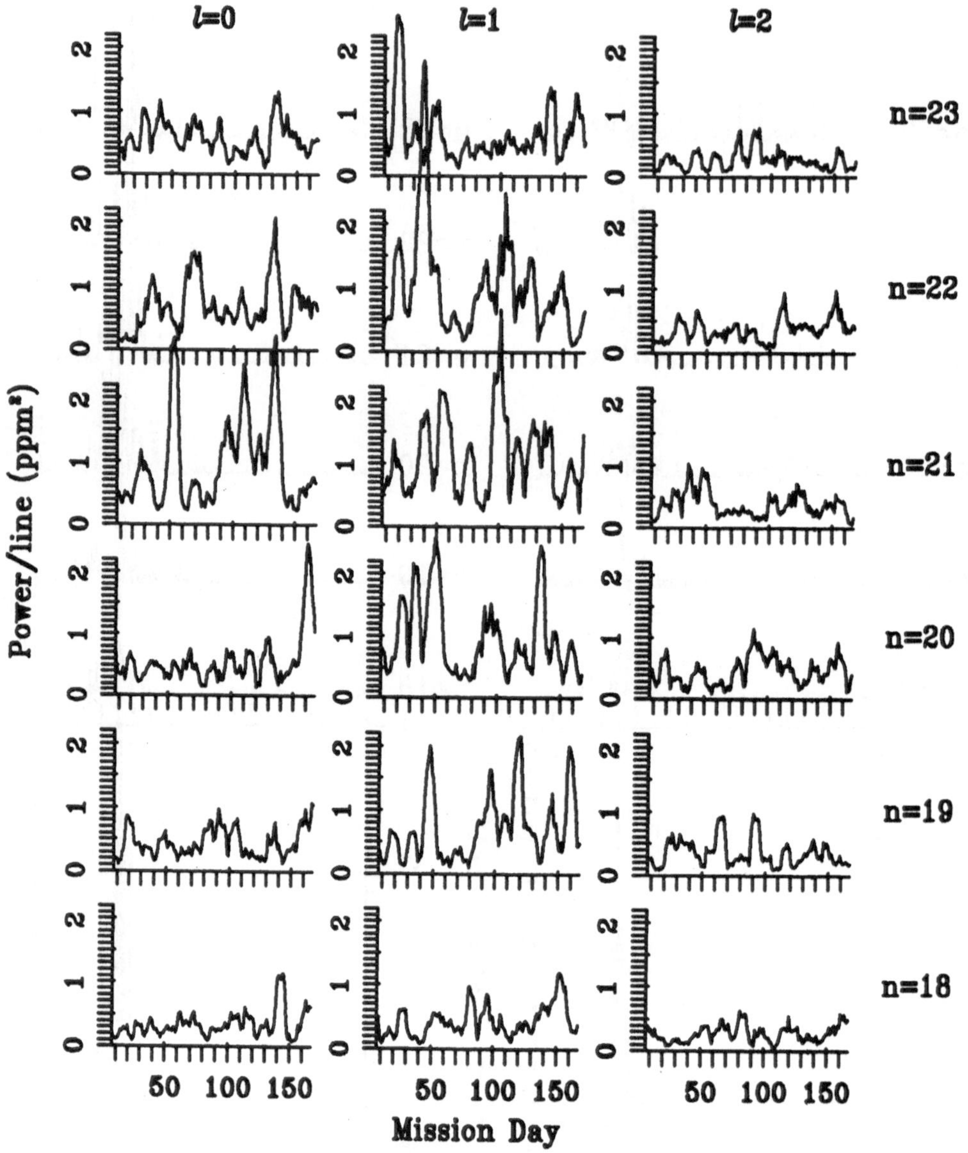

Fig. 3. Time Variation of p-mode line power for $l = 0$, 1, and 2 and $n = 18, ..., 23$.

are performed. The constants a, b and c and the standard deviation of the fit are summarized in the following:

	a	b	c	σ
$l = 0$	0.0741	135.067	3101.53	0.09
$l = 1$	0.0052	135.145	3098.57	0.10

As input the frequencies from Lorentz-fitting have been used in contrast to the table given in Fröhlich *et al.* (1990) where the mean value of the Lorentz-center and the one determined as bary-center have been given. Information about the line width is scarce: the individual lines are too noisy possibly due to excitation statistics and the superposition is further influenced by the increase of the width with increasing n. From the superposed lines the average width for $l = 0$ amounts to 1.5 μHz and for the $l = 1$ doublet to 2.0 μHz. These values are mainly determined by the highest n and compare well with the widths of Woodard and Hudson (1983) and Elsworth *et al.* (1988). In order to asses the widths of the individual lines a more adequate fitting should be used (other than just least-squares) by taking into account the probability of the distribution of the peaks around the centroid due to excitation. This would not only improve the reliability of the fitted Lorentzians, but also allow to estimate accurate uncertainties of the centroid frequencies and widths.

The separation $\delta_{2,1} = \nu_{0,n+1} - \nu_{2,n}$ is calculated from the $l = 0, 2$ and $n = 19, ..., 23$ frequencies by fitting a straight line as function of frequency. δ amounts to 9.33 μHz at 3.0 mHz and exhibits a slope of -2.2μHz/mHz. This compares well with 9.30 μHz (Woodard and Hudson 1983) from ACRIM (Active Cavity Radiometer on the Solar Maximum Mission Satellite, Willson 1982) and is somewhat higher than 9.04 μHz calculated from the observations at Tenerife (Jiménes *et al.* 1988). Indeed, the separation $\delta_{2,1}$ seems to be slightly lower than predicted by a standard solar model (9.54 μHz, Christensen-Dalsgaard 1986).

4 Time Variation of p-Mode Line-Power

Linked to the excitation is also the variability of the p-mode amplitudes with time. Figure 3 shows the line-power variation over the 160 days calculated as 5-day running spectra which look quite erratic with no obvious correlation between the different l and n. It is interesting to note that the $l = 1$ modes show much more variation than the $l = 0$ modes, which could be an indication of beating between the unresolved $m = \pm 1$ modes (unresolved due to their lifetime). It seems, however, that there is some relation between the number of peaks in the power spectrum of a line and the number of peaks in its time series of the power. Unfortunately the $l = 2$ modes are too weak, and thus too much influenced by noise, that conclusions about their behavior could not be drawn. Comparison of the p-mode amplitude variation with the simultaneous variation of solar activity parameters – 10.7 cm flux, projected sunspot index (PSI) and the total irradiance from ACRIM in Fig. 4 – does not reveal any obvious correlation. If the excitation is by turbulent convection this result is not unexpected as the convection itself is not changed on the time scale of months.

Acknowledgments

Without the continuous efforts of the IPHIR team – scientific and technical – and the efficient co-operation with the Space Science Institute at Moscow during the preparation of the experiment and the mission this work would not have been possible. T.Toutain performed this work during a stay at the Institut d'Astrophysique Spatiale which is

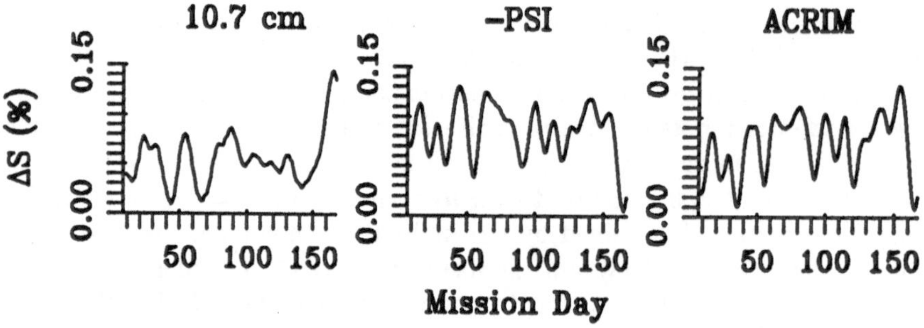

Fig. 4. Time Variation of the 10.7 cm flux, the Projected Sunspot Index (PSI, plotted negative to simulate the irradiance change) and the total solar irradiance from ACRIM.

gratefully acknowledged. Thanks are extended to the Swiss National Science Foundation and the Centre National d'Etudes Spatiales for financial support of the experiment.

References

Christensen-Dalsgaard, J. 1986, in *Seismology of the Sun and Distant Stars*, ed. by D.O.Gough (Reidel, Dordrecht), p.23.

Elsworth, Y., Isaak, G.R., Jefferies, S.M., McLeod, C.P., New, R., Pallé, P.L., Régulo, C., and Roca Cortés, T. 1988, in *Seismology of the Sun and Sun-like Stars*, ed. E. Rolfe, ESA SP-286 (ESA Publication Division, Noordwijk), p.27.

Fröhlich, C., Bonnet, R.M., Bruns, A.V., Delaboudinière, J.P., Domingo, V., Kotov, V.A., Kollath, Z., Rashkovsky, D.N., Toutain, T., Vial, J.C., and Wehrli, Ch. 1988, in *Seismology of the Sun and Sun-like Stars*, ed. E. Rolfe, ESA SP-286 (ESA Publication Division, Noordwijk), p.359.

Fröhlich, C., Toutain, T., Bonnet, R.M., Bruns, A.V., Delaboudinière, J.P., Domingo, V., Kotov, V.A., Kollath, Z., Rashkovsky, D.N., Vial, J.C., and Wehrli, Ch. 1990, in *Proc. IAU Colloq. No. 121, Inside the Sun*, ed. G. Berthomieux and M. Cribier (Klewer, Dordrecht), p.279.

Jiménez, A., Pallé, P.L., Pérez, J.C., Régulo, C., Roca Cortés, T., Isaak, G.R., McLeod, C.P., and van der Raay, H.B. 1988, in *Proc. IAU Symp. No. 123, Advances in Helio- and Asteroseismology*, ed. J. Christensen-Dalsgaard and S. Frandsen, (Reidel, Dordrecht), p.205.

Lazrek, M, Delache, P., and Fossat, E. 1988, in *Seismology of the Sun and Sun-like Stars*, ed. E. Rolfe, ESA SP-286 (ESA Publication Division, Noordwijk), p.673.

Willson, R.C. 1982, *Solar Phys.*, **74**, 217.

Woodard, M., and Hudson, H.S. 1983, *Nature*, **305**, 589.

Search for g-Modes in the IPHIR Data

Claus Fröhlich

Physikalisch-Meteorologisches Observatorium Davos, World Radiation Center, CH-7260 Davos-Dorf, Switzerland

Abstract: The results of the IPHIR experiment on the USSR planetary mission to Phobos presented here are from data gathered during 160 days of the cruise phase of PHOBOS II to Mars, launched on 12 July 1988. The search for g-modes is based on a cross-spectral analysis of the low frequency spectrum from IPHIR (20 to 120 μHz) with theoretical g-mode spectra. These spectra are calculated using a second order asymptotic theory for the frequencies and a visibility function for the amplitudes which depends on degree and frequency; the basic period spacing of the g-modes and the rotational rate are varied in the ranges $26 < P_0 < 45$ minutes and $0.4 < \nu_R < 2.0$ μHz respectively. Comparison with artificial noise spectra indicates that the solar g-modes – if they exist – are buried in solar noise with an upper limit for their amplitudes of the order of 1.3 ppm at 20 μHz. More sophisticated methods to extract a possible g-mode signal from noise are proposed.

1 Introduction

Searches for g-modes have been performed using results from Doppler measurements of the global Sun (e.g. Kotov *et al.* 1984, García *et al.* 1988, Henning *et al.* 1988, van der Raay 1990) and from irradiance data (e.g. Fröhlich and Delache 1984, Fröhlich 1987, Kroll *et al.* 1988). The results are still controversial and a wide range for the basic period spacing P_0 of solar g-modes between 28 to 45 minutes is proposed. The opportunity for a dedicated helioseismology experiment from space came up in 1985 when the final payload of the USSR mission to the martian satellite Phobos was planned. The cruise phase of such a mission offers the unique chance to gather truly uninterrupted time series of several months duration which reduces the observational noise substantially. The experiment IPHIR (*I*nter*P*lanetary *H*elioseismology by *IR*radiance measurements) uses a very simple sunphotometer (SPM) to observe the global Sun in three wavelength channels (at 335, 500, and 865 nm). It was realized by an international consortium lead by PMOD/WRC and consisting of the Laboratoire de Physique Stellaire et Planétaire (LPSP, France), the Space Science Department (SSD of ESA), the Crimean Astrophysical Observatory (KrAO, USSR), and the Central Research Institute for Physics (CIRP, Hungary). A description of IPHIR has been presented by Fröhlich *et al.* (1988).

2 Solar Gravity Modes

G-mode periods for oscillations with radial order n much higher than the degree l can be described by asymptotic theory. To the second order the g-mode periods are as follows (Provost and Berthomieu 1986):

$$P_{n,l} = P_0 \cdot \frac{n + \frac{l}{2} - \frac{1}{4} - \epsilon_0}{\sqrt{l(l+1)}} + \frac{P_0^2}{P_{n,l}} \cdot \frac{l(l+1)V_1 + V_2}{l(l+1)} \tag{1}$$

The fundamental period, P_0, depends mainly on the Brunt-Väisälä frequency N near the center and is very sensitive to the structure of the core. The parameter ϵ_0 equals $0.09 \ldots 0.21$ for $l = 1 \ldots 4$, and $V_{1,2}$ depend in a intricate way on P_0 and $\lim_{\epsilon \to 0}(\int_\epsilon^R \frac{dr}{N \cdot r} - \frac{1}{N(\epsilon)})$. The importance of the second order term in the asymptotic approximation decreases with increasing n, crossing the 1% level at $n = 25$ for a standard solar model (Provost and Berthomieu 1986). At lower frequencies when the period of g-modes approaches the rotational period, coupling due to the Coriolis force takes place and the frequencies are slightly increased (Berthomieu *et al.* 1978). With a rotational rate of $\nu_R = 1\mu$Hz this latter deviation amounts to almost 5% at 20 μHz. The prediction of P_0 from different standard models of the Sun is 33 to 36 minutes (e.g. Christensen-Daalsgard 1986). A model with a core mixed by turbulent diffusion, as proposed by Schatzman *et al.* (1981) has a P_0 of 40 to 57 minutes depending on the amount of mixing assumed. A WIMP model (weakly interacting massive particles) in which the central temperature and thus the neutrino flux is lowered by increased thermal transport near the center by hypothetical WIMPs has P_0 of 29 to 31 minutes (Faulkner *et al.* 1986, Däppen *et al.* 1986).

The fact that g-modes are equidistant in period means that their density is rapidly increasing ($1/\nu^2$) with decreasing frequency. With a length of the time series of 160 days the individual $l = 2$ g-modes are resolved down to frequencies of about 10 μHz (≈1 day period). The spectrum is further complicated by rotational splitting, which is a fixed amount in frequency and equals:

$$\Delta\nu = m \left[1 - \frac{1}{l(l+1)}\right] \nu_R. \tag{2}$$

The density of the rotationally split $l = 1, 2$ modes reaches almost 60 lines/μHz at 10 μHz.

3 Data Set

The data used in this analysis are from 160 days of cruise phase starting 2 days after launch of PHOBOS II on 12 July 1988. The data coverage is excellent with virtually no gaps due to transmission or operations. Two problems have been encountered with the sensors: degradation of the sensitivity and an unexpected influence of the offset pointing on the signal of all three channels. The degradation depends strongly on wavelength and the present analysis uses the least influenced channel at 865 nm. The unexpected influence of the pointing, however, is a more serious problem. During offset pointing the signal increases due to straylight in the baffle and the 'cleaning' method used for the low frequency spectrum is somewhat more demanding than the one used for p-mode analysis, where a highpass filter is sufficient. For the low frequency spectrum the 'cleaning' is based on an interactive correction procedure which tries to keep the low frequency variations unaffected, but might be somewhat subjective. The result is shown in Fig. 1 and demonstrates the effectiveness of the method used.

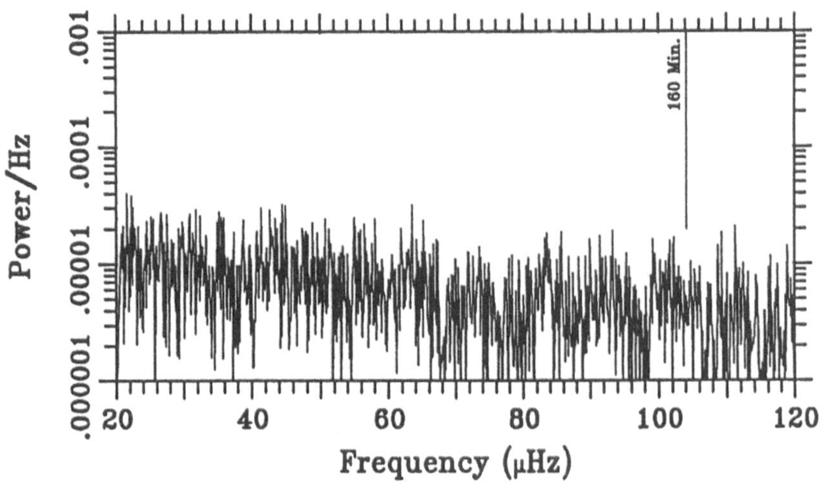

Fig. 1. Power spectrum of the time series of the 865 nm channel of IPHIR during 160 days of the cruise phase of PHOBOS II. Note the location of the 160-minute oscillation observed from ground (e.g. Kotov *et al.* 1984).

4 Method for g-Modes Search

Figure 1 shows clearly that a direct identification of the modes is not possible as the spectrum is dominated by solar noise. Thus a statistical method has been used for the search which allows to detect g-modes in the presence of noise, similar to the synchronous detection in electrical signal processing. In the latter case the frequency of the signal to be extracted has to be known; for the 'synchronous detection' of g-modes in a power spectrum a theoretical g-mode spectrum is needed (e.g. Fröhlich and Delache 1984). The frequencies can be calculated from second order asymptotic theory with a given P_0 and rotational rate ν_R taking into account the mode coupling according to Berthomieu *et al.* (1978) and Eqs. (1) and (2). As V_1 and V_2 in Eq. (1) cannot be expressed analytically in terms of P_0 a linear interpolation is performed between the V's of a WIMP, standard and mixed model respectively. Not only the frequencies of the theoretical spectrum have to be known, but also the relative amplitudes as function of degree and frequency. The visibility of g-modes in brightness has been calculated by Berthomieu and Provost (1990) assuming a non-adiabatic solar atmosphere and energy equipartition of the different modes. The visibility in brightness increases toward higher and lower frequencies being zero at 61 μHz for $l = 1$ and at 85 μHz for $l = 2$. At 20 and 120 μHz the amplitudes for $l = 1, 2$ are 12, 8 ppm and 15, 8 ppm respectively for an energy per mode of $E_0 = 2 \cdot 10^{37}$erg. The cited amplitudes are for $m = 1$, $l = 1$ and $m = 0$, $l = 2$; due to the projection into the line of sight the $m = \pm 2, l = 2$ modes have ≈20% higher amplitudes than those of $m = 0$.

The search for g-modes is performed along to the following steps:

1. calculate for a given P_0 and ν_R and for $l = 1$, $m = \pm 1$ and $l = 2$, $m = 0, \pm 2$ the frequencies of g-modes in the range 20-50 μHz;
2. produce a g-mode power spectrum f_j with the resolution of the smoothed IPHIR spectrum (0.093μHz) and with the amplitudes as described;
3. calculate the coherence c_{ij} between f_j and the spectrum f_i under investigation from the complex cross-spectrum f_{ij}. In order to get reasonably unbiased estimates for f_i, f_j and f_{ij} smoothing is performed before the coherence is evaluated. The c_{ij}^2 are summed over the frequency range under consideration and the mean $\bar{c}_{ij}^2$ for each pair (P_0, ν_R) is calculated;
4. steps 1...3 are repeated for other P_0 (26–45 min with 0.1 min increments) and ν_R (0.4-2.0 μHz with 0.04 μHz increments) yielding a 191×41 matrix of $\bar{c}_{ij}^2(P_0, \nu_R)$ values. Before plotting this matrix as a grey tone map, a two dimensional linear detrending is performed and the map is smoothed with a running mean over 3×3 elements.

The method is illustrated in Fig. 2 for three different signal-to-noise ratios. A g-mode spectrum with $P_0 = 35.5$ min and $\nu_R = 1.0\,\mu$Hz is used for the simulation with the $l = 1$ amplitude normalized to 1 at 120 μHz. The noise power spectra added to the g-mode spectrum have a χ^2-distribution with expectation values of 0.1, 0.3 and 1.0 respectively at 120 μHz and increasing as $1/\nu$ towards lower frequencies. The ratios of the g-mode power-to-noise (both integrated from 20 to 50 μHz) for the three spectra are 0.51, 0.17 and 0.05. The peak of the g-modes at 35.5 min and 1.0 μHz can be recognized in all plots; in the bottom picture, however, there are other, even higher peaks. This seems to be the limit for g-mode detection by simply looking for a peak in the g-mode map.

5 Results and Conclusions

The result of the analysis of the IPHIR spectrum is shown in Fig. 3. It looks very similar to the bottom picture of the simulations in Fig. 2. From this and the visibility data the upper limit of the $l = 1$ g-mode amplitude at 20 μHz can be estimated as 1.3 ppm in relative brightness. From Fig.10 of Berthomieu and Provost (1990) this corresponds to ≈3.5 cm s^{-1} (at 40 μHz) for velocity observations of the global Sun. More sophisticated methods for the analysis of the g-mode maps are needed. Investigations searching with patterns including the variability in the vicinity of the 'peak' instead of only the 'peak' are under way. The typical patterns are searched from simulations with different P_0 and ν_R g-mode spectra corrupted with different noise spectra. The method looks quite promising, although the 'best' pattern is not yet found.

Acknowledgments

Without the continuous efforts of the IPHIR team – scientific and technical – and the efficient co-operation with the Space Science Institute at Moscow during the preparation

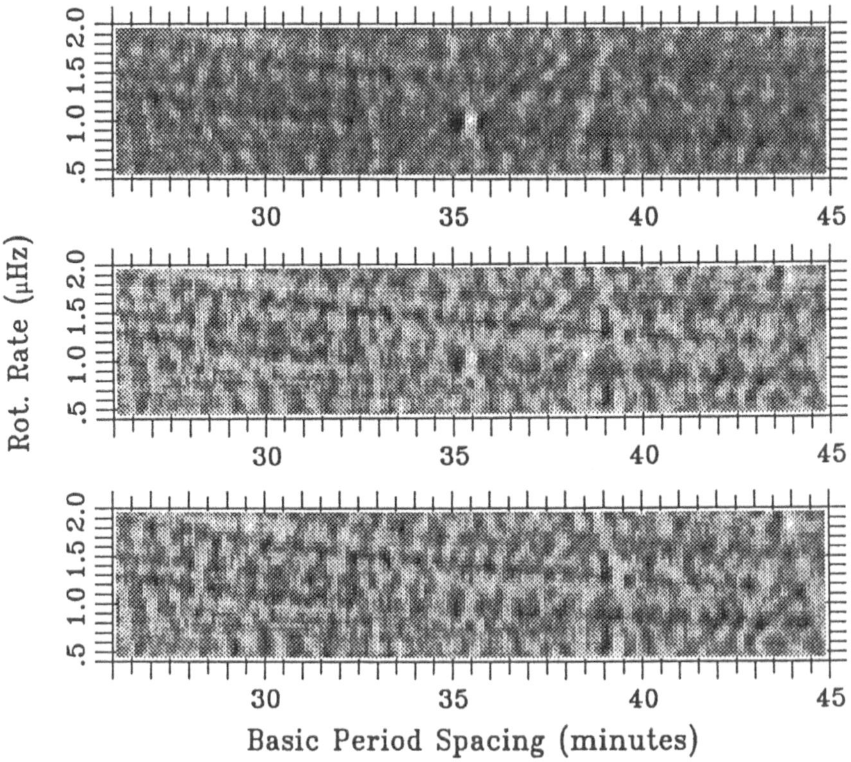

Fig. 2. Analysis of a simulated g-mode spectrum with $P_0 = 35.5$ min and $\nu_R = 1.0\mu$Hz corrupted by different amount of noise. Noise-to-g-mode power amounts to 2, 6 and 20 (top to bottom). Bright corresponds to high, dark to a low c_{ij}^2 with a 16 level grey scale; the ranges between maximum and minimum of c_{ij}^2 are (top to bottom): .05–.34, .06–.23, .06–.23.

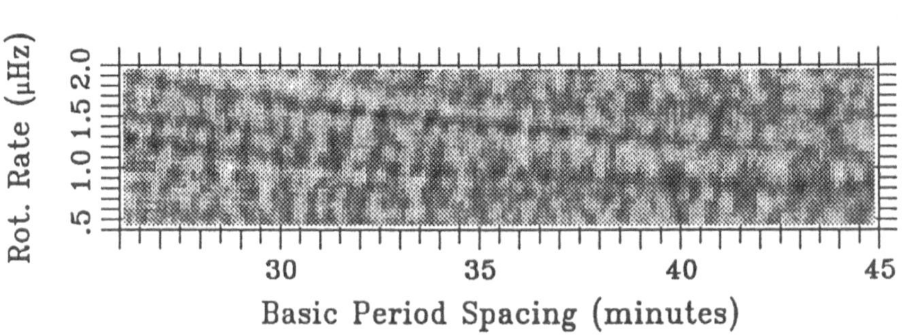

Fig. 3. G-mode map from IPHIR spectrum of Fig. 1.

of the experiment and the mission this work would not have been possible. Part of the present study has been performed during a two-month stay in the fruitful research environment of the Observatoire de Nice, which is gratefully acknowledged. Thanks are extended to G.Berthomieu and J.Provost, Observatoire de Nice, for the development of the second order g-mode calculation algorithm and many helpful discussions, and to the Swiss National Science Foundation for financial support of the experiment.

References

Berthomieu, G., Gonczi, G., Graff, Ph., Provost, J., and Rocca, A. 1978, *Astron. Astrophys.*, **70**, 579.

Berthomieu, G., and Provost, J. 1990, *Astron. Astrophys.*, **227**, 563.

Christensen-Dalsgaard, J. 1986, in *Seismology of the Sun and Distant Stars*, ed. D.O. Gough (Reidel, Dordrecht), p.23.

Däppen, W., Gilliland, R.L., and Christensen-Dalsgaard, J. 1986, *Nature*, **321**, 229.

Faulkner, J., Gough, D.O., and Vahia, M.N. 1986, *Nature*, **321**, 226.

Fröhlich, C. 1987, in *New and Exotic Phenomena*, ed. O. Fackler and J. Tran Thanh Van (Editions Frontières, Gif sur Yvette), p.395.

Fröhlich, C., Bonnet, R.M., Bruns, A.V., Delaboudinière, J.P., Domingo, V., Kotov, V.A., Kollath, Z., Rashkovsky, D.N., Toutain, T., Vial, J.C., and Wehrli, Ch. 1988, in *Seismology of the Sun and Sun-like Stars*, ed. E. Rolfe, ESA SP-286 (ESA Publication Division, Noordwijk), p.359.

Fröhlich, C., and Delache, Ph. 1984, in *Solar Seismology from Space*, ed. R.K. Ulrich, J. Harvey, E.J. Rhodes, and J. Toomre (JPL Publ.84-84, Pasadena), p.183.

García, C., Pallé, P.L., and Roca Cortés, T. 1988, in *Seismology of the Sun and Sun-like Stars*, ed. E. Rolfe, ESA SP-286 (ESA Publication Division, Noordwijk), p.353.

Henning, H.M., and Scherrer, P. 1988, in *Seismology of the Sun and Sun-like Stars*, ed. E. Rolfe, ESA SP-286 (ESA Publication Division, Noordwijk), p.419.

Kotov, V.A., Severny, A.B., and Tsap,T. 1984, *Mem. Soc. Astron. Ital.*, **55**, 117.

Kroll, R.J., Chen, J., and Hill, H.A. 1988, in *Seismology of the Sun and Sun-like Stars*, ed. E. Rolfe, ESA SP-286 (ESA Publication Division, Noordwijk), p.415.

Provost, J., and Berthomieu, G. 1986, *Astron. Astrophys.*, **165**, 218.

Schatzmann, E., Maeder, A., Angrand, F., and Glowinski, R. 1981, *Astron. Astrophys.*, **96**, 1.

van der Raay, H.B. 1990, these proceedings.

Solar g-Modes

H. B. van der Raay

School of Physics and Space Research, The University of Birmingham,
Birmingham B15 2TT, UK

Abstract: An analysis of long data stretches covering the years 1982-1987, obtained from a whole disc optical resonant scattering spectrometer, indicate the presence of significant signals in the 25-75 μHz frequency range. The period spacings of these signals are found to be in four groups, which, determined independently for each year, are not only consistent but correspond with the predictions of the Tassoul (1980) asymptotic theory. Further, an investigation of the phase coherence of specific signal frequencies indicate life-times of several hundred days. These signals therefore show all the anticipated characteristics of solar g modes.

1 Introduction

The field of helioseismology has shown significant advances since the first experimental demonstration of the global nature of the 5 minute signal (Claverie *et al.*, 1979). The precise frequency determinations (Jiménez *et al.*, 1988a) and evidence for rotational splitting (Duvall and Harvey, 1984; Libbrecht, 1986; Jefferies *et al.*, 1988) have placed severe constraints on the possible theoretical models. However an investigation of the core of the sun requires information obtainable from the deeply penetrating solar g modes.

The experimental verification of the existence of g modes is more complex than that for p modes, due mainly to the following three factors. The anticipated amplitudes are lower, the number of modes per unit frequency interval is large, especially at very low frequencies, and the measured 'solar' noise spectrum (Jiménez *et al.*, 1988b) indicates a rapid increase in this region. The actual pattern produced is also relatively complicated, corresponding to a series of equally spaced signals in period space and a repetative pattern in frequency space occasioned by rotational splitting.

The close spacing in frequency space of individual modes and the interspersed components resulting from rotational splitting necessitate the use of long coherent data sets in order to obtain the required resolution. Fortunately the predicted life times of g modes meet this requirement.

The problem of g mode identification thus turns into one of specific pattern recognition in either period space, to select the modes or frequency space to determine the rotational splitting. Checks on correct selection are provided by firstly the consistent appearance of discrete peaks in successive data sets, a limited number (≤ 4 for full disc observations) of constant period groupings and, for the higher order modes, a fit to the Tassoul (1980) asymptotic relation. As a further test the phase coherence of individually selected modes which fit the above criteria may be checked to verify the expected long life-times associated with g modes.

2 Analysis

Previously data obtained in Tenerife over the years 1984-87, were analysed over the frequency range 50-75 μHz (van der Ray, 1988). The method of exact fractions was successfully utilized to select three period groupings centred on 9.9, 11.4, and 16.9 minutes. These were attributed to ℓ_4, ℓ_3, and ℓ_2 g-modes. The ℓ_1 grouping was not evident as only 4 components are predicted over this frequency range. A consistent period spacing was found from the independent analysis of all 4 sets of data.

This analysis has been extended to cover the 25-50 μHz frequency region and also to include 122 days of data obtained in 1982. The previous findings have been confirmed and for each year period groupings have been identified independently which are consistent with those found for the 50-75 μHz region. A plot of all these data is shown in Fig. 1.

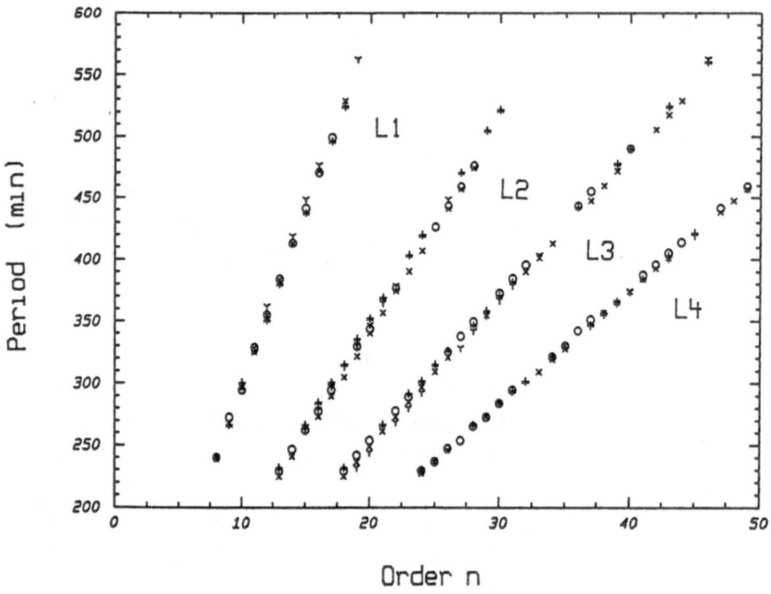

Fig. 1. Plot of the period against order of the identified g modes over the frequency range 25-75 μHz for the years 82(x), 84(+), 85(λ), 86(Y) and 87(o).

The Tassoul (1980) asymptotic formula

$$T = T_0(n + \ell/2 - k)/\sqrt{\ell(\ell+1)} \tag{1}$$

represents a linear relation between the observed period and order for a particular mode which is clearly fitted by the data in Fig. 1. Having assigned values for ℓ and n, two constants T_0 and k may be evaluated for each ℓ value as listed in Table 1 for the 1985 data.

Table 1. Constants in Tassoul relation evaluated for 1985 data

	ℓ_1	ℓ_2	ℓ_3	ℓ_4
T_0 (min)	41.47±0.07	40.62±0.05	41.00±0.21	41.57±0.09
k	0.31±0.03	0.36±0.03	0.30±0.17	0.55±0.09

The value for T_0 is in agreement with that found previously from data in the 50-75 μHz range (van der Raay, 1988) and now with the extended range a realistic evaluation of k, consistent with predictions, is obtained.

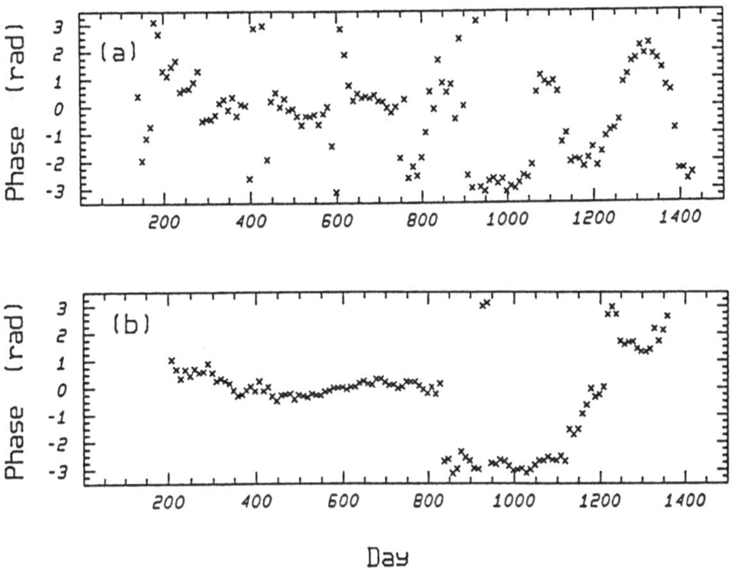

Fig. 2. Phase variation of the 52.64 μHz signal from 1/1/84 to 31/12/87 a) 60 day mean, b) 200 day mean.

If the selected signals are g modes then the anticipated long life-time would predict a phase coherent signal from year to year. The 4 years of data are considered as a single data string and using iterative sinewave fitting with a frequency increment of 0.01 μHz the phase variation of the 52.64 μHz component is followed over the entire data span. The results, where each data point represents the mean phase over a 60 day stretch, are shown in Fig. 2a. These data clearly indicate that this particular mode ($\ell = 2$, $n = 18$) has existed for at least 600 days and also show the effects of a beat with a neighbouring component ($\ell = 3$, $n = 25$) at 52.58 μHz. The beat shows up as a rapid phase change after which the phase recovers to its previous value. This emphasizes the need to use long data stretches so as to obtain the requisite frequency resolution. In Fig. 2b results obtained by considering the mean phase over 200 day stretches are shown. A detailed investigation of this position of the frequency spectrum revealed the two closely spaced components (Fig. 3).

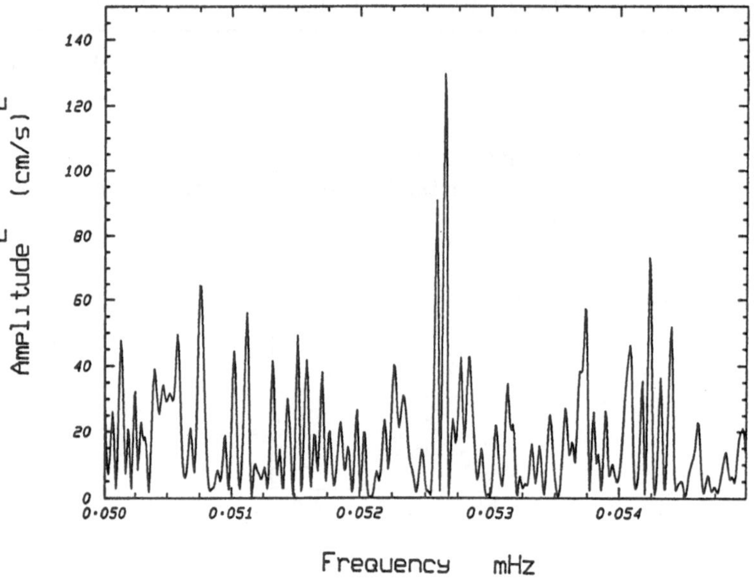

Fig. 3. Power spectrum of 1985 data revealing two closely space modes.

Similar analyses of signals at the two extremes of the frequency range considered, namely 26.11 μHz and 73.75 μHz yield the data illustrated in Fig. 4. These and other data suggest a possible decrease in life-time as the frequencies increase.

As the life-times, particularly of the lower frequency signals, are measured to be several years, a power spectrum over four years should increase the signal to noise ratio in this region whereas at the higher frequencies both the signal and the random background noise will decrease due to random phase and/or frequency changes. This is illustrated by the power spectrum for the 84-87 data string over the 25-50 μHz range shown in Fig. 5. The overall decrease in power towards the lower frequencies results from an effective high pass filter introduced into the analysis as a consequence of fitting the basic daily data with a sinewave to remove the effects of the earth's rotation on the measured line of sight velocity. The two dominant peaks are the 3rd and 4th harmonics of the day occasioned by the night time interruption of the data string.

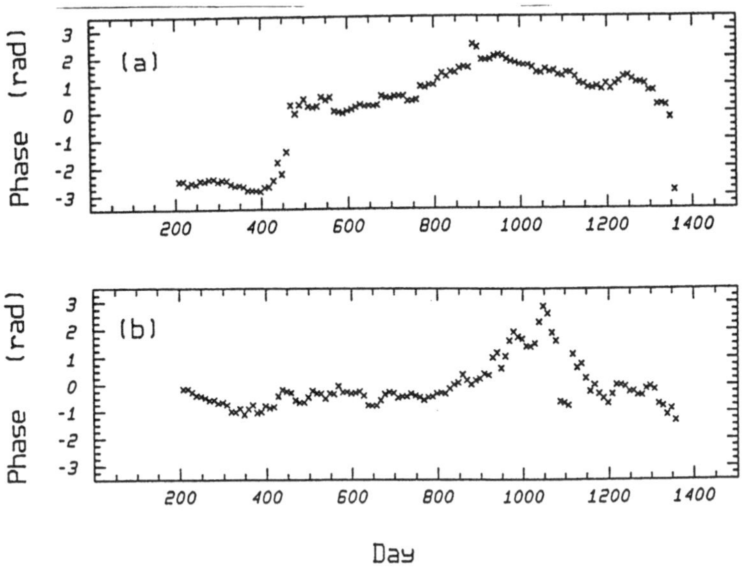

Fig. 4. Phase variation using 200 day stretches for signals of a) 26.11 μHz and b) 73.5 μHz.

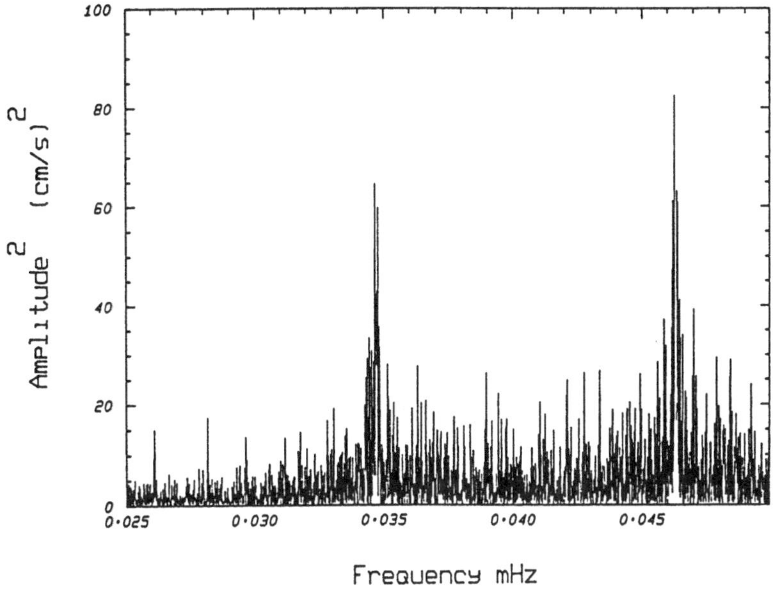

Fig. 5. Power spectrum over the frequency range 25-50 μHz for the data string 84-87 inclusive.

Rotational splitting of the various modes is predicted to follow the relation

$$\nu_s = m\nu_r \left[1 - \frac{1}{\ell(\ell+1)}\right] \tag{2}$$

where ν_s is the observed frequency splitting and ν_r is the rotational frequency. Hence for each ℓ value a particular pattern of equally spaced peaks is generated in the frequency spectrum. A search for a common frequency difference between any two peaks in the spectrum of each year considered is sought. This procedure overcomes the problem that the frequency spacings of orders of a particular mode are not equal. Certain preferred common differences are found in each of the four years and if very tentatively interpreted in terms of Eq. (2), yield the rotational splittings indicated in Fig. 6. The results all tend to indicate an increase in rotational splitting with ℓ and suggest a variation with the solar cycle. As the g modes are concentrated in the solar core these preliminary results suggest that the core is rotating 3-4 times more rapidly than the observed solar surface.

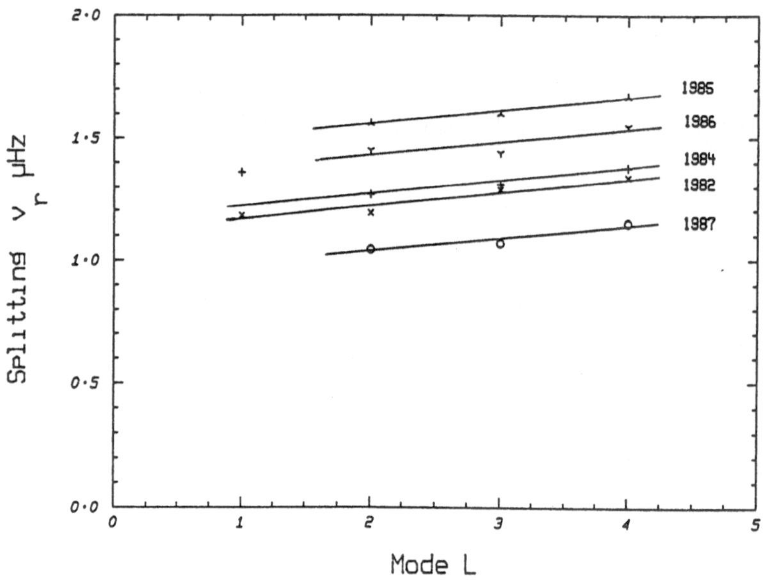

Fig. 6. Tentative rotational g mode splitting.

3 Conclusion

Solar line of sight velocity data for the full disc obtained in 1982 and 1984-1987 from a single station, reveal peaks in the 25-75 μHz range which re-occur in each year's data set. The period structure of these peaks show groupings around four time interval spacings which are consistent with those predicted by the Tassoul (1980) asymptotic relation with a single value for the constant T_0. Assigning n and ℓ values to the observed signals in accordance with this relation, allows an evaluation of the second constant, k, which is found to be in the expected range.

Specific signals are found to be phase coherent over long periods varying from ~ 900 days at the lower frequencies to ~ 600 days at the upper frequency limit considered.

Tentative allocations of rational splitting frequencies indicate values of ~ 1.5 μHz which slightly increase with ℓ value and suggest a variation with the solar cycle.

Consequently the persistent solar velocity signals which have been detected in the low frequency range over a 6 year span, exhibit characteristics which are consistent with those attributed to solar g modes.

Acknowledgements

The assistance of all members of the Birmingham and I.A.C. solar oscillation groups both past and present is gratefully acknowledged, especially the support of the technical staff. This work was partly funded by the SERC (UK) and the CAICYT (Spain).

References

Claverie, A., Isaak, G.R., McLeod, C.P., van der Raay, H.B., and Roca Cortés, T. 1979, *Nature*, **282**, 591.

Duvall, T.L., Jr., and Harvey, J.W. 1984, *Nature*, **310**, 19.

Jefferies, S.M., McLeod, C.P., van der Raay, H.B., Pallé, P.L., and Roca Cortés, T. 1988, in *Advances in Helio- and Asteroseismology, IAU Symp. No. 123*, eds. J. Christensen-Dalsgaard and S. Frandsen (Reidel, Dordrecht), p.25.

Jiménez, A., Pallé, P.L., Pérez, J.C., Régulo, C., Roca Cortés, T., Isaak, G.R., McLeod, C.P., and van der Raay, H.B. 1988a, in *Advances in Helio- and Asteroseismology, IAU Symp. No. 123*, eds. J. Christensen-Dalsgaard and S. Frandsen (Reidel, Dordrecht), p.205.

Jiménez, A., Pallé, P.L., Péréz Hernández, F., Régulo, C., and Roca Cortés, T. 1988b, *Astron. Astrophys.*, **192**, L7.

Libbrecht, K.G. 1986, *Nature*, **319**, 753.

Tassoul, M. 1980, *Astrophys. J. Suppl.*, **43**, 469.

van der Raay, H.B. 1988, in *Seismology of the Sun and Sun-like Stars*, ed. E. Rolfe, ESA SP286 (ESA Publication Division, Noordwijk), p.339.

A Problem with the 160-Minute Pulsation of the Sun

V. A. Kotov

Crimean Astrophysical Observatory, Nauchny, Crimea 334413, USSR

Abstract: Differential measurements (1974-1988) of the solar velocity and photospheric brightness show the presence of long-term pulsation of the Sun with a period of P_0=160.0101(±1) min. Its nature is poorly known since it seems hardly possible to explain the periodicity in terms of solar g-modes. But the most fascinating appears to be the recent discovery of the same 160-min periodicity in the light-flux variations of several AGN (active galactic nuclei). This finding strongly supports the hypothesis advanced much earlier about cosmological origin of the oscillation.

The two important suppositions published in the past can now be regarded as direct precursors of the discovery (Severny *et al.* 1976; Brookes *et al* 1976; Kotov *et al.* 1978) of the solar oscillation with a period of near 160 min. Firstly, as long ago as 1974 Roxburgh (1974) has made an intriguing and a bit provocative remark: "The central region spins at a phenomenal rate with a period of about 50 min — it would be of considerable interest to look for phenomena on the Sun with a period of about 30 min to 1 hr that might reflect this rapid motion...." Secondly, much earlier (in 1946), Sevin (1946) has explicitly stated that the "infra-sound" of solar vibration must have a 1/9-of-a-day (160 min) period.

The best value of the period determined via statistical analysis of the time sequence (1947-1980) of solar flares is equal to 160.0101(±1) min (Kotov and Tsap, 1989). The oscillation is characterized by stable, in average, initial phase (the phase everywhere is set to be zero at UT 00^h00^m 1974 January 1).

The Doppler shift measurements were made from 1974 through 1988 with the use of differential (center-to-limb) technique (Kotov *et al.*, 1978). Being performed with almost null spatial resolution, the data might contain a great deal of information about the deep interior of the Sun, — plausibly right up to the energy generating core.

The first observations, 1974-1982, have shown the presence of 160.010-min periodicity in the low frequency part of the power spectrum (PS) which was significantly above the noise level (Kotov *et al.*, 1984). This finding was strongly supported then by independent measurements carried out in 1977-1980 at Stanford and South pole (Scherrer and Wilcox, 1983; Grec *et al.*, 1980) and also by the solar diameter observations (Hill *et al.*, 1986). Later on, however, the authors (van der Raay, 1988; Zhugzhda, 1989) have questioned the reality and solar nature of the P_0-oscillation, mainly due to possible influence of the terrestrial atmosphere and to the closeness of P_0 to 9th daily harmonic. The new evidences in favour of true solar nature of the oscillation were presented recently in (Kotov and Tsap, 1989).

The PS of the Crimean 1974-1988 velocity observations computed for the narrow frequency range around 104.16 μHz, is shown in Fig. 1. A total number of days is N_d=987; N=74359 is a number of individual measurements made with 5-min integration time. The major peak corresponds to a period of 160.009 (±1) min. Also, because the data have gaps at winter, the yearly sidelobe structure introduces several other apparently significant peaks: 159.967, 160.057, and 160.110 min. The 9th daily harmonic peak (160.00 min), which is for some reason a subject of great interest for opponents, is negligible.

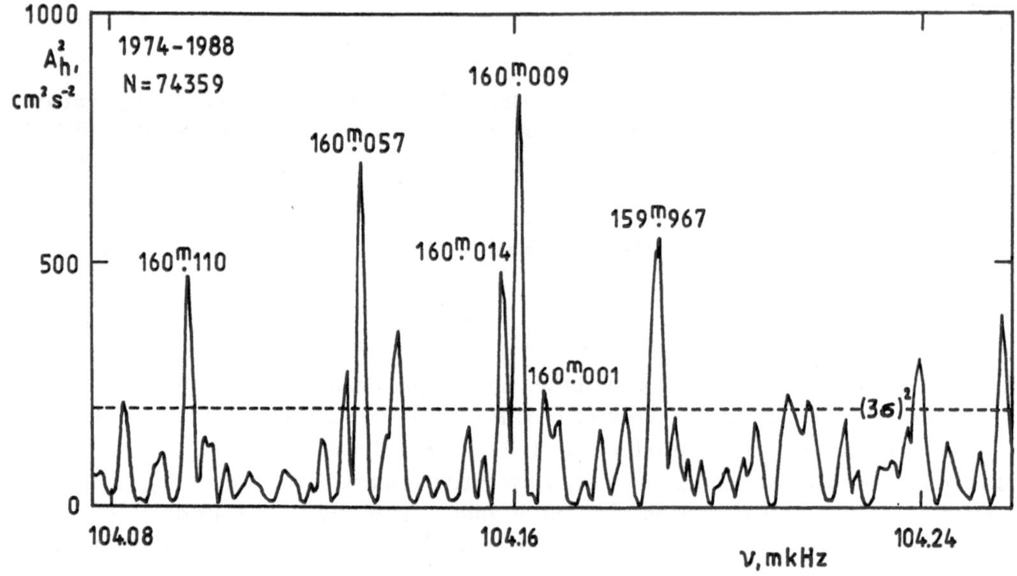

Fig. 1. The power spectrum of solar oscillations, 1974-1988.

The 160.01-min peak emerges as a doublet feature. The spacing between two peaks (160.009 and 160.014 min) corresponds to a beat period of 10±3 yr and thus fairly coincides, within the error limits, with the average length of 11-yr cycle. One may tentatively conclude, therefore, that there might be some physical connection between 160-min oscillation and 11-yr cycle. One can also suggest that the observed fine structure might be produced by nonlinear interaction between solar global pulsation and fast rotation (hypothetical) of the central solar core. But the primary role in appearance of the splitting should be attributed perhaps to considerable change of the internal general magnetic field (GMF) of the Sun during the 22-yr magnetic cycle.

The mean P_0-curves for velocity, brightness (Kotov and Tsap, 1989) and GMF of the Sun (seen as a star) are plotted in Fig. 2 (dotted lines are the best-fitted sinusoids; the "+"-sign corresponds to "expansion" velocity and to the northern polarity of GMF). The brightness curve shows statistically significant P_0-oscillation with the mean harmonic amplitude $A_h \sim 2 \cdot 10^{-5}$ in relative units. The latter, being rescaled to the full-disk observations, corresponds to $\sim 4 \cdot 10^{-6}$ and, thus, does not contradict to the upper limit for solar irradiance variation set by the ACRIM/SMM bolometric measurements (Woodard and Hudson, 1983).

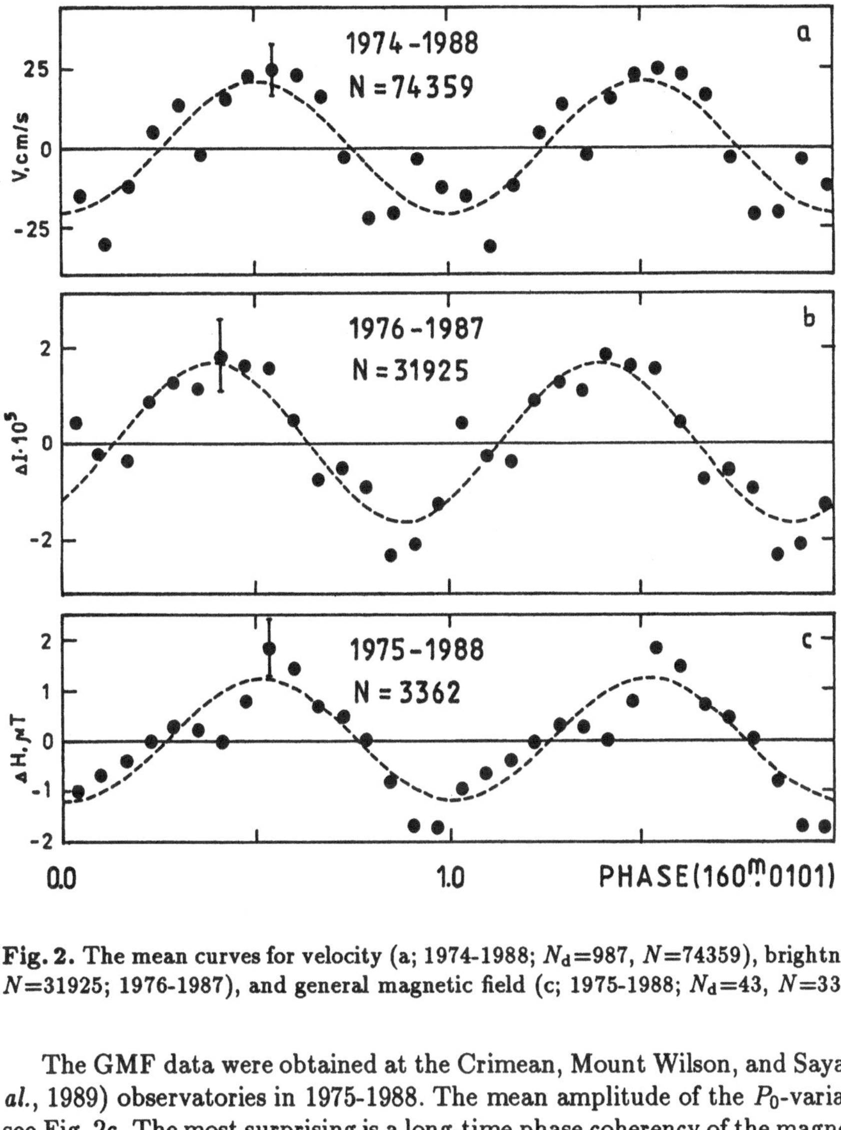

Fig. 2. The mean curves for velocity (a; 1974-1988; N_d=987, N=74359), brightness (b; N_d=487, N=31925; 1976-1987), and general magnetic field (c; 1975-1988; N_d=43, N=3362) of the Sun.

The GMF data were obtained at the Crimean, Mount Wilson, and Sayan (Demidov *et al.*, 1989) observatories in 1975-1988. The mean amplitude of the P_0-variation $\sim 1.2\,\mu$T, see Fig. 2c. The most surprising is a long-time phase coherency of the magnetic oscillation: it appears to be paradoxically independent of the GMF polarity. This effect, perhaps, indicates the existence of strong, $\sim 10^7$ - 10^9 G, magnetic field in deep solar interior. [Notice that Dziembowski and Goode (1989), on the basis of the frequency splitting of acoustic oscillations, inferred a $\sim 10^6$ G magnetic field for the outermost part of the radiative zone.]

It seems impossible to treat P_0-oscillation in terms of solar gravity modes (Deubner and Gough, 1984; Kotov, 1985). Accordingly, a number of different suggestions, including rather exotic ones, have already been advanced for its explanation [for the references see

Kotov (1985)]. In addition, it was recently hypothesized (Kotov, 1986) that the oscillation must have a cosmological status. In consequence of it, the P_0-periodicity was soon discovered in light-flux variations of several AGN (Kotov and Lyuty, 1989). The finding was immediately confirmed by analysis of the X-ray data available from the ARIEL-5 and EXOSAT satellites, 1975- 1986. Figure 3 shows the PS for the Seyfert galaxy NGC 4151 where one of the two highest peaks corresponds exactly to P_0 (the second one is artifact due to a 1-day regularity of substantial portion of observations). The same period was also found in the optical data for the famous quasar 3C 273 (Kotov and Lyuty, 1989).

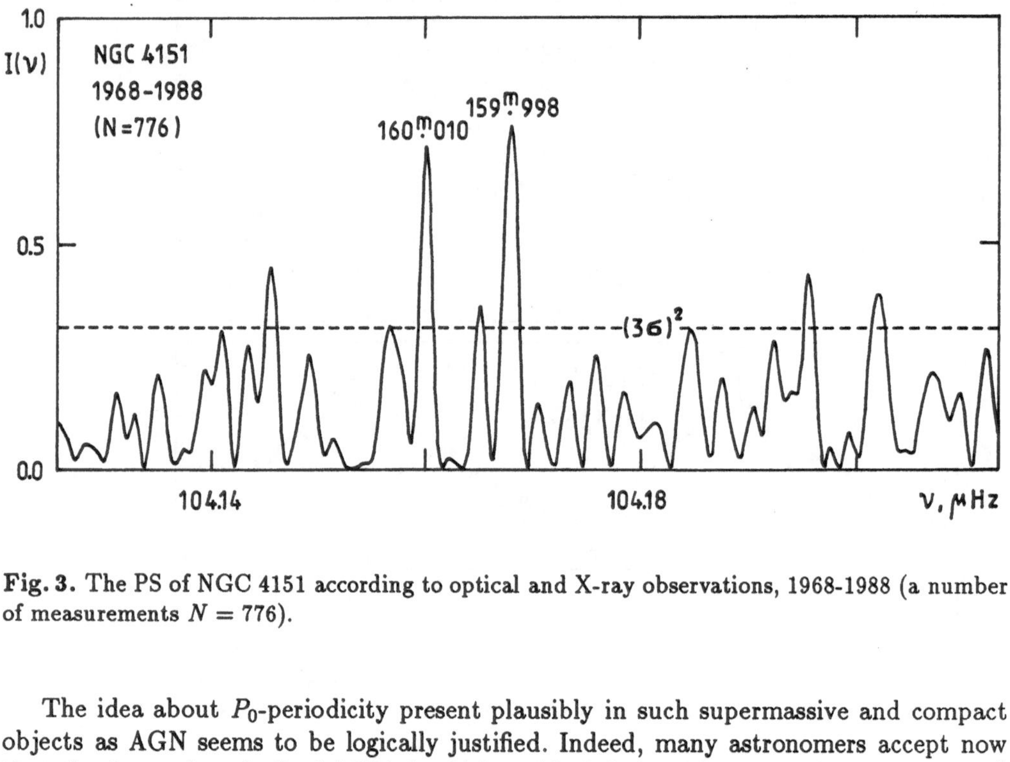

Fig. 3. The PS of NGC 4151 according to optical and X-ray observations, 1968-1988 (a number of measurements $N = 776$).

The idea about P_0-periodicity present plausibly in such supermassive and compact objects as AGN seems to be logically justified. Indeed, many astronomers accept now that the "central engine" of AGN should be a black hole with a typical mass M $\sim$ 10^6 - 10^{10} $M_\odot$ ($M_\odot$ - solar mass). Then, if M $\sim$ 10^9 $M_\odot$, for the gravitational radius we get (in usual notations): $r_g = 2GM/c^2 \sim 3 \cdot 10^{14}$ cm and, therefore, transporting time for light $\sim$ 160 min.

But the most fascinating is the circumstance that P_0-oscillation observed in about half AGN appears to be almost synchronous with the Sun (see Fig. 2), whereas for the rest of sources the resultant P_0-curves are found to be in antiphase with respect to the Sun. This distinction (for a complete set of 14 extragalactic objects) is significant at about 3-sigma confidence level (Kotov and Lyuty, 1990); the average curves for the two groups of AGN are plotted in Fig. 4 (the X-ray data were rescaled to rms-values of the optical measurements; δ - a relative change of the emission).

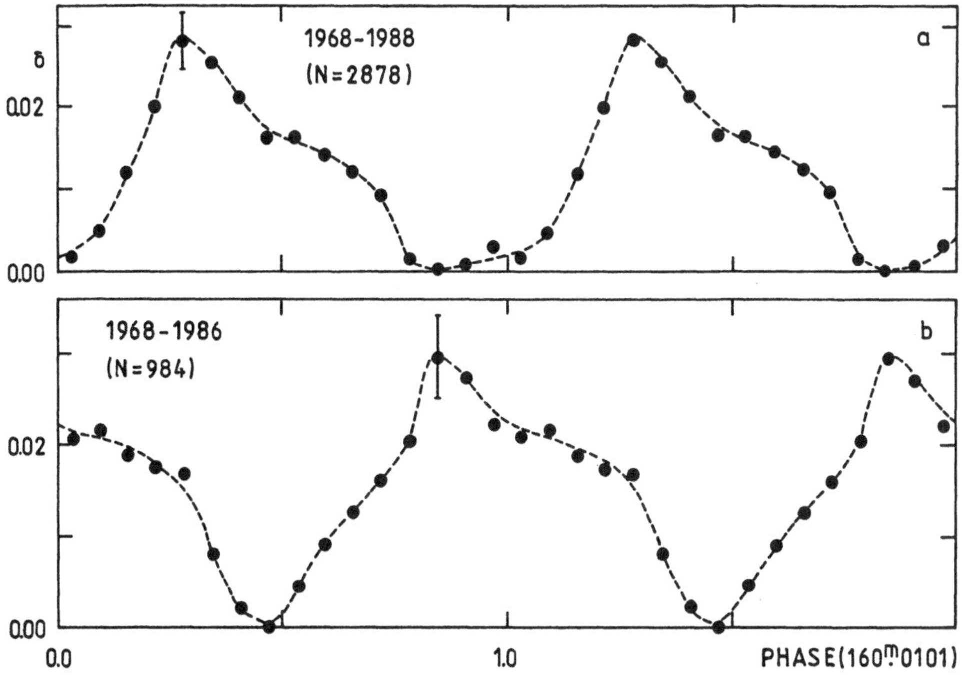

Fig. 4. Mean curves for 9 "bright" (a; NGC 4051, NGC 4151, NGC 6814, NGC 7469, 3C 371, Mrk 335, Mrk 421, OJ 287, MGC-6-30-15) and 5 "dark" (b; NGC 1275, NGC 3516, NGC 5506, 3C 273, EXO 1126+691) AGN. Dotted lines are drawn "by hand." The time of each observation was reduced to the Sun; N - a total number of optical and X-ray data.

We suggest that the first and second AGN-groups consist of matter and antimatter ("worlds" and "antiworlds" of the Universe) respectively. This possible explanation of the distinction can be provided perhaps by simple and logic analogy between 180°-phase shift of AGN and that observed for right- and left-hand rotations among particles and antiparticles in the fundamental physics [for details, see Kotov and Lyuty (1990)].

In conclusion it is worth to quote (Hawking, 1989): "Humanity's deepest desire for knowledge is justification enough for our continuing quest." In the case of AGN we face indeed quite intriguing questions which may stimulate research in the field of astero- and AGN-seismology: why we measure one and the same period in AGN, no matter how fast these extragalactic objects are moving? And how very distant AGN can keep constant phase if we know that nothing may travel in the Universe more rapidly than light?

References

Brookes, J.R., Isaak, G.R., and van der Raay, H.B. 1976, *Nature,* **259**, 92.
Demidov, M.L., Kotov, V.A., and Grigoryev, V.M. 1989, *Izv. Krymsk. Astrofiz. Obs.*, **83** in press.
Deubner, F.-L., and Gough, D.O. 1984, *Ann. Rev. Astron. Astrophys.*, **22**, 593.
Dziembowski, W.A., and Goode, P.R. 1989, in list of abstracts of *Inside the Sun, IAU Colloq. No. 121*, p.67.
Grec, G., Fossat, E., and Pomerantz, M. 1980, *Nature,* **288**, 541.
Hawking, S.W. 1989, *A brief history of time* (Bantam, London).
Hill, H.A., Tash, J., and Padin, C. 1986, *Astrophys. J.*, **304**, 560.
Kotov, V.A. 1985, *Solar Phys.*, **100**, 101.
Kotov, V.A. 1986, *Izv. Krymsk. Astrofiz. Obs.*, **74**, 69.
Kotov, V.A., and Lyuty, V.M. 1989, *Compt. Rend. Acad. Sci. Paris*, in press.
Kotov, V.A., and Lyuty, V.M. 1990, in preparation.
Kotov, V.A., Severny, A.B., and Tsap, T.T. 1978, *Monthly Notices Roy. Astron. Soc.*, **183**, 61.
Kotov, V.A., Severny, A.B., and Tsap. T.T. 1984, *Mem. Soc. Astron. Ital.*, **55**, 117.
Kotov, V.A., and Tsap, T.T. 1989, *Solar Phys.*, submitted.
Roxburgh, I.W. 1974, *Nature,* **248**, 209.
Scherrer, P.H., and Wilcox, J.M. 1983, *Solar Phys.*, **82**, 37.
Severny, A.B., Kotov, V.A., and Tsap, T.T. 1976, *Nature,* **259**, 87.
Sevin, E. 1946, *Compt. Rend. Acad. Sci. Paris*, **222**, 220.
van der Raay, H.B. 1988, in *Seismology of the Sun and Sun-like Stars*, ed. E. Rolfe, ESA SP-286 (ESA Publication Division, Noordwijk), p.339.
Woodard, M., and Hudson, H. 1983, *Solar Phys.*, **82**, 67.
Zhugzhda, Y. D. 1989, in list of abstracts of *Inside the Sun, IAU Colloq. No. 121.*

The IRIS Network for Full Disk Helioseismology. Present Status of the Programme

Francois-Xavier Schmider [1], Eric Fossat [2], Bernard Gelly [2], and Gérard Grec [2]

[1]Astro Unit, Queen Mary College, Mile End Road, E1 4NS, London, UK
[2]Département d'Astrophysique, Université de Nice, Parc Valrose, 06034 Nice Cedex, France

Abstract: I.R.I.S.(International Research on the Interior of the Sun) is the name of a worldwide network of 8 stations of observation in full disk helioseismology. The I.R.I.S. scientific community is organizing a yearly workshop in one of the 8 sites. This paper presents the status of the network as it was for the second IRIS workshop, held at Tashkent, Uzbekistan SSR, in September 1989. It presents a brief history, the structure of the international cooperation, the membership rule, the list of sites and members, the scientific working teams additional structure and, as an appendix, the report of the first meeting

1 Introduction

The need of continuous data for obtaining good quality measurements in helioseismology has been felt for a long time. The first separation of the spectrum of full-disk helioseismology in individual mode has been obtained from 5 days of continuous observations at geographic South Pole (Grec *et al.*, 1980). After the dramatic increase of precision which has been made both in theory and in observations, it is now necessary to have longer sets of continuous data. In particular, the search for rotational splitting on low degree modes, still open, can be addressed only by obtaining a better signal to noise ratio and a better resolution.

Indeed, obtaining better sensitivity and better frequency resolution are not the only reasons for getting longer and longer observational time series. It has now been established that the low degree p-mode frequencies are slightly changing along the solar cycle (Woodard and Noyes, 1985, Fossat *et al.*, 1987, Gelly *et al.*, 1988, Fossat, 1988, Palle *et al.*, 1988). In the last reference, it has also been shown that the mean oscillation amplitudes are also function of the solar cycle phase. It is then quite possible that the damping time of the p-modes is also a function of the activity cycle, as well as could be its frequency dependence or the amplitude frequency dependence. The detailed study of all these possibilities requires a network to be operated during the complete period of 11 years. It will perhaps help to finally locate where is the real seat of this magnetic cycle.

Three main ideas have been used for obtaining continuous observations on times longer than the typical 12-hour run of a mid-latitude single site. The first one was to go to Antarctica, where the summer sky makes possible to obtain uninterrupted sequences as long as one week. The second possibility is to go to space. An instrument suitably located on a full sunshine orbit could provide uninterrupted sequences of data over periods as long as the lifetime of the spacecraft mission. This idea was first developed on the "DISCO" project of the European Space Agency, which was finally not selected. It was restarted on the "SOHO" mission, which is now definitely selected, and which will set near one of the Lagrange points of the Earth-Sun system three (among 12) instruments of helioseismology for a 2 to 6 year mission starting around 1995. The third idea consists in setting a network of observing sites around the world, suitably located in complementary longitudes and latitudes. It was also pioneered by our British colleagues (Claverie *et al.*, 1984) operating a two-station network during the summer season at Izana and Haleakala.

In 1982, the IAU Commission 12 voted a resolution (C7) "recognizing the extreme importance of the observation of solar seismology," and "strongly supporting international cooperation in establishing a worldwide network of observing stations." It was a starting point for both the GONG and the IRIS network.

It has been shown by simulations (Hill and Newkirk, 1985, Hill *et al.*, 1988) that a well selected 6-site network should easily be able to provide more than 90% duty cycle, which is generally regarded as acceptable for the sort of data helioseismology is dealing with. With such a duty cycle, there are many possibilities of using the statistical properties of the data, as defined by the 90% available, to efficiently fill the gaps or deconvolve the power spectra.

Now, three main networks are under construction. Two of them are based upon the full disk measurements using the optical resonance spectrophotometers. They will measure the eigenmodes of very low degree, between 0 and 3, which are the modes penetrating deep inside the solar core. One is authored by the Birmingham group, using Potassium resonance cells, and the other one is "IRIS," using Sodium resonance cells. The third network, "GONG," is an important American project, the headquarter of which is at N.S.O., Tucson, Arizona (GONG, 1984). It will provide two-dimensional Doppler solar images onto a 256 × 256 pixels CCD camera, and intends to give access to all eigenmodes of degree lower than about 200.

2 IRIS Calendar

IRIS (for International Research on the Interior of the Sun) was first presented to the French Astronomical Agency "INSU" in October 1983, as a 6 to 8 site network project. The proposed instrument was the sodium cell spectrophotometer making full-disk Doppler measurements (Grec *et al.*, 1976), and the intention was to run the network during at least one eleven year solar cycle. The IRIS project was approved for funding in 1984.

In 1986, the first prototype of the IRIS instrument was set at La Silla. This instrument was almost identical to the one previously used with success at the Geographic South

Pole, and during one year in Chile, it was used to define all the modifications made necessary for such a long term project.

In 1987, the first real IRIS station, still not using the definitive instrumental design, was started at John Wilcox Stanford Observatory in California.

In 1988, the first complete IRIS instrument, including its own heliostat, was started at Kumbel, in a remote mountain site of the Soviet Republic of Uzbekistan. The same year was organized the first IRIS workshop at Peillon, near Nice.

In 1989, all sites were selected. Another instrument was started in a remote site, in l'Oukaimeden, Morocco, and in December 1989, the fourth instrument was started in Izaã, Tenerife, Spain. The second IRIS workshop was organized at Tashkent, Uzbekistan in September 1989. During this meeting, the first data available has been distributed to the site representatives.

In February 1990, the instrument of La Silla will be replaced by a definitive IRIS instrument. At this moment, the network will have five station running. The next instruments should be started at the end of 1990 and in 1991 for the last one. The network will then be routinely operated until the year 2001 at least.

The next IRIS workshop will be held in Marrakech, Morocco, in September 1990.

3 Structure of the International Cooperation. Membership

The success of such a long term international cooperation is not very easy to guarantee. The basic idea of the IRIS programme is a full participation of every local scientific team. In each site, one local scientist has the responsibility of the instrument. He has first visited Nice during a few months in order to participate to the instrumental integration phase to have a reasonable knowledge of this instrument, in case of on-site hardware problems.

The total volume of data produced by our one-pixel instruments is not tremendous and does not require a specific computing center, as it is the case for the GONG programme. One year of IRIS data should be of the order of 100 Mbytes (Gelly, 1990), and this can easily be accessible to all participating teams, even if it requires some level of specific management. The full network data can be distributed to all participants who can then efficiently share the work of data analysis. On the other hand, each instrument is provided with a serial output, so that the local team can have an immediate access to its own data, for partial but faster analysis.

The structure of the IRIS scientific community has been defined as follows.

Project Scientist: Eric Fossat, Nice.
Instrument Scientist: Gerard Grec, Nice.
Data bank responsibility: Bernard Gelly, Nice.

In each site, there are three key persons:

1 The Responsible for Cooperation (RC) deals with all questions regarding international cooperation, bilateral agreements, travel, funding, ...
2 The Site Representative (SR) has the responsibility of the instrument in the observing sites. He is also responsible for deciding who, in his team, is an IRIS member. The IRIS members have the priority of participating to the scientific analysis before the data will be made available to the broader scientific community.

3 One scientist is member of the IRIS Scientific Committee (ISC).

The questions of raw data distribution policy, raw data calibration software package production and qualification, scientific data distribution policy, publication policy and organization of the research in scientific teams are the responsibility of an IRIS Scientific Committee, which was created during the first workshop in 1988 and which first met during the second workshop in 1989 (see the report of this Committee meeting in the appendix).

The whole IRIS community meets once per year during the yearly workshop, which is organized sequentially by one of the participating teams. The workshop is also the opportunity for the IRIS Scientific Committee meetings.

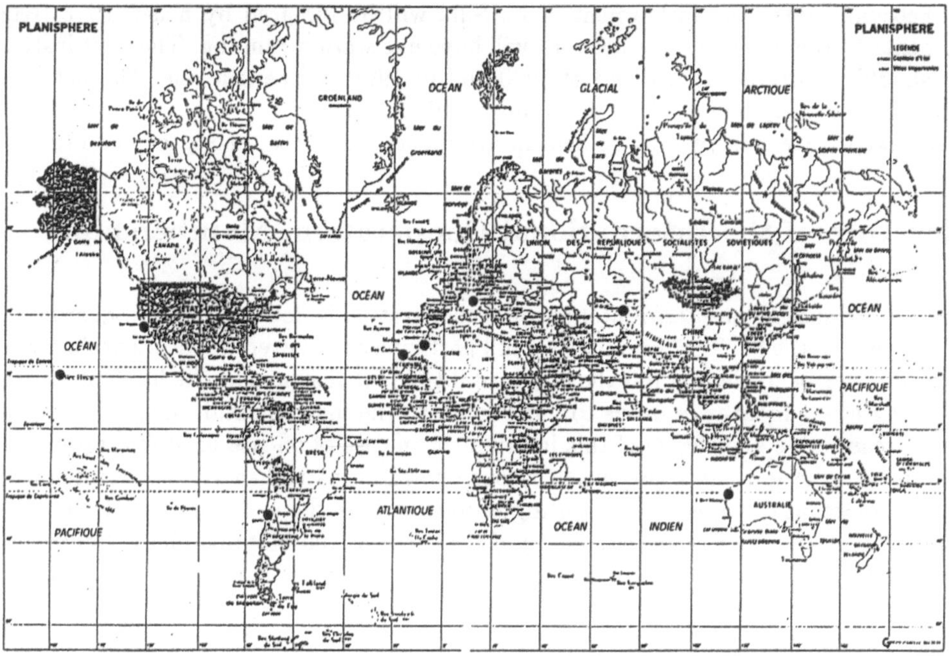

Fig. 1. The 8 sites finally selected for the IRIS network are shown on this map.

4 IRIS Sites and Teams

At the date of the second workshop, 8 observing sites have been definitely selected (Fig. 1), and two additional groups are officially participating to the programme. The 10 following teams are then involved in the IRIS programme.

4.1 University of Nice.
RC: E. Fossat.
SR: E. Fossat.

ISC: E. Fossat and G. Grec.
Members: E. Fossat, B. Gelly, G. Grec, F. X. Schmider.
No instrument will be operated there, but all instruments are built, calibrated and integrated with the participation, in each case, of one scientist of the local team. The data bank management is centralised at Nice.

4.2 Nice Observatory.
RC: Ph. Delache.
SR: B. Gelly.
ISC: Ph. Delache.
Members: G. Berthomieu, Ph. Delache, G. Gonczi, J. Provost.
One instrument will be set in 1990, mostly for technical work and studies of instrumental improvements to be tested and implemented.

4.3 John Wilcox Solar Observatory, Stanford, California, USA.
RC: J. T. Hoeksema.
SR: J. T. Hoeksema.
ISC: J. T. Hoeksema.
Members: J. T. Hoeksema, P. Scherrer.
Instrument routinely operated there since September, 1987. At present, the duty cycle is not optimal because of some interference with other observing programmes. Some recent instrumental improvements remain to be implemented.

4.4 Kumbel, Uzbekistan, Soviet Union.
RC: T. Yuldashbaev.
SR: S. Ehgamberdiev.
ISC: T. Yuldashbaev.
Members: M. Baijumanov, S. Ehgamberdiev, S. Iljasov, A. Kamaldinov, S. Khalikov, I. Khamitov, G. Menshikov, S. Raubaev, T. Yuldashbaev.
Instrument started in September, 1988 in the first remote site where nothing existed before. Routinely operated since summer, 1989.

4.5 Institute for Nuclear Research, Moscow, Soviet Union.
RC: G. Zatsepin.
SR: no site.
ISC: E. Gavryuseva.
Members: V. Gavryusev, E. Gavryuseva, A. Rosljakov, G. Zatsepin.
Group of theorists specially involved in the low frequency g-modes and in the physics of the solar core. This group is working more specifically in the solar neutrino experiments, which is the alternative and complementary method for the study of the solar core.

4.6 Oukaimeden, Morocco.
RC: S. Kadiri.
SR: M. Lazrek.
ISC: S. Kadiri.
Members: Z. Benkhaldoun, S. Kadiri, M. Khatami, M. Lazrek, H. Touma.
Instrument started in December, 1988 in the second remote site. Routinely operated since summer, 1989.

4.7 Izaña Observatory, Tenerife, Spain.
RC: P.L.Palle.
SR: L.Sanchez.

ISC: P.L.Palle.

Members: M. Anguerra, A. Jimenez, P. L. Palle, F. Perez, C. Regulo, T. Roca-Cortes, L. Sanchez.

Instrument started in December, 1989, near the so called MARK-1 potassium cell instrument of the Birmingham group.

4.8 La Silla, ESO Observatory, Chile.

RC: NA.

SR: D. Hofstadt.

ISC: D. Hofstadt.

Members: No scientific team is involved at La Silla.

Instrument setup scheduled for January, 1990.

4.9 Haleakala, Hawaii.

Only preliminary contacts have been taken so far. At the date of the second IRIS workshop, B. Labonte has been the scientific contact person. Instrument setup scheduled for the second half of 1990.

4.10 Learmonth, Western Australia.

Only preliminary contacts too. J. Kennewell has been the scientific contact person. Instrument setup scheduled for 1991.

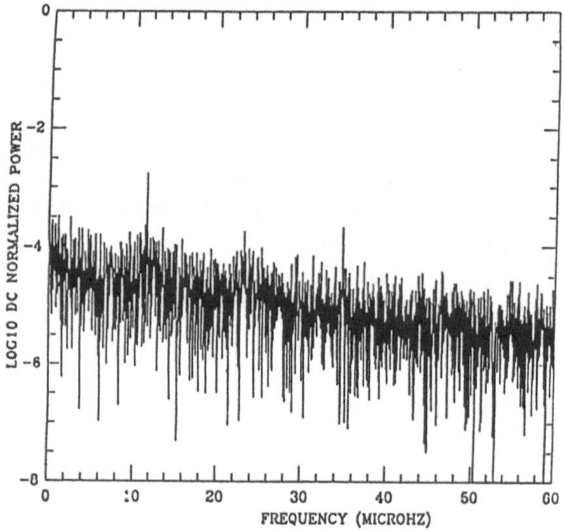

Fig. 2. In the case of a 6-site network not including Hawaii, this spectral window function obtained by a simulation using the GONG site survey routine (Hill *et al.*, 1988) shows a residual effect of the daily sidelobes at a level above 10^{-3} in power. This is due to the too large longitude gap across the Pacific Ocean.

At the time of the first IRIS workshop in 1988, no plan existed to select a site at Hawaii. However, the various possibilities which existed by then for a 6 or 7-site not including Hawaii have been simulated by means of the programme developed for the GONG site survey. As an example, Fig. 2 shows that a residual 24 hour effect cannot be

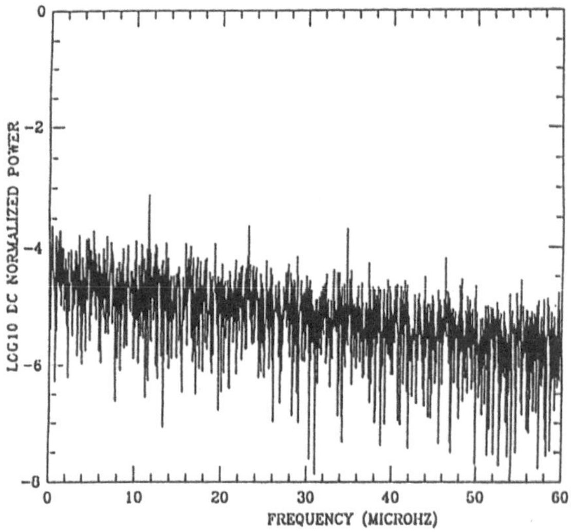

Fig. 3. The same simulation made for the finally selected IRIS network sites, including Hawaii, shows a decrease of about a factor 3 of the daily sidelobe structure and of the mean level of the window function.

avoided in this case, because of the too large longitude gap represented by the Pacific Ocean. For comparison, the same programme of simulation shows (Fig. 3) the slightly better efficiency of our final selection.

Acknowledgements

The duty cycle simulations and the last two figures were kindly provided by F. Hill.

Appendix

IRIS Scientific Committee

Final report, Meeting of September 8th, 1989
Tashkent, Uzbekistan

The IRIS Scientific Committee was created in October 1988, during the first IRIS workshop held at Peillon near Nice, following a suggestion made by Taymas Yuldashbaev.

This committee had his first meeting at Tashkent in September 1989, on the last day of the second IRIS workshop. This is a brief report of the decisions taken during this meeting.

1 Composition of the Committee

As a principle, the IRIS Scientific Committee is composed of:

–One scientist of each participating team, who is designated inside his team, with an exception for the University of Nice, author of the project, which has two members.
–Three outside experts, who are designated by the members of the participating teams.

All members are designated for one year, until the next workshop. The membership can be prolongated without restriction. The number of outside experts can also be adjusted once per year, during the annual workshop.

Are members of the Committee, in September 1989:

1. Team members
   Philippe Delache, Nice Observatory, France
   Eric Fossat, Nice University,France
   Elena Gavryuseva, INR Moscow, Soviet Union
   Gerard Grec, Nice University, France
   Todd Hoeksema, Stanford University, USA
   Daniel Hofstadt, ESO - La Silla, Chile
   Samir Kadiri, Rabat/Oukaimeden, Morocco
   John Kennewell, Learmonth, Australia
   Barry Labonte, Haleakala, Hawaii, USA
   Pere Palle, Izana, Tenerife, Spain
   Taymas Yuldashbaev, Kumbel/Tashkent, Soviet Union
2. Outside experts
   Frank Hill, GONG, Tucson, USA
   Douglas Gough, Cambridge University, England
   Ian Roxburgh, Queen Mary College, London, England

2 Chairman

The Committee is chaired by one of his members, for one year. The new chairman is elected during the yearly meeting. For the first year, Eric Fossat is acting as chairman.

3 Role of the Committee

The IRIS Scientific committee is responsible for all questions regarding the raw data distribution and release policy, the scientific data distribution and release policy, the software distribution policy and the publication policy. It will meet once per year during the yearly workshop, and it will work mostly by electronic mail, or telex where e-mail is not available, during the rest of the year.

4 Scientific Working Teams

The IRIS scientific community is naturally distributed in geographical teams, mostly located in or near the observing sites. It was felt necessary to also organise working teams, each one having a specific scientific responsibility. The number, role, and composition of these teams will be established yearly. In September 1989, four teams are created with the following members and responsibility.

4.a Instrument team

Chairman: Gerard Grec (Instrument Scientist)
Members: the Nice University group and the site representatives: Eric Fossat, Bernard Gelly, Jean-Francois Manigault, Guy Rouget, Francois-Xavier Schmider, Todd Hoeksema, Shuhrat Ehgamberdiev, Mohamed Lazrek, Daniel Hofstadt, Pere Palle, Barry Labonte, and John Kennewell.
Responsibility: To know the instruments in detail and to define any modification that could result in an improvement of its performance and/or an easier use.

4.b Raw Data Software Team

Chairman: Shuhrat Ehgamberdiev
Members: Bernard Gelly, Pere Palle, Shukur Khalikov, Eric Fossat, Luis Sanchez.
Responsibility: To produce within one year a complete software package, tested, agreed and qualified, which makes the characterization of the raw data, and the calibration in order to produce for each day and each site a velocity versus time signal.

4.c Data Characterization Team

Chairman: Samir Kadiri
Members: open
Responsibility: to have someone checking systematically the result of the data characterization software, in order to possibly improve it on a long term time basis.

4.d Data Merging Team

Chairman: Eric Fossat
Members: Vladimir Gavryusev, Todd Hoeksema, Philippe Delache, Bernard Gelly, F. X. Schmider, and we expect some input from the GONG members.
Responsibility: To produce, tentatively within one year, a complete software package tested, agreed and qualified, which produces optimal solutions to the data merging question, as a function of the required data quality.

5 Raw Data Distribution

Under the responsibility of Bernard Gelly, the raw data from the complete operated network will be distributed, not later than one year after the original observation, to the 9 scientists who are responsible for international cooperation in the 9 IRIS teams. This will be possible if the original cassettes are forwarded to Nice not later than 6 months after the original observations.

During the Tashkent meeting, the first data available have been distributed to the scientific responsible of each site.

At least one year after the raw data calibration software package will have been distributed to the 9 IRIS teams, this raw data will be made available to a broader scientific community.

6 Software Production and Distribution

The scientific teams B and D have the responsibility of producing, within one year or as soon as possible if later, a software able to provide calibrated velocity data from the whole network. The final approval of these two packages will be the responsibility of the Committee. Once approved, these two softs will be distributed to the 9 IRIS teams.

After a delay of one to two years (that remains to be precisely defined), the calibrated data will be made available to the broad scientific community.

7 Publications

It was strongly felt that the questions regarding the publication policy will find easy solutions when they raise during the development of the general IRIS programme. At the still initial stage of development which has been reached so far, the following rules have already been defined:

7.a The single site data analysis can be made by every observing team, which is allowed to publish its own results, with the mention of being IRIS members.
The full network data cannot be used for a scientific publication before the approval of the calibration and of the data merging softwares.

7.b All papers using IRIS data should be submitted to the IRIS Committee chairman for clearance of IRIS rules and for scientific comments.
This can be done at the same time as the paper is submitted to a refereed journal. The Committee is supposed to have enough time to react before the paper is definitely accepted by the journal referee.

7.c The statistical informations regarding the network itself will be published by the whole IRIS community.

7.d The first scientific result obtained from the analysis of a new data set, directly like a list of precise frequencies or indirectly like a theoretical deduction of the molecular mass inside a given layer, will always be published by the whole IRIS community.

This report was approved by the Committee members who attended the September 8 meeting: E. Fossat, Ph. Delache, T. Hoeksema, E. Gavryuseva, T. Yuldashbaev, S. Kadiri, P. Palle, D. Gough, and I. Roxburgh.

References

Claverie, A., Isaak, G.R., McLeod, C.P., and Van der Raay, H.B. 1984, *Mem. Soc. Astron. Ital.*, **55**, 63.

Fossat, E. 1988, *Adv. Space Res.*, **8**, 107.

Fossat, E., Gelly, B., Grec, G., and Pomerantz, M.A. 1987, *Astron. Astrophys.*, **177**, L47.

Gelly, B. 1990, *Solar Phys.*, submitted.

Gelly, B., Fossat, E., and Grec, G. 1988, *Astron. Astrophys.*, **200**, L29.

GONG (The Global Oscillation Network Group) 1984, Project, National Solar Observatory, Tucson, AZ.

Grec, G., Fossat, E., and Pomerantz, M.A. 1980, *Nature*, **288**, 541.

Grec, G., Fossat, E., and Vernin, J. 1976, *Astron. Astrophys.*, **50**, 221.

Hill, F. and the GONG site survey team, 1988, in *Seismology of the Sun and Sun-like Stars*, ed. E. Rolfe, ESA SP-286 (ESA Publication Division, Noordwijk), p.209.

Hill, F., and Newkirk, G. 1985, *Solar Phys.*, **95**, 201.

Palle, P.L., Regulo, C., and Roca Cortes, T. 1988, in *Seismology of the Sun and Sun-like Stars*, ed. E. Rolfe, ESA SP-286 (ESA Publication Division, Noordwijk), p.285.

Woodard, M.F., and Noyes, R.W. 1985, *Nature*, **318**, 449.

Helioseismology Observations by Solar-A Satellite

Takashi Sakurai

National Astronomical Observatory, Mitaka, Tokyo 181, Japan

Abstract: Helioseismological observations that can be made by using the aspect telescope on board the Solar-A satellite are discussed.

1 Introduction

The Solar-A satellite (Ogawara 1987) will be launched by the Institute of Space and Astronautical Science, Japan, in August 1991. Main scientific objective of this satellite is to observe solar activity, particularly solar flares, by using X-ray telescopes in soft and hard X-ray energies. The Soft X-ray Telescope (SXT) on board Solar-A will be equipped with the aspect telescope, which is made of a 5 cm lens mounted on the central axis of the X-ray grazing incidence mirror system. The primary purpose of the aspect telescope is to provide optical images of the Sun, which can be used as a reference in registering soft and hard X-ray images obtained by Solar-A. The CCD detector located at the focal plane of the SXT observes the soft X-ray images of the Sun through X-ray filters (opaque to the visible light). The same CCD records the optical images of the Sun when optical filters made of glass (opaque to X-rays) are in the light path. This aspect telescope can be a highly capable instrument for observing solar (brightness) oscillations. In the following we will describe the observation of helioseismology that is planned by using this instrument.

2 Description of the Instrument

The aspect telescope is made of a 5 cm primary lens, a pre-filter in front of the primary, and a narrow-band interference filter near the focal plane. The pass band of the last filter is centered at 4308Å (at the so-called G-band, the band head of CN) with 30Å band width. The overall transmission of the optics is such that an optimum exposure is 0.03 s.

The CCD detector at the focal plane has 1024×1024 pixels covering a field of 40 arcmin. (A single pixel corresponds to $2.4''$.) We will combine 4×4 pixels into a macro-pixel, which covers the area of $10'' \times 10''$ on the Sun. This binning is due to the limitation in the data memory of the satellite. Each image data will consist of 256×256 macro-pixels, and the solar image has the diameter of about 200 macro-pixels. In order to observe solar oscillations in the 5 minute band, we choose the sampling interval of 32 s. The orbital period of the satellite is 96 minutes, and 80 images are taken in each orbit. One day is made of 15 orbits, and 5 orbits are accessible from Japan. Most (but not all) of the data taken in the other orbits are transmitted to NASA's deep space network stations and are shipped to Japan later.

The full well capacity of a macro-pixel is about 2×10^5 electrons. The relative accuracy in a single measurement of brightness is therefore roughly 3×10^{-3}. The digital output from the CCD camera is in 12 bits (4096 levels), fully covering this accuracy. The data stored in the memory are however in 8 bit format, and for the X-ray images the conversion from 12 to 8 bits is applied via square root function. (Since one X-ray photon produces tens of electrons in the CCD, this conversion is compatible with the X-ray photon statistics.) For optical images the full accuracy of CCD output should be retained. For this purpose we will record the lower 8 bits by discarding the top 4 bits. Therefore the solar optical data are represented in 4096 levels modulo 256. The solar disk will look like Newton rings, but the original brightness distribution can easily be reconstructed because we know the limb darkening curve fairly accurately.

3 Expected Sensitivity of the Instrument

If each macro-pixel generates noise whose power spectrum is $P(\nu)$, coherent oscillations with the amplitude $\delta I/I_0$ will be detected when

$$\delta I/I_0 > \sqrt{P(\nu)/(N_x N_y T)}\ , \tag{1}$$

where $N_x = N_y = 200$ is the number of macro-pixels and T is the time span of observing run. Possible sources of $P(\nu)$ will include the photon noise, granulation, and the fluctuation of satellite pointing. The noise spectrum due to the granulation may be written as (Harvey 1985)

$$P_g(\nu) = \frac{2\tau_g}{1+(\omega\tau_g)^2} \frac{d^2}{l_x l_y} \left(\frac{\delta I_g}{I_0}\right)^2 . \tag{2}$$

Here $\tau_g = 6$ minutes is the characteristic time scale of the granulation, $\omega = 2\pi\nu$, and $l_x = l_y = 10''$ is the size of macro-pixels. The quantity $2\delta I_g/I_0$ is the brightness contrast of granules observed under the resolution of 5cm lens (i.e. $d = 2.5''$) and is taken to be 0.04.

The fluctuation in the satellite pointing produces additional noise by bringing the granulation pattern into and out of the CCD pixels. In order to evaluate this effect, the signal from the CCD macro-pixel traversing a fixed brightness pattern with a constant velocity v is calculated to be

$$P_d(\nu) = 4\left(\frac{\delta I_g}{I_0}\right)^2 \frac{d}{v} \frac{d}{l_y} \frac{1}{1+(k_x l_x/2)^2} \frac{1}{1+(k_x d/2)^2} \qquad (k_x = \omega/v). \tag{3}$$

The expected pointing drift is $v = 7'' \mathrm{min}^{-1}$. Based on these formulae, the noise levels are evaluated as in Fig. 1. The detection limit of our instrument according to (1) is given in Table 1. The photon noise is generally smaller than P_g and P_d.

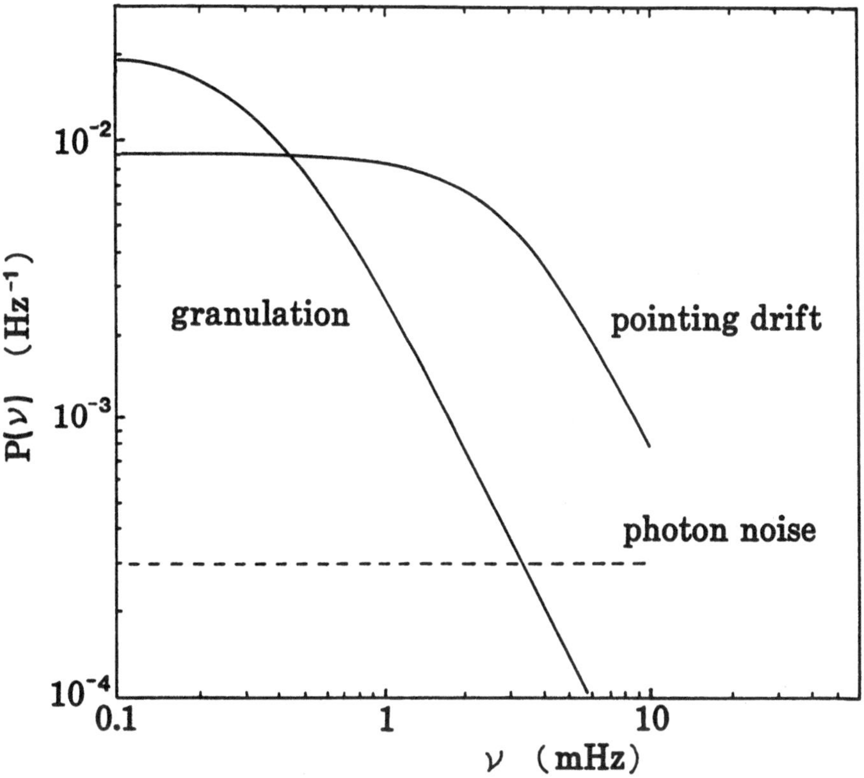

Fig. 1. Spectra of noise sources.

Table 1. Expected sensitivity of the instrument

observing run	number of exposures	detection limit ($\delta I/I_0$ per mode) at 3 mHz	at 1 mHz
week	8,400	4.9×10^{-7}	6.9×10^{-7}
month	36,000	2.3×10^{-7}	3.3×10^{-7}
year	438,000	6.8×10^{-8}	1.0×10^{-7}

4 Discussion

Table 1 shows that one week observing run will detect coherent oscillations whose amplitude is smaller than 1 ppm. In terms of velocity oscillations this is equivalent to the sensitivity better than $10\,\mathrm{cm\,s^{-1}}$.

A few supplementary points are in order here concerning the instrumental specification. To be ideal, the exposure should extend the whole sampling time (32 s). Such a long exposure will, however, blur the image due to the pointing drift. To make several exposures and add the data on board is not practically possible. Therefore our data will be like snapshot pictures rather than the time integration over 32 s. We do not expect severe aliasing in the power spectrum because the solar eigenmodes have no significant power in the frequency range higher than 5 mHz.

The effect due to the nonuniformity in the filter transmission is estimated to be less than 10^{-3} in terms of $\delta I/I_0$ and will not degrade the accuracy in the measurement. A remaining uncertainty is the effect of scattered light and ghost images, which may cause a spurious large scale brightness pattern that will move in responding to the pointing drift.

During this helioseismology program, the SXT can take only a limited amount of soft X-ray images. Therefore the first trial run will be after sufficient X-ray data are accumulated (say one year after launch), and will continue for a week or so. A longer observation run may be made after a preliminary analysis of the trial run gives a promising result.

References

Harvey, J.W. 1985, in *Proc. ESA Workshop on Future Missions in Solar, Heliospheric, and Space Plasma Physics*, eds. E. Rolfe and B. Battrick, ESA SP-235 (ESA Publication Division, Noordwijk), p.199.

Ogawara, Y. 1987, *Solar Phys.*, **113**, 361.

The Soho Project and Helioseismology

Vicente Domingo

Space Science Department of ESA, ESTEC,
2200AG Noordwijk, The Netherlands

Abstract: The Solar and heliospheric observatory (Soho) space mission being developed by ESA and NASA will carry together with other instruments devoted to the study of the solar atmosphere and solar wind, a set of instruments that will provide a comprehensive set of measurements of solar oscillations. Two investigations in Soho aim primarily at the study of g-modes and low l p-modes and a third one will have the central interest in high degree oscillations while aiming to extend the validity of its measurements to low l modes overlapping with the other two investigations. The Soho mission is being designed having into account the needs of the helioseismology experiments and therefore should be able to provide the best possible infrastructure for the production of good quality data. For the data analysis the three experiments will coordinate their operation and the data handling. It is expected that Soho will greatly profit of the experience gained with GONG, as a large fraction of the co-investigators in the Soho helioseismology investigations form part of the GONG project. The investigations in Soho have finished the definition phase of their instruments and the contractor selected by ESA has started on 1 December 1989 the industrial Phase B, or design phase, of the spacecraft and mission for a launch in March 1995.

1 Soho Mission. Status

The Solar and Heliospheric Observatory (Soho) is a joint ESA and NASA endeavour. The main aims of the mission are to study: a) The structure and dynamics of the Sun's interior and the variability of its output. b) The physical processes that form and heat the solar corona and originate the solar wind.

The mission consists of a spacecraft that carries a payload module in which 12 scientific instruments or groups of detectors are mounted (see table 1). These instruments, defined to accomplish the aims of the mission, are designed and built by scientific institutes, or universities, mainly from Europe and the United States of America. Three of them have helioseismology as their main objective, the remaining nine are devoted to the study of the solar corona and the solar wind. The spacecraft will be continuously pointed to the Sun's center within 10 arc sec, with a pointing stability better than 1 arc sec per 15 minutes. (A description of the Soho investigations can be found in ESA SP-1104).

The Soho project is now at the stage of industrial definition (Phase B). Current plans call for Soho to be launched in March 1995. About four months later the spacecraft will

Table 1. Soho investigations

SOHO INVESTIGATIONS

Investigation	Measurements	Investigation	Measurements
Global Oscillations at Low Frequencies (GOLF)	Global Sun velocity and magnetic field oscillations Harmonic degree L = 0–3	Ultraviolet Coronagraph Spectrometer (UVCS)	Electron & ion temperatures, densities, velocities in corona (1.3 - 10 R_0)
Variability of Solar Irradiance (VIRGO)	Low degree (L=0–7) irradiance oscillation and solar constant	White Light and Spectrometric Coronagraph (LASCO)	Structures evolution, mass, momentum and energy transport in corona (1.1–30 R_0)
Michelson Doppler Imager (MDI)	Velocity oscillations high degree modes (up to L = 4500)	Solar Wind Anisotropies (SWAN)	Solar wind mass flux anisotropies. Temporal variations
Solar Ultraviolet Measured Emitted Radiation (SUMER)	Plasma flow characteristics (temperature, density, velocity) chromosphere through corona	Charge, Element and Isotope Analysis (CELIAS)	Energy distribution & composition (mass, charge & charge state) ions 0.1 - 1000 keV/e
Coronal Diagnostic Spectrometer (CDS)	Temperatures and density: transition region & corona	Suprathermal and Energetic Particle Analyzer (COSTEP)	Energy distribution & composition, ions 1.2 - 330 MeV/n & electrons 0.06 - 25 MeV
Extreme Ultraviolet Imaging Telescope (EIT)	Evolution of chromospheric and coronal structures	Energetic Particle Analyzer (ERNE)	

be injected in a halo orbit around the L_1 Sun-Earth Lagrangian point, about 1.5 × 10 km sunward from Earth. Soho is being designed for a lifetime of 2 years, but it will be equipped with sufficient on-board consumables for an extra 4 years. Therefore Soho will, in principle, be in operation during the next solar minimum and may well be operational during the rise to the following solar maximum.

The scientific instruments aboard will be continuously operated. Regularly, Soho telemetry will be received by ground stations of NASA's Deep Space Network (DSN) during three short (1.3 hours) and one long (8 hours) periods each day, but during at least two consecutive months every year the telemetry will be received continuously (24 hours per day). Scientific data acquired outside these periods will be stored on magnetic tape aboard the spacecraft and transmitted to the ground during the short daily sessions. The Soho payload will produce a continuous stream of 40 kb/s; however, the bit rate will be increased by 160 kb/s during the regular 8-hr direct telemetry periods and during the two months campaign of direct telemetry. The additional 160 kb/s are generated by the solar oscillations imaging instrument in addition to its regular 5 kb/s bit rate.

2 Soho Helioseismology Investigations

2.1 GOLF

The Global Oscillations at Low Frequencies (GOLF) investigation will perform uninterrupted velocity oscillations measurements of the full solar disk. It is designed for attaining the highest sensitivity (1 mm/s and 1 milligauss) in the observation of the global solar velocity and magnetic field oscillations at very low frequencies. The investigation is being developed by a consortium of laboratories led by the Institut d'Astrophysique Spatiale (IAS), CNRS, Verrières-le-Buisson (F). Principal Investigator is A. Gabriel (IAS).

The instrument is based on the resonant-scattering spectrometric technique used on ground based instruments. The heart of the experiment is a resonance cell filled with sodium vapor which, by use of the the Zeeman effect, enables the selective absorption of light in the two wings of the sodium D lines (Na D_1, 589.6 nm and Na D_2, 589 nm). One fixed quarter wave plate and two rotation mechanisms (polarizing cube and quarter wave plate) allow the successive polarization of the beam into right and left circular polarizations when coupled with a small modulation of 200 gauss, in a 8 position sequence of 40 seconds on the permanent magnetic field (4750 gauss); this provides a precise determination of the global solar velocity and magnetic fields (with a calibration of the instantaneous sensitivity of the instrument by measuring the slopes of the profile). The detection makes use of 4 photomultiplier tubes, working in photon counting mode. The use of this technique allows the instrument to be photon noise limited, which guarantees high sensitivity in the long period range.

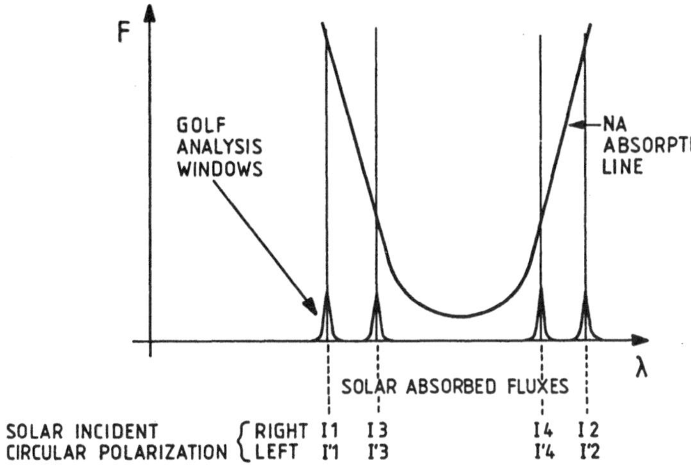

Fig. 1. Principle of the GOLF Doppler shift measurements.

The nominal operating mode is the following (Fig. 1): to two successive positions of the polarizer, correspond two absorbed fluxes I_1 and I_2, and $V_1 = (I_1 - I_2)/(I_1 + I_2)$ is an elementary velocity measurement. By changing the magnetic field, this elementary measurement becomes $V_2 = (I_3 - I_4)/(I_3 + I_4)$, and comparison between V_1 and V_2 gives the essential information on the solar profile slope.

In summary GOLF is designed to achieve low noise, high stability measurements of line-of-sight velocity integrated over the solar disk. It is particularly well suited to measure low l ($l = 0, 1, 2, 3$) oscillations (p- or g-modes). It aims at obtaining oscillations in velocity with a sensitivity of 1 mm/s, frequencies between 8 mHz (i.e. about 2 minutes) and 0.1 mHz (about 3 hours) and a resolution better than 10 μHz (about 20 days of observation).

2.2 VIRGO

Variability of IRradiance and Gravity Oscillations (VIRGO) is an experiment to investigate solar irradiance variability and oscillations. The consortium that develops VIRGO is led by the Physikalisch-Meteorologisches Observatorium/World Radiation Center at Davos (CH), C. Fröhlich being the principal investigator.

The instrumentation of VIRGO comprises two active cavity radiometers (PMO6 and CROM) observing the total irradiance, sunphotometers (SPMs) measuring the spectral irradiance at three wave-lengths in the near UV, the visible and near IR (335, 500, and 865 nm). In addition narrow band (500 nm) radiance measurements are carried out with 12 resolution elements on the solar disc by a high precision luminosity oscillation imager (LOI).

Two types of absolute radiometers are incorporated to guarantee the high stability needed for the measurement of the variability of the solar constant by assessment of possible degradation by internal comparison. Moreover, the PMO6 radiometers have means to measure, in flight, the reflectivity of their cavities, the most probable candidate for slow and small changes in performance.

The three-channel sunphotometers are simple instruments using interference filters for the wavelength selection and silicon diodes as detectors. They are very stable and have a precision which is well below 1 ppm needed for the detection of small amplitude solar oscillations. One of the instruments is exposed at most once per week for checking degradation.

The LOI comprises a telescope and a custom made multi-element silicon detector with 16 individual detector of similar noise characteristics to the ones in the SPMs. 12 elements are used for the measurement of the oscillations and 4 annular elements for the guiding (Fig. 2).

The data acquisition is performed by voltage-to-frequency converters. A microprocessor controls all the functions of the experiment and the S/C interface. For the scientific aims in general and to allow noise frequency analysis in particular it is essential to obtain time series with no or only a few gaps. Thus, a large memory is incorporated in the experiment. This guarantees the continuity of the data, even if some data were lost for technical reasons.

In summary the photometers and LOI will allow VIRGO to measure solar irradiance oscillations of low degree ($l = 0 - 7$) with a precision better than 1 ppm, low noise (solar limited) and a time resolution better than 10 μHz.

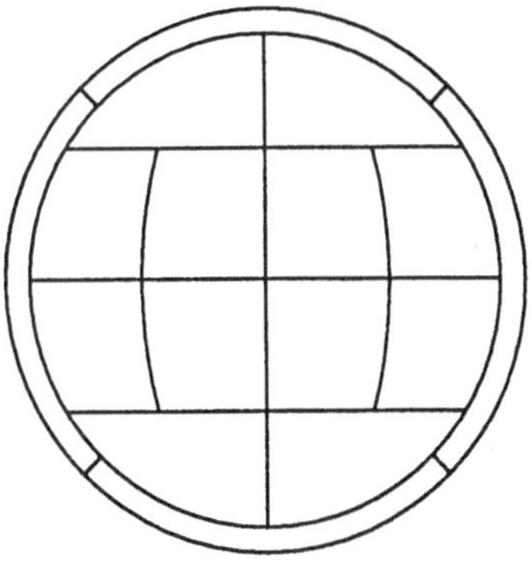

Fig. 2. Detector configuration of the low resolution oscillations imager (LOI) in VIRGO. 12 pixels to measure the oscillations and 4 circular sector pixels to guide the solar image.

2.3 MDI/SOI

The Michelson Doppler Imager (MDI) or Solar Oscillations Imager (SOI) investigation is developed by a consortium led by P. Scherrer, Stanford University (Cal.) as Principal Investigator.

MDI is essentially a flight version of the Fourier Tachometer. The Ni I 6768 Å line position is determined by making eight intensity measurements at four wavelengths evenly spaced across the absorption line. The wavelength is selected by a fixed Lyot filter and tunable Michelson interferometers. Two linear combinations of the intensities give the fundamental components of a four point discrete Fourier transform in wavelength from which the Doppler shift of the spectral line can be determined. This technique yields line-shift estimates with a linear response over a range of 4000 m/s, needed as a result of multiple effects like solar rotation, granulation, spacecraft orbital motions, etc. The telescope is a 12 cm f/15.6 classical refracting telescope, focal length 187 cm. Dual final optical paths provide 1.4 arc-sec-resolution images of a sub-region of the disk or 4 arc-sec resolution images of the full disk; either can be read out of the same CCD camera.

Part of the instrument is an Image Processing Unit (IPU) that receives directly the 1024 × 1024 pixel information from the CCD to perform the data extraction and processing needed to make best use of the available telemetry.

The MDI has several observing configurations; a high-resolution ($1.4''$) 717 arc-sec square image or full disk images ($4''$ resolution) can be made with linear or circular polarization in either sense. They allow to measure all components of the velocity and magnetic field vectors, and continuum and line intensities.

Diverse types of scientific investigations will be served by four complementary modes of operation:

- Full-disk observations of Doppler velocity in the primary line and continuum intensity will be carried out continuously for the duration of the mission. These observations will be processed on-board, and fixed parameters describing low-degree and selected high-degree oscillation modes transmitted in the continuous 5 kbps channel. They will be directed at understanding the global structure of the Sun.

- The same observations will be transmitted with minimal processing over the 160 kbps channel for two months at a time, providing the continuous time series of 4 arc-sec resolution Dopplergrams needed to determine the complete set of solar modes.
- During some campaigns a variety of observations will be made during the eight hours or more per day when the 160 kbps channel is available. These observations, of varying duration and repetition, will constitute individual experiments aimed at exploring the many small-scale and time-varying phenomena of the solar surface.
- Full-disc magnetograms will be supplied at least once every day for the duration of the mission.

3 Soho Helioseismology — Overview

The investigations in Soho will cover a large fraction of the oscillations degree - frequency (or k-ω) domain, but it will be in two regions where the Soho observations are expected to be more unique. One, in the low degree-low frequency oscillations area, where the g-modes are expected to be observable, the other, in the high degree area, where information about the upper boundary layer of the convection zone is missing (Fig. 3).

l	0-2	-3	-7	-20	-200	-750	-4500	OTHER
GOLF	V, B							
VIRGO	I							S
VIRGO	I							
MDI 2"	V, I, R				V, I *			B
MDI 0.7"						V, I **		B
GONG	V							

* PART OF THEM, CONTINUOUS
** DURING CAMPAIGNS

Fig. 3. Comparative table of the oscillations measurements made by the different instruments aboard Soho and by GONG.

Although both, GOLF and VIRGO, claim that the search for g-modes is a major aim in their investigation, today there is no clear indication of the fact that they will be able to see and identify g-modes. There are no firm observation predictions about the expected amplitudes, either in velocity or in intensity. Another problem is the mode identification; it may prove very elusive until one finds a data analysis method, probably guided by theoretical models, that favours the isolation of the g-modes. The collaboration between the different experiments in Soho will be important to help identify modes of oscillation, particularly because each of them has a particular domain for which it is best suited, and there will be partial overlap with each of the others in other domains. GOLF will provide the best, stable, velocity measurements for $l = 0 - 3$, MDI, for high l, perhaps less stable, and VIRGO, will have good stability for intensity, and will overlap for the degrees $l = 0 - 3$ with GOLF and extend its observations to $l = 7$, thus facilitating, perhaps, the mode identification. To have a good overlap between MDI and VIRGO, MDI will use a logical mask that will simulate the VIRGO 12-pixel measurements with integrated measurements from its 1024×1024 detector.

4 Soho in Relation to Other Helioseismology Programs

Figure 4 shows a time bar schedule of the known helioseismology programs from space and from the ground that attempt to operate in continuous mode. Only networks are mentioned for ground systems (in addition there is a network being developed by the University of Birmingham, that is already being operated partially).

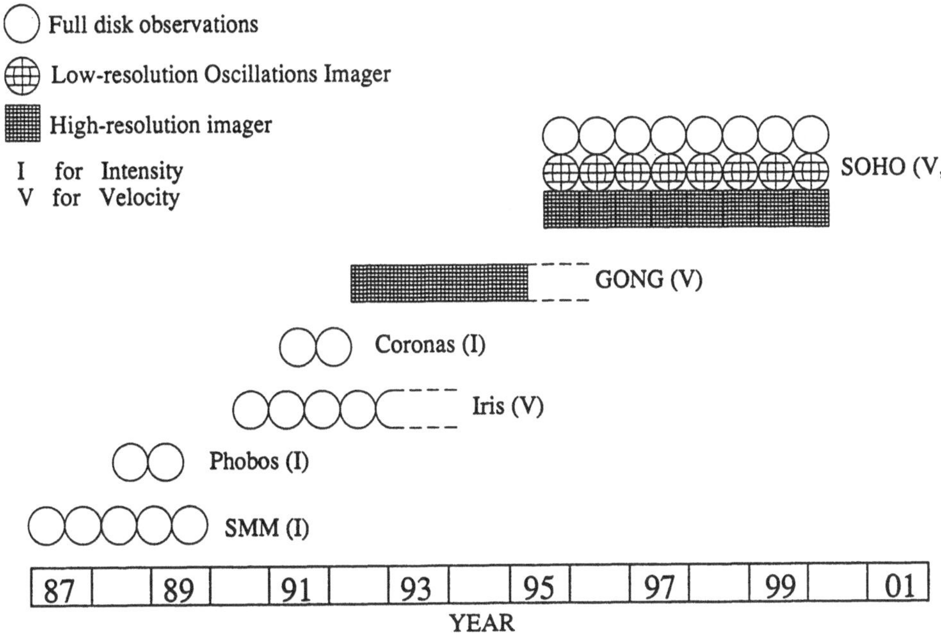

Fig. 4. Time chart of the different helioseismology investigations (space projects and networks). SMM = Solar Maximum Mission earth orbiting spacecraft, Phobos = Space probe to Mars and Phobos, Iris = International Research on the Interior of the Sun network, CORONAS = earth orbiting, sun synchronous satellite, GONG = Global Oscillations Network Group.

It is important to notice that Soho will start to operate after the Iris network (integrated solar disk velocity in six stations around the world) will have been in operation more that 5 years, and GONG (256×256 pixels velocity images of the sun, from 6 stations around the world) some 3 years, and hopefully both of them will still be operating when Soho will be launched. When Soho starts to operate much experience will have been obtained from the previous experiments and it is expected that it will profit from data analysis tools that will have been developed for these experiments, since they are sort of half way between the past generation, single station, short continuous periods of observation, and the Soho level, of 1 through 10^6 pixels, continuous operation during 2 to 6 years.

5 Data Handling and Analysis Aspects

The three investigations in Soho will coordinate their measurements (i.e. synchronized sampling) and the data analysis to make best use of their capabilities. GOLF and VIRGO will take special steps to insure close to 100 data recovery. MDI will insure close to 100 of parameters, probably some 95 some 2-months periods of high resolution, high degree oscillations continuous measurements, and other shorter periods (at least 8 hours) of selected high resolution information.

The data produced by MDI is a category of its own. MDI will produce about 300 gigabytes of raw telemetry data per year for at least 2 and perhaps 6 years. After decompression and production of over 20 sets of reduced and calibrated data required for analysis, the total data volume will expand by a factor of 10. Access to this unique data set of 6 to 18 terabytes will be available to members of the helioseismology community worldwide. The large quantity and variety of data will require special efforts and significant capabilities in data management. In particular, the MDI team intends to develop a data processing plan which takes the GONG experience into account, with possible overlap in both software design and hardware.

6 Conclusion

The helioseismology investigation in Soho will provide a coherent set of measurements, in the whole range of spatial and temporal ranges of oscillations in the Sun, that should allow for a comprehensive study of the solar interior structure and dynamics. The operation of Soho will take place at a time when the first generation of continuous data sets obtained with ground networks will have been analysed, or will be in the process of being analysed, and should benefit from the experience gained with them.

References

The Soho Mission, Scientific and Technical Aspects of the Instruments, ESA SP-1104, November 1988.

Magnetic Field Modulation Issues for Improving Global Solar Oscillation Measurements from Space

L. Damé [1], R.K. Ulrich [2], M. Martić [3], and P. Boumier [3]

[1]Direction des Etudes de Synthèse, ONERA, 92190 Meudon, France
[2]Astronomy Department, UCLA, Los Angeles, CA 90024, USA
[3]Institut d'Astrophysique Spatiale, 91371 Verrières-le-Buisson, France

Abstract: The measurement of global oscillations of the Sun from space will provide the ultimate means by which we will assess the existence, and hopefully observe some of the expected gravity modes. The SOHO-GOLF experiment, with a 4-point measurement in the line profile (resonance scattering method with a variable magnetic field applied to a sodium cell), may be able to distinguish between magnetic effects and true velocities. In this paper we characterize the effects of the magnetic fields and active regions on all aspects of the solar D lines in order to determine the best way to extract this signal from the solar background noise. These preliminary findings are then used to quantify the precision requirements for the GOLF Magnetic Field Modulation measurement method.

1 Rational for the Study

The observation and identification of solar g-modes is a major goal of helioseismology since the detection of only a few of them would constrain inversion models, and benefit our understanding of the solar interior. The SOHO mission addresses directly this goal by means of a stabilized satellite at the Lagrange point. This mission includes three complementary types of oscillation experiment each of which focuses on a different aspect of the problem of detecting low amplitude global oscillations. The resolved sun oscillations in velocity will be measured by the Michelson Doppler Imager (MDI), global and resolved Sun oscillations in intensity will be measured by VIRGO, and global oscillations in velocity will be measured by the Global Oscillations at Low Frequencies experiment (GOLF).

The MDI does not have an adequate system to control the Michelson's tuning and passband of the prefilter and can only observe g-modes through their differential effects on the separate parts of the solar image. VIRGO by measuring both the global oscillations in intensity and by resolving the solar image into 12 pixels may be able to use observations of surface structures to assess long term drifts due to magnetic effects. However, the intensity is more sensitive to solar noise such as that from plages than is the velocity. The frequency dependence of the low amplitude solar intensity noise is not yet known, and there might be a fundamental limitation of the temperature amplitude due to radiative processes.

The noise sources are either **solar** in origin: photon noise, active regions magnetic effects, supergranulation or giant cells; or **instrumental**: filter stability, polarizer positioning, detector stability, or sampling accuracy (for Fourier analysis). In the case of MDI, the limit is instrumental, since a stabilized laser (10^{-10} stability required) is not part of the payload. In the case of GOLF and VIRGO the limit will probably come from the solar noise itself; both instruments propose specific developments to better address the g-mode detection: GOLF through the 4-point measurement and VIRGO through the limited resolution mode.

The g-modes are indeed at the limit of our present detecting capabilities on ground [$\leq 7\ \mathrm{cm\,s^{-1}}$ according to the velocity observations of van der Raay (1989, 1990); of an amplitude $\leq 5\ \mathrm{cm\,s^{-1}}$ according to Fröhlich (1990) deduced from IPHIR intensity observations). And since the "solar noise" background (Jiménez *et al.*, 1988) is comparable to these values, in order to get a confident detection of the g-modes (i.e. with at least a 3σ level over the noise), the solar background has to be reduced to the $1\ \mathrm{cm\,s^{-1}}$ level in the frequency range of interest, i.e. for periods of, or longer than, one hour. This supposes a better knowledge of the nature and content of the "solar noise," and means to calibrate it out.

2 The GOLF Strategy

For the purpose of dealing with "solar noise," the GOLF instrument (Damé *et al.*, 1987) was proposed with a 4-point measurement of the line profile rather than the classical 2-point measurement usually performed by resonance scattering spectrometers. The principles of the instrument and implementation of the 4-point measurement are described in Damé (1988). In brief, an extra magnetic field is applied to the sodium cell in order to slightly change the position of the absorption bands in the line profile ; this allows a measurement of the slope around the working points and a better determination of the line profile. In this paper we concentrate our effort on evaluating the required amplitude and precision of this magnetic modulation in order for it to address the spurious velocities induced by magnetic effects/active regions.

Accordingly, one of the prime objectives of our study is to clarify the nature of the magnetic effects line shifts and shape changes. The correlation of the downdraft effect and regions of high magnetic field suggests that the changes are somehow caused directly by radiative transfer in regions of high magnetic field. We have investigated this possibility by separately measuring the line profiles in the two polarization states which are normally used for the magnetic field maps. We find that the line profiles in detail are independent of the polarization state. This suggests that the line shift of the downdraft effect is not directly caused by the magnetic field which would also produce different profiles in the two different states of circular polarization. Another question is relationship between the apparent velocities and the line profile changes. Is the apparent shift simply a result of the profile change or is the line as a whole shifted? We find that the NaD lines mostly have their profile changed with the line center relatively unshifted while $\lambda5237$

is shifted with a profile that is virtually unchanged. The case of λ5250 is intermediate. The λ5237 line is from Cr II and is formed most deeply in the solar atmosphere while the core of the Na resonance D lines is formed highest in the atmosphere. The fact that λ5237 is most sensitive to the downdraft effect while the D line cores are the least sensitive suggests that the shifts and profile changes result from true velocities induced by modifications to the energy transport brought about by the magnetic fields. A reduction of the convective energy flux would reduce the replenishment of photospheric energy and cause those regions to become cooler than average and settle inward. The deeper layers are the ones most affected by convection and therefore are the layers which should experience the strongest downdraft effect.

This argument suggests that the line shifts are a secondary result from the magnetic fields and that the correlation between the shifts and the magnetic field strength may not be perfect. An additional indication that the magnetic field is not the best parameter for detecting the regions of non-oscillatory velocities comes from the presence of strong Evershed flows in spot groups. These regions are most evident when they are at intermediate center to limb angles and typically have a near side which is approaching with a velocity of 200 to 400 $\mathrm{m\,s^{-1}}$ and a far side which is receding at a similar velocity. A simple approach which anticipates only receding velocities cannot deal with such structures.

Cavallini *et al.* (1988) have studied line bisectors in several spectral lines of Fe I, Fe II and Ca I as a function of the level of solar activity. They come to conclusions that are very similar to ours and ascribe the effects "to the modulation of the convective effect by the magnetic field." In a related work Solanki (1989) has examined theoretical Stokes profiles of four spectral lines with widely different properties. He concludes that the observed asymmetries are a result of magnetic elements embedded in cool downflowing intergranular lanes and that large amplitude non-stationary mass motions are present within the magnetic elements. In view of the agreement among the investigators, we refer to these line shifts as a downdraft effect and will base our discussion on this physical picture.

Another aspect of the downdraft is the solar cycle dependence. Is the general level of sunspot activity a good indicator of the importance of the downdraft effect? Explicit models attempting to reproduce the velocity shift of integrated sunlight have been developed based on sunspot numbers (Jiménez *et al.*, 1986). These models are only reliable if the downdraft effect is in fact closely tied to the sunspots and such a relationship is not yet established empirically. The most important property of the downdraft effect is its time dependence. Previous effort to measure the velocity of downdraft areas on the solar surface depended on a selective isolation of the region from unaffected regions nearby and a computation of the relative velocities inside the two areas. This approach indicated a large enough amplitude of variation in the critical 15 to 40 minute period band that most of the solar noise is due to this process (Ulrich *et al.*, 1989). However, the supergranulation may be a more significant contributor to the solar noise and if this is true there is little hope of compensating for the noise from this source in the integrated sunlight velocities. If on the other hand, the solar noise is mostly due to active regions then we can hope to develop an algorithm to deduce its amplitude to the $\mathrm{cm\,s^{-1}}$ level where confident g-modes detection could be carried. Evidence that the supergranular noise dominates is found in the paper by Jiménez *et al.* (1988) where it is shown that the observed background power matches the theoretical frequency spectrum predicted by Harvey (1985), and dominates in the crucial range of frequency. This match is not convincing evidence that the supergranulation is the cause however since most stochastic processes will produce a similar frequency dependence. A more discriminating test might be to look for a solar cycle dependence of the noise – the supergranulation noise should be independent of solar activity. We point out below that even this method of discriminating may be inconclusive because the downdraft effect appears to be somewhat less dependent on the overall level of activity than might have been expected. This paper is a progress report on our efforts which are not yet conclusive.

3 Solar Cycle Variations

The Mt. Wilson 150-foot tower system since 1982 has had the capability of observing two spectral lines simultaneously through a dual stage assembly in the exit slit enclosure. Since 1988, as part of the GOLF investigation, we have used this capability to study the downdraft effect through the comparison of the velocities in $\lambda5250$ and $\lambda5237$, observations which are now carried on a daily basis. Fortunately, prior to our program, and for a period of 6 weeks during 1985, the second stage was also used to measure $\lambda5237$ and we are now able to investigate the downdraft effect for those observations. At that time there were still a few spot groups near the solar equator from solar cycle 21 while the new spot groups from solar cycle 22 had not yet appeared at the higher latitudes. In fact at no time during sunspot minimum was the solar surface much less covered with spots than during that interval. The spot groups were primarily in just one E-W hemisphere of the solar surface.

Accordingly to the Doppler differencegrams, one from the more active hemisphere, and one from the inactive hemisphere, the inactive hemisphere is completely uniform whereas the more active hemisphere is not noticeably less covered with downdraft areas than is typically the case in 1989 during the period near sunspot maximum. Although the coverage is less at solar minimum than at solar maximum, the low frequency noise from active regions is still present and significant; yet, the issue is still to determine which, from the active region noise and supergranulation one, is the relevant one in the low frequency band of the solar g-modes.

4 Resolved Sun Measurements of Magnetic Effects

We first give here an evaluation of the level of line profile distortions that can be expected from magnetic effects, and then we compare this evaluation with observed resolved sun profiles of the sodium lines (D1 and D2 since both will be observed by GOLF), and global sun profiles, either by a spectrograph set up (Mt. Wilson), or a GOLF type experiment (Tenerife). Depending on the time dependence of these effects, the level of profile distortion observed could be harmful to the measurement of long period oscillations. In order to address this potential problem we are searching for the parameters whose measurement may yield a reasonable indicator of these magnetic effects. The slope parameter, chosen for GOLF as previously mentioned (2), is measured at Tenerife by a simplified GOLF breadboard instrument. Another indicator is the intensity of the center of the line. In active regions this parameter has a different value than in quiet regions so that it may permit a more precise resolution of the observed signal into magnetic and oscillatory components. The limited length of this proceeding did not allow to include these results herein (even though presented orally).

To determine the contribution of downdraft regions to the integrated sunlight velocity as measured at the working points of the GOLF experiment (±0.1Å from the center) in the Na D lines, a necessary first step is the measurement of the line profiles in a typical downdraft region and a nearby quiet region. From a Doppler difference-gram measured on August 11, 1989, we were able to select (in interactive mode) a downdraft region which was far enough from sunspot groups that the intensity variations on the solar surface would not introduce additional noise from image motion on the entrance slit. A series of 25 scans was made. The average of 4 of these is shown in Fig. 1a. A 12th order polynomial was fitted through the points between ±0.3Å. The dots at the top of the figure show the deviation of each point from the fitting curve with a scale as shown on the left of the figure. For comparison, a similar line profile from a non-downdraft region is shown in Fig. 1b. These profiles are both for the sum of the two states of circular polarization. The data were taken with the polarization analysis system operating and could have been displayed in either state of circular polarization but would have been indistinguishable from those shown.

At the working point level in the downdraft line profile, a significant change in the slope can be noticed, which account for 20 to 25 % difference when compared to the non-downdraft region. Another significant difference in between the profiles is the core intensity, which is sensibly affected in the downdraft region. Similar changes were reported by the Bordeaux Observatory team (Robillot and Bocchia, 1989) when performing preliminary tests of their resolved sun scattering spectrometer (with a 5-point measurement of the profile).

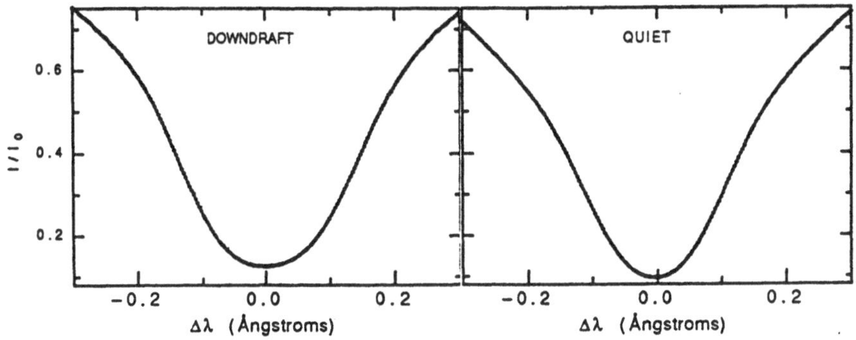

Fig. 1. Na D1 line profile in downdraft (a) and quiet (b) solar regions.

The time evolution of these downdraft profiles show variations of the slope parameter (the average slope, both the red and blue slope being correlated) of ~ 5 %. Note that the same sensitivity has been observed for the D1 and D2 lines. From these downdraft regions observations, a crude estimate of the resulting effect they should have on the integrated solar profile can be evaluated. Altogether with the downdraft regions coverage (~ 5 %), the mean slope change (~ 20 %), and the slope variations observed, the variations of the slope of the full solar disk profile should be about $0.05 \times 0.8 \times 0.05 \simeq 2 \cdot 10^{-3}$ (0.2 %); the 0.8 value reflects the fact that the mean downdraft slope is inferior to the mean quiet slope (and thus has a reduced effect in the weighted average). Note that these values are very preliminary and that a systematic program to better evaluate the coverage and to calculate the effect of integrating the profiles from very different latitudes has to be worked out to account for the different working points positions (and thus slope sensitivity) due to offset velocity (the solar rotation in this case being the predominant effect: $\pm$ 2 $\mathrm{km\,s^{-1}}$, ~ $\pm$ 40 mÅ in the position in the line profile).

5 Integrated Sun Measurements of Magnetic Effects

Two series of observations were performed in order to quantify the magnetic effects on the integrated line profile. These observations are recent and not all systematic effects might have been removed yet. In particular the series of observations performed at Tenerife, had the D1 and D2 lines at once, and might present a variable telluric lines influence in D2. Also, in the case of this GOLF simplified breadboard, the position of the working points in the line profile is affected by the mean velocity, which result in large sensitivity changes (further accentuated by the fact that D2 has two absorption bandwidths, one nearest from the bottom of the line).

The Mt. Wilson observations have been carried during several months and several systematic effects have already been removed or better understood (like vignetting). The December observing days were obtained under good conditions. The noise level is $\simeq \pm 5$ $\mathrm{m\,s^{-1}}$. Even though instrumental problems still are present, the day to day changes and the variations with time are believed to be reliable. Figure 2 shows 3 successive days of observations reported to the same UT hour. One can easily notice the changes from day to day of the average slope level, ~ 0.3 to 0.5 % (which will need in future work to be correlated with the downdraft regions coverage), and the slow variations during the day.

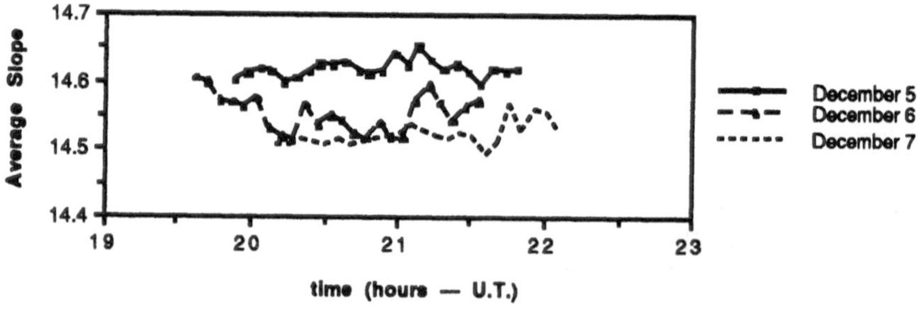

Fig. 2. Mt. Wilson integrated Sun Na D1 line – time evolution of the line profile slope.

The $\pm$ 1 σ amplitude of these fluctuations calculated for a representative day, the 5th, is 0.18 %. This is in good agreement with the level of variations estimated from the resolved Sun observations and this prompt us to regard this number as correct and, further, to expect that *most of the solar noise background might therefore be due to the magnetic effects/downdraft regions noise*, as expected by Ulrich *et al.* (1989) previous results.

The Tenerife results, obtained with a resonance scattering experiment in which a modulation of ± 100 gauss is added to the fixed (4300 gauss) magnetic field, show similar variations. However, as mentioned, the working points position changes slightly from day to day and the D2 line influence is still largely unknown. As an example Fig. 3 shows the red (right) and blue (left) slopes recorded for several hours on September 18 (the data gap around noon corresponds to a problem with the celostat at this period of the year). One can see that even though the two slopes measurements are indeed correlated, the linear trend seen in the left slope is not as well followed by the right one, and the amplitudes of the two do present differences, mainly due to the asymmetry of the working points.

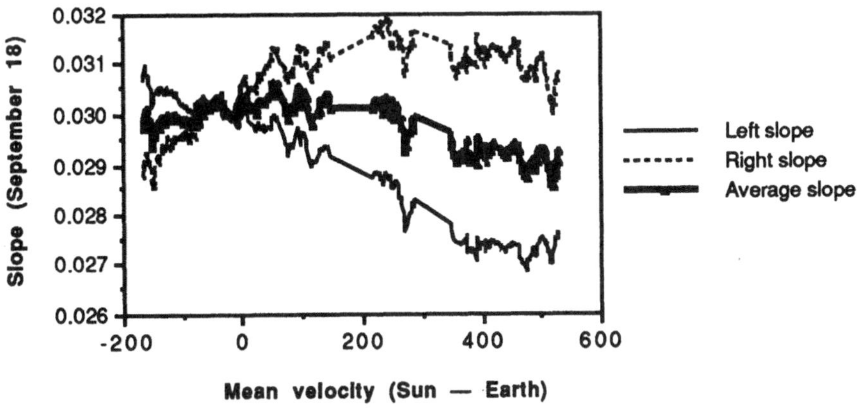

Fig. 3. Time evolution of the left and right line profile slopes – Tenerife observations.

From 3 days of observations, September 18, 19 and 20 (cf. Fig. 4), where we plotted the average slope only, day to day changes, and variations during the day are also observed. Note that in this case the slope is simply given by the ratio $(I_+ - I_-)/(I_+ + I_-)$. The Mt. Wilson slopes are normalized by the half $\Delta\lambda$ amplitude of the modulation in between the points used for slope measurements. In same units, the 14.55 Mt. Wilson average slope would convert to 0.0315.

To do this calculation the sensitivity of D1 and D2 were assumed equivalent. Even though the absolute slope values are comparable, the day to day changes and amplitudes of variations are different in the two cases. The day to day changes measured at Tenerife are around 3 to 5 % and the slope variations $\simeq$ 1.6 % (only in the best period of September 19, they get below 1 %). There is clearly a sensitivity difference of 8 to 10 with Mt. Wilson data which is hardly explainable and deserves further investigations. The long term stability of the Tenerife instrument can certainly be questionable but the fact that the mean slopes are comparable and that the same ratio exits both at Tenerife and Mt.

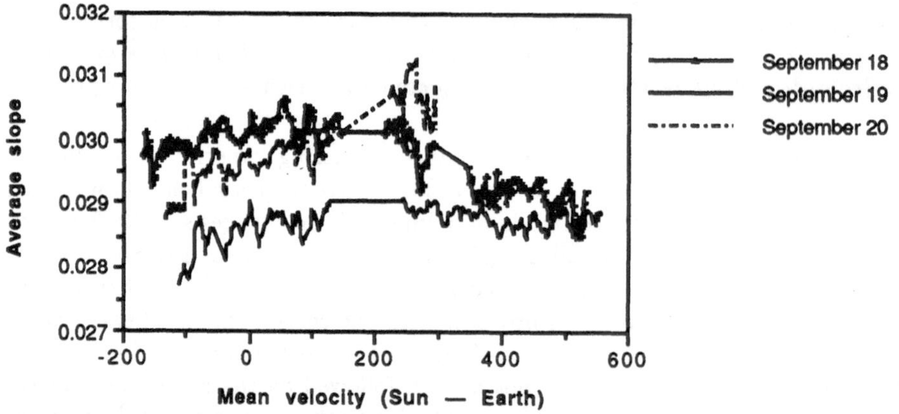

Fig. 4. Time evolution of the average slope in the Na lines – Tenerife 4-point measurement.

Wilson in between the day to day changes and the amplitudes of the fluctuations during the day makes difficult presently to think of a systematic effect which could explain such sensitivity differences (also the bandwidths are comparable: 80 mÅ for Mt. Wilson, three times 25 mÅ for Tenerife). Photon noise could account only for less than 0.4 %, and the problem might rather be in the long time in between the modulation measurements : 40 seconds.

6 Precision Requirements on the 4-Point Measurement

6.1 Required Precision on the Slope Measurement

According to the preliminary observations that have been made, the slope fluctuations expected on the integrated solar profile are of $\simeq 0.2$ %, i.e. $2 \cdot 10^{-3}$.

This 0.2 % represents the amplitude of the phenomenon. To correctly address and measure it, we need to define a precision on this measure. It has to be based on the expected noise of the long-term fluctuations. For coherence with the frequency band of interest, a corrective signal at least every hour with a precision of $\simeq 1$ $\mathrm{cm\,s^{-1}}$ is needed. Since the noise in this frequency range is $\simeq 20$ $\mathrm{cm\,s^{-1}}$ (Ulrich *et al.*, 1989), to achieve 1 $\mathrm{cm\,s^{-1}}$ out of 20 representing 0.2 % variation, a precision of $2\ 10^{-3}/20$, i.e. $\simeq 10^{-4}$ is foreseen on the slope measurement every hour.

The slope measurement gets its precision from the amplitude of the modulation on which two errors participate (cf. Fig. 5): the photon noise and the precision on the ΔB ($\delta\Delta$B).

For GOLF, a slope measurement is done every 40 s, with a duty cycle of 4/5 (useful observing time), and a velocity sensitivity of 4000 $\mathrm{m\,s^{-1}}$ (conversion factor between the measured intensity and the associated velocity), which leads to a photon noise of:

$$< \sigma_\nu^2 >= \frac{5}{4}\frac{S^2}{N} = 1.6\ (\mathrm{m\,s^{-1}})^2\mathrm{Hz}^{-1} \tag{1}$$

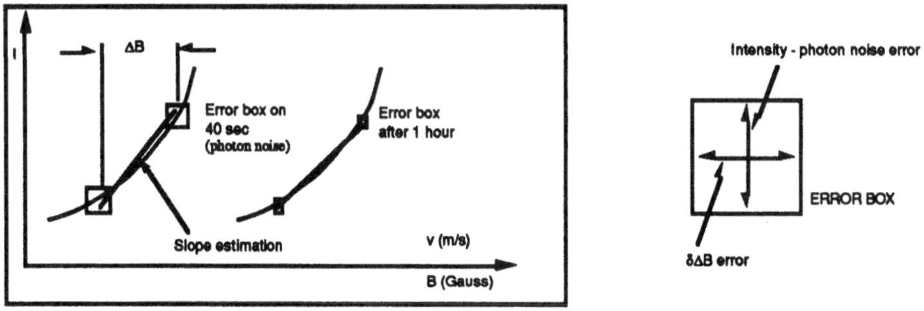

Fig. 5. Errors on the profile's slope measurement.

which over 40 s reduces to: $\sigma_\nu = \sqrt{1.6/40} = 20\ \mathrm{cm\,s^{-1}}$.

6.2 Required Amplitude for the Magnetic Field Modulation

After an hour, the error box on the measurement is reduced by $\sqrt{90}$, i.e. to 2.1 $\mathrm{cm\,s^{-1}}$. To achieve a 10^{-4} precision on the slope with two measurements with an error of 2.1 cm/s, the required amplitude is :

$$\pm\frac{2.1}{\sqrt{2}}\ \mathrm{cm\,s^{-1}} \times 10^4 \simeq \pm 150\ \mathrm{m\,s^{-1}} \tag{2}$$

which, according to the experiment characteristics, will correspond to an amplitude of the modulation of ± 150 gauss.

6.3 ΔB Precision

The same precision of 10^{-4} applies to ΔB which implies a 0.03 gauss σ on 1 hour and, respectively, 0.4 gauss on an individual measurement, if one assumes white noise.

Note that these calculations are based on a slope measurement using only one wing of the line profile, and that a cross measurement using both sides would require a smaller amplitude of modulation (the reference amplitude being given simply in this case by the one in between the working points). This has to be investigated thoroughly even though the profile distortions and working point slow drifts would probably prevent using this option.

7 Conclusions

The results are still preliminary. For the integrated Sun measurements the net effect is small and the noise performance of the Mt. Wilson system is not yet adequate to permit the unambiguous detection of oscillatory changes in the line profiles. Even though the observation of 0.2 % slope variations both deduced from resolved Sun and integrated Sun observations, gives us some confidence in the result. The Tenerife experiment with the GOLF type experiment shows similar trends than the Mt. Wilson observations, despite a larger amplitude of variations which might result from some still unexplained effects resulting from the D2 use or other instrumental instabilities to be identified. Joint periods of observations are required for the two experiments to properly compare them and account for Solar Activity changes.

The preliminary profile distortion reported, leads to precision requirements on the slope measurement of 10^{-4}, which is in reach of the GOLF current design and should allow a calibration point for the non-magnetic velocities, every hour, at the 1 $\mathrm{cm\,s^{-1}}$ level.

After additional improvements to the systems are completed, we expect to use the line profile measurements of Mt. Wilson and Tenerife, to obtain further indications of the potential of the GOLF 4-point measurement in addressing the magnetic effects calibration.

Acknowledgements

We are grateful to H.B. van der Raay for providing us with the basic of the instrumental hardware for the Tenerife experiment, and for the time spend in the instrumental set up and its calibrations. We further thank J.M. Robillot and R. Bocchia for provision of the magnet assembly and G. Grec and E. Fossat for the sodium filter. Special thanks go to P. Pallé for support in the Tenerife observations and data analysis, and to T. Roca Cortés for allowing this new instrument at the Izana Observatory, and for providing support for its operation.

References

Cavallini, F., Ceppatelli, G., and Righini, A. 1988, *Astron. Astrophys.*, **205**, 278.

Damé, L. 1988, in *Seismology of the Sun and Sun-like Stars*, ed. E. Rolfe, ESA SP-286 (ESA Publication Division, Noordwijk), p.367.

Damé, L. 1989, *Comparison of the Measurement Sensitivity to Global Solar Oscillations of GOLF and the MDI*, Helioseismology Symposium, Annales Geophysicæ, Special Issue EGS XIV General Assembly, p.226.

Damé, L., Cezarsky, C., Delache, Ph., Deubner, F.-L., Foing, B., Fossat, E., Fröhlich, C., Gabriel, M., Gorisse, M., Gough, D., Grec, G., Pallé, P., Paul, J., Roca Cortés, T., and Stenflo, J.L. 1987, *Global Oscillations at Low Frequencies (GOLF)– an Investigation of the Solar Interior*, Proposal submitted to ESA and NASA in response to the Announcement of Opportunity SCI (87) 1/OSSA-1-87 for the Solar and Heliospheric Observatory.

Fröhlich, C. 1990, these proceedings.

Harvey J. 1985, in *Future Missions in Solar, Heliospheric and Space Plasma Physics*, eds. E. Rolfe and B. Battrick, ESA SP-233 (ESA Publication Division, Noordwijk), p.199.

Jiménez, A., Pallé, P.L., Pérez Hernandez, F., Régulo, C., and Roca Cortés, T. 1988, *Astron. Astrophys.*, **192**, L7.
Jiménez, A., Pallé, P.L., Régulo, C., Roca Cortés, T., Isaak, G.R., McLeod, C.P., and van der Raay, H.B. 1986, *Adv. Spac. Res.*, **6**(8), 89.
Robillot, J.M., and Bocchia, R. 1989, private communication.
Solanki, S.K. 1989, *Astron. Astrophys.*, **224**, 225.
Ulrich, R.K., Boyden, J.E., Webster, L., and Shieber, T. 1989, *Magnetically Induced Spectral Line Redshifts Full Disk Measurements*, Helioseismology Symposium, Annales Geophysicæ, Special Issue EGS XIV General Assembly, p.325.
van der Raay, H.B. 1989, *g-Modes Identification Problems and Possibilities*, Helioseismology Symposium, Annales Geophysicæ, Special Issue EGS XIV General Assembly, p.226.
van der Raay, H.B. 1990, These proceedings.

Construction of Long-Life Magneto-optical Filters for Helioseismology Observations

T. Sakurai [1], K. Tanaka [1], H. Miyazaki [1], K. Ichimoto [1],

A. Sakata [2], and S. Wada [2]

[1]National Astronomical Observatory, Mitaka, Tokyo 181, Japan
[2]University of Electro-Communications, Chofu, Tokyo 181, Japan

Abstract: A design of magneto-optical filters we are developing is described. By heating the cell to about 200°C, a lifetime more than a year has been achieved.

1 Introduction

The magneto-optical filter (MOF, or Cacciani filter) is made of a transparent cell placed in a strong magnetic field (Agnelli *et al.* 1975; Cacciani and Fofi 1978). The cell contains the vapor of material like sodium. By utilizing the magneto-optical effect near the wavelength of a spectral line, MOF provides a very narrow and stable passband, which is particularly suitable for helioseismological observations. We started basic experimental studies on this type of filters in 1984 (Koyama 1986). In the design of most MOF's, only the sodium reservoir is heated, so that the sodium vapor attaches onto the wall when it cools there. The cell therefore becomes less transparent as it ages, and can only be used in observations for several months. Our aim was to construct a filter that can be continuously used for several years without losing the sodium vapor at the cell wall. This can be achieved by heating the whole cell to about 200°C.

2 Design of the Filter

Since the sodium vapor is chemically highly active, ordinary glass when heated to ~200°C reacts with the vapor. Namely silicon atoms in the glass are replaced with sodium atoms and the glass is gradually deteriorated in transmission. In addition if the cell is placed in the air of room temperature, large temperature gradient in the cell window introduces mechanical stress in the window. The stressed window shows a retardation effect which hinders the performance of the filter. Therefore the window material should have small thermal expansion coefficient.

Glass which is relatively resistant to sodium vapor (Schott 8436) was adopted for the body of the cell. The loss of sodium into the glass is only of a minor quantity, but the darkening of the window makes it inappropriate to use this glass as the window of the cell. After several attempts, we found a satisfactory material, which is Gadolinium Gallium Garnet (GGG). This material is not deteriorated by sodium, and showed negligible retardation effects due to thermal distortion. The thermal distortion of the cell window can be minimized by putting the cell in a vacuum enclosure. Then the cell window is not cooled by the air, and the uniformity in the temperature is significantly improved. The configuration of heater coils is also important in realizing uniform temperature in the cell window.

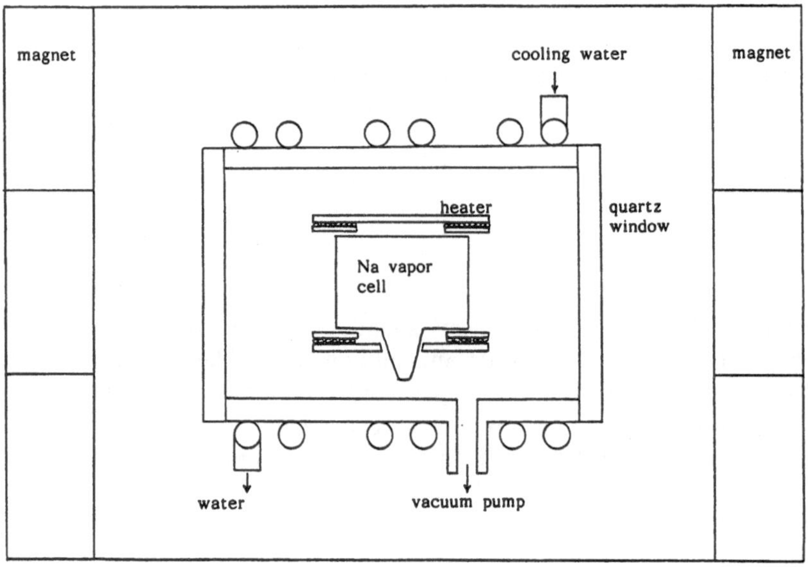

Fig. 1. Schematic structure of our magneto-optical filter.

A schematic structure of our MOF is given in Fig. 1. The magnetic field is applied either parallel or perpendicular to the optical axis by arranging a permanent magnet. The field strength in the MOF cell is adjustable to some degree (1000 to 2000 G) by inserting an iron spacer which alters the amount of leak magnetic flux. The window of the vacuum chamber is made of quartz. The cell is supported in the vacuum chamber by a mechanism made by thermal insulator. The vacuum chamber is cooled by circulating the water around it, in order to keep the magnet cool.

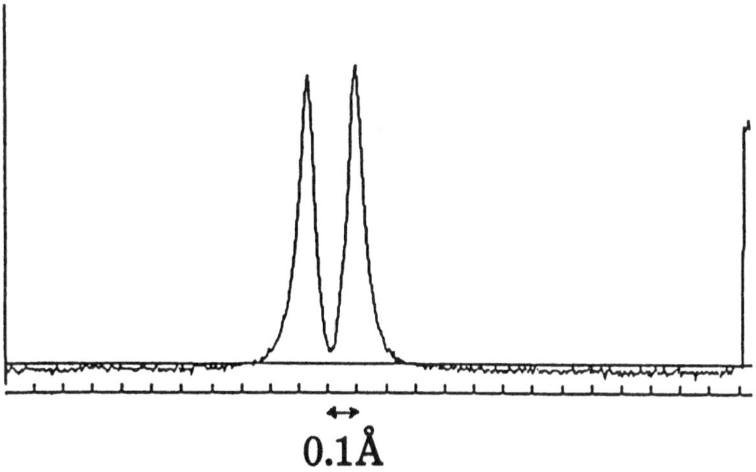

Fig. 2. An example of transmission curves at the wavelength of Na D_2 line. The temperature in the cell is 170°C.

3 Current Status and Prospects

By using the filter with a usable diameter of 2 cm, we measured the transmission characteristics by a spectrograph (Fig. 2). The dependence of the transmission profile on the temperature in the cell was as is expected from theoretical consideration. We also succeeded in taking images of the sun through MOF with satisfactory resolution (Fig. 3). The cells have been used for more than a year without any degradation in performance. Measurements of solar velocity fields by combining MOF with a KDP modulator will start shortly, and the construction of filters with larger aperture is now under way.

References

Agnelli, G., Cacciani, A., and Fofi, M. 1975, *Solar Phys.*, **44**, 509.

Cacciani, A., and Fofi, M. 1978, *Solar Phys.*, **59**, 179.

Koyama, K. 1986, *Hydromagnetic and Magnetohydrodynamic Problems in the Sun and Stars*, ed. Y. Osaki (University of Tokyo, Tokyo), p.339.

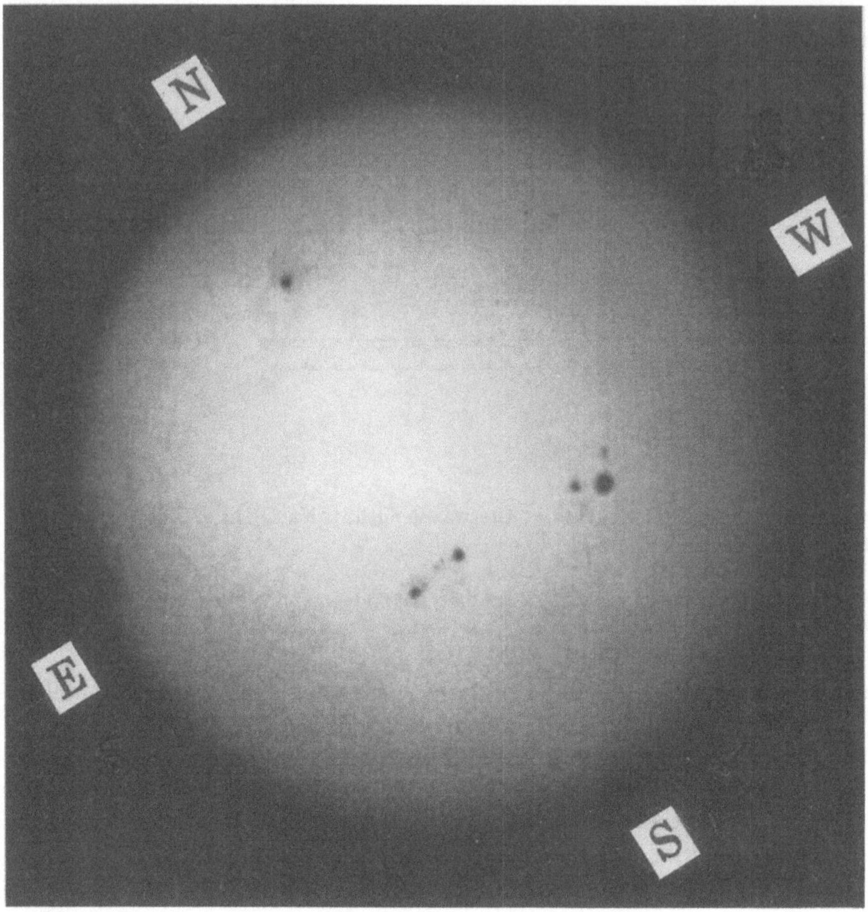

Fig. 3. A solar image taken through our magneto-optical filter at the wavelength of Na D_2 line. The transmission profile of the filter has double passbands in the wings of the line as is shown in Fig. 2. A notable feature is that facular regions (near the east limb) appear darker than the surroundings.

V

Inverse Problems of Solar Oscillations

Comments on Helioseismic Inference

Douglas Gough

Institute of Astronomy and
Department of Applied Mathematics and Theoretical Physics,
University of Cambridge, Madingley Road, U.K.
and
Institute for Theoretical Physics, University of California,
Santa Barbara, CA 93106, U.S.A.

Abstract: Helioseismic inference can be made within a wide spectrum of sophistication, from arguments based on the results of very simple and highly idealized model problems which depend on specific limited aspects of the data to a variety of formal numerical inversions of all the data that are available. The idealized problems are relatively simple to analyze, and provide a tool for making immediate qualitative and sometimes even quantitative estimates of certain aspects of the structure of the sun. If well chosen, they are likely to add substantially to our understanding of the situation; indeed, they can be an extremely useful guide to designing the more formal techniques which, though numerically more precise, are frequently also more opaque. Therefore it is often prudent to utilize methods throughout the entire spectrum. In this lecture a selection of the techniques for making immediate inferences will be discussed, and illustrated with examples of topical interest.

1 Introduction

Simple models offer an easy way to obtain a qualitative sense of the behaviour of the oscillations of more realistic and therefore more complicated models of stars. Provided they incorporate the essence of the physics one wishes to describe, they are extremely useful tools for any preliminary investigation. The understanding they provide enables one to appreciate the generality of the qualitative conclusions one may derive from them, particularly when addressing some new physical issue for the first time. Thus they are an invaluable complement to detailed numerical models, which provide but isolated, albeit more accurate examples of the phenomena under investigation.

In this lecture I shall consider two simple physical models, which I shall introduce at the outset. One is spherical, but unstratified, and therefore incorporates the geometry of the star. The other is plane, but stratified. Experience of the two provides an extremely useful introduction to any study of stellar oscillations.

My main motivation is solar diagnostics, and therefore I shall concentrate on adiabatic acoustic modes. I consider only the frequency spectrum, and shall not discuss excitation and damping. I shall start by considering the basic properties of the frequency spectra of my two models, and then relate them to the asymptotic properties of the oscillations of a real star. The ideas are particularly useful for considering the influence of small perturbations, and so can be used to anticipate the results of inversions of frequency data, degeneracy splitting and temporal variations. I shall also use a variational principle, coupled with asymptotic argument, to discuss an oscillatory property of the distribution of oscillation eigenfrequencies that might be used to determine the locations of the edges of convection zones in stars, and which might possibly also provide a seismic calibration of the helium abundance. Finally, I shall apply the ideas that I shall have developed to discuss what can be inferred about the new solar-cycle variations reported at this seminar by Libbrecht and Woodard, Pallé *et al.* and by Jefferies and his colleagues. It is quite evident that there is an acoustically sensitive variation taking place in the very outer layers of the sun, but what that variation is is at present a mystery.

2 Simple Models

I shall discuss two simple models: the isothermal sphere and the complete plane-parallel polytrope. Both are useful for illustrating some of the more basic properties of stellar oscillations. They can each be made more elaborate to address more subtle issues.

2.1 The Isothermal Sphere

Some of the most basic properties of high-order stellar p modes of low and intermediate degree come about because the star is essentially spherical. It is therefore sometimes useful to represent the star by the simplest of spherical models: an isothermal sphere of perfect gas with radius R in the absence of gravity. Its adiabatic acoustic oscillations have been discussed by Rayleigh (1896). A scalar wave variable ψ, such as the pressure perturbation (because the basic state is uniform there is no difference in linear theory between Lagrangian and Eulerian perturbations), satisfies the simple wave equation

$$\nabla^2\psi + \frac{\omega^2}{c^2}\psi = \nabla \ln \rho \cdot \nabla\psi = 0, \tag{2.1}$$

where ω is the frequency of oscillation, ρ is the (constant) density of the basic state and c is the adiabatic sound speed. This equation has separable solutions of the form

$$\psi = r^{-1}\Psi(r)P_l^m(\cos\theta)\cos m\phi \cos\omega t \tag{2.2}$$

with respect to spherical polar coordinates (r, θ, ϕ), where t is time and P_l^m is the associated Legendre function (of the first kind) of degree l and order m. The radial eigenfunction $\Psi(r)$ satisfies

$$\Psi'' + \left(\frac{\omega^2}{c^2} - \frac{L^2 - \frac{1}{4}}{r^2}\right)\Psi = 0, \tag{2.3}$$

where $L = l + \frac{1}{2}$ and a prime denotes differentiation with respect to the argument, and can thus be expressed in terms of a Bessel function of the first kind:

$$\Psi = r^{\frac{1}{2}} J_L\left(\frac{\omega r}{c}\right). \tag{2.4}$$

If pressure is presumed to be constant at the outer boundary, which is a good first approximation for a star, then

$$\omega = j_{n,L}\pi^{-1}\omega_0, \tag{2.5}$$

where

$$\omega_0 = \pi c/R \tag{2.6}$$

and $j_{n,L}$ is the nth zero of the Bessel function J_L. In particular, for low-degree modes satisfying $l \ll n$,

$$\omega \sim \left(n + \frac{1}{2}l + \epsilon\right)\omega_0 + [Al(l+1) - B]\omega_0^2/\omega + \ldots, \tag{2.7}$$

where $\epsilon = 0, A = (2\pi^2)^{-1}$ and $B = 0$. By varying the boundary condition at $r = R$ one finds that of the coefficients in the asymptotic expression (2.7) only ϵ and B vary; they depend on the phase jump suffered by a wave on reflection at the surface of the sphere. The quantity A is invariant; it depends on the geometry of the waves near the centre of the star, though this is more easily seen by expressing the solution of Eq. (2.1) as a superposition of plane waves (*cf.* Keller and Rubinow, 1960).

As is well known, the basic structure of the asymptotic expression (2.7) for the eigenfrequencies ω is preserved even when the sphere is stratified under gravity. This was first demonstrated by Tassoul (1980) and was discussed by Shibahashi in the introduction to this seminar. I shall not yet venture further into consideration of more realistic stellar models; I mention Tassoul's result here simply to establish that there is a practical domain of validity of my simple model.

One of the reasons for being interested in so simple a model is that one can quite easily consider what happens to the eigenfrequencies if a perturbation to the simple isothermal sphere is imposed. Suppose, for example, the core of the sun were to have been mixed. What would be the influence on the eigenfrequencies? To get an impression of the general kind of change that would take place one first asks what the effect of the mixing would be. Broadly speaking, it would modify the density ρ and the sound speed c in the core, and produce steep gradients in ρ and c at the interface between the mixed core and its unmixed surroundings. Let us consider its implications.

Because the equilibrium pressure is uniform one can write

$$\rho = \rho_0[1 - \epsilon f(r)], \quad \frac{1}{c^2} = \frac{1}{c_0^2}[1 - \epsilon f(r)] \tag{2.8}$$

and if ϵ [which here is unrelated to the phase factor ϵ appearing in Eq. (2.7)] is small one need consider only perturbations linear in ϵ. Thus one expresses ψ and ω as the sums of zero-order solutions, which I now call ψ_0 and ω_0, obtained from expressions (2.2), (2.4) and (2.5) with c replaced by the unperturbed sound speed c_0, plus small perturbations ψ_1 and ω_1. The perturbation equation, neglecting terms quadratic in ϵ, is then

$$\nabla^2\psi_1 + \frac{\omega_0^2}{c_0^2}\psi_1 = -\frac{\omega_0^2}{c_0^2}\left(2\frac{\omega_1}{\omega_0} - f\right)\psi_0 - \frac{df}{dr}\frac{\partial\psi_0}{\partial r} \tag{2.9}$$

in which the solubility condition [writing ω_0 as ω to avoid confusion with ω_0 of Eq. (2.6)]

$$\omega_1 = \frac{1}{2}\omega \frac{\int \left\{ \psi_0^2 + \frac{c_0^2}{2\omega^2}\nabla^2\psi_0^2 \right\} f dV}{\int \psi_0^2 dV} \tag{2.10}$$

$$= \frac{\omega \int_0^R \left\{ \Psi^2 + \frac{c_0^2}{\omega^2}\frac{d}{dr}\left[r\Psi \frac{d}{dr}\left(r^{-1}\Psi\right)\right]\right\} f(r) dr}{2\int_0^R \Psi^2 dr} \tag{2.11}$$

must be satisfied. In Eq. (2.10) the integrals are over the volume V of the star. Thus we have been able to write down the form of the perturbed eigenfrequency in terms of an arbitrary small spherically symmetrical perturbation to the sound speed. Although this expression cannot be used to make precise quantitative estimates of the frequency perturbation, it can give one a good idea of the general form the perturbation must take: how it depends on n and l. Kosovichev and I used this model in a preliminary investigation of the influence of core mixing in an attempt to explain an anomaly that appeared to have been exhibited in some observations of relatively low-frequency modes with $l = 5$ (Gough and Kosovichev, 1988). Its advantage over the usual asymptotic analyses of more realistic models is that it is valid for low-order modes. What is important is that it does not require the scale of variation of the sound-speed perturbation to be much greater than the characteristic wavelength of the mode, and can therefore be applied to the core. However, it does, of course, require the magnitude of the perturbation to be small. Under some circumstances asymptotic expressions for eigenfunctions of a more realistic solar model can be used in conjunction with perturbation integrals such as those in Eq. (2.10) to obtain more accurate estimates of frequency changes. I shall employ such a hybrid method later to discuss small-scale structure in the outer layers of the sun.

One can also use the isothermal model to study perturbations that are not necessarily small. The price one must pay to retain analytical simplicity, however, is a much tighter restriction on the functional form of the sound-speed variation. If, for example, $c = c_0$ in the envelope: $a < r \leq R$, and $c = c_1$ in the core: $r < a$, then in the core Ψ is given by expression (2.4) with $c = c_1$ but in the envelope there are contributions from Bessel functions of both kinds:

$$\Psi = r^{\frac{1}{2}} J_L\left(\frac{\omega a}{c_1}\right) \frac{J_L\left(\frac{\omega r}{c_0}\right) + K Y_L\left(\frac{\omega r}{c_0}\right)}{J_L\left(\frac{\omega a}{c_0}\right) + K Y_L\left(\frac{\omega a}{c_0}\right)}, \qquad a \leq r \leq R \tag{2.12}$$

for some constant K. I have written the solution in such a way as to ensure continuity of the pressure fluctuation ψ at $r = a$. It is now a straightforward matter to determine K in terms of ω by demanding continuity of the vertical component ξ of the displacement, using the vertical component of the momentum equation to relate ξ to ψ. The eigenvalue ω can then be obtained, as before, by applying the boundary condition at $r = R$. The expression for ω is a little cumbersome, so I refrain from presenting it explicitly. Needless to say, its value reduces to that obtained from Eq. (2.11) when $|c_1/c_0 - 1|$ is small, even though the function Y_L does not appear explicitly in the solubility condition. The perturbed radial eigenfunction ψ_1 which satisfies Eq. (2.9) does, of course, contain the appropriate contribution from Y_L.

In Fig. 1 I illustrate the relative perturbation ω_1/ω to the frequencies ω of several low-degree modes arising from a 10-per-cent augmentation of the sound speed in a core $r < a = \lambda R$ where $\lambda = 0.25$. The results are plotted against $\alpha \equiv \omega R/\pi c_0$ which is approximately $n + \frac{1}{2}l$ when $n \gg l$; it is related to the lower turning point r_t, defined by Eq. (6.3), according to $r_t/R = L/\pi\alpha$. For $r < r_t$ the eigenfunction is evanescent; consequently if $a < r_t$ the perturbation to ω is relatively small. When $a > r_t$, the perturbation is approximately proportional to $(a - r_t)/(R - r_t) \to \lambda = 0.25$ as $\alpha \to \infty$, as one might expect, upon which is an oscillatory contribution associated with the oscillatory structure of the eigenfunction experiencing the discontinuity in the sound speed.

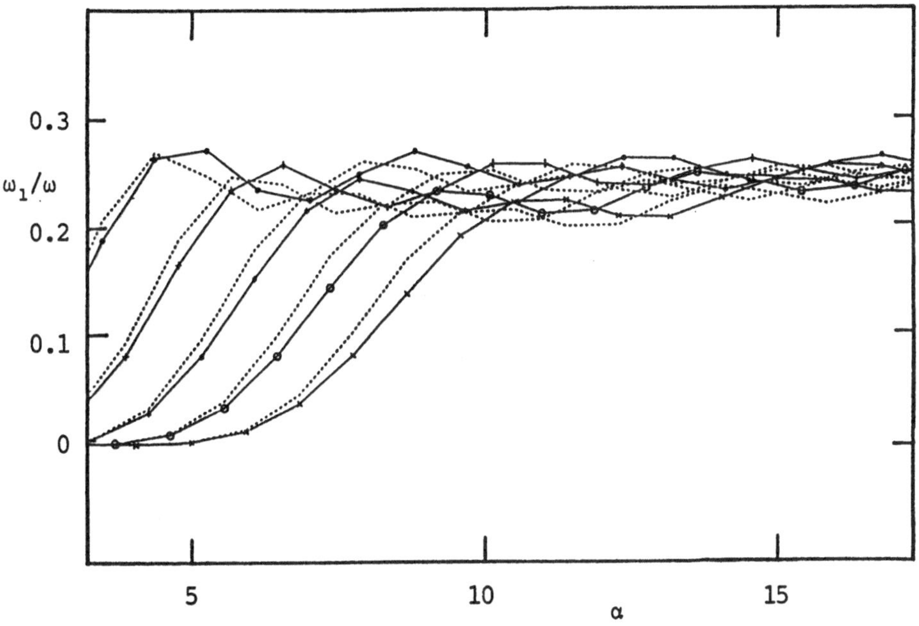

Fig. 1. Scaled relative frequency perturbation ω_1/ω resulting from augmenting the sound speed by ϵc_0 in $r < a = \lambda R$. Here $\epsilon = 0.1$ and $\lambda = 0.25$. The zero-order eigenfrequencies are given by Eq. (2.5) and the symbols joined by continuous lines represent perturbations ω_1 obtained by subtracting them from the exact eigenvalues of Eq. (2.1) and dividing by ϵ. The symbols denote the degree of the mode: $\bullet(l = 1)$, $+(l = 2)$, $*(l = 3)$, $\circ(l = 4)$, $\times(l = 5)$. The dashed lines join linearized perturbations, computed from Eq. (2.11) retaining only the first term in the numerator, which arises from the sound-speed perturbation. If ϵ were 0.01, the dashed and continuous lines would be barely distinguishable.

The simple example of a sound-speed perturbation I have chosen to illustrate is evidently not the best approximation to the outcome of the material mixing in a stellar core that motivated the discussion, but the behaviour of the eigenfrequencies that has been found is generic, and provides a good basis from which to generate more realistic representations. It must also be remembered that I have ignored a sharp spike in the buoyancy frequency N (defined on p. 292) which is produced at the interface between the mixed and unmixed layers. Strictly speaking, it is necessary to introduce gravity to describe it. However, if the interface is very thin, it can be represented as a discontinuity.

Equation (2.1) can still be used above and below the interface, the density discontinuity resulting from the composition discontinuity being accounted for merely in the matching conditions at $r = a$: the vertical component ξ of the displacement remains continuous, but now there is a jump of magnitude $-g\Delta\rho\xi$ in the (Eulerian) pressure perturbation at the unperturbed interface, where $\Delta\rho$ is the difference between the density below and above the interface and g is the acceleration due to gravity. The outcome is the addition of a further oscillatory contribution to ω_1. Such oscillatory components to ω are discussed in more detail in Section 8.

2.2 The Plane-Parallel Polytrope

Near the surface of the star the scale of variation of the background state is comparable with the wavelength of the oscillations. To describe acoustic oscillations in this region it is necessary to take the stratification into account. A plane-parallel polytrope is a convenient model for accomplishing this. In its simplest manifestation, the density ρ and pressure p of the undisturbed state are given by

$$\rho = \rho_0 z^\mu \quad , \quad p = (\mu + 1)^{-1} g \rho_0 z^{\mu+1}, \tag{2.13}$$

where ρ_0, g and μ are constants, μ being the polytropic index and g the gravitational acceleration. The coordinate z is depth below some fiducial level, which might be regarded as the surface of the star. This is the so-called complete polytrope. The square of the sound speed c increases linearly with depth

$$c^2 = \gamma(\mu + 1)^{-1} g z, \tag{2.14}$$

where γ is the adiabatic exponent $(\partial \ln p / \partial \ln \rho)_s$, the thermodynamic derivative being taken at constant specific entropy s.

Adiabatic oscillations of this model have been discussed by Lamb (1932). Since the background state is dependent only on z, separable solutions that are wavelike in the horizontal direction with wave number k can be found. Lamb used the dilatation div$\boldsymbol{\xi}$ for the principal dependent variable, obtaining the confluent hypergeometric equation

$$\zeta \chi'' + (\mu + 2 - \zeta)\chi' + 2\beta\chi = 0 \tag{2.15}$$

for the scaled amplitude χ defined by div$\boldsymbol{\xi} = \mathrm{Re}[\chi(\zeta)\exp(-\zeta - ikx)]$, x being a horizontal Cartesian coordinate in the direction of variation of the wave and $\zeta = kz$. In this equation

$$2\beta = \frac{\mu+1}{\gamma}\sigma^2 - (\mu + 2) + \left(\mu - \frac{\mu+1}{\gamma}\right)\sigma^{-2}, \tag{2.16}$$

and $\sigma^2 = \omega^2/gk$. Lamb's interest was in modelling waves in the terrestrial atmosphere whose horizontal wavelength $2\pi/k$ is much greater than the depth of the atmosphere. Accordingly he needed to approximate the solutions to Eq. (2.15) only for $\zeta \ll 1$. Our interest, however, is in modelling acoustic oscillations of high degree which, as will be seen later [see Eq. (2.20), and Table 1 (p. 302)] are confined to the outer layers of the star where the plane-parallel approximation is valid. Thus we can effectively regard the envelope as being infinitely deep. This is essentially the situation considered by Spiegel and Unno (1962) in their study of convective instability; the analysis of oscillations is essentially the same, save for a few sign changes.

From two independent solutions of Eq. (2.15) a single linear combination can be found for which χ does not diverge as $\zeta \to \infty$. It is normally called $U(-\beta, \mu+2, 2\zeta)$. Moreover, $U(-\beta, \mu+2, 2\zeta) \sim (2\zeta)^{\beta}$ as $\zeta \to \infty$ (*e.g.*, Abramowitz and Stegun, 1964), implying that div$\boldsymbol{\xi}$ decays exponentially to zero at great depths, which justifies regarding the model as being infinitely deep; high-degree p modes do not sense directly conditions at the centre of the star. Indeed, it is evident from the asymptotic expansion that the magnitude of U is substantial only for $\zeta \lesssim C\beta$, where C is of order unity, which implies that the waves are confined to a cavity of depth

$$z_t \simeq C\beta k^{-1}. \tag{2.17}$$

The condition that U does not diverge at the singular point $\zeta = 0$ is that β be a non-negative integer, which implies the eigenvalue equation

$$\omega^4 - 2\gamma(\mu+1)^{-1}(n+\alpha)gk\omega^2 + [\gamma\mu(\mu+1)^{-1} - 1]g^2k^2 = 0, \tag{2.18}$$

where $\alpha = \frac{1}{2}\mu$.

There are two classes of solutions to Eq. (2.18): those corresponding to p modes, with $\omega^2 > \gamma(\mu+1)^{-1}(n+\alpha)gk$, and those corresponding to g modes, with $\omega^2 < \gamma(\mu+1)^{-1}(n+\alpha)gk$. The latter may have $\omega^2 > 0$, when they represent stationary gravity waves, or $\omega^2 < 0$, when they represent convective modes, depending on whether $\mu - (\mu+1)/\gamma$ is positive or negative. I shall discuss only the p modes, which satisfy

$$\frac{\omega^2}{gk} \simeq \frac{2\gamma}{\mu+1}(n+\alpha) - \frac{1}{2}\left(\mu - \frac{\mu+1}{\gamma}\right)(n+\alpha)^{-1}, \tag{2.19}$$

at least when n is large. The second term in the expression on the right-hand side of this equation is the buoyancy correction to acoustics. It is proportional to the square of the buoyancy frequency, defined by Eq. (3.4), through the factor $\mu - (\mu+1)/\gamma$ which is proportional to the subadiabatic temperature gradient. In the case of the sun the outer layers are convective, and except very close to the surface the stratification is very close to being adiabatic, so one would expect to obtain a good approximation to the p-mode eigenfrequencies by retaining just the leading term of the right-hand side of Eq. (2.19), even when n is not large. Notice that in a convectively stable region, where the buoyancy frequency is real, Eqs. (2.18) and (2.19) indicate that the buoyancy acts in opposition to the acoustic restoring force and decreases the frequency of the waves. On the other hand, the acoustic influence on g modes augments ω.

A convenient way of specifying the depth of the acoustical cavity more precisely than expression (2.17) is to adopt the lower turning point of the wave equation, which occurs where the characteristic vertical wave number κ, defined in the case of the isothermal sphere by the square root of the coefficient of the undifferentiated Ψ in the simple wave equation (2.3) and which is given more generally by Eq. (3.9) with $k^2 = l(l+1)/r^2$, vanishes. For p modes in a plane parallel polytrope, this yields

$$z_t \simeq 2(n+\alpha)k^{-1}, \tag{2.20}$$

which shows that high-degree modes are concentrated in the outer layers of the star. It can be verified from the high-degree entries in Table 1 (p. 302) that this result is also approximately true of a more realistic solar model.

It was from the simple eigenvalue equation (2.18) that perhaps the first seismic diagnostic of the structure of the outer layers of the sun was made. From a systematic discrepancy between observed solar oscillation frequencies (Deubner, 1975) and the eigenfrequencies of a model of the solar envelope (Ando and Osaki, 1975) it was realized that the stratification of the upper boundary layer of the convection zone should be more nearly adiabatic than in the theoretical model. Consequently, the mixing length in the model should be increased, implying an increase in the depth of the convection zone to about 2000 Mm (Gough, 1976). Crude as this argument is, the inference is basically correct, and has been confirmed by subsequent more careful analysis. It is important to realize, however, that the rough argument based on this highly simplified model could not have been carried out without the prior existence of numerical computations of a much more realistic model. The simple model served merely as a basis for an approximate perturbation calculation to establish how the theoretical model should be modified to remove the small discrepancy between theory and observation.

A more reliable estimate might have been made by using a hybrid model, composed of a polytropic base supporting an isothermal atmosphere. I shall discuss that only briefly, but before doing so I must consider the isothermal atmosphere.

2.3 Oscillations of an Isothermal Atmosphere

The basic state of a plane-parallel isothermal atmosphere is given by

$$\rho = \rho_a e^{(z-z_0)/H} \quad , \quad p = gH\rho_a e^{(z-z_0)/H} , \tag{2.21}$$

where ρ_a, z_0 and H are constants, H being the scale height of both density and pressure. Once again, the coordinate z measures depth. It has been assumed that the atmosphere is a perfect gas with mean molecular mass μ_0, the scale height being related to the temperature T by

$$H = \frac{\Re T}{\mu_0 g}, \tag{2.22}$$

where $\Re$ is the gas constant.

Adiabatic oscillations of this atmosphere have also been discussed by Lamb (1932). As with the plane-parallel polytrope, one may express oscillation variables in separated form with constant horizontal wave number k. The Lagrangian pressure perturbation δp, for example, can be written

$$\delta p = \mathrm{Re}\left[\Psi(z)\exp(-z/2H + ikx - i\omega t)\right], \tag{2.23}$$

and the amplitude function Ψ satisfies

$$\Psi'' + \kappa^2 \Psi = 0, \tag{2.24}$$

where

$$\kappa^2 = \frac{\omega^2}{c^2} - \frac{1}{4H^2} - k^2\left[1 - \frac{\gamma - 1}{\omega^2}\left(\frac{c}{\gamma H}\right)^2\right]. \tag{2.25}$$

Evidently, if $\kappa^2 < 0$, the oscillation is evanescent. The atmosphere, if it is considered to extend to $z = -\infty$, cannot transmit waves, and therefore acts as a perfect reflector of waves incident from below.

In the case of purely vertical motion, having $k = 0$, which was first considered by Lamb in 1908, evanescence occurs for frequencies ω below $\omega_c = c/2H$, which is sometimes called Lamb's acoustic cutoff frequency. When ω exceeds this value, vertical oscillations are no longer confined by the atmosphere. Wave energy can propagate outwards until the displacement amplitude, which is proportional to $\rho^{-\frac{1}{2}}$, is so large that the waves become nonlinear and shock.

The critical frequency is modified by horizontal variation of the wave. Its value is given by

$$\omega_c = \left\{ \frac{1}{2}\left(k^2 + \frac{1}{4H^2}\right) + \left[\frac{1}{4}\left(k^2 + \frac{1}{4H^2}\right)^2 - \frac{(\gamma - 1)k^2}{\gamma^2 H^2}\right]^{\frac{1}{2}} \right\} c, \tag{2.26}$$

which exceeds $c/2H$ when $k \neq 0$. Thus, as ω increases, vertical oscillations are the first to leak through the atmosphere.

Since κ is constant, the eigensolutions of Eq. (2.24) are either exponential or sinusoidal.

2.4 Simple Hybrid Models

Somewhat more realistic yet simple models of the outer layers of a star can be modelled with an isothermal atmosphere supported by a plane-parallel polytrope. Thus, p and ρ are given by Eq. (2.13) for $z > z_0$ and by Eq. (2.21) for $z < z_0$, with the condition $H\rho_a = (\mu + 1)^{-1}\rho_0 z_0^{\mu+1}$ demanded by continuity of pressure at $z = z_0$. Temperature, and hence density, need not be continuous at $z = z_0$, the discontinuity being introduced to represent the thin boundary layer at the top of the convection zone in which the structure of the star is neither polytropic nor isothermal. The atmospheres of stars with coronae, such as the sun, can be modelled with two superposed isothermal regions, the upper high-temperature region extending to $z = -\infty$.

A model of this kind (without the corona) has been used to estimate the importance of the atmosphere on low-degree oscillations at frequencies below the cutoff frequency in the atmosphere (Christensen-Dalsgaard and Gough, 1980) and in assessing the influence on acoustic eigenfrequencies of an atmospheric magnetic field (Campbell and Roberts, 1989). In this case the atmosphere oscillates everywhere in phase with the motion of the fitting surface at the junction with the polytropic interior, and so far as the oscillations of the polytrope are concerned can be regarded simply as a boundary with inertia that resists the pressure fluctuations. Since the displacement eigenfunction increases with height, the atmosphere oscillates with a greater mean amplitude than does the fitting surface, and therefore its apparent inertia is greater than its mass. At frequencies above the cutoff, phase varies with height; waves propagate upwards and energy leakage damps the modes, but the interior of the star continues to exhibit discrete albeit broadened resonant frequencies as a result of reflection at the discontinuities (Balmforth and Gough, 1990).

3 Simple Asymptotics

Asymptotic approximations to the adiabatic acoustic oscillations of a star can be obtained if the characteristic vertical wavelength of the eigenfunction is small compared with the scale of variation of the background state. In that case two approaches present themselves, both of which depend on representing the oscillation locally as a superposition of waves.

The first stage for either method is to transform the equation into a form suitable for easy analysis. This is a very important step, because a blind application of either method to raw equations can sometimes require a considerable amount of unnecessary extra labour to achieve the required accuracy; moreover, it can also shrink the domain of validity of the approximation.

For simplicity, I shall adopt the so-called Cowling approximation, in which the Eulerian perturbation Φ' to the gravitational potential is ignored. The analysis can be generalized to include Φ', and has been carried out for a spherically symmetrical background state by Vorontsov (these proceedings) and by Dziembowski and myself (1990). My independent variables will be spherical polar coordinates (r, θ, ϕ) about the centre of the star. I shall work with a dependent variable which is proportional to the Lagrangian pressure perturbation δp. This is equivalent to the approach taken by Lamb, who based his analysis on div$\boldsymbol{\xi}$, which is also proportional to δp. I find it more convenient to start with a scalar, rather than construct one from a vector, particularly when the full spherical geometry of the problem is taken into account.

Different choices of either dependent or independent variables can lead to somewhat different representations of the solution. For example, Smeyers and Ruymaekers (these proceedings) compare the use of the scalar δp and the radial component of the displacement vector, and Christensen-Dalsgaard *et al.* (1983) obtain a different formula for the acoustic cutoff frequency in a stratified envelope when the more natural acoustical radius $\int c^{-1} dr$ is used for the radial coordinate. A systematic comparison of the different representations has never been published.

To make my presentation yet simpler I ignore local curvature effects. To include them is tedious (*cf.* Gough, 1990), though it probably does not introduce any fundamental difficulties. I then determine the appropriate dependent variable ψ by multiplying δp by that function u which eliminates first derivatives of ψ from the governing differential equation. The outcome, for high-frequency modes, is

$$\psi \sim \rho^{-\frac{1}{2}} \delta p, \tag{3.1}$$

which satisfies

$$\frac{\partial^2}{\partial t^2}\left(\frac{\partial^2}{\partial t^2} + \omega_c^2 - c^2\nabla^2\right)\psi - c^2 N^2 \nabla_h^2 \psi = 0, \tag{3.2}$$

where the acoustic cutoff frequency ω_c and the buoyancy frequency N are given by

$$\omega_c^2 = \frac{c^2}{4H^2}\left(1 - 2\frac{dH}{dr}\right) = \frac{1}{2}c^2\left[\frac{3}{2}\left(\frac{d\ln\rho}{dr}\right)^2 - \frac{1}{\rho}\frac{d^2\rho}{dr^2}\right], \tag{3.3}$$

$$N^2 = g\left(\frac{1}{H} - \frac{g}{c^2}\right) = -g\left(\frac{d\ln\rho}{dr} + \frac{g}{c^2}\right), \tag{3.4}$$

$H = (-d\ln\rho/dr)^{-1}$ being the density scale height of the equilibrium state, and ∇_{h}^2 is the horizontal Laplacian operator:

$$\nabla_{\mathrm{h}}^2 = \nabla^2 - \frac{1}{r^2}\frac{\partial}{\partial r}r^2\frac{\partial}{\partial r} = \frac{1}{r^2\sin\theta}\frac{\partial}{\partial\theta}\sin\theta\frac{\partial}{\partial\theta} + \frac{1}{r^2\sin^2\theta}\frac{\partial^2}{\partial\phi^2}. \tag{3.5}$$

Since the background state is considered to be independent of time, one can anticipate eigensolutions with sinusoidal time dependence of frequency ω. Equation (3.2) then reduces to

$$\nabla^2\psi + \left(\frac{\omega^2-\omega_{\mathrm{c}}^2}{c^2} - \frac{N^2}{\omega^2}\nabla_{\mathrm{h}}^2\right)\psi = 0. \tag{3.6}$$

Strictly speaking, Eqs. (3.2) and (3.6) are valid only for spherically symmetrical stars, though if there were a slight asphericity, with horizontal gradients of the background state being everywhere very much smaller than radial gradients, one might expect to be able to use these relatively simple equations as a guide.

If the star can be considered to be spherically symmetrical, one can separate the solution further, into the form

$$\psi \propto \mathrm{Re}\left[r^{-1}\Psi(r)P_l^m(\cos\theta)e^{i(m\phi-\omega t)}\right] \tag{3.7}$$

as was done for the isothermal sphere. Substituting into Eq. (3.6) yields

$$\Psi'' + \kappa^2\Psi = 0, \tag{3.8}$$

where

$$\kappa^2 = \frac{\omega^2-\omega_{\mathrm{c}}^2}{c^2} - \frac{l(l+1)}{r^2}\left(1-\frac{N^2}{\omega^2}\right). \tag{3.9}$$

One sees that Eqs. (3.8) and (3.9) reduce immediately to Eq. (2.3) in the absence of stratification ($\omega_{\mathrm{c}} = 0, N = 0$), and that Eq. (3.9) is equivalent to Eq. (2.25) for a plane-parallel isothermal atmosphere, for which one sets $k^2 = l(l+1)/r^2$.

One can easily solve Eq. (3.8) in the JWKB approximation. Except in the vicinity of the radii at which κ vanishes one can well approximate Ψ by the form

$$\Psi = A(r)\exp\left(i\int\kappa dr\right) \tag{3.10}$$

in which it is assumed that the characteristic scale of variation Λ, say, of A and κ greatly exceeds the local inverse wave number κ^{-1}. On substituting this expression into Eq. (3.8) and satisfying the resulting equation separately at each order of $(\kappa\Lambda)^{-1}$, one finds that the leading order is automatically satisfied, as it was designed to do by the choice of the argument of the exponential function in Eq. (3.10), and that the subsequent order implies $A \propto \kappa^{-\frac{1}{2}}$. In regions where $\kappa^2 > 0$ the solution is wavelike; here waves can propagate in the radial direction. Where $\kappa^2 < 0$ the solution is evanescent. Matching across the turning point, where $\kappa = 0$, is accomplished by Olver's method, in which the solution is approximated by an Airy function (*e.g.*, Vandakurov, 1967; Shibahashi, 1979; Tassoul, 1980; Unno *et al.*, 1989). By choosing that solution which, where $\kappa^2 < 0$, decays away from the region of propagation, the integral in the representation (3.10) can be made definite, leading to the expression

$$\delta p \sim \Psi_0 r^{-1} \left(\frac{\rho}{\kappa}\right)^{\frac{1}{2}} \sin\left[\int_{r_t}^{r} \kappa dr + \frac{\pi}{4}\mathrm{sgn}(r - r_t)\right] \tag{3.11}$$

for the pressure fluctuation in the region of propagation, where Ψ_0 is a constant and r_t is a turning point. Identifying the two representations (3.11) obtained by equating r_t separately to the lower and upper turning points r_1 and r_2 of a region of propagation, assuming for the moment that r_2 exists, then yields the eigenvalue equation

$$\int_{r_1}^{r_2} \left[1 - \frac{\omega_c^2}{\omega^2} - \frac{L^2 c^2}{\omega^2 r^2}\left(1 - \frac{N^2}{\omega^2}\right)\right]^{\frac{1}{2}} \frac{dr}{c} \sim \frac{\left(n - \frac{1}{2}\right)\pi}{\omega}, \tag{3.12}$$

where now $L^2 = l(l+1)$. This expression indicates that the eigenfunction is such as to fit an integral number of waves within the cavity (r_1, r_2), save for a phase factor $-\frac{\pi}{2}$ which arises from the elongation of the wave resulting from the fact that it penetrates somewhat into the two bounding evanescent regions.

It is instructive to recall the significance of the quantities N and ω_c which appear in the eigenvalue relation (3.12). Let us first note that once again two classes of solution are admitted: the p modes with high frequency for which the first term in the square brackets dominates, and the g modes with low frequency for which the third term dominates. For g modes of moderate order, κ is moderate (I speak loosely, in search of clarity) and therefore as $l \to \infty$ the factor $1 - N^2/\omega^2$ must become small. Thus $\omega \to N$; the buoyancy frequency is the value to which the frequencies of g modes tend as the degree increases at fixed order. The motion becomes almost vertical nearly everywhere, in the form of elongated oscillatory cells. This is evident from the asymptotic forms for the vertical and horizontal amplitude components $(\Xi(r), H(r))$ defining the displacement eigenfunction

$$\boldsymbol{\xi}(\boldsymbol{r}, t) = \mathrm{Re}\left[\left(\Xi P_l^m, L^{-1} H dP_l^m/d\theta, imL^{-1} H \operatorname{cosec}\theta P_l^m\right) e^{im\phi - i\omega t}\right], \tag{3.13}$$

which are given approximately by

$$\Xi \sim \frac{\Psi_0 L}{\omega r^2}(\rho\kappa)^{-\frac{1}{2}} \sin\left(\int_{r_1}^{r} \kappa dr + \frac{\pi}{4}\right), \tag{3.14}$$

$$H \sim \frac{\Psi_0}{\omega r}(\kappa/\rho)^{\frac{1}{2}} \cos\left(\int_{r_1}^{r} \kappa dr + \frac{\pi}{4}\right) \tag{3.15}$$

well away from the turning points in regions where $\kappa^2 > 0$. The pressure fluctuation, which diverts the vertical flow into the horizontal direction near the top and bottom of each cell, is relatively small. Thus, N is the characteristic frequency of a fictitious element of fluid imagined to be moving vertically under the action of buoyancy and otherwise unimpeded by pressure fluctuations. Of course, strictly speaking, buoyancy is produced by a pressure imbalance with gravity, as Archimedes must have known, and which is necessarily dependent on there being horizontal fluctuations, and a consequent horizontal component of the flow; but in fluid dynamics it isn't usually thought of quite in that way. It is more commonly thought of as the force of gravity acting upon the local density deviation from its horizontal average. It follows immediately that there can be no buoyancy in a fluid where the only spatial variation is with height. Therefore the value of N cannot influence the dynamics of radial modes, as is obvious from Eqs. (3.8) and (3.9).

The nature of the acoustic cutoff frequency is quite different. As its name indicates, it is concerned principally with acoustic-wave propagation, and not with the dynamics of gravity waves. It is sometimes associated with the buoyancy frequency, however, and has even been replaced by it (*e.g.*, Shibahashi, 1990), perhaps because in a plane-parallel isothermal atmosphere the numerical values of the two frequencies are similar. [$N^2 = (1-\gamma^{-1})g/H$ and $\omega_c^2 = \frac{1}{4}\gamma g/H$, the two frequencies differing by less than 1% when $\gamma = \frac{5}{3}$. In an adiabatically stratified plane-parallel polytrope, however, $\omega_c^2 = [(2\gamma-1)/4(\gamma-1)]g/z$ whereas $N^2 = 0$.] Equating the two can sometimes be convenient mathematically, but it is important to realize that it is no more than a convenience for calculation. The essential difference should not be forgotten. The influence of the acoustic cutoff becomes important when the scale height H of density becomes comparable with the wavelength of the wave. If density declines (or augments) too abruptly there is too great a mismatch between the inertia of the material in neighboring regions of compression and rarefaction for them to interact continually in opposition. It is evident that this mismatch has nothing to do with the presence of gravity, which is an essential ingredient of buoyancy. Indeed, Eq. (3.2) is still satisfied when $g = 0$; in that case $N = 0$, but if the density of the background state is not uniform ω_c remains nonzero.

The role of the acoustic cutoff can be illustrated by the purely vertical oscillations of a plane-parallel polytrope of index μ. In this case there is no buoyancy, and Eq. (3.8) reduces to

$$\Psi_{zz} + \left[\frac{\omega^2}{c_0^2 z} - \frac{\mu(\mu+2)}{4z^2}\right]\Psi = 0, \tag{3.16}$$

where $c_0^2 = \gamma g/(\mu+1)$. The second term in square brackets is ω_c^2/c^2, and increases in magnitude with μ as the stratification becomes more severe. The solution is

$$\Psi = \Psi_0\ z^{\frac{1}{2}} J_{\mu+1}\left(\frac{2\omega z^{\frac{1}{2}}}{c_0}\right), \tag{3.17}$$

where Ψ_0 is again a constant amplitude factor. This illustrates that the wave is evanescent near the surface, for $\Psi \propto z^{\mu/2+1}$ where $\omega z/c \ll 1$.

The vertical component ξ of the displacement eigenfunction can be obtained by substituting the solution (3.17) into the momentum equation, yielding

$$\xi = -\xi_0\ z^{-\mu/2} J_\mu\left(2\omega\sqrt{\frac{(\mu+1)z}{\gamma g}}\right)\cos\omega t, \tag{3.18}$$

where $\xi_0^2 = (\mu+1)\Psi_0^2/(\omega^2\gamma g\rho_0)$. It can be illustrated practically by oscillating horizontally a rope hanging under gravity whose mass ρ per unit length is given by the first of Eqs. (2.13), where now z is height above the free end. In that case the horizontal displacement is given by Eq. (3.18) with $\gamma = 1$, and is illustrated in Fig. 2. It can clearly be seen that near the free end, where the tension is too low to provide adequate propagation, the solution is not wave-like. Above the turning point the wavelength decreases in proportion to the propagation speed $c \propto z^{\frac{1}{2}}$, and the amplitude diminishes in proportion to $(\rho c)^{-\frac{1}{2}}$.

Notice in Fig. 2 that when ξ is expressed with respect to the more natural independent variable $\tilde{\tau} = \int c^{-1}dz$, its spatial variation is very weak at the top of the evanescent region close to the free surface. The behaviour of div$\boldsymbol{\xi}$, which is proportional to $\delta p/p$, can be seen from Eqs. (3.1), (2.13) and (3.17) to be qualitatively similar. I record also that at

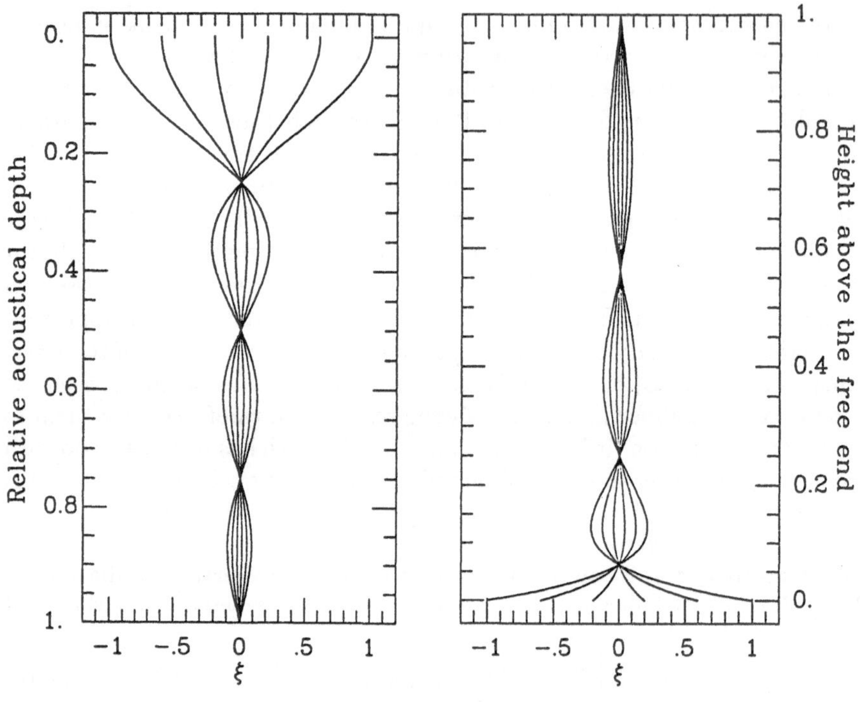

Fig. 2. The left-hand panel illustrates the displacement ξ, given by Eq. (3.18), of a vertically oscillating plane-parallel polytrope plotted as a function of acoustical depth $\tilde{\tau} = \int c^{-1}dz$, normalized such that the acoustical depth of the layer is unity. The displacement is shown at several different instances in time. In this example $\gamma = 5/3$ and $\mu = 0.5$. A small value of μ was chosen because otherwise ξ declines too rapidly with $\tilde{\tau}$ to be clearly visible. The right-hand panel shows the shape of a laterally oscillating rope of unit length, given by Eq. (3.18), hanging under its own weight from a fixed point. The mass per unit length is proportional to z^{μ} where $\mu = 0.5$. Aside from the orientations and the independent variables against which they are plotted, the displacements in the two panels are identical.

the free surface div$\boldsymbol{\xi} \propto \omega^2\xi$. These properties will be invoked in Section 9 when discussing the solar-cycle frequency variations reported by Libbrecht and Woodard (1990).

The depth (in the case of the polytrope) of the first node in the eigenfunction, which provides some indication of where propagation has begun, increases with μ, as does the extent of the region near the surface where the Lagrangian pressure fluctuation is insignificant. In the almost inert evanescent region the fluid is hardly compressed; the region is simply periodically lifted up bodily and subsequently lowered by the oscillating medium beneath. As can be seen in Fig. 2, however, the displacement is not constant, but decreases linearly with z away from $z = 0$. Thus, as in the case of the isothermal atmosphere, the centre of mass of the evanescent zone oscillates with a greater displacement amplitude than does the material near the turning point: the inertia the zone presents to the wave is again greater than its mass.

It is sometimes useful to think of the oscillations as a resonant superposition of locally plane propagating waves interfering coherently to form a standing wave pattern. This is

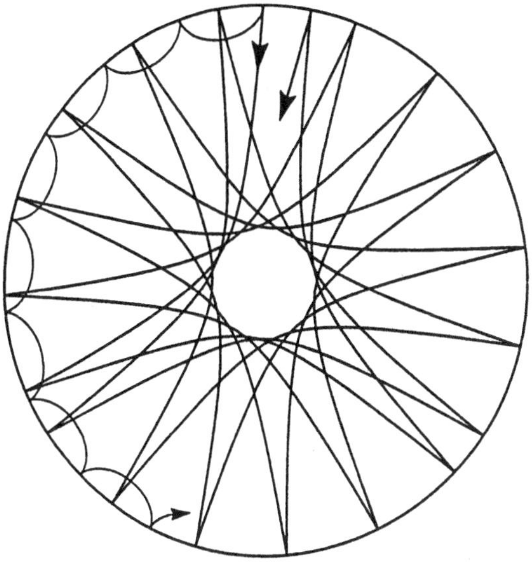

Fig. 3. Ray paths of components of two acoustic modes of an idealized model of the sun. As is almost always the case, the paths are not closed. The number of reflections near the surface per revolution is therefore not integral, and is not, as has sometimes been stated, the value of L (or l); it depends on the radius r_t of the lower turning point, given by Eq. (6.3), at which the waves travel horizontally and beneath which they cannot propagate. Thus, using Table 1 (p. 302) as a guide, one finds that the more deeply penetrating ray path represents a consituent of a mode with $\nu/L \simeq 560\mu\text{Hz}$, where $\nu = \omega/2\pi$ is the cyclic frequency; it could be a mode with $l = 2$ and $n = 8$ or perhaps a mode with $l = 6$ and $n = 23$. Similarly one deduces that for the shallower ray $\nu/L \simeq 30\mu\text{Hz}$, which corresponds to modes with $L/n \simeq 5$.

the second of the two approaches to which I alluded at the beginning of this section. The paths of all possible rays associated with a particular mode, examples of two of which are shown in Fig. 3, determine the domain within which the mode has substantial amplitude, and satisfying the condition that constructive interference takes place everywhere within that domain, in circumstances in which that is possible, both assures the asymptotic existence of the mode and determines its eigenfrequency. The procedure by which the interference conditions are computed is explained by Keller and Rubinow (1960). They are basically quantization conditions of the form

$$\oint \boldsymbol{K} \cdot d\boldsymbol{r} = 2(n' + m'/4)\pi, \tag{3.19}$$

where $\boldsymbol{K}$ is the total wave number and the integral is taken over an arbitrary closed contour C within the domain of propagation which suffers m' simple grazings with caustic surfaces; n' is an integer quantum number which in the case of a spherically symmetric star is related to the order n, the degree l and the azimuthal order m of the mode in a way that depends on C. For spherical stars the conditions (3.19) imply the existence of the integral quantum numbers l, m and n, and recover the relation (3.12) derived above by JWKB expansion of the radially varying factor $\Psi(r)$ in the separated eigenfunction (3.7) with the sole difference that L is now $l+\frac{1}{2}$. Since the approximation uses the asymptotic wave description in all three dimensions it can be formally valid only for $l \gg 1$, so the difference is immaterial.

This method is potentially more powerful than the JWKB analysis of separable solutions which I described above (at least for high-degree modes) because it does not require separable solutions to exist. In particular, one can formally write down conditions (3.19) that incorporate rotation and other aspherical components of the structure of the star. However, to evaluate them it is usually expedient to make use of the fact that the aspherical influences on the oscillations are small, and to perturb about the spherical state. In that case the results reduce to what one would obtain by applying degenerate perturbation theory about the separable solutions (3.7). Within that approximation, therefore, the two methods are essentially equivalent.

Finally, I record for future use the asymptotic displacement eigenfunctions for high-order p modes. They are given approximately by

$$\Xi \sim \Psi_0 \omega^{-2} r^{-1} (\kappa/\rho)^{\frac{1}{2}} \cos\left(\int_{r_1}^{r} \kappa dr + \frac{\pi}{4}\right), \tag{3.20}$$

$$H \sim \Psi_0 L \omega^{-2} r^{-2} (\rho\kappa)^{-\frac{1}{2}} \sin\left(\int_{r_1}^{r} \kappa dr + \frac{\pi}{4}\right), \tag{3.21}$$

in the propagating region well away from the turning points.

4 Low-degree p Modes

The frequencies of high-order p modes of low degree are given asymptotically by Eq. (2.7), where now

$$\omega_0 = \pi \left(\int_0^R \frac{dr}{c}\right)^{-1}, \tag{4.1}$$

which is an obvious generalization of Eq. (2.6), and ϵ, A and B are now functionals of the equilibrium state. The result was first derived to this order by Tassoul (1980), and the essential features of it are recovered by expanding Eq. (3.12) in inverse powers of ω/ω_0 at fixed l.

An important feature of Eq. (2.7) is that it is influenced separately by conditions in different regions of the star. In particular, ϵ and B depend predominantly on conditions in the surface layers of the star, just as is the case for the isothermal sphere (however, now $\epsilon \simeq \frac{1}{4} + \frac{1}{2}\mu$, where μ is an effective polytropic index in the vicinity of the upper turning point of the mode; B is rather more complicated), whereas A, which according to Tassoul is given by

$$A = \frac{1}{2\pi\omega_0}\left[\frac{c(R)}{R} - \int_0^R \frac{1}{r}\frac{dc}{dr} dr\right], \tag{4.2}$$

is dominated by the second term in the square brackets and hence depends to a large extent on conditions near the centre of the star. Thus by comparing observed low-degree eigenfrequencies with the asymptotic formula one can hope to estimate the parameters ω_0, A, ϵ and B, and so calibrate theoretical models. In particular, because of the different ways in which these parameters depend on the structure of the star, one can infer from discrepancies between theory and observation in what way a theoretical model would need to be adjusted in order to remove those discrepancies.

I mention this simple example first because low-degree modes can in principle be measured in stars other than the sun, so comparison of these simple parameters is likely to be the basis of the earliest asteroseismic inferences.

I should perhaps stress again that such inferences will be made not from asymptotic analysis alone, but from carefully calculated eigenfrequencies of hopefully realistic stellar models which are computed numerically. The role of the asymptotic formulae is to guide the way in which the comparisons are made. In particular, the modelling of conditions near the surfaces of stars is very uncertain, and because all p modes are reflected near the surface they must all suffer from this uncertainty to some degree. However, combining data in such a way as to estimate the parameter A, for example, provides a measure of conditions in the core from which, one hopes, the surface uncertainties have been largely eliminated. Even though such a comparison needs to be carried out entirely numerically, the asymptotic analysis has played an important role in the design of the test and the interpretation of the results. This is one of the justifications for developing the asymptotic analysis yet further.

5 A Variational Principle, and its Application

If a frame of reference exists in which the basic structure of a star, which in general is presumed to be rotating, is independent of time, then one can seek linearized adiabatic oscillations in that frame whose time dependence is sinusoidal with frequency ω. That frequency is related to the displacement eigenfunction $\boldsymbol{\xi}(\boldsymbol{r})e^{-i\omega t}$ through a variational principle of the form

$$\omega^2\mathcal{I} - \omega\mathcal{R} - \mathcal{K} = 0 \tag{5.1}$$

(Lynden-Bell and Ostriker, 1967) provided ω does not exceed the critical cutoff frequency in the atmosphere, where $\mathcal{I}, \mathcal{R}$ and $\mathcal{K}$ are integrals over the volume of the star of functions which are bilinear in the components of $\boldsymbol{\xi}$ or their derivatives and which, of course, depend on the structure of the star but which do not contain ω explicitly. I shall not write down those integrals in their general form, but instead I shall consider some simple special cases. The variational principle is that, considered as a solution of the quadratic Eq. (5.1), ω is stationary with respect to arbitrary variations $\delta\boldsymbol{\xi}$ (satisfying appropriate boundary conditions, which I also refrain from quoting) to $\boldsymbol{\xi}$ when $\boldsymbol{\xi}$ is an eigenfunction and ω an eigenvalue of the adiabatic oscillation problem. This principle can conveniently be used to estimate eigenfrequencies from approximate eigenfunctions, or as the basis of perturbation theory.

In the case of a nonrotating star with no internal motion, $\mathcal{R} = 0$ and

$$\omega^2 = \frac{\mathcal{K}}{\mathcal{I}}. \tag{5.2}$$

Moreover, $\mathcal{I}$ and $\mathcal{K}$ are real. Hence ω^2 is real, and it is possible to choose $\boldsymbol{\xi}$ to be real. Then $\mathcal{I}$ and $\mathcal{K}$ can be written

$$\mathcal{I} = \int \rho \boldsymbol{\xi} \cdot \boldsymbol{\xi} dV, \tag{5.3}$$

where dV is a volume element, and

$$\mathcal{K} \simeq \int \left[\rho c^2 (\mathrm{div}\boldsymbol{\xi})^2 + 2(\boldsymbol{\xi} \cdot \nabla p)\mathrm{div}\boldsymbol{\xi} + \boldsymbol{\xi} \cdot \nabla p \, \boldsymbol{\xi} \cdot \nabla \ln \rho\right] dV \tag{5.4}$$

(*cf.* Ledoux and Walraven, 1958; Chandrasekhar, 1964). For simplicity I have not written $\mathcal{K}$ exactly; in particular, I have omitted terms accounting for oscillating perturbations to the gravitational potential which make a relatively small contribution, but which in accurate computations must, of course, be included.

One of the important uses of Eqs. (5.2)–(5.4) in helioseismology is in constructing kernels for computing the small differences between the solar frequencies and the eigenfrequencies of a good theoretical model of the sun. Because the quotient $\mathcal{K}/\mathcal{I}$ is stationary to variations in $\boldsymbol{\xi}$, small perturbations to $\mathcal{K}/\mathcal{I}$ resulting from variations in the model can be computed without requiring explicit knowledge of the resultant perturbation to $\boldsymbol{\xi}$. Thus one can compute, for example, kernels for spherically symmetric structure perturbations for the purposes of inverting frequency discrepancies. Alternatively one can compute kernels for aspherical perturbations, which provide formulae for degeneracy splitting. Similarly, one can introduce rotation as a small perturbation, which yields the formula

$$\delta\omega \simeq \frac{\mathcal{R}}{2\mathcal{I}} \tag{5.5}$$

for rotational splitting $\delta\omega$, where once again $\mathcal{R}$ and $\mathcal{I}$ are evaluated with the unperturbed eigenfunctions. It is important to appreciate that the substitution of unperturbed eigenfunctions into equations such as (5.5) does not imply that perturbations to the eigenfunctions are assumed not to exist; it is merely that the variational principle has permitted the evaluation of the frequency perturbation without requiring the eigenfunction perturbation to be known. Indeed, Eq. (5.5) can be derived without explicit use of the variational principle by expanding the basic oscillation equations and their eigensolutions about the spherically symmetrical nonrotating state, analogously to the expansion that led to Eq. (2.9) resulting from a spherically symmetric perturbation to the isothermal sphere. Equation (5.5) then arises as a solubility condition, analogous to Eq. (2.10), and is thus actually the condition that the perturbation to the eigenfunction exists. That perturbation has not only a different radial dependence but also a different angular dependence from the unperturbed eigenfunction.

6 Asymptotic p-mode Frequency Perturbations

For high-order p modes the term N^2/ω^2 can be ignored in Eq. (3.9) for the vertical component κ of the wave number characterizing the spatial oscillation of the eigenfunction. Moreover, ω_c is comparable with ω only in the vicinity of the upper reflecting layer, so I shall ignore that too. Then I can approximate κ by the very simple acoustic dispersion relation

$$\kappa^2 + \frac{L^2}{r^2} \simeq \frac{\omega^2}{c^2} \tag{6.1}$$

[*cf.* Eq. (2.3)], where $L^2 = l(l+1)$ or $\left(l+\frac{1}{2}\right)^2$. This can be substituted into Eq. (3.12) to yield a simpler eigenvalue equation, which I write as

$$F\left(\frac{\omega}{L}\right) \equiv \int_{r_t}^{R} \left(1 - \frac{L^2c^2}{\omega^2 r^2}\right)^{\frac{1}{2}} \frac{dr}{c} = \frac{(n+\alpha)\pi}{\omega}, \tag{6.2}$$

where r_t is an approximation to r_1 obtained by setting the integrand to zero:

$$\frac{c(r_t)}{r_t} = \frac{\omega}{L}. \tag{6.3}$$

As a result of removing ω_c from the formula, the integrand no longer vanishes near the surface, and I have been forced to replace r_2 by some radius R which represents the radius of the star. The error so introduced can hardly depend on L, because $Lc/\omega r \ll 1$ near the surface [I am ignoring modes of very high degree, with $l \gtrsim 10^3$, for which that is not the case], so I can absorb the transformation into the factor α which could depend on ω but not on L. It is straightforward to show that if the outer layers of the star can be approximated by a polytrope of index μ, then $\alpha = \mu/2$, and in that case is even independent of ω. Indeed, Eq. (6.2) with α constant was first shown by Duvall (1982) to approximate the solar frequencies when $\alpha \simeq 3/2$; a better representation is obtained with a function α that varies weakly with ω. Notice, however, that so far as studying the properties of the oscillations in the vicinity of $r = r_2$ is concerned, if the outer layers of the sun are to be represented by a polytrope one should choose $\mu = 3$; I shall use this result in Section 9 when discussing temporal frequency variations associated with the solar cycle. Values of r_t/R in a standard solar model corresponding to three representative frequencies are listed in Table 1 (p. 302).

One can formally perturb Eq. (6.2) to estimate the frequency difference $\delta\omega$ between two solar models (having the same radius) whose sound speeds differ by δc:

$$\frac{\delta\omega}{\omega} \simeq \left\langle\frac{\delta c}{c}\right\rangle \equiv S^{-1} \int_{\tau_t}^{T} \left(1 - \frac{L^2c^2}{\omega^2 r^2}\right)^{-\frac{1}{2}} \frac{\delta c}{c} d\tau, \tag{6.4}$$

where

$$S = \int_{\tau_t}^{T} \left(1 - \frac{L^2c^2}{\omega^2 r^2}\right)^{-\frac{1}{2}} d\tau, \tag{6.5}$$

$$\tau(r) = \int_0^r c^{-1} dr \tag{6.6}$$

is acoustic radius, $\tau_t = \tau(r_t)$ and $T = \tau(R)$. In obtaining this relation I have ignored the variation of α with both ω and $c(r)$, which appears to be a fairly good approximation (Christensen-Dalsgaard *et al.*, 1988).

Equation (6.4) can also be obtained, more laboriously, by computing a kernel for $c^{-1}\delta c$ from the variational principle (5.2)–(5.4) using the asymptotic eigenfunctions (3.20) and (3.21). When carried out in this way it is necessary to average over the rapidly oscillating part of the eigenfunction, which is valid only when the scales of variation of the models are much greater than κ^{-1}, which is, of course, a necessary condition for the asymptotic equation (6.2) to be valid. It becomes apparent from that analysis that S is proportional to the inertia $\mathcal{I}$ of the mode, defined by Eq. (5.3). One can similarly evaluate the formula

Table 1. Formal lower turning points r_t, in units of the photospheric radius R, of a standard solar model. The turning point is defined by Eq. (6.3) with $L = l + \frac{1}{2}$, and is evaluated at cyclic frequencies ($\nu = \omega/2\pi$) 2, 3 and 4 mHz. Notice that because the sun is roughly isothermal in the core, r_t/R is roughly proportional to $(l + \frac{1}{2})/\nu$ when l is small. It can also be verified that when l is large, the depth $1 - r_t/R$ is approximately proportional to ν^2/L^2 as predicted by the plane-parallel polytropic relations (2.19) and (2.20).

l	$\nu = 2$	$\nu = 3$	$\nu = 4$
1	0.0869	0.0581	0.0436
2	0.1401	0.0963	0.0725
3	0.1856	0.1318	0.1009
4	0.2252	0.1638	0.1275
5	0.2605	0.1926	0.1522
6	0.2929	0.2189	0.1749
7	0.3236	0.2432	0.1960
8	0.3526	0.2660	0.2157
9	0.3799	0.2876	0.2343
10	0.4058	0.3084	0.2519
15	0.5199	0.4016	0.3310
20	0.6124	0.4810	0.3994
25	0.6862	0.5495	0.4604
30	0.7359	0.6096	0.5147
40	0.8036	0.7037	0.6082
50	0.8501	0.7605	0.6830
60	0.8821	0.8027	0.7338
70	0.9052	0.8358	0.7715
80	0.9221	0.8614	0.8022
90	0.9347	0.8816	0.8279
100	0.9444	0.8979	0.8491
150	0.9701	0.9443	0.9138
200	0.9813	0.9642	0.9442
250	0.9867	0.9745	0.9604
300	0.9897	0.9812	0.9700
350	0.9918	0.9854	0.9764
400	0.9935	0.9879	0.9812
450	0.9948	0.9897	0.9845
500	0.9958	0.9912	0.9867
550	0.9967	0.9924	0.9884
600	0.9973	0.9935	0.9897
650	0.9979	0.9944	0.9908
700	0.9984	0.9952	0.9918
750	0.9988	0.9958	0.9927
800	0.9991	0.9964	0.9935
850	0.9994	0.9969	0.9942
900	0.9996	0.9973	0.9948
950	0.9997	0.9977	0.9953
1000	0.9998	0.9981	0.9958

Table 2. Acoustical structure of a solar model: $x = r/R$ where $R = 6.96 \times 10^{10}$cm; τ is acoustical radius, defined by Eq. (6.6) and $T = \tau(R) = 3518$ s; the sound speed c is quoted in $\mathrm{Mm\,s^{-1}}$, and $c/2\pi r$ in μHz; $\zeta(r) = c^2/r^2$ is the asymptotic depth co-ordinate appearing in Eq. (7.3) and $\zeta_0 = \zeta(R) = 1.29 \times 10^{-10}\,\mathrm{s^{-2}}$; ψ is defined by Eq. (7.4) and $\psi_0 = \psi(R) = 1.31 \times 10^{-12}\,\mathrm{s^{-3}}$; $-x^{-1}dc/dx$ is proportional to the integrand in Eq. (4.2) for the coefficient A in the asymptotic relation (2.7) applied to the sun, and is quoted in $\mathrm{Mm\,s^{-1}}$.

τ/T	r/R	c	$c/2\pi r$	ζ/ζ_0	ψ/ψ_0	$-x^{-1}dc/dx$
0.000	0.0000	0.5067	–	–	–	−3.681
0.025	0.0641	0.5078	1.81×10^3	1.00×10^6	2.25×10^6	0.293
0.050	0.1278	0.4957	8.87×10^2	2.40×10^5	2.93×10^5	3.340
0.075	0.1885	0.4621	5.61×10^2	9.59×10^4	8.43×10^4	3.445
0.100	0.2445	0.4247	3.97×10^2	4.82×10^4	3.28×10^4	2.718
0.125	0.2960	0.3923	3.03×10^2	2.80×10^4	1.51×10^4	1.944
0.150	0.3439	0.3671	2.44×10^2	1.82×10^4	8.04×10^3	1.436
0.175	0.3889	0.3455	2.03×10^2	1.26×10^4	4.80×10^3	1.171
0.200	0.4314	0.3274	1.74×10^2	9.19×10^3	3.03×10^3	0.942
0.225	0.4718	0.3114	1.51×10^2	6.96×10^3	2.06×10^3	0.819
0.250	0.5102	0.2969	1.33×10^2	5.41×10^3	1.46×10^3	0.723
0.275	0.5469	0.2837	1.19×10^2	4.29×10^3	1.06×10^3	0.645
0.300	0.5819	0.2715	1.07×10^2	3.48×10^3	7.94×10^2	0.583
0.325	0.6155	0.2602	9.67×10^1	2.85×10^3	6.14×10^2	0.547
0.350	0.6477	0.2493	8.80×10^1	2.37×10^3	4.89×10^2	0.532
0.375	0.6785	0.2383	8.03×10^1	1.97×10^3	4.08×10^2	0.560
0.400	0.7079	0.2260	7.30×10^1	1.63×10^3	3.71×10^2	0.687
0.425	0.7355	0.2112	6.57×10^1	1.32×10^3	3.11×10^2	0.744
0.450	0.7613	0.1971	5.92×10^1	1.07×10^3	2.46×10^2	0.726
0.475	0.7854	0.1837	5.35×10^1	8.73×10^2	1.96×10^2	0.710
0.500	0.8078	0.1713	4.85×10^1	7.17×10^2	1.56×10^2	0.686
0.525	0.8287	0.1596	4.40×10^1	5.92×10^2	1.27×10^2	0.685
0.550	0.8481	0.1484	4.00×10^1	4.88×10^2	1.06×10^2	0.698
0.575	0.8662	0.1375	3.63×10^1	4.02×10^2	8.82×10^1	0.711
0.600	0.8829	0.1270	3.29×10^1	3.30×10^2	7.32×10^1	0.723
0.625	0.8983	0.1170	2.98×10^1	2.71×10^2	6.14×10^1	0.747
0.650	0.9125	0.1072	2.69×10^1	2.20×10^2	5.19×10^1	0.783
0.675	0.9254	0.0977	2.41×10^1	1.78×10^2	4.38×10^1	0.826
0.700	0.9372	0.0884	2.16×10^1	1.42×10^2	3.70×10^1	0.882
0.725	0.9478	0.0793	1.91×10^1	1.12×10^2	3.12×10^1	0.952
0.750	0.9572	0.0703	1.68×10^1	8.62×10^1	2.60×10^1	1.038
0.775	0.9655	0.0616	1.46×10^1	6.50×10^1	2.17×10^1	1.156
0.800	0.9728	0.0532	1.25×10^1	4.77×10^1	1.70×10^1	1.240
0.825	0.9790	0.0456	1.07×10^1	3.46×10^1	1.25×10^1	1.249
0.850	0.9843	0.0385	8.95×10^0	2.45×10^1	1.11×10^1	1.582
0.875	0.9887	0.0307	7.11×10^0	1.54×10^1	8.59×10^0	1.957
0.900	0.9922	0.0242	5.57×10^0	9.47×10^0	5.02×10^0	1.861
0.925	0.9949	0.0191	4.40×10^0	5.91×10^0	3.16×10^0	1.878
0.950	0.9970	0.0151	3.47×10^0	3.67×10^0	1.96×10^0	1.882
0.975	0.9987	0.0118	2.70×10^0	2.23×10^0	1.38×10^0	2.175
1.000	1.0000	0.0079	1.81×10^0	1.00×10^0	1.00×10^0	3.522

(5.5) for rotational splitting. In the special case when the angular velocity Ω of the star is a function of radius alone, the result is

$$\delta\omega = mS^{-1}\int_{\tau_t}^{T}\left(1-\frac{L^2c^2}{\omega^2r^2}\right)^{-\frac{1}{2}}\Omega d\tau. \tag{6.7}$$

Alternatively, Eqs. (6.4) and (6.7) can be derived from ray theory, by representing the eigenfunctions as resonant superpositions of plane waves as described at the end of Section 3. In that case the relative frequency perturbations are seen to be averages of $c^{-1}\delta c$ and $m\omega^{-1}\Omega$ respectively, the averages being weighted by the time a wave spends in a given interval $d\tau$ of τ. The formulae are then seen to be quite natural.

The variational principle and ray theory can still be carried out when δc is aspherical or when Ω depends on the polar angles θ and ϕ. Associated with such perturbations is likely to be a meridional flow, but I shall ignore its influence on ω. I shall also here restrict attention to axisymmetric perturbations, and take the axis of my spherical polar coordinates to be the axis of symmetry. In that case the functions (3.7) with unique l and m (rather than linear combinations of such functions with different values of m) continue to represent the zero-order eigenfunctions, which simplifies the analysis considerably. One way to proceed with the variational principle is to expand δc or Ω in Legendre functions:

$$c^{-1}\delta c = \sum_k \beta_{2k}(r)P_{2k}(\cos\theta) \tag{6.8}$$

or a similar expression for Ω, and evaluate the angular integrals analytically. Only even terms are included in the sum because, as can easily be seen from the symmetry of the integrals, odd terms provide no contribution. In the case of acoustic asphericity, the frequency splitting between two modes of like n and l is given by

$$\omega^{-1}(\omega_m-\omega_0) \simeq \sum_k Q_{2k,lm}\langle\beta_{2k}\rangle, \tag{6.9}$$

where, provided $k \ll l$, $\langle\beta_{2k}\rangle$ is the average of β_{2k} weighted by $S^{-1}(1-L^2c^2/\omega^2r^2)^{-\frac{1}{2}}$ as in Eq. (6.4) and

$$Q_{2k,lm} = \frac{1}{2}(2l+1)\frac{(l-m)!}{(l+m)!}\int_{-1}^{1} P_{2k}(\mu)[P_l^m(\mu)]^2 d\mu. \tag{6.10}$$

The subscripts on ω in Eq. (6.9) denote the value of the azimuthal order, and are included only where they are needed. We note that $Q_{2k,lm}=0$ when $k>l$. When $k \ll l$,

$$Q_{2k,lm} \simeq \frac{(-1)^k(2k)!}{2^{2k}(k!)^2}P_{2k}(m/L); \tag{6.11}$$

this approximate expression can be used as a first guide even when k quite close to l.

If ray theory is employed, the relative frequency perturbation is once again found to be the average of $c^{-1}\delta c$ weighted by the relative time a sound wave spends in a given element of the propagating region. The splitting is given by

$$\omega^{-1}(\omega_m-\omega_0) \simeq \frac{2}{\pi}\int_0^M (M^2-\cos^2\theta)^{-\frac{1}{2}}\left\langle\frac{\delta c}{c}\right\rangle d\cos\theta \tag{6.12}$$

(Gough, 1990), where $M^2 = 1 - m^2/L^2$ and, once again, the angular brackets denote the radial average defined by Eq. (6.4). The formula is valid only for modes of oscillation which vary with θ on a scale much less than the perturbation δc, which is equivalent to the condition that only coefficients with $k \ll l$ contribute substantially to the sum in Eq. (6.9). Equation (6.12) can also be obtained by combining Eq. (6.9) and (6.11) and using an asymptotic expression for $P_{2k}(\mu)$ (Kosovichev and Parchevskii, 1988). Equations similar to (6.9) and (6.12) hold for rotational splitting. Analagous asymptotic formulae for splitting due to magnetic fields with simple geometries are presented by Gough and Thompson (1990).

7 Inversion of Asymptotic Formulae: the Abel Calculus

It is evident that the average $\langle f \rangle$, defined as in Eq. (6.4), of any function $f(r)$, such as $c^{-1}\delta c$, is a function of the parameters defining the mode of oscillation only in the combination ω/L, as also is the integral S. The splitting data must therefore also be functions of only ω/L; they define weighted averages of f over domains beneath the surface which penetrate down to the lower turning points of the modes. By considering combinations of different averages one can deduce the function f.

As is well known, the equation can be transformed into Abel's integral equation [as can Eq. (6.12)] and, provided sufficient data are available, inverted to give f in terms of the data. To carry that out one represents the data, essentially the averages $\langle f \rangle$, by a sufficiently smooth function of ω/L [which one associates with the independent variable τ_t via Eqs. (6.3) and (6.6)] and formally inverts the equation. Although the inversion is apparently an elementary procedure, it is useful to try to acquire some intuition about it, for then one might be able to assess the implications of new data even when one has no computer at hand.

Perhaps the first and crudest approach one might contemplate is to ignore the Abel kernel save for the lower turning point it implies, and approximate Eq. (6.4) by

$$\langle f \rangle \simeq (T - \tau_t)^{-1} \int_{\tau_t}^{T} f \, d\tau. \tag{7.1}$$

As a result of ignoring the singularity at $\tau = \tau_t$ the average of f is rather smoother than it should be, though the gross features of the general form are probably still discernable (*cf.* Christensen-Dalsgaard *et al.*, 1988). Evidently some appreciation of the data can therefore be obtained immediately by plotting them as a function of the lower turning point τ_t associated with ω/L. One sees immediately that a slight change in the value of $\langle f \rangle$ with τ_t must be the result of a more severe change in $f(\tau)$. Indeed, f depends on the derivative of $\langle f \rangle$, as can be seen by solving Eq. (7.1) for f:

$$f(\tau_t) \simeq \langle f \rangle - (T - \tau_t) \frac{d\langle f \rangle}{d\tau_t}. \tag{7.2}$$

Thus, in the case of rotational splitting, for example, one sees that the value of Ω at some position τ is not simply the value of the rotational splitting divided by m for modes with turning points at τ, as has sometimes been suggested in the literature. Indeed, as was illustrated by Gough (1984), the sun could have rotational splitting always less than the surface angular velocity yet have a rapidly rotating core, provided the splitting were to increase fast enough with ω/L at low L. However, because the average (7.1) provides excessive smoothing, its inverse (7.2) must overestimate the variation of f. Indeed, it can be badly in error in regions where $\langle f \rangle$ varies rapidly. The differentiation in Eq. (7.2) is too severe an operator. As I shall now demonstrate, it is basically the square root of the derivative operator that should be brought into play.

I begin by setting $\xi = \omega^2/L^2$ and introducing the depth coordinate $\zeta(\tau) = r^{-2}c^2$, neither of which should be confused with the depth coordinates used to describe the plane-parallel polytrope and the isothermal atmosphere in Section 2. Then Eq. (6.4) becomes

$$F(\xi) \equiv (\pi\xi)^{-\frac{1}{2}} S(\xi)\langle f \rangle = \pi^{-\frac{1}{2}} \int_{\xi_0}^{\xi} (\xi - \zeta)^{-\frac{1}{2}} \psi^{-1} f d\zeta, \tag{7.3}$$

where $\xi_0 = \zeta(T) = R^{-2}[c(R)]^2$ and

$$\psi = \frac{2c^3}{r^3} \left(1 - \frac{d\ln c}{d\ln r}\right). \tag{7.4}$$

This can be rewritten

$$F(\xi) = \mathcal{A}\psi^{-1} f, \tag{7.5}$$

where $\mathcal{A}$ is the Abel operator:

$$\mathcal{A}u = \pi^{-\frac{1}{2}} \int_{\xi_0}^{\xi} (\xi - \zeta)^{-\frac{1}{2}} u(\zeta) d\zeta. \tag{7.6}$$

Notice now that

$$\begin{aligned} \mathcal{A}^2 u &= \pi^{-1} \int_{\xi_0}^{\xi} (\xi - \zeta)^{-\frac{1}{2}} d\zeta \int_{\xi_0}^{\zeta} (\zeta - \eta)^{-\frac{1}{2}} u(\eta) d\eta \\ &= \pi^{-1} \int_{\xi_0}^{\xi} u(\eta) d\eta \int_{u}^{z} [(\xi - \zeta)(\zeta - \eta)]^{-\frac{1}{2}} d\zeta \\ &= \int_{\xi_0}^{\xi} u(\eta) d\eta. \end{aligned} \tag{7.7}$$

In other words, $\mathcal{A}^2$ is the integration operator. Consequently $\mathcal{A}$ is the square root of integration. Indeed, $\mathcal{A}$ is a special case of the Riemann-Liouville fractional integral. The solution to Eq. (7.5) is now evident. It simply requires inverting the operator $\mathcal{A}$, which yields what one should logically call the square root of differentiation:

$$f = \psi \frac{d^{\frac{1}{2}}}{d\xi^{\frac{1}{2}}} F, \tag{7.8}$$

the right-hand side being evaluated at $\xi = \zeta$. The halfth derivative in Eq. (7.8) can be regarded as the derivative of the halfth integral. Thus

$$\begin{aligned} f(\zeta) &= \psi \frac{d}{d\xi} \mathcal{A} F \\ &= \pi^{-\frac{1}{2}} \psi \frac{d}{d\zeta} \int_{\xi_0}^{\zeta} (\zeta - \xi)^{-\frac{1}{2}} F(\xi) d\xi. \end{aligned} \tag{7.9}$$

Eq. (7.9) can be regarded as a formal explanation of Eq. (7.8). It is what one must use to evaluate f; but perhaps the concept behind Eq. (7.8) is more illuminating. As is well known, differentiation exaggerates the variations in a typical function. The square root of differentiation also exaggerates variation, but less severely so, and can be regarded as performing just half the operation of transforming a function to its derivative. Thus, for example, if D represents differentiation, $D^{\frac{1}{2}}x^{\mu} = [\Gamma(\mu+1)/\Gamma(\mu+\frac{1}{2})]x^{\mu-\frac{1}{2}}$, where Γ is the gamma function. Fractionally differentiating the exponential function is somewhat more complicated, but when the argument is large the result simplifies: $D^{\frac{1}{2}}e^{kx} \sim k^{\frac{1}{2}}e^{kx}$ and $D^{\frac{1}{2}}\cos kx \sim k^{\frac{1}{2}}\cos(kx+\pi/4)$ as $kx \to \infty$. A more complicated function F and its derivatives $D^{\frac{1}{2}}F$ and DF are illustrated in Fig. 4.

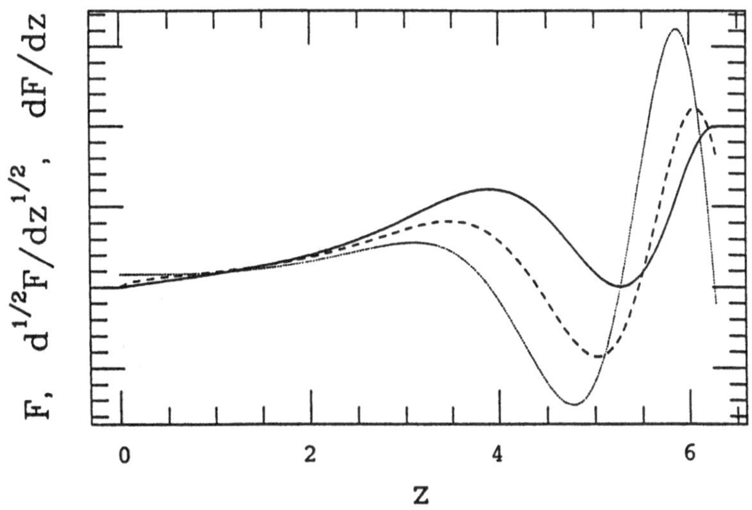

Fig. 4. The first derivative dF/dz (dotted curve) and the halfth derivative $d^{\frac{1}{2}}F/dz^{\frac{1}{2}}$ (dashed curve) of the function $F(z)$ represented by the continuous curve.

Evidently, if one is to assess data in this way one needs as part of one's inverter's tool kit the functions τ, ζ and ψ for the sun. A good solar model provides an adequate substitute, and in Table 2 (p. 303) I supply these, together with the sound speed c and the integrand $-x^{-1}dc/dx$, where $x = r/R$ is normalized radius, in the formula (4.2) for the coefficient A that appears in the asymptotic expression for low-degree p-mode eigenfrequencies.

I conclude this discussion by pointing out that fitting a smooth curve through the data is not a wholly straightforward operation. For example, if one tries to infer from the data the function $F(\omega/L)$ defined by Eq. (6.2) by finding that function $\alpha(\omega)$ which causes $\pi(n+\alpha)/\omega$ to be in some sense most closely a function of ω/L alone, one obtains a result such as that illustrated in Fig. 5. It is evident from the figure that $\pi(n+\alpha)/\omega$ depends also on some other parameter, because the points do not lie on a single curve; this renders it difficult to infer $F(\omega/L)$ precisely because it is not obvious how to draw

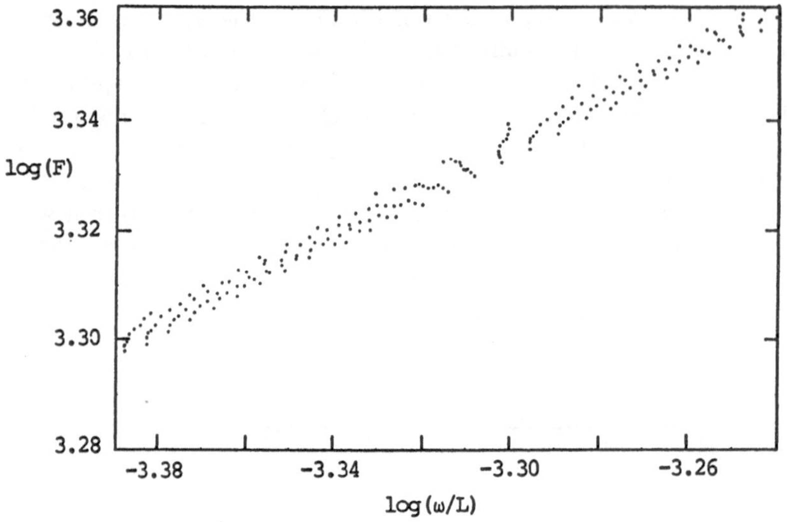

Fig. 5. Log $[\pi(n+\alpha)/\omega]$ computed from adiabatic oscillations of a standard solar model plotted against log $[\omega/(l+\frac{1}{2})]$ over a typical small range, where $\alpha(\omega)$ is chosen such as to minimize the scatter about a single curve for those modes whose lower turning points are in the convection zone.

a curve through the points. However, the deviations of the points from a single curve are systematic, and therefore contain extractable information about the structure of the star. The deviations arise partly from the neglect of N^2 and the inadequate treatment of ω_c^2 in approximating Eq. (3.12) by Eq. (6.2) (and also from the neglect of some relatively small curvature terms), partly from the neglect of the perturbation to the gravitational potential in the differential equations from which the eigenvalue Eq. (3.12) was derived, and partly from the asymptotic truncation adopted in deriving Eq. (3.12) from Eq. (3.6). With a better understanding of these approximations it should be possible to obtain an improved estimate of $F(\omega/L)$ than has hitherto been accomplished merely by adopting some arbitrary procedure to draw a representative curve through the points. This is so also of the formula for the differences in frequency between two similar models used for differential inversions and for the formulae for degeneracy splitting by symmetry-breaking agents. Indeed, the comparison between the asymptotic formula for differential inversions and the corresponding function of exact frequency differences presented by Christensen-Dalsgaard *et al.* (1988) suggests that some envelope of the data rather than a representative mean is more appropriate. More detailed asymptotic analysis will be necessary before such an envelope can be estimated from the frequency data alone.

8 The Signature of Rapid Variation

The JWKB approximation upon which most of the asymptotic analysis is based is valid only where the scale of variation of the background state is very much greater than the wavelength of the oscillations. This is not the case throughout the entire interior of the sun, and therefore in reality there are significant errors in the eigenvalue equation (3.12). One such error, as I have already discussed, comes from the inadequate treatment of the outer layers of the region of propagation where $\omega_c \simeq \omega$. Other regions of rapid variation in the background state are the He II ionization zone and the base of the convection zone. In the He II ionization zone there is a rapid variation of the adiabatic exponent γ which enters directly in the formula for the sound speed. In typical solar models the base of the convection zone is essentially a discontinuity in the second derivative of the sound speed, which produces a discontinuity in ω_c.

One can estimate the influence of localized rapid variation in two ways, and I illustrate both here. The first is via the variational principle discussed in Section 5. The reason I specialize to rapid variation of the background state which is localized is that then I can assume that because the JWKB analysis is valid almost everywhere, the values of the asymptotic p-mode eigenfunctions given by Eqs. (3.20) and (3.21) are adequate for substituting into Eqs. (5.3) and (5.4), as Fig. 1 illustrates. I shall discuss explicitly the localized variation of γ in the He II ionization zone. Noting that away from the turning points the vertical component of $\boldsymbol{\xi}$ dominates the displacement, and that for low-degree modes vertical derivatives of the high-order eigenfunctions are much greater in magnitude than horizontal derivatives, it is a good approximation to represent div$\boldsymbol{\xi}$ by the radial derivative of the radial component ξ of $\boldsymbol{\xi}$. I now consider the contribution from the first term in Eq. (5.4) and integrate it by parts twice, obtaining

$$\int \gamma p(\mathrm{div}\boldsymbol{\xi})^2 dV \simeq 2\pi \int p\frac{d^2\gamma}{dr^2}\xi^2 r^2 dr + \text{integrated parts.} \tag{8.1}$$

In obtaining this equation I have ignored derivatives with respect to r of r,p and ξ compared with the corresponding derivative of γ.

The reason for performing this integration by parts is to express the integral in terms of the undifferentiated eigenfunction ξ, which is much more reliably represented by the JWKB formula than is its derivative. It is evident now that the relatively rapid variation of γ in the He II ionization zone leads to a high value of $d^2\gamma/dr^2$, and hence a contribution to the total integral $\mathcal{K}$ approximately proportional to

$$\mathcal{K}_0 = \cos^2\left(\int_{r_i}^{r_2} \kappa dr + \frac{\pi}{4}\right) \simeq \frac{1}{2}[1 - \sin 2\omega(T - \tau_i)] \tag{8.2}$$

for low-degree modes, where r_i is the radius and $T - \tau_i$ the acoustical depth of the ionization zone. [In deriving the right-hand side of Eq. (8.2) I ignored the small l-dependent terms and removed ω_c from the integrand, incorporating it into the phase by replacing $\pi/4$ by $-\alpha\pi/2$ as in Eq. (6.2) and substituting the observed value $\alpha \simeq 3/2$.] Actually, the contribution $\mathcal{K}_0$ is an appropriate average of the expression (8.2) over the region of variation of γ, but if ω is not too great that average is similar to a point value. Thus there is a contribution to the integral $\mathcal{K}$, and hence to ω given by Eq. (5.2), which is itself an oscillatory function of ω with 'frequency' $2(T - \tau_i)$.

This contribution to $\mathcal{K}$ is quite small in magnitude, but can be recognized by its oscillatory behaviour. Thus, if the major contributions to the frequencies ω_n of order n and their first differences are eliminated by considering the second differences

$$\delta_2\omega \equiv \omega_{n+1} - 2\omega_n + \omega_{n-1}, \tag{8.3}$$

the oscillating term becomes apparent. This is illustrated in Fig. 6, where second differences of low-degree eigenfrequencies of a solar model are plotted against ω.

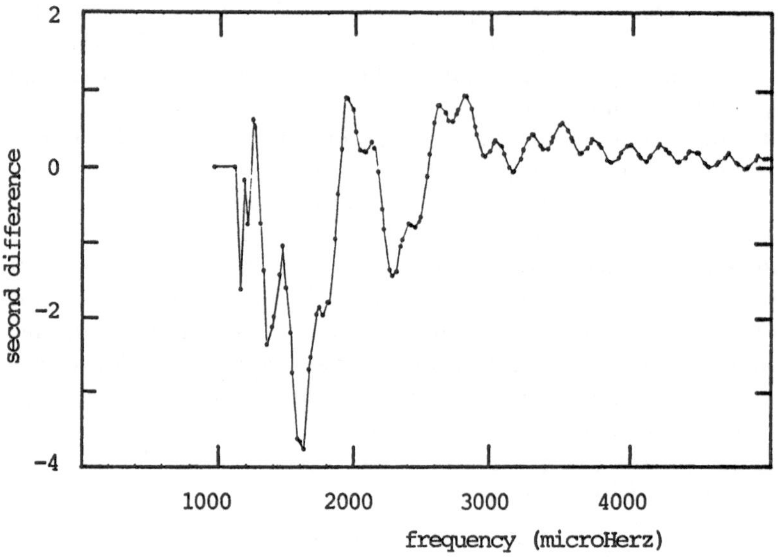

Fig. 6. Second cyclic frequency differences $\delta_2\nu$ defined as in Eq. (8.3) with ω_n replaced by $\nu_n = \omega_n/2\pi$, of modes with $l = 0, 1, 2, 3$ and 4 of a standard solar model, plotted against ν_n. The units are μHz.

The second method of estimating the effect of rapid variation is to regard the localized region as a discontinuity, and at that discontinuity match asymptotic eigenfunctions appropriate to conditions on either side of it. This procedure is perhaps more appropriate for modelling the discontinuity of ω_c^2 at the base of the convection zone, which leads to a discontinuity in the coefficient κ^2 of the undifferentiated term in Eq. (3.8). Rather than present the details of the stellar asymptotics explicitly, I treat a much simpler but mathematically similar problem which exhibits the essence of the influence of the discontinuity that I wish to illustrate. I consider the longitudinal acoustic oscillations of gas in a pipe governed by the simple equation

$$\frac{d^2\Psi}{dz^2} + \frac{\omega^2}{c^2}\Psi = 0, \tag{8.4}$$

in which $c = c_1$ if $0 \leq z < \tilde{\lambda}a$ and $c = c_2$ if $\tilde{\lambda}a < z \leq a$, where $\tilde{\lambda}$, c_1 and c_2 are constants. Adopting, for simplicity, the boundary conditions $\Psi = 0$ at $z = 0$ and $z = a$, with Ψ and $c^2 d\Psi/dz$ continuous at $z = \tilde{\lambda}a$, then yields the eigenvalue equation

$$c_2 \tan \omega\tau_1 + c_1 \tan \omega\tau_2 = 0, \tag{8.5}$$

in which $\tau_1 = \tilde{\lambda}a/c_1$ and $\tau_2 = (1-\tilde{\lambda})a/c_2$ are the acoustical lengths of the two uniform regions in the pipe. It is straightforward to show from the eigenvalue equation (8.5) that if $c_2 = (1+\epsilon)c_1$ with $|\epsilon| \ll 1$, then again there is an oscillatory component to ω, such that

$$\delta_2\omega \propto \epsilon \sin\left(2\omega\tau_1 + \tan^{-1}\sin\frac{2\pi\tau_1}{\tau_1+\tau_2}\right), \tag{8.6}$$

which also has a 'frequency' that is twice the acoustical length of one of the uniform regions. [Since, in this case, $\omega \simeq n\pi/(\tau_1+\tau_2)$, τ_1 and τ_2 are interchangeable in the expression on the right-hand side of Eq. (8.6).]

Alternatively, one can consider a mathematical model given by

$$\frac{d^2\Psi}{dz^2} + V\Psi = 0 \tag{8.7}$$

with $V(z;\omega) \equiv (\omega^2 - \omega_c^2)/c^2$ being piecewise constant with respect to z. The model can be considered to be isothermal throughout, with a constant density scale height, the discontinuity in V arising from a discontinuity in the density scale height. This simple model exhibits the character of the base of the convection zone in a standard solar model more closely. Its eigenvalue equation is

$$\tilde{\omega}_2 \tan \tilde{\omega}_1\tau_1 + \tilde{\omega}_1 \tan \tilde{\omega}_2\tau_2 = 0, \tag{8.8}$$

where $\tilde{\omega}_i^2 = \omega^2 - \omega_{ci}^2$, ω_{c1} and ω_{c2} being the values of the critical cutoff frequency in regions 1 and 2 respectively. In much the same way as before one can set $\omega_{c2} = (1+\epsilon)\omega_{c1}$ and expand the eigenfrequencies to first order in ϵ. Again an oscillatory component is found, whose 'frequency' tends to $2\tau_1$ for high-order modes for which $\omega^2 \gg \omega_{c1}^2$.

Oscillations due to both the discontinuity at the base of the convection zone and the variation of γ in the He II ionization zone are evident in Fig. 6, where $\delta_2\omega/2\pi$ (evaluated at fixed degree l) is plotted against $\omega/2\pi$ for modes with $0 \le l \le 5$. Several low values of l are plotted in order to give a great enough density of points to exhibit the oscillations; high-degree modes are not included, because my approximations demand that the discontinuity be far above the lower turning point, where the actual value of l is immaterial. The most prominent feature of the data is a large-amplitude oscillation, with a 'period' of about 750μHz, which corresponds to a localized variation in the background state at an acoustical depth $T - \tau \simeq (2 \times 750\mu\text{Hz})^{-1} \simeq 670\,\text{s}$. Thus, since $T \simeq 3518\,\text{s}$, $\tau/T \simeq 0.81$. One can now consult Table 2 (p. 303) to learn that this corresponds to a radius $r \simeq 0.975R$. The oscillation is therefore presumably a consequence of the variation of γ due to He II ionization. (50 per cent of He is doubly ionized at the radius $0.975R$.) Notice that its amplitude decreases with increasing frequency. This must be a result of the variation of γ occurring over an extended region, of thickness d, say, of the star. As ω increases, the wavelength λ of the eigenfunction decreases, as is evident from Eq. (8.2), and once λ becomes comparable with d cancellation of the contribution to the integral (8.1) from the oscillatory component of ξ becomes significant. The value of the amplitude of the large-scale oscillation in the points plotted in Fig. 6 depends on the magnitude of the variation of γ, which in turn depends on the abundance Y of helium. Thus perhaps we are offered a potential diagnostic to determine Y through a measurement of the equation of state which does not depend on the overall structure of the star, and therefore ought to be insensitive to many of the assumptions of the theory of stellar evolution and errors in the uncertain values of the opacity and the nuclear reaction rates. This particular diagnostic is therefore very exciting because, unlike the previous suggestion for making an almost direct seismic measurement of Y in the convection zone (Däppen and Gough, 1986), with sufficiently accurate data it might perhaps be possible to carry out model calibrations using only modes of low degree that can be obtained from integrated whole-disk measurements, which might therefore also be applicable to data from other stars too. A careful analysis of the sensitivity of the oscillation to the value of Y and to uncertainties in the theoretical models will be required before we can judge whether the diagnostic is likely to be of practical use in the foreseeable future.

The other obvious feature in Fig. 6 is the small-scale oscillation with a 'period' of about 230μHz. This has a smaller amplitude than the He II oscillation, at least at low frequency, which does not diminish with increasing ω, indicating that the variation in the background state that causes it is highly localized. It occurs at an acoustical depth of $(2 \times 235\mu\text{Hz})^{-1} \simeq 2130\,\text{s}$, corresponding to $\tau/T \simeq 0.40$ which is situated at a radius $r \simeq 0.71R$, precisely at the base of the convection zone. Notice that this particular property of the oscillation data arises from a discontinuity in ω_c arising from the discontinuity in the second derivative of density in the theoretical solar model, and not from some perturbing agent such as a confined magnetic field which would have had a superficially similar signature. Unlike a magnetic field, however, this property is spherically symmetrical, and therefore does not contribute to degeneracy splitting. Thus there might perhaps be some means of differentiating between the two phenomena.

9 Solar-cycle Frequency Variations

I close with a few remarks about the spectacular new observations reported at this meeting by Libbrecht and Woodard. I shall concentrate on the frequency dependence of the overall frequency change from 1986 to 1988, which was found to be approximately inversely proportional to the inertia $\mathcal{I}$ of the mode, where $\mathcal{I}$ is given by Eq. (5.3) for a displacement eigenfunction $\boldsymbol{\xi}$ normalized such that $\Xi(R) = 1$. What does this imply?

It is useful to discuss the result in terms of the variational principle of Section 5. The effect of a small perturbation to the structure of the sun on the oscillation eigenfrequencies can then be obtained by perturbing Eq. (5.2):

$$\delta\omega = \frac{\delta\mathcal{K} - \omega^2\delta\mathcal{I}}{2\omega\mathcal{I}}, \tag{9.1}$$

where $\delta\mathcal{K}$ and $\delta\mathcal{I}$ are the perturbations to $\mathcal{K}$ and $\mathcal{I}$ at constant $\boldsymbol{\xi}$. It is immediately clear that for $\delta\omega$ to be proportional to $\mathcal{I}^{-1}$, $\delta\mathcal{K} - \omega^2\delta\mathcal{I}$ must be proportional to ω.

Rather than using the variational principle in the form given explicitly in Section 5, it is safer to write it with respect to independent variables that remain constant on the surface of the star, for then it is not necessary to consider variations of the limits of integration when the solar surface is perturbed (*cf.* Gough and Thompson, 1990). It will soon be evident that I shall be discussing only perturbations near the surface of the sun, where all p modes of low and moderate degree have essentially the same spatial structure. Moreover, the perturbation will be predominantly spherically symmetrical. Therefore for the purposes of calculating $\delta\mathcal{K}$ I shall sometimes find it convenient to simplify the discussion to radial modes of a spherical star, replacing $\mathcal{K}$ by (*cf.* Ledoux and Walraven, 1958)

$$\int_0^M \left\{ rc^2 \left(\frac{d\tilde{\xi}}{dr}\right)^2 - \frac{1}{\rho}\frac{d}{dr}\left[(3\gamma-4)p\right]\tilde{\xi}^2 \right\} rdm \tag{9.2}$$

with respect to the Lagrangian variable m, which is the mass enclosed in a sphere of radius r: $dm = 4\pi r^2 \rho dr$. Also $\tilde{\xi} = r^{-1}\xi$, ξ being the radial component of $\boldsymbol{\xi}$, and M is the total mass of the star.

Before continuing I should point out that the frequency dependence of the inertia $\mathcal{I}$ of the mode is quite steep, and depends predominantly on conditions in the outer layers of the sun where the mode is evanescent, and not on the structure of the deep interior. This is because the normalization of the eigenfunction is carried out at the surface of the star, where the motion can be observed, rather than deep in the interior where most of the energy resides. Indeed, $\mathcal{I}$ depends simply on the depth $z_2 = R - r_2$ of the upper turning point. To see this, note that according to the asymptotic relations (3.20) and (3.21), $r^2\rho c\boldsymbol{\xi}\cdot\boldsymbol{\xi}$ is approximately constant in the region of propagation, aside from rapidly oscillating sinusoidal terms, so that the contribution to $\int \rho\boldsymbol{\xi}\cdot\boldsymbol{\xi}r^2 dr$ from that region is proportional to $\int c^{-1}dr$, which is the sound travel time between the turning points. For p modes of low and moderate degree, this depends only quite weakly on the radius of those turning points, because the sound speed is relatively high near the lower turning point and conditions vary so rapidly near the upper turning point that the frequency variation of the acoustical radius τ_2 at that point, relative to the total acoustical radius T of the sun, is quite small. The value of $\mathcal{I}$ therefore depends predominantly on the normalization $\Xi(R) = 1$, which requires relating the magnitude of Ξ in the acoustic cavity with the value at the surface. As we learned from the discussion of vertical oscillations of a plane-parallel polytrope, Ξ varies slowly near the very top of the evanescent region where $\omega < \omega_c$, and for a polytrope $\omega_c \propto z^{-\frac{1}{2}}$. Thus, at the depth z_2 at which $\omega_c = \omega, c \propto z^{\frac{1}{2}}$ which is proportional to ω^{-1} and $\rho \propto z^{\mu}$ which is proportional to $\omega^{-2\mu}$, where μ is a polytropic index characterizing the stratification near $z = z_2$. Consequently, $\mathcal{I} \propto \omega^{-2\mu-1}$, and therefore $\mathcal{I} \propto \omega^{-7}$ if $\mu = 3$, the value suggested by Duvall's analysis (see p. 301). It is straightforward to confirm this result by direct computation using the oscillation eigenfunctions of the plane-parallel polytrope discussed in Section 2. The outer layers of the sun are rather more complicated than a simple polytrope, and consequently $\mathcal{I}$ is not a simple power of ω, but the value of the polytropic exponent $-(2\mu+1) \simeq -7$ is typical of $d\ln\mathcal{I}/d\ln\omega$ for a realistic solar model.

What we have learned from this discussion is that, relative to the surface displacement $\Xi(R)$, the amplitude of the displacement is approximately $\Xi(R)$ well above the upper turning point, and is proportional to $\omega^{-\mu}\Xi(R)$ well below it. Thus, any perturbation to the structure of the star well above the turning points of the modes leads to a weak frequency dependence of $\delta\mathcal{K} - \omega^2\delta\mathcal{I}$, and hence, by Eq. (9.1), to a frequency perturbation $\delta\omega$ which is roughly proportional to $\mathcal{I}^{-1}$. A perturbation well beneath the turning points of the modes, on the other hand, introduces a factor $\omega^{-2\mu}$ into the numerator of Eq. (9.1) which largely cancels the factor $\mathcal{I}$ in the denominator, and the frequency perturbation $\delta\omega$ then varies only weakly with ω. We can conclude with Libbrecht and Woodard, therefore, that the variation in the structure of the sun responsible for the observed frequency changes must lie in the very outer layers where the modes are evanescent. Their effect can be regarded simply as a modification to the reflecting boundary condition that the inert outer layers present to the wave propagating in the cavity beneath.

The precise frequency dependence of $\delta\omega$ depends on the nature of the variation of the background state. Continuing to use the plane-parallel polytrope as a guide, we note from the discussion on pp. 295-296 that in the vicinity of the free surface div$\boldsymbol{\xi} \propto \omega^2\xi$. Therefore, from Eqs. (9.1), (5.3) and (5.4) or (9.2), it appears that a superficial variation in the solar structure leads to a frequency change of the form

$$\delta\omega \propto \frac{Q(\omega^2)}{\omega\mathcal{I}}, \tag{9.3}$$

where $Q(x)$ is a quadratic function of x. Some care must be taken in interpreting this result, however, because the stratification of the sun is essentially hydrostatic, and one cannot arbitrarily modify ρ and c^2 without addressing how that modification might come about. Let us consider a few examples.

If there were a magnetic field introduced into the evanescent layers, which for the moment I shall suppose is concentrated into fibrils by the convective motion, then the characteristic propagation speed of acoustic waves is modified, without necessarily making a serious impact on the mean density stratification. This could be modelled by a change in c^2 appearing in $\mathcal{K}$, and hence $Q \propto \omega^4$ and $\delta\omega \propto \omega^3/\mathcal{I}$, which is the result suggested by Woodard and Libbrecht. If, alternatively, the dominant effect is to change the efficacy of convection and thereby modify principally the scale heights in the rapidly varying superadiabatic boundary layer, then one might expect the modification of the last term in the expression (9.2) for $\mathcal{K}$ to dominate the perturbation, and $\delta\omega \propto (\omega\mathcal{I})^{-1}$. More subtle perturbations, such as modifications to the density that leave the density jump across the superadiabatic boundary layer unmodified, lead to intermediate variation: if the solar radius is unperturbed, they are sensitive mainly to the second term in the integrand in the expression (5.4) for $\mathcal{K}$. Then $\delta\omega \propto \omega/\mathcal{I} \propto \omega^{2(\mu+1)} = \omega^8$ when $\mu = 3$. These results are exhibited by the analyses of the vertical oscillations of the plane-parallel polytrope with a superposed isothermal atmosphere mentioned in Section 2. The variation $\delta\omega \propto \omega^{2\mu}$ arises from variations in a temperature (and density) discontinuity modelling the superadiabatic convective boundary layer, and can be considered to result from the change in the reflection coefficient at the discontinuity. The variation $\delta\omega \propto \omega^{2(\mu+1)}$ arises from variations in the value of the temperature of an atmosphere that matches continuously onto the polytrope (Christensen-Dalsgaard and Gough, 1980), and arises from a variation in the inertia of the oscillating atmosphere.

Of the three examples I have mentioned, the second and third have frequency dependences that are closest to the observations. Indeed, it might be the case that either $\delta\omega \propto (\omega\mathcal{I})^{-1}$ or $\delta\omega \propto \omega/\mathcal{I}$ is not a significantly poorer representation of the data than the $\mathcal{I}^{-1}$ dependence reported by Libbrecht and Woodard. The former might arise from variations in the efficacy of convection, brought about, perhaps, by magnetic-field variations, which modify the structure of only the very outer layers of the sun within and above the superadiabatic boundary layer. Indeed, this was the first idea that came to mind (Gough and Thompson, 1988) to explain solar acoustic asphericity implied by the even component of the degeneracy splitting originally reported by Duvall *et al.* (1986).

The asphericity was modelled by a latitudinal variation in the mixing length used to determine the convective heat and momentum transport, and the structure was computed by integrating the equations governing a spherically symmetrical star inwards from the photosphere, adjusting the mass, radius and effective temperature at the photosphere to ensure that conditions beneath the convection zone were invariant, in order to maintain the horizontal component of hydrostatic balance at greater depths. Of course, this implied a latitudinally varying heat flux, which was presumed to have been accommodated by a small relative horizontal component of the tiny superadiabatic temperature gradient in the essentially adiabatically stratified interior of the convection zone where the thermal capacity of the convecting fluid is very great, thereby permitting the convection zone to be supported by a spherically symmetrical radiative interior.

The response of the radial stratification of the convection zone to the value of the mixing length depends critically on how the zone abuts onto the radiative interior. The computations by Thompson and myself to mimic the latitudinal variation were carried out before any temporal variations had been established, and a model in thermal balance was sought (Gough and Thompson, 1988). The effect of reducing the efficacy of convection in the equatorial regions is to contain the heat more effectively within the star, causing in this case at almost all depths within the convection zone an increase in the sound speed and a consequent expansion, inducing an oblateness of surfaces of constant temperature which is compensated by a mass transfer from the poles towards the equator. The perturbation to the value of the sound speed c varies weakly with depth beneath the superadiabatic boundary layer, leading to a variation of c^{-1} (upon which the acoustical depth, and hence also the oscillation frequencies depend) which is concentrated near the surface of the model. In the superadiabatic boundary layer the temperature gradient is substantially greater in the equatorial regions, to counteract the reduction in the efficacy of convective transport; this leads to a reduction in the photospheric temperature, despite the fact that the material beneath the boundary layer is hotter.

A temporal change of magnetic activity over all latitudes causing a modulation in convective efficacy is likely also to modify the stratification predominantly in the outer superadiabatic boundary layer, though the overall reaction of the convection zone would be somewhat different. In particular, the mass beneath the photosphere is conserved. Conditions at the base of the convection zone would not be temporally invariant, and the perturbed structure should really be matched onto an essentially adiabatically perturbed radiative interior in hydrostatic equilibrium (*cf.* Gough, 1981). Nevertheless, the amplitude of the perturbation of the interior of the star is very small compared with the perturbation at the top of the convection zone, and the latitudinal variation in the surface layers computed by Gough and Thompson (1988) therefore probably mimics the temporal variation, and can therefore be used for a first estimate of the frequency changes. It is therefore instructive to consider these frequency changes in more detail.

In Fig. 7 is plotted frequency variations for several values of the degree l. Reducing the efficacy of convection increases the frequencies of the modes. The magnitude of the reduction of the mixing length, some 2 per cent, has been chosen to yield frequency variations of about the same magnitude as those reported by Libbrecht and Woodard in these proceedings. The functional form is very similar to the observations too, confirming that a perturbation to the convective boundary layer is likely to be responsible for the frequency changes. At low degree l the frequency changes are independent of l, but as l increases the magnitude of the changes increases. This is largely due to the reduction in the modal inertia $\mathcal{I}$. The frequency changes of 3 mHz modes are plotted as a function of l by Gough and Thompson (1988), and fit the values plotted in Fig. 3 of Libbrecht and Woodard (1990) somewhat better than the inverse modal inertia, though not significantly so.

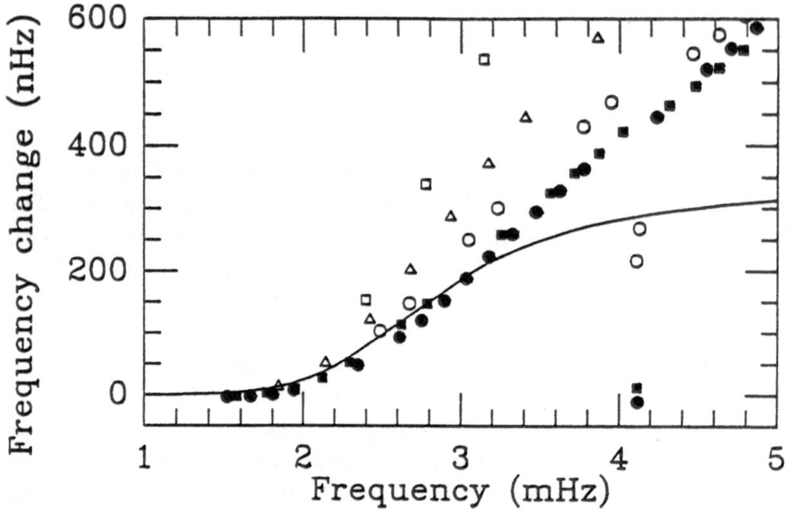

Fig. 7. Frequency dependence of the frequency differences $\Delta\nu$ resulting from a change in mixing length of a solar envelope model (from the computations discussed by Gough and Thompson, 1988) for a selection of modes with $l = 8(\bullet)$, $l = 15(\blacksquare)$, $l = 30(\circ)$, $l = 100(\triangle)$ and $l = 200(\square)$. The units are nHz. The continuous line is proportional to the inverse inertia $\mathcal{I}^{-1}$ corresponding to modes with $l = 0$ of a standard solar model, where $\mathcal{I}$ is defined by Eq. (5.3) normalized such that $\Xi = 1$ at an optical depth of 10^{-2}. This curve does not lie precisely on the low-degree points because $\mathcal{I}$ would be somewhat different, particularly at high frequencies, if it were normalized (more appropriately, for this discussion) at the location where the change in the model is greatest. The modes with $\nu \simeq 4.1$ mHz with small frequency differences resonate with the chromosphere.

Despite the ability of the model to reproduce the observed frequency shifts, it cannot be correct as it stands. The model requires a reduction in effective temperature from 1986, at sunspot minimum, to 1988 of 0.7 per cent. This is contradicted by the observed rise in irradiance reported by Willson and Hudson (1988) from the ACRIM data, implying that an appropriate spatial average of the effective temperature increased during that interval, by only 10^{-2} per cent. Moreover, the latitudinal temperature variation is similarly at variance with the brightness-temperature measurements of Kuhn *et al.* (1988) which is what led to the rejection of the model as an explanation of the acoustic asphericity (Gough and Thompson, 1988). However, the response of the convection zone to a solar-cycle perturbation, varying on a timescale much shorter than the characteristic cooling time of the entire zone (about 10^5 years), might perhaps be rather different from the structure which would result if the zone were given time enough to achieve thermal balance with its surroundings. Therefore, magnetic inhibition of convection is not yet ruled out by this model as a candidate for the frequency variations.

It should perhaps be remarked that Kuhn's (1988a, b) subsequent suggestion of a depth-independent relative temperature variation of about 0.1 per cent extending down to some radius r_0 (where $r_0 \lesssim 0.95R$, and might be as small as $0.2R$ or less) substantially beneath the superadiabatic convective boundary layer also cannot account for Libbrecht and Woodard's observations. As is evident from the discussion above, the frequency dependence of the splitting coefficients predicted by that hypothesis is much weaker than $\mathcal{I}^{-1}$, which is at variance with that reported by Libbrecht and Woodard (1990).

Of course, there are other possibilities for modifying the wave-propagation speed and the location of the upper reflecting boundary of the acoustical cavity, such as the direct influence of the Lorentz force, and these too need to be investigated. Indeed, Gough and Thompson (1988) suggested that the frequency changes may be a result of changes to fibril magnetic fields which might change the propagation speed of large-scale acoustic waves without substantially modifying the mean thermal stratification of the convection zone. Numerical simulations of Boussinesq convection in a magnetic field have shown that at high Rayleigh number the magnetic field is expelled to the interstices between convective cells without modifying greatly the convective flow elsewhere, except possibly to change the preferred horizontal length scale. There is certainly evidence that the scale of solar granulation is different in active regions from that in the quiet sun. Of course, the energy flux is altered, but so too is the magnitude of the inhomogeneities, and since the mean heat flux and the mean acoustic propagation speed depend differently on the inhomogenieties, p-mode oscillation frequency changes could be induced without such large energy-flux changes as the current mixing-length models seem to require. These new observations will certainly provoke considerable further thought.

Acknowledgements

A.G. Kosovichev computed the points plotted in Fig. 1, and N.J. Balmforth produced Fig. 2 and helped to produce Figs. 4 and 7. The discussion in Section 8 was prompted by E. Novotny, and subsequent conversations with N.J. Balmforth and W.J. Merryfield, with whom I have also discussed the contents of Section 9. I thank them all. I acknowledge support in part by the National Science Foundation under Grant PHY89-04035, supplemented by funds from the National Aeronautics and Space Administration, at the University of California at Santa Barbara.

References

Abramowitz, M., and Stegun, I.A. 1964, *Handbook of Mathematical Functions* (National Bureau of Standards, Washington DC).

Ando, H., and Osaki, Y. 1975, *Publ. Astron. Soc. Japan*, **27**, 581.

Balmforth, N.J., and Gough, D.O. 1990, *Astrophys. J.*, in press.

Campbell, W.R., and Roberts, B. 1989, *Astrophys. J.*, **338**, 538.

Chandrasekhar, S. 1964, *Astrophys. J.*, **139**, 664.

Christensen-Dalsgaard, J., Cooper, A.J., and Gough, D.O. 1983, *Monthly Notices Roy. Astron. Soc.*, **203**, 165.

Christensen-Dalsgaard, J., and Gough, D.O. 1980, *Nature*, **288**, 544.

Christensen-Dalsgaard, J., Gough, D.O., and Pérez-Hernández, F. 1988, *Monthly Notices Roy. Astron. Soc.*, **235**, 875.
Däppen, W., and Gough, D.O. 1986, in *Seismology of the Sun and the Distant Stars*, ed. D.O. Gough (Reidel, Dordrecht), p.275.
Deubner, F.-L. 1975, *Astron. Astrophys.*, **44**, 371.
Duvall, T.L., Jr. 1982, *Nature*, **300**, 242.
Duvall, T.L., Jr., Harvey, J.W., and Pomerantz, M. 1986, *Nature*, **321**, 500.
Dziembowski, W.A., and Gough, D.O. 1990, in preparation.
Gough, D.O. 1976, in *The Energy Balance and Hydrodynamics of the Solar Chromosphere and Corona*, ed. R.M. Bonnet and P. Delache (G. de Bussac, Clermont-Ferrand), p.3.
Gough, D.O. 1981, in *Variations of the Solar Constant*, ed. S. Sofia (NASA Publ., Washington DC), p.185.
Gough, D.O. 1984, *Phil. Trans. R. Soc. London*, **A 313**, 27.
Gough, D.O. 1990, in *Dynamiques des Fluides Astrophysiques*, ed. J.-P. Zahn and J. Zinn-Justin (Elsevier), in press.
Gough, D.O., and Kosovichev, A.G. 1988, in *Seismology of the Sun and Sun-like Stars*, ed. E.J. Rolfe (ESA SP-286, Noordwijk), p.47.
Gough, D.O., and Thompson, M.J. 1988, in *Advances in Helio- and Asteroseismology*, ed. J. Christensen-Dalsgaard and S. Frandsen (Reidel, Dordrecht), p.175.
Gough, D.O., and Thompson, M.J. 1990, *Monthly Notices Roy. Astron. Soc.*, **242**, 25.
Keller, J.B., and Rubinow, S.I. 1960, *Ann. Phys.*, **9**, 24.
Kosovichev, A.G., and Parchevskii, K.V. 1988, *Soviet Astron. Letters*, **14**(3), 201.
Kuhn, J. 1988a, *Astrophys. J. Letters*, **331**, L131.
Kuhn, J. 1988b, in *Seismology of the Sun and Sun-like Stars*, ed. E.J. Rolfe (ESA SP-286, Noordwijk), p.87.
Kuhn, J.R., Libbrecht, K.G., and Dicke, R.H. 1988, *Science*, **242**, 908.
Lamb, H. 1908, *Proc. London Math. Soc.*, **7**, 122.
Lamb, H. 1932, *Hydrodynamics (Sixth Edition)* (Cambridge Univ. Press, Cambridge).
Ledoux, P., and Walraven, T. 1958, in *Handbuch der Physik Bd. 51*, ed. S. Flügge (Springer, Berlin), p.353.
Libbrecht, K.G., and Woodard, M.F. 1990, these proceedings.
Lynden-Bell, D., and Ostriker, J. 1967, *Monthly Notices Roy. Astron. Soc.*, **136**, 293.
Rayleigh 1896, *The Theory of Sound* (Cambridge Univ. Press, Cambridge).
Shibahashi, H. 1979, *Publ. Astron. Soc. Japan*, **31**, 87.
Shibahashi, H. 1990, these proceedings.
Smeyers, P., and Ruymaekers, E. 1990, in these proceedings.
Spiegel, E.A., and Unno, W. 1962, *Publ. Astron. Soc. Japan*, **14**, 28.
Tassoul, M. 1980, *Astrophys. J. Suppl.*, **43**, 469.
Unno, W., Osaki, Y., Ando, H., Saio, H., and Shibahashi, H. 1989, *Nonradial Oscillations of Stars (Second Edition)* (Univ. of Tokyo Press, Tokyo).
Vandakurov, Yu. V. 1967, *Astron. Zh.*, **44**, 786.
Willson, R.C., and Hudson, H.S. 1988, *Nature*, **332**, 810.

Structure of the Solar Core Inferred from Inversion of Frequencies of Low-Degree p-Modes

A. G. Kosovichev

Crimean Astrophysical Observatory, 334413, Nauchny, Crimea, USSR

Abstract: Results of estimations of density and a parameter of convective stability in the central regions of the Sun from observed frequencies of 5-min modes are presented.

1 Introduction

Helioseismic inversions of the observed p-mode frequencies have shown significant deviations of the solar core parameters from that predicted by a standard model Christensen-Dalsgaard (1988). In particular, Shibahashi and Sekii (1988), Vorontsov (1988), Christensen-Dalsgaard *et al.* (1988), and Dziembowski *et al.* (1989) found a sharp rise of the sound speed toward the centre; Korzennik and Ulrich (1989) and Gough and Kosovichev (1988, 1989) deduced that the Sun appears to be denser at the centre than the standard solar model. However the results obtained by different authors are not in perfect agreement. For instance, according to Dziembowski *et al.* (1989), the density at the centre of the Sun is lower than in the standard model. The cause of the discrepancy is not understood.

The most important direct information about the central regions of the Sun is provided by the lowest angular degree p modes which propagate closest to the solar centre. However, the frequencies of the modes in the 5-min band are more sensitive to the structure near the surface than in the interior. Therefore an important problem is to separate information about the solar core from the surface structure which depends on the complicated and ill understood physics of convection and magnetic fields.

2 Inversion Technique

A useful helioseismic inversion technique can be constructed from a variational principle for the eigenfrequencies and Backus-Gilbert optimal averaging procedure (Backus and Gilbert 1968, Gough 1985). A linearization of the variational principle in the difference between the equilibrium state of the Sun and a reference solar model leads to a system of linear integral equations connecting the frequency changes $\delta\omega_{n,l}$ with differences of the density $\delta\rho$ (Backus and Gilbert 1968; Gough and Kosovichev 1989):

$$\frac{\delta\omega_{n,l}^2}{\omega_{n,l}^2} = \int_0^R K_\rho^{(n,l)} \frac{\delta\rho}{\rho} dr \qquad (1)$$

where $\omega_{n,l}$ is the eigenfrequency of a mode of degree l and order n, R is the radius of the Sun. Spherical symmetry and hydrostatic equilibrium of the internal structure are assumed; moreover, changes of the adiabatic exponent γ, which are significant only in the ionization zones near the solar surface, are neglected. The kernels (or sensitivity functions) $K_\rho^{(n,l)}(r)$ depend on the equilibrium state and the eigenfunctions of oscillations of the reference solar model.

The kernels in Eq. (1) characterize the sensitivity of the frequencies to perturbations to the structure in different parts of the Sun. The deviation of the solar density from the reference solar model can be estimated by a standard inversion technique using differences between the observed frequencies and those computed for the model. Similar equations can be derived for other hydrostatic parameters [for example, the squared sound speed $c^2(r)$ and a parameter of convective stability $A^*(r) = \gamma^{-1} d\ln P/d\ln r - d\ln\rho/d\ln r$] using the equations of hydrostatic support. Corresponding kernels $K_{c^2}^{(n,l)}(r)$, $K_{A^*}^{(n,l)}(r)$ are obtained as solutions of the adjoint linearized structure equations (Gough and Kosovichev 1989).

It should be noted that inversions using the constraints (1) or the similar constraints for c^2 and A^* can be carried out separately to determine deviations from a reference solar model: $\delta\rho/\rho$, $\delta c^2/c^2$ and δA^*. Another approach (Dziembowski *et al.* 1989) is to invert for only one parameter (e.g. $\delta c^2/c^2$) and then calculate the others ($\delta\rho/\rho$, δA^* etc.) using the linearized hydrostatic equations. Of course, for artificial error-free data and for infinite resolution of the inverted solar structure these two approaches should be equivalent. However, for observed frequencies which can be contaminated by random and systematic errors separate inversions for different parameters provide a consistency check. The first results of the separate inversions for $\delta\rho$ and δc^2 seem not to be consistent (Gough and Kosovichev 1989). The inconsistency may be due to poor stability of the determination of c^2 near the centre as a result of the heavy concentration of $K_{c^2}^{(n,l)}(r)$ near the surface, or it could arise from possible deviations of the Sun from a simple spherically symmetric hydrostatic model (Kosovichev and Perdang 1988).

3 Results

The inversions have been carried out using 119 frequencies with $l < 5$ (Duvall *et al.* 1988). To test the procedure and resolving power of the data set the theoretical eigenfrequencies of the corresponding modes of a chemically homogeneous model of the Sun have been inverted. The results are compared with the exact functions $\delta\rho/\rho$ and δA^* in Fig. 1. It can be concluded that the procedure provides a good agreement even for the relatively large deviations from the standard solar model. The structure of a small inner region at $r < 0.05R$ cannot be resolved using the p-mode data. The test inversion for $\delta c^2/c^2$ is also in agreement with the exact deviation. However this inversion was less stable to random frequency errors added to the artificial data set. To model unknown surface perturbations systematic frequency shifts were included to the artificial data by adding smooth functions of frequency. The results of the inversions were insensitive to the shifts.

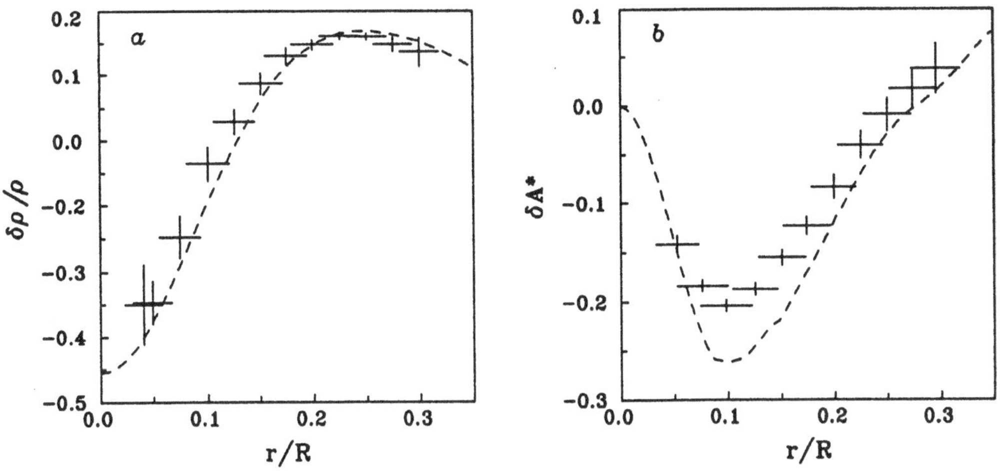

Fig. 1. The dashed lines are the relative differences between a homogeneous model and a standard model: a) density ρ; b) a parameter of convective stability A^*. The crosses represent optimal averages deduced from the differences between eigenfrequencies of the two models corresponding to those modes that were used for inversion of the solar data. The horizontal components of the crosses indicate the widths of the optimal averaging kernels and represent the resolution of the inversion; the vertical components the standard errors. The latter were computed assuming the standard errors for the corresponding observed frequencies.

Inversions of the real data are shown in Fig. 2. One deduces that the inner core of the Sun is denser than the standard solar model, which is in qualitative agreement with the previous results of Gough and Kosovichev (1988, 1989) and Korzennik and Ulrich (1989). However the results are not consistent with the inversion of Dziembowski *et al.* (1989). A precise estimation of the central density depends on a regularization parameter in Backus-Gilbert procedure. A revised analysis of trade-off between resolution and error magnification led to a maximal value of $\delta\rho/\rho$ about 20%, which is twice as large of the previously reported estimation (Gough and Kosovichev 1988, 1989). A recent inversion by Thompson (1990) confirms the result.

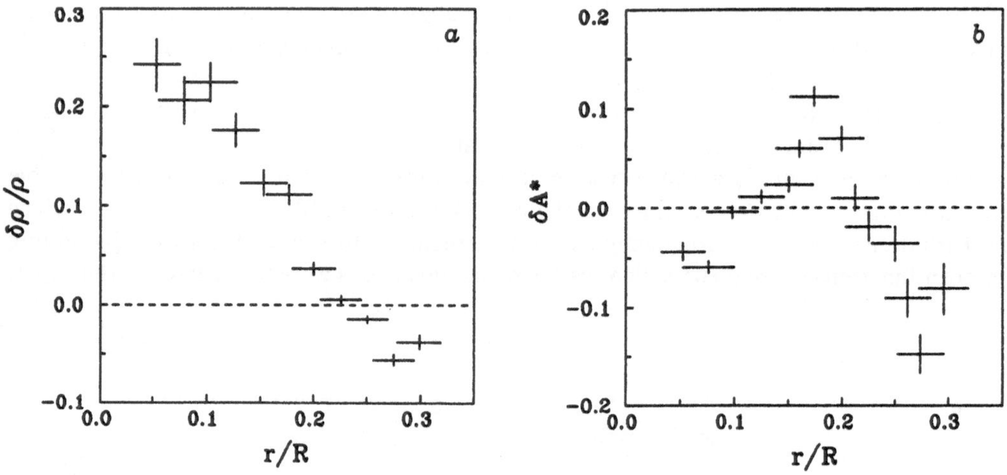

Fig. 2. Optimal averages of the relative deviation of the Sun from the reference solar model: a) density ρ; b) a parameter of convective stability A^*, obtained by inversion of the low-degree p-mode frequencies. The notation is otherwise the same as in Fig. 1.

The parameter of convective stability A^* slightly decreases in the central region at $r < 0.1R$. However the decrease is not sufficient for the formation of a convective core in the Sun, which has been suggested by Dziembowski *et al.* (1989). A local maximum of A^* at $r \approx 0.2R$ seems to be consistent with a sharp decreasing of $\delta\rho/\rho$ in this region.

References

Backus, G., and Gilbert, F. 1968, *Geophys. J. Roy. Astron. Soc.*, **16**, 169.

Christensen-Dalsgaard, J. 1988, in *Seismology of the Sun and Sun-like Stars*, ed. E. Rolfe, ESA SP-286 (ESA Publication Division, Noordwijk), p.431.

Christensen-Dalsgaard, J., Gough, D.O., and Thompson, M. 1988, in *Seismology of the Sun and Sun-like Stars*, ed. E. Rolfe, ESA SP-286 (ESA Publication Division, Noordwijk), p.493.

Duvall, T.L., Jr., Harvey, J.W., Libbrecht, K.G., Popp, B.D., and Pomerantz, M. 1988, *Astrophys. J.*, **324**, 1158.

Dziembowski, W., Pamyatnykh, A.A., and Sienkiewicz, R. 1989, preprint.

Gough D.O. 1985, *Solar Phys.*, **100**, 65.

Gough, D.O., and Kosovichev, A.G. 1988, in *Seismology of the Sun and Sun-like Stars*, ed. E. Rolfe, ESA SP-286 (ESA Publication Division, Noordwijk), p.195.

Gough, D.O., and Kosovichev, A.G. 1989, in *Inside the Sun*, eds. G. Berthomieu and M. Cribier, IAU Colloq. No. 121 (Kluwer, Dordrecht), in press.

Korzennik, S.G., and Ulrich, R. 1989, *Astrophys. J.*, **339**, 1144.

Kosovichev, A.G., and Perdang, J. 1988, in *Seismology of the Sun and Sun-like Stars*, ed. E. Rolfe, ESA SP-286 (ESA Publication Division, Noordwijk), p.539.
Shibahashi, H., and Sekii T. 1988, in *Seismology of the Sun and Sun-like Stars*, ed. E. Rolfe, ESA SP-286 (ESA Publication Division, Noordwijk), p.471.
Thompson, M.J. 1990, submitted to *Astrophys. J.*
Vorontsov, S.V. 1988, in *Seismology of the Sun and Sun-like Stars*, ed. E. Rolfe, ESA SP-286 (ESA Publication Division, Noordwijk), p.475.

Second-Order Asymptotic Inversions of the Sound Speed Inside the Sun

S. V. Vorontsov [1] and H. Shibahashi [2]

[1]Institute of Physics of the Earth, Moscow 123810, USSR
[2]Department of Astronomy, University of Tokyo, Tokyo 113, Japan

Abstract: We report the results of nonlinear inversions of the sound speed using the second-order asymptotic theory of solar acoustic oscillations. Three data sets are used for the inversions: the eigenfrequencies of solar model SNG1 of Shibahashi, Noels, and Gabriel, the eigenfrequencies of solar model 1 of Christensen-Dalsgaard, and the observational frequencies.

The second-order asymptotic equation for the frequencies of solar p-modes is (Vorontsov 1990)

$$F(w) + \frac{1}{\omega^2} P(w) = \frac{\pi[n + \alpha(\omega)]}{\omega} , \tag{1}$$

where $w = \omega/(l + 1/2)$, $F(w)$ is determined by the sound speed, $P(w)$ contains all the second-order terms, and $\alpha(\omega)$ is surface phase shift. We use the inversion technique for this equation as described in Vorontsov (1988). Although a number of terms in $P(w)$ were not specified in Vorontsov (1988), the functional form of (1) was the same, so that the sound-speed inversion was actually the second-order inversion. The only improvements we use to the boundary conditions are the additional boundary conditions established by Vorontsov (1990)

$$\lim_{1/w \to 0} \frac{d^3 F}{d(1/w)^3} = 0 , \tag{2}$$

and a different extrapolation of $F(w)$ to the surface value of $w = w_s$. Specifically, $dF/d(1/w)$ was extrapolated to $dF/d(1/w) = \text{const.}(w - w_s)$ from the last breakpoint position but one. In the approximation of $F(w)$ and $P(w)$, 25 cubic B-splines were used, and the number of splines and distribution of breakpoints being determined by the noise level in the observational data. The eigenfrequencies of model SNG1 of Shibahashi, Noels, and Gabriel (1983) and model 1 of Christensen-Dalsgaard (1982) were computed using a fourth order difference scheme with 1021 and 705 mesh points, which gives an accuracy of about 0.1 μHz. We used the observational frequencies published in Duvall *et al.* (1988) and Libbrecht and Kaufman (1988).

The phase shift $\alpha(\omega)$ was determined as described in Brodsky and Vorontsov (1988). The results are shown in Fig. 1, where $\beta(\omega) = -\omega^2 d(\alpha/\omega)/d\omega$. The stable evaluation of the phase shift from the observational data is possible in frequency range 1.8 – 3.8 mHz, which limits the data sets used in the inversions (about 1300 modes). Note that

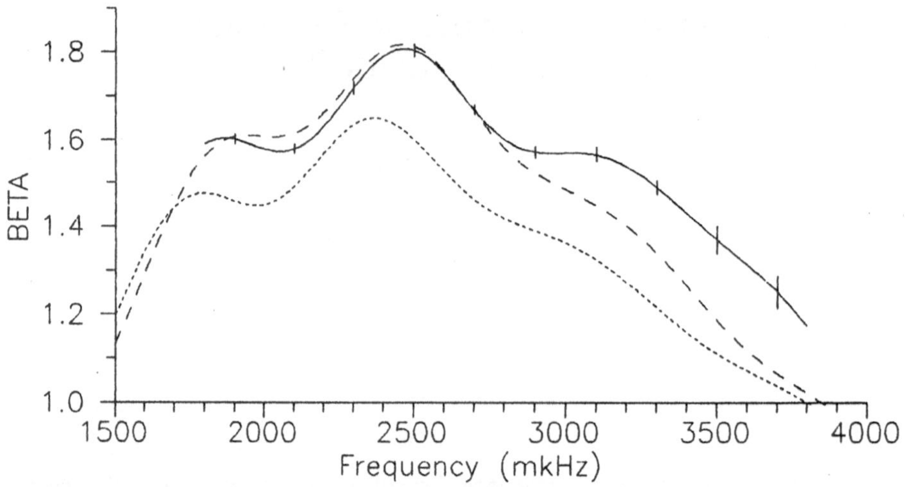

Fig. 1. The frequency dependence of the phase shift $\beta(\omega)$, obtained from the observational frequencies (solid line with error bars), eigenfrequencies of model 1 (dotted line), and eigenfrequencies of model SNG1 (dashed line).

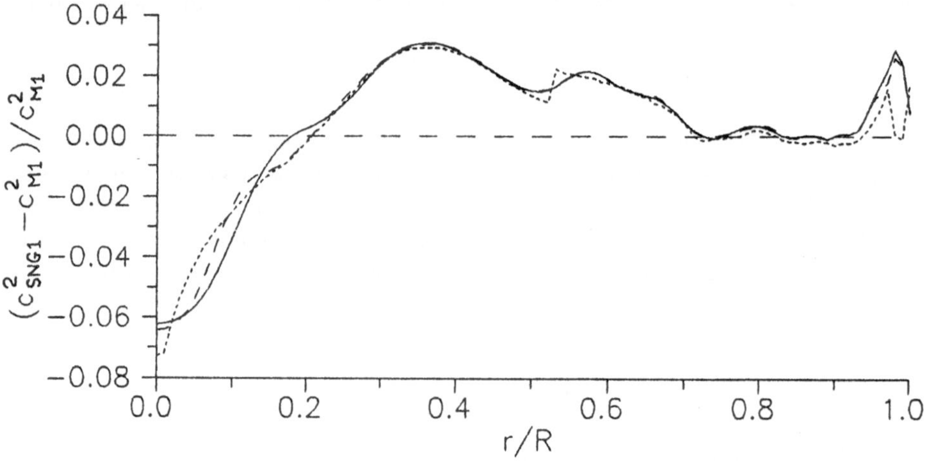

Fig. 2. Relative difference between the squared sound speed in model SNG1 and model 1. Dotted line shows the exact difference. Solid line was obtained by inverting the eigenfrequencies of those p-modes which are available in observational data set in frequency range 1.8 – 3.8 mHz, dashed line by inverting data added with higher frequencies.

$\beta(\omega)$ obtained for model SNG1 is much closer to observational curve, indicating better description of the outer solar layers. This improvement is certainly connected with the improvements in the equation of state (Pamyatnykh, Marchenkov, and Vorontsov 1990).

The relative difference between $c^2(r)$ in two models is shown by dotted line in Fig. 2. The discontinuity at $r \approx 0.53R$ is due to a small discontinuity in the model SNG1. The difference between $c^2(r)$ inverted from the eigenfrequencies is shown by solid line. This result is improved, if higher-frequency modes up to 4.5 mHz are added to the data.

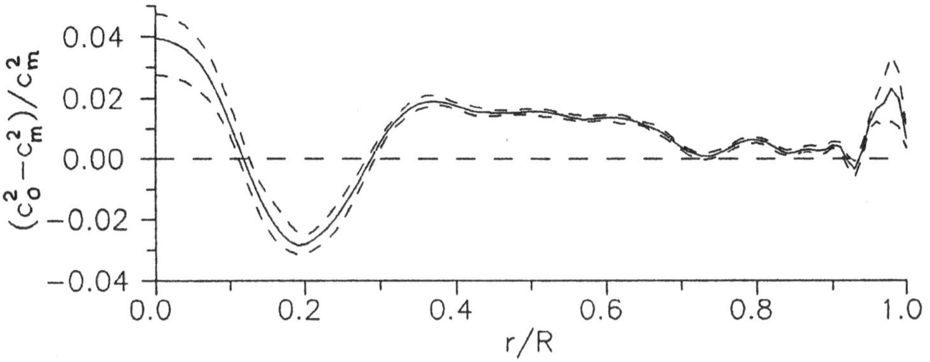

Fig. 3. Relative difference between the squared sound speed inverted from the observational frequencies and from the eigenfrequencies of model 1. Dashed lines show the envelope of the solutions obtained by adding white noise to observational frequencies, with limiting amplitude corresponding to reported observational errors, in 100 realizations.

The relative difference between $c^2(r)$ inverted from the observational frequencies and from the eigenfrequencies of model 1 is shown in Fig. 3. The result is essentially the same as in Vorontsov (1988), deviating slightly in the core; this is due to the application of additional boundary conditions (2), which require $dc^2/dr = 0$ at $r = 0$.

The similar comparison with model SNG1 is shown in Fig. 4. No significant discrepancies are now seen for the outer half of the solar radius. This seems to be connected with the use of more recent opacities. The wave-like feature around $r = 0.53R$ is produced by the discontinuity in the model. For $0.2R < r < 0.4R$ the sound speed in the Sun appears to be lower than in the model; a similar conclusion was drawn in Shibahashi and Sekii (1988) by using the first-order asymptotic inversion.

In the solar core, both models meet the same problem when compared with observational data. This problem is seen also in the second-order terms (Vorontsov 1988) and may be interpreted as an evidence for some sort of mixing in the solar core.

Acknowledgements

We thank J. Christensen-Dalsgaard for his solar model 1. This work was partly supported by "Zodiac" Scientific Methodological Council and by Itoh Science Foundation.

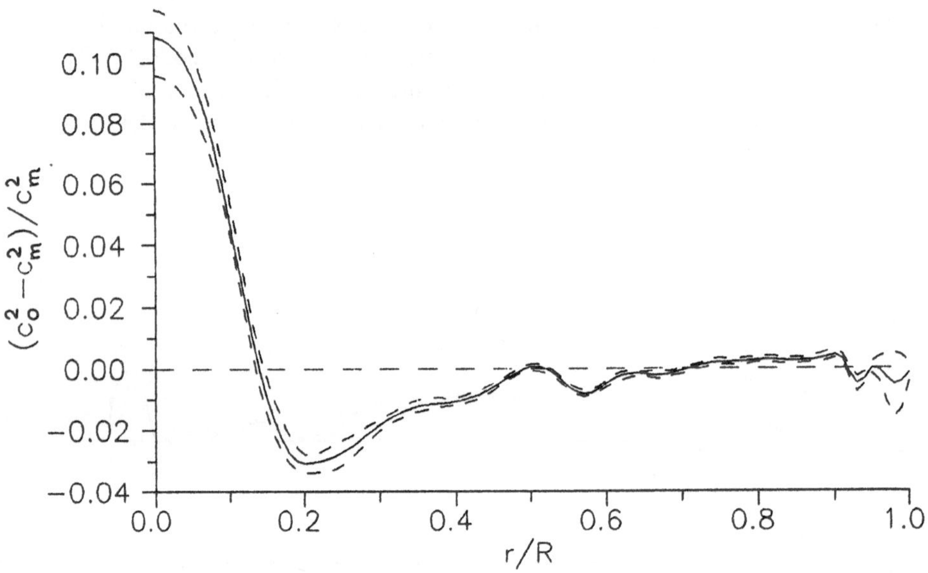

Fig. 4. Relative difference between the squared sound speed inverted from the observational frequencies and from the eigenfrequencies of model SNG1.

References

Brodsky, M.A., and Vorontsov, S.V. 1988, in *Seismology of the Sun and Sun-Like Stars*, ed. E. Rolfe, ESA SP-286 (ESA Publishing Division, Noordwijk), p.487.

Christensen-Dalsgaard, J. 1982, *Monthly Notices Roy. Astron. Soc.*, **199**, 735.

Duvall, T.L., Jr., Harvey, J.W., Libbrecht, K.G., Popp, B.D., and Pomerantz, M.A. 1988, *Astrophys. J.*, **324**, 1158.

Libbrecht, K.G., and Kaufman, J.M. 1988, *Astrophys. J.*, **324**, 1172.

Pamyatnykh, A.A., Marchenkov, K.I., and Vorontsov, S.V. 1989, These proceedings.

Shibahashi, H., Noels, A., and Gabriel, M. 1983, *Astron. Astrophys.*, **123**, 283.

Shibahashi, H., and Sekii, T. 1988, in *Seismology of the Sun and Sun-Like Stars*, ed. E. Rolfe, ESA SP-286 (ESA Publishing Division, Noordwijk), p.471.

Vorontsov, S.V. 1988, in *Seismology of the Sun and Sun-Like Stars*, ed. E. Rolfe, ESA SP-286 (ESA Publishing Division, Noordwijk), p.475.

Vorontsov, S.V. 1989, These proceedings.

Testing Solar Envelope Models Using Intermediate-Degree p-Mode Frequencies

A. A. Pamyatnykh [1], K. I. Marchenkov [2], and S. V. Vorontsov [2]

[1]Astronomical Council, Moscow 109017, USSR
[2]Institute of Physics of the Earth, Moscow 123810, USSR

Abstract: The accurately measured frequencies of intermediate-degree acoustic oscillations permit to infer the frequency dependence of the phase shift, corresponding to the reflection of the trapped acoustic waves from the solar surface. This function is sensitive to the structure of the outermost solar layers down to to the depth of second helium ionization zone. It deviates significantly from the predictions of standard solar models, representing the main source of discrepancies of oscillation frequencies in all the degree range. We report the results of studying this problem in the framework of adiabatic approximation, using both linearized inversions for the reflecting acoustic potential and direct tests of a variety of solar envelope models.

In the inverse problem of solar acoustic oscillations, the frequency dependence of the phase shift $\alpha(\omega)$ is determined from the oscillation frequencies in a form $\beta(\omega) = -\omega^2 d(\alpha/\omega)/d\omega$ (see, e.g., Vorontsov and Zharkov 1989). For a given envelope model, $\beta(\omega)$ can be easily calculated from the corresponding acoustic potential (Brodsky and Vorontsov 1988, Vorontsov and Zharkov 1989), which permits to calibrate the envelope model without any recourse to interior solar structure. The functions $\beta(\omega)$ obtained from observational frequencies and for model 1 of Christensen-Dalsgaard (1982) are shown in Fig. 1 by solid and dotted lines. The significant discrepancy is clearly seen. The acoustic potential in the model 1 is shown by dotted line in Fig. 2.

The problem can be considered as the inverse scattering problem of reconstructing the acoustic potential from the observational $\beta(\omega)$, but this problem is extremely ill-defined due to the complex structure of the potential and limited frequency range of observational data. We made an attempt to solve this problem by linearized inversion technique limiting the variation of the potential by adiabaticlly stratified layers of the convection zone. The technique is described in Marchenkov and Vorontsov (1989) and the result obtained after six iterations is shown in Fig. 2 by the solid line. The phase function $\beta(\omega)$ computed for the resulting potential is shown in Fig. 1 by line marked with circles. The discrepancy with observational data is almost removed.

Although such a simplified solution is not unique, it gives some feeling of how the envelope model should be modified to reduce the discrepancy with observations. The significant increase of potential which is obtained needs lower specific entropy in the convection zone, while its detailed behavior is governed by the equation of state (Marchenkov and Vorontsov 1989).

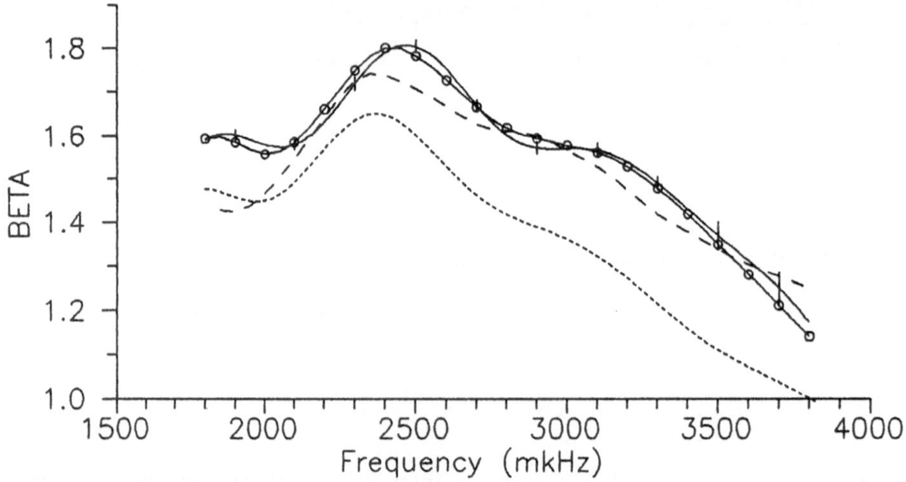

Fig. 1. Frequency dependence of the phase shift $\beta(\omega)$. Solid line with error bar was obtained from observational data, dotted line corresponds to model 1. Line marked with circles was computed from the acoustic potential, inverted from the observational $\beta(\omega)$. Dashed line corresponds to a "best-fit" model.

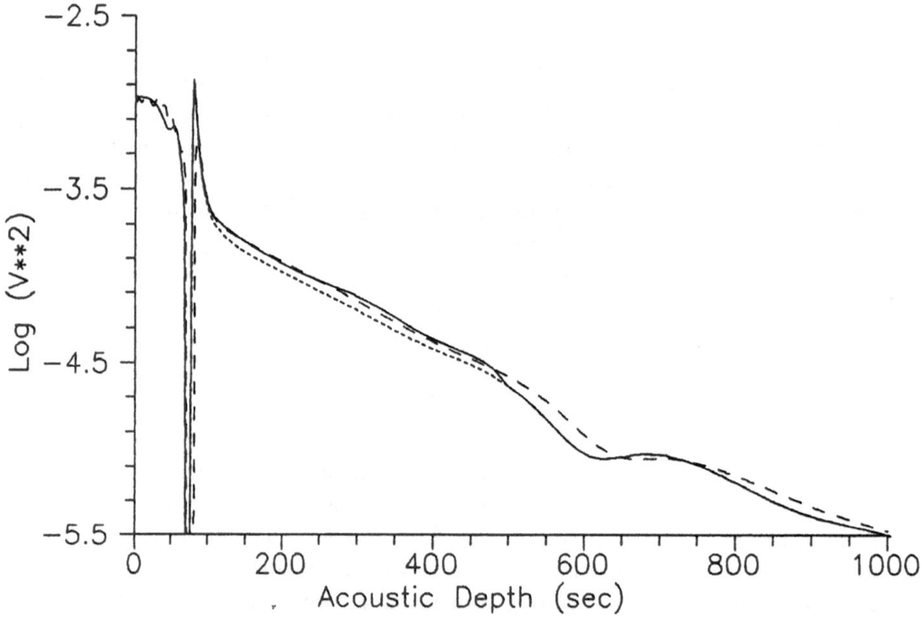

Fig. 2. Acoustic potential versus acoustic depth. Dotted line corresponds to model 1, solid line is the result of the inversion of observational data. Dashed line corresponds to a "best-fit" model.

Another approach consists in the direct verification of a variety of solar envelope models. The computations were done for a grid of models constructed as described in Pamyatnykh (1988), but in a wide range of convection efficiency and helium abundance. Figure 3 demonstrates the potential capability of calibrating the model with respect to this two parameters if $\beta(\omega)$ is known. Comparison with observations shows, however, that the discrepancy can only be reduced, but not removed (Fig. 4). Results for some "best-fit" model are shown in Figs. 1 and 2 by the dashed lines. At lower frequencies, the discrepancy in $\beta(\omega)$ remains the same.

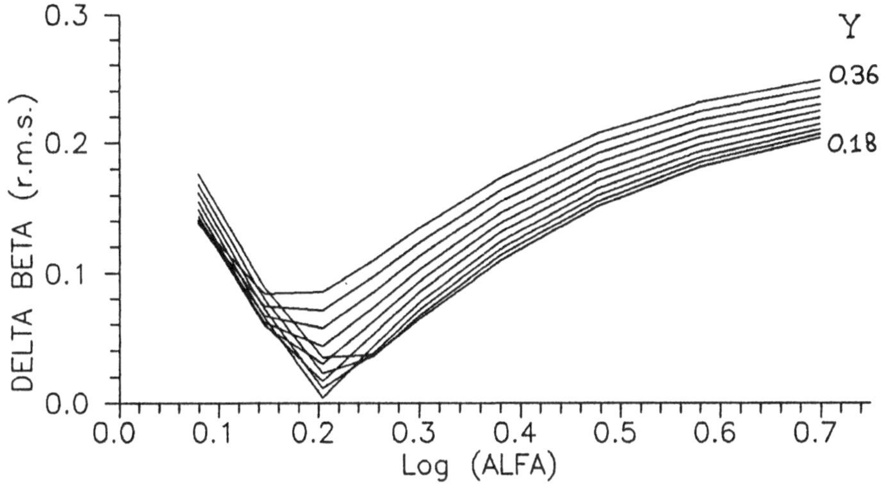

Fig. 3. Root-mean-square deviation of $\beta(\omega)$ for the grid of envelope models from that of the "reference" model. Models in the grid differ in helium abundance (from 0.18 to 0.36) and in convection efficiency (ratio of the mixing length to pressure scale height is from 1.2 to 5, on the horizontal axis). The "reference" model is the standard envelope model described in Pamyatnykh (1988).

A lot of similar grids of models have been tested, with different $T-\tau$ relation in the atmosphere (Holweger and Mueller 1974), different opacities (Huebner *et al.* 1977), and modified version of mixing-length theory (Deupree 1979, Deupree and Varner 1980). It was not found possible to improve the results significantly.

All the results indicate that the discrepancy cannot be removed without the improvement of the equation of the state. If so, the equation of state can be calibrated using this technique. Influence of non-adiabatic effects also remains to be studied.

Acknowledgements

We thank J. Christensen-Dalsgaard for his solar model 1 used in the computations. This work was partly supported by "Zodiac" Scientific-Methodological Council.

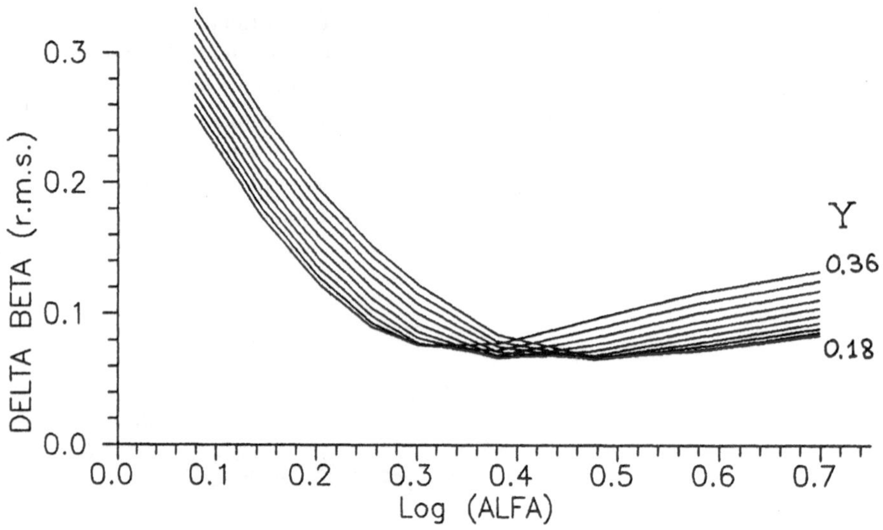

Fig. 4. Root-mean-square deviation of $\beta(\omega)$ for the grid of envelope models from $\beta(\omega)$ inferred from the observational data.

References

Brodsky, M.A., and Vorontsov, S.V. 1988, in *Seismology of the Sun and Sun-like Stars*, ed. E. Rolfe, ESA SP-286 (ESA Publication Division, Noordwijk), p.487.

Christensen-Dalsgaard, J. 1982, *Monthly Notices Roy. Astron. Soc.*, **199**, 735.

Deupree, R.G. 1979, *Astrophys. J.*, **234**, 228.

Deupree, R.G., and Varner, T.M. 1980, *Astrophys. J.*, **237**, 558.

Holweger, H., and Mueller, E.A. 1974, *Solar Phys.*, **39**, 19.

Huebner, W.F., Merts, A.L., Magee, N.H., Jr., and Argo, M.F. 1977, *Astrophysical Opacity Library* (Los Alamos Scientific Laboratory report LA-6760).

Marchenkov, K.I., and Vorontsov, S.V. 1989, *Solar Phys.*, in press.

Pamyatnykh, A.A. 1988, in *Advances in Helio- and Asteroseismology*, eds. J. Christensen-Dalsgaard and S. Frandsen (Reidel, Dordrecht), p.99.

Vorontsov, S.V., and Zharkov, V.N. 1989, *Soviet Sci. Rev. E, Astrophys. Space Phys. Rev.*, **7**, 1.

On the Inverse Problem of High-Degree Solar Acoustic Oscillations: Local Analysis of p-Mode Ridges

M. A. Brodsky [1] and S. V. Vorontsov [2]

[1]Department of Statistics, University of California, Berkeley, CA 94720, USA
[2]Institute of Physics of the Earth, Moscow 123810, USSR

Abstract: The simple technique of analyzing the information contained in the frequencies of high-degree p-modes is proposed. Its application demonstrates some difficulties in studying the helium ionization zone by using the asymptotic inversion. It indicates also the explicit requirements to the accuracy of frequency measurements needed for these studies, interesting for helium abundance determination.

In the asymptotic inversions of p-mode oscillation frequencies, the radial distribution of the sound speed $c(r)$ is obtained by the application of Abel inversion to the function

$$\frac{dF(w)}{dw} = \frac{1}{w^3}\int_{r_1}^{R}\left(\frac{r^2}{c^2}-\frac{1}{w^2}\right)^{-1/2}\frac{dr}{r}, \tag{1}$$

which is to be inferred from the frequencies. The simple relations exist between this function and the partial derivatives of $\omega(n,l)$, when the frequency ω is considered as a continuous function of radial order n and degree l (Brodsky and Vorontsov 1988a). Combining the expressions given in Brodsky and Vorontsov (1988a) to exclude $\partial\omega/\partial n$ and neglecting higher-order asymptotic terms, we obtain

$$\frac{dF(w)}{dw} = \frac{\pi[n+\beta(\omega)]}{w^2}\,\frac{\partial\omega/\partial l}{\omega - L\partial\omega/\partial l}, \tag{2}$$

where $L \equiv l+1/2$, $w \equiv \omega/L$, $\beta(\omega) \equiv -\omega^2 d(\alpha/\omega)/d\omega$, and $\alpha(\omega)$ is the surface phase shift. This expression is particularly useful for high-degree modes, because $\partial\omega/\partial l$ can be easily estimated along the p-mode ridges. The phase shift is determined from the frequencies of intermediate-degree modes. The polynomial representation of $\beta(\omega)$ inferred from the observational frequencies is given in Brodsky and Vorontsov (1988b).

The anomalous behavior of the sound-speed gradient in the second helium ionization zone (Däppen and Gough 1986, Däppen, Gough, and Thompson 1988) is reflected as a hump in the function dF/dw, shown for the standard solar model by the solid line in Fig. 2. Also shown is the result of application of Eq. (2) to the eigenfrequencies of this model. The frequency set was used with step 10 in degree l, and each point in Fig. 2 corresponds to a particular mode, $\partial\omega/\partial l$ being estimated by centered difference between

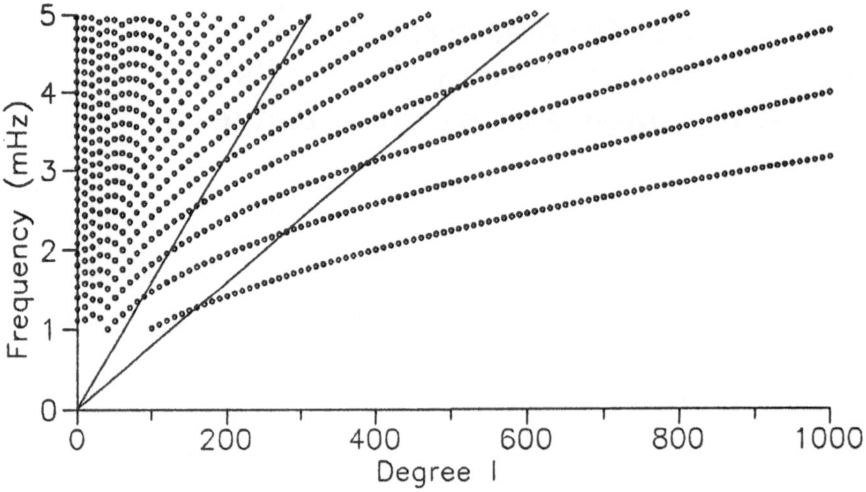

Fig. 1. Theoretical dispersion curves of high-degree solar acoustic oscillations. The step of 10 in degree is used for the presentation. Lower curve corresponds to f-mode. Modes between two straight lines ($w = 0.5 \cdot 10^{-4}\,s^{-1}$ and $w = 10^{-4}\,s^{-1}$) have turning points around the second helium ionization zone.

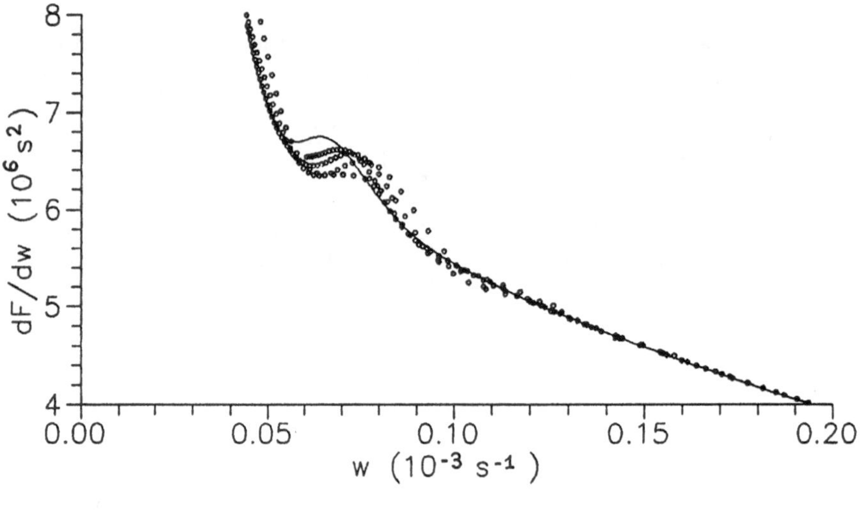

Fig. 2. The function dF/dw computed directly for the standard solar model using Eq. (1) (solid line) and the result obtained from the eigenfrequencies of this model using Eq. (2).

the neighbouring modes. Below the ionization zone (at higher values of w) the function dF/dw is reproduced by this technique with high accuracy.

The discrepancies are seen, however, near the helium hump, and shown in a larger scale in Fig. 3. Their origin is due to the "surface phase shift," which is no longer a function of frequency alone. The phase shift absorbs all the deviations from the simple asymptotic description of a vertically propagating acoustic wave. Both the inaccuracy of the asymptotic description (due to the short-scale variation of seismic parameters) and the curvature of the ray paths become significant in this region. The contribution of the helium ionization zone to the phase shift, producing the significant periodic component in $\beta(\omega)$, was shown in Vorontsov and Zharkov (1989). The use of p-modes of higher radial order (and of higher frequencies) improves the results, which is seen in Fig. 3.

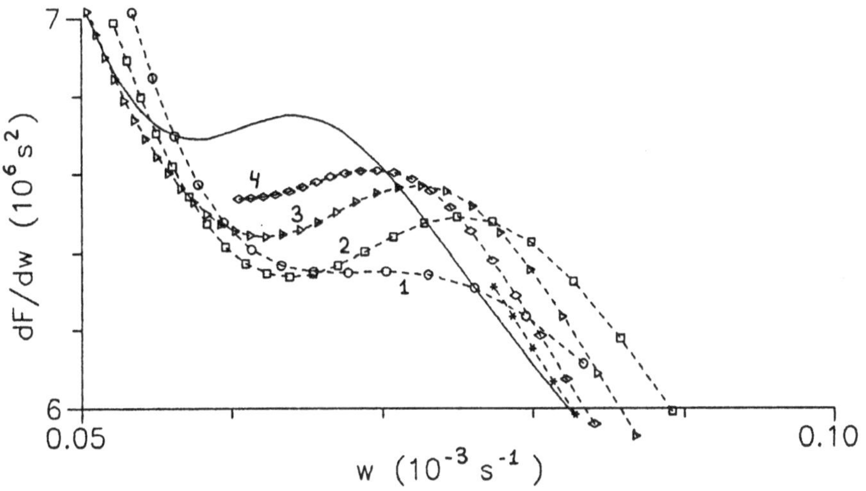

Fig. 3. The same as Fig. 2, but in enlarged scale. The dashed lines marked with radial order n connects data obtained from the corresponding dispersion curves.

These problems inevitably affect the sound-speed inversions in helium ionization zone, producing the systematic errors. It is possible that the significant improvements of the inversions will be achieved by the application of the second-order asymptotic theory.

An attempt to reproduce the function dF/dw from the observational frequencies (Libbrecht and Kaufman 1988) is demonstrated in Fig. 4. It is clear that the data available now are too noisy. It should be noted that the technique described above provides by itself the simple and explicit requirements to the accuracy of the observational data needed for the sounding of the outer solar layers.

Acknowledgements

We thank J. Christensen-Dalsgaard for his solar model 1 used in the computations. This work was partly supported by "Zodiac" Scientific-Methodological Council.

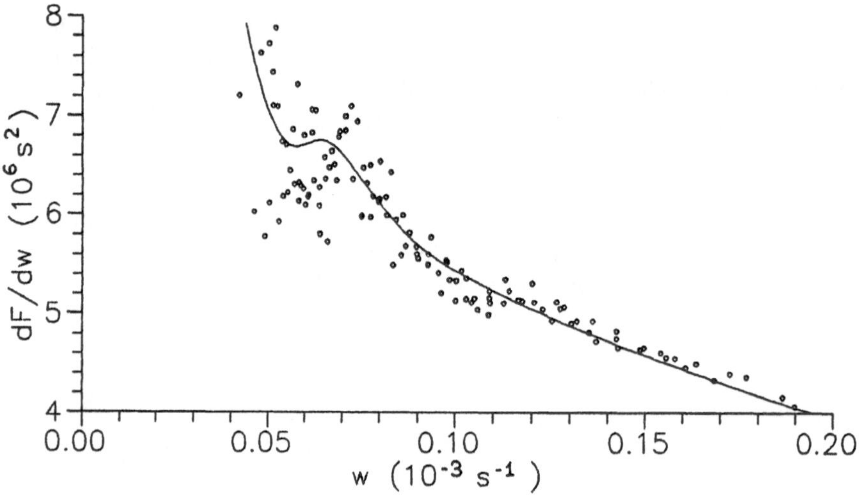

Fig. 4. The same as Fig. 2, but when using the observational frequencies.

References

Brodsky, M.A., and Vorontsov, S.V. 1988a, in *Advances in Helio- and Asteroseismology*, eds. J. Christensen-Dalsgaard and S. Frandsen (Reidel, Dordrecht), p.137.

Brodsky, M.A., and Vorontsov, S.V. 1988b, in *Seismology of the Sun and Sun-like Stars*, ed. E. Rolfe, ESA SP-286 (ESA Publication Division, Noordwijk), p.487.

Däppen, W., and Gough, D. O. 1986, in *Seismology of the Sun and the Distant Stars*, ed. D. O. Gough (Reidel, Dordrecht), p.275.

Däppen, W., Gough, D. O., and Thompson, M.J. 1988, in *Seismology of the Sun and Sun-like Stars*, ed. E. Rolfe, ESA SP-286 (ESA Publication Division, Noordwijk), p.505.

Libbrecht, K.G., and Kaufman, J.M. 1988, *Astrophys. J.*, **324**, 1172.

Vorontsov, S.V., and Zharkov, V.N. 1989, *Soviet Sci. Rev. E, Astrophys. Space Phys. Rev.*, 7, 1.

Two-Dimensional Inversion of Rotational Splitting Data

Takashi Sekii

Department of Astronomy, University of Tokyo, Bunkyo-ku, Tokyo 113, Japan

Abstract: We solve a two-dimensional inverse problem of the rotational frequency splitting to infer the rotation rate in the sun as a function of both the radius and the latitude. We use Libbrecht's (1989) observational data of the solar p-mode frequency splitting. We discretize a set of linear integral equations for rotational splittings and reduce them a set of linear algebraic equations. We solve the resultant algebraic equations by imposing an error-weighted least squares condition cooperated with boundary constraint at the surface. In order to stabilize the solution to observational and numerical errors, we discard small singular values of the coefficient matrix, and this keeps some parameters undetermined. To determine these parameters we impose a flatness condition. The inverted results show the solar internal rotation becomes slower at low latitudes and faster at high latitudes with increasing depth. The most significant deviation from this trend is that rotation is slow at low latitudes in the convection zone.

1 Introduction

The internal rotation of the sun is the main source of the observed frequency splittings of solar p-mode oscillations. We invert the splitting data of Libbrecht (1989) to infer the internal rotation of the sun. The data contains frequency splittings for $5 \leq l \leq 60$. The inversion is two dimensional, since the rotation rate is supposed to depend both on depth and latitude. We assume that standard solar models are sufficiently good that we can replace splitting kernels of the real sun by those of a standard model. We use the model 1 of Shibahashi *et al.* (1983) as the reference. The inversion technique adopted here is basically identical with Sekii and Shibahashi (1988). The technique is essentially the least squares method of solving underdetermined problems. We discretize the integral equations for frequency splittings

$$\delta\omega_{nlm} = \int_0^R \int_0^\pi K_{nlm}(r,\theta)\Omega(r,\theta)drd\theta \quad , \tag{1}$$

where $\delta\omega_{nlm}$ is the frequency splitting of the mode with quantum numbers (n,l,m)., $\Omega(r,\theta)$ is the rotation rate, and $K_{nlm}(r,\theta)$ stands for the splitting kernel of the mode. The discretization yields an algebraic equation of the form

$$\boldsymbol{K\Omega} = \boldsymbol{e} \quad ,$$

where $\boldsymbol{K}$, $\boldsymbol{\Omega}$ and $\boldsymbol{e}$ are the discretized expression of K_{nlm}, Ω and $\delta\omega_{nlm}$, respectively. We impose an error-weighted least squares condition firstly, cooperated with boundary constraint at the surface. We carry out this process by utilizing singular value decomposition of the coefficient matrix $\boldsymbol{K}$. Here we introduce a trade-off parameter ε and discard singular values less by a factor ε than the largest singular value. As a consequence some parameters remain undetermined, and we impose a flatness condition to fix them. If we adopt a large value of ε, we obtain only broad features of the rotation rate. Adopting a small ε will bring us finer structures at the cost of high sensitivity to the data error. It should be noted that imposing a flatness condition is necessary even if we pick all the singular values since we solve an underdetermined problem as is described below.

2 Inversion of Frequency Splittings

We first calculated eigenfunctions of the reference model. Then splitting kernels were then computed following Hansen *et al.* (1977). In the present inversion analysis, the number of radial and angular meshes were chosen to 100 and 20, respectively. We decided to concentrate mainly on low to mid-latitude zones, and distributed angular meshes accordingly. We estimated discretization errors by performing some radial and angular integrals, and rejected those modes which were not properly treated in the present mesh system. As was expected, eigenmodes with small m were rejected, and high overtones of high degree modes were also excluded. After all, 1053 eigenmodes of various quantum numbers (l, m, n) were chosen. Thus the present problem became underdetermined, since the number of meshes is about twice greater than the number of modes. As a flatness measure, we adopted the first derivative of the rotation rate. We also added boundary constraint on the rotation rate. The rotation rate at the surface was required to be close to the observed surface rotation rate of Snodgrass (1983). In his work the relation between the rotation rate Ω_s and the latitude ϕ were studied from the motion of magnetic field pattern. Surface constraint and equations for splittings were treated equally.

3 Results

From numerical experiments of inverting artificial data, which preceded inversions of the observed data, we found resolution was poor at deep interior and at high latitudes. The poor resolution at the deepest region is a consequence of deficiency in accurate splitting data for low degree modes which penetrate to that depth, while omitting low m modes resulted poor resolution at high latitudes. It is also reasonable that the best resolution was obtained at the depth range $0.6 \lesssim r/R \lesssim 0.9$ and at the latitude range $0° \leq \phi \lesssim 45°$. The penetration depth of $l = 60$ mode is around 0.8. In the shallower region, all the eigenmodes used here have substantial amplitudes there. Since the number of meshes distributed in the shallower region is limited, independence of eigenfunctions is poorer there than that in deeper region, and hence the resolution power is also poor there.

Then we inverted the observed splitting data. Figures 1 and 2 show the result of inversion with the trade-off parameter $\varepsilon = 5 \times 10^{-4}$. Broadly speaking, the rotation is found to be slower at low latitudes and faster at high latitudes, with increasing the depth.

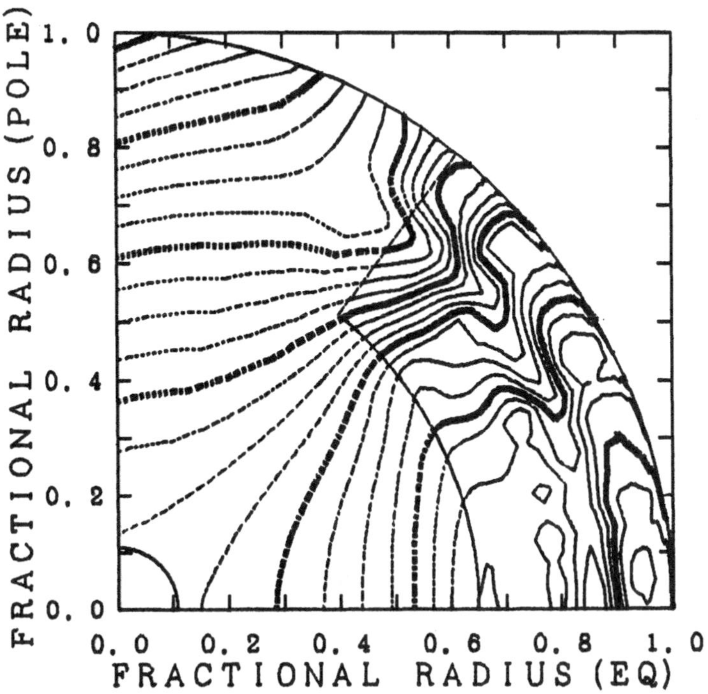

Fig. 1. The contour map of the rotation rate inverted from the observed frequency splittings (Libbrecht 1989) with $\varepsilon = 5 \times 10^{-4}$. The contour lines are drawn for 26 levels: from 340 nHz to 470 nHz, with a step of 5 nHz. Thick curves are for from 340 nHz to 460 nHz with a step of 20 nHz. Outside the region enclosed by the dashed curve, the solution is mainly determined by the flatness condition.

Contour lines are distorted in the vicinity of the depth $r/R \simeq 0.85$. This corresponds to a dip in the rotation rate. We also find that in the low latitude region the rotation rate reaches its maximum at $r/R \simeq 0.95$. This feature is seen even in results with larger trade-off parameters giving a broad feature of the rotation rate. Inversions with smaller ε result similar features, though spurious oscillatory variation appears and the results respond more wildly to data errors.

Christensen-Dalsgaard and Schou (1988) and Dziembowski *et al.* (1989) also inverted the same data set. Both of them expanded the rotation rate by $\cos^2\theta$ and inverted the observational data to obtain the first three coefficients of the series as functions of the depth. The present results are qualitatively similar to theirs in the sense that the rotation is slower in the low latitude with increasing the depth while it is faster in the high latitude. However, there are some differences in detail. For example, Christensen-Dalsgaard and Schou's (1988) and Dziembowski *et al.*'s (1989) results show a fairly steep gradient of the rotation rate at the base of the convection zone, while the present result does not. These differences may be caused by the differences of sensitivity of the inversion methods to the r- and the θ-dependence of Ω: While Christensen-Dalsgaard and Schou (1988) and

Dziembowski *et al.* (1988) fix the θ-dependence of the rotational rate, the present method deal with the r-dependence and the θ-dependence of Ω equally. A detailed comparison of inversion techniques is desirable.

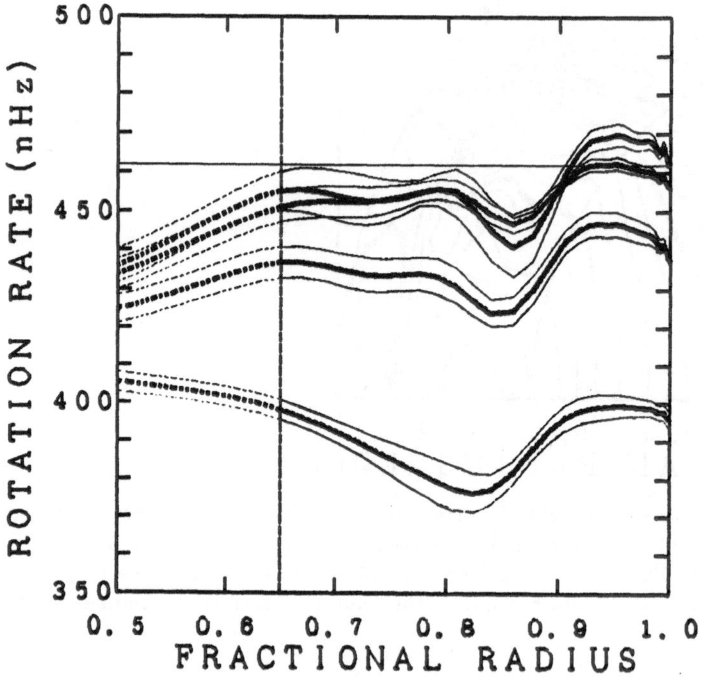

Fig. 2. The thick curves show the rotation rate presented in Fig. 1, versus radius at latitudes $\theta = 0°$ (uppermost), $15°, 32°$, and $52°$ (lowermost). The thin curves indicate $\pm 1\sigma$ confidence levels. In the region $r/R \lesssim 0.65$, the solution is mainly determined by the flatness condition. The thin horizontal line shows the surface equatorial rotation rate.

References

Christensen-Dalsgaard, J., and Schou, J. 1988, in *Seismology of the Sun and Sun-like Stars*, ed. E. Rolfe, ESA SP-286 (ESA Publication Division, Noordwijk), p.149.

Dziembowski, W.A., Goode, P.R., and Libbrecht, K.G. 1989, *Astrophys. J. Letters*, **337**, L53.

Hansen, C.J., Cox, J.P., and van Horn, H.M. 1977, *Astrophys. J.*, **217**, 151.

Libbrecht, K.G. 1989, *Astrophys. J.*, **336**, 1092.

Sekii, T., and Shibahashi, H. 1988, in *Seismology of the Sun and Sun-like Stars*, ed. E. Rolfe, ESA SP-286 (ESA Publication Division, Noordwijk), p.521.

Shibahashi, H., Noels, A., and Gabriel, M. 1983, *Astron. Astrophys.*, **123**, 283.

Snodgrass, H.B. 1983, *Astrophys. J.*, **270**, 288.

Contribution of High-Degree Frequency Splittings to the Inversion of the Solar Rotation Rate

Sylvain G. Korzennik [1], Alessandro Cacciani [2],

Edward J. Rhodes, Jr. [3], and Roger K. Ulrich [1]

[1]University of California, Los Angeles, CA 90024, USA
[2]University "La Sapienza", Rome, Italy
[3]University of Southern California, Los Angeles, CA 90089, USA, and
Jet Propulsion Laboratory, California Institute of Technology,
Pasadena, CA 91109, USA

Abstract: We present the contribution of high degree rotational splittings to the inversion of the internal rotation rate around the equator. The extention of the input data set to ℓ of 500, allow us to improve the resolution of the solution mainly in the outermost 15% of the solar radius. The rotational profile obtained in the regions below the surface leads to an attractive picture that could reconcile different non-seismic estimates of the "surface" rotation rate.

1 Introduction

One of the most directly accessible internal properties of the Sun from helioseismologic data is the solar internal rotation rate. Indeed, the Sun's slow rotation splits the azimuthal degeneracy of a spherically symmetric model, and induces for each eigenmode a frequency shift proportional to some mode-dependent average of the rotation rate and also proportional to the azimuthal order m. Such frequency shifts, commonly referred as the rotational frequency splittings, are well within today's observational capabilities, and have been measured with reasonable accuracy by numerous groups using data sets ranging from a few days to a few months of observations and for spherical harmonic degree ℓ ranging from low ($\ell = 5$) to intermediate ($\ell = 120$) values.

Since the rotational frequency splitting for each mode is related through an integral equation to the internal rotation rate, accurate measurement of the splittings for a large set of modes that adequately samples the solar interior would allow a determination of the internal rotation rate using a procedure commonly referred as an inversion. Various authors have discussed the merits of the different inversion techniques from a terrestrial perspective (Parker 1977) and from a solar perspective (Gough 1985).

Recent accurate measurements of the frequency splittings have prompted several attempts to determine the solar internal rotation rate and to test with real data the merit of different techniques (Duvall *et al.* 1984; Christensen-Dalsgaard and Schou 1988; Korzennik *et al.* 1988; Dziembowski *et al.* 1989; Goode *et al.* 1989). While the present level of accuracy of the rotational frequency splittings has allowed us to deduce the gross features of the internal rotation rate, the limited accuracy of the low degree modes has prevented the inversion procedure from providing significant results in the deepest regions, while the restriction to intermediate degrees has limited the resolution near the surface. Also, discrepencies between different data sets have indicated the potential for systematic effects in the determination of the splittings and henceforth in the inverted results (Korzennik *et al.* 1988).

More recent observations at the 60-Foot Solar Telescope of the Mt. Wilson Observatory, using high spacial resolution measurements, have allowed us to extend the determination of rotational frequency splittings to high degree modes ($\ell \leq 500$). While the reduction of these data is still in progress, using preliminary results based on the sectoral modes only, we present here the contribution of those high degree modes on the determination of a rotation profile.

2 The Rotational Frequency Splittings

High resolution full-disk Dopplergrams were obtained with a 1024×1024 CCD camera and a Na magneto-optical filter, at the 60-Foot Solar Tower of the Mt. Wilson Observatory, on 20 consecutive days, beginning July 1st 1988. For each Dopplergram, spherical harmonic coefficients for each even azimuthal order m and each degree ℓ have been computed up to $\ell = 500$, and from the 20-day-long time series, a 65536-point power spectrum for each coeficient has been computed (Rhodes *et al.* 1990).

From the 250 pairs of even-ℓ sectoral spectra ($m = \pm\ell$), estimates of the rotational frequency splitting have been computed using a cross-correlation technique between the prograde and the retrograde spectra. Estimates of the splittings from all the tesseral spectra using a Legendre polynomial parametrization of the azimuthal dependency, similar to the one carried out on the 1984 CID data (Tomczyk 1988), is now in progress (Rhodes et al. 1990), although it has not yet been completed for the purpose of this work on the 1988 CCD data.

Since the cross-correlation method is based entirely upon the sectoral spectra, its results are inherently less accurate and therefore the individual estimates were averaged over 10-ℓ-wide bins; an estimate of the uncertainty was computed from the scatter in each bin.

Initial work on the distortion of eigenfunctions by the differential rotation and its effect on the determination of high degree rotational frequency splittings has indicated a potential for systematic errors (Woodard 1989). However, numerical evaluation of equation (13) from Woodard (1989) as a function of the azimuthal order m, for relevant values of the degree ℓ and the coefficient p, indicates that the effect of the distortion by differential rotation of the sectoral eigenfuctions (i.e. $m = \pm\ell$) can be neglected. Indeed, for high degree sectoral modes, the eigenfunctions are confined to a narrow region centered around the equator, and are therefore fairly insensitive to any distortion due to differential rotation.

In order to assess the contribution of the high degree modes to the inversion of the angular velocity profile, we have combined estimates of the frequency splittings obtained at high ℓ ($120 < \ell \leq 500$), using the cross-correlation analysis on the 1988 sectoral modes, with estimates obtained at low and intermediate ℓ ($5 \leq \ell \leq 120$), using the Legendre polynomial representation from an earlier work based on the 1984 CID observations (reduced to their sectoral equivalent, i.e. $a_1 + a_3 + a_5$).

3 The Inversion Procedure

The rotationally induced frequency splitting, reduced to the "equatorial" approximation, can be writen as

$$\frac{\nu_{n,l,l} - \nu_{n,l,-l}}{2l} = -\frac{1}{2\pi} \int_0^R K_{n,l}(r)\Omega(r)dr \tag{1}$$

where $\nu_{n,l,m=\pm l}$ represents the sectoral modes, $\Omega(r)$ represents the solar rotation rate around the equator, and $K_{n,l}$ are normalized rotational kernels. Since the methods used to estimate the rotational frequency splittings provide us with an average over radial order n, the rotational kernels were also averaged over n to reduce (1) to its n-averaged form. The average of the kernels was weigthed by a Lorentzian adjusted to represent the power envelope of the 5-minute band.

To solve this system of integral equations, we have used an iterative version of the spectral expansion method, and two versions of the constrained least-squares method, using respectively a first and a second derivative constraint.

The spectral expansion method performs an eigenvalue and eigenvector decomposition (orthogonal basis), and gives the least-norm, least-squares solution to the problem. Since small eigenvalues introduce large uncertainties, the decomposition is truncated to limit the error magnification. We have used an iterative procedure to optimize the radial discretization of the problem as described in Korzennik *et al.* (1988).

The constrained least-squares method regularizes the ill-conditioned standard least-squares problem by adding a smoothness constraint to the solution in the minimization criterion. The constraint is weighted by some arbitrary parameter to adjust the degree of smoothness and therefore the error magnification. Two forms of a smoothness constraint have been used, using respectively the square of the norm of the first and the second derivative (Korzennik *et al.* 1988).

4 The "Equatorial" Rotation Rate

Where the inversion procedure indicates significance in the deduced profile, all three methods give consistent results as presented in Fig. 1.

Figure 2 shows inverted profiles, obtained with the second derivative constrained least-squares method only, but computed with and without the high degree modes to evaluate the effect of including the high degree modes in the inversion procedure.

As expected from the radial dependence of the rotational kernels, the major difference between the inverted profiles obtained with and without the high degree modes is confined

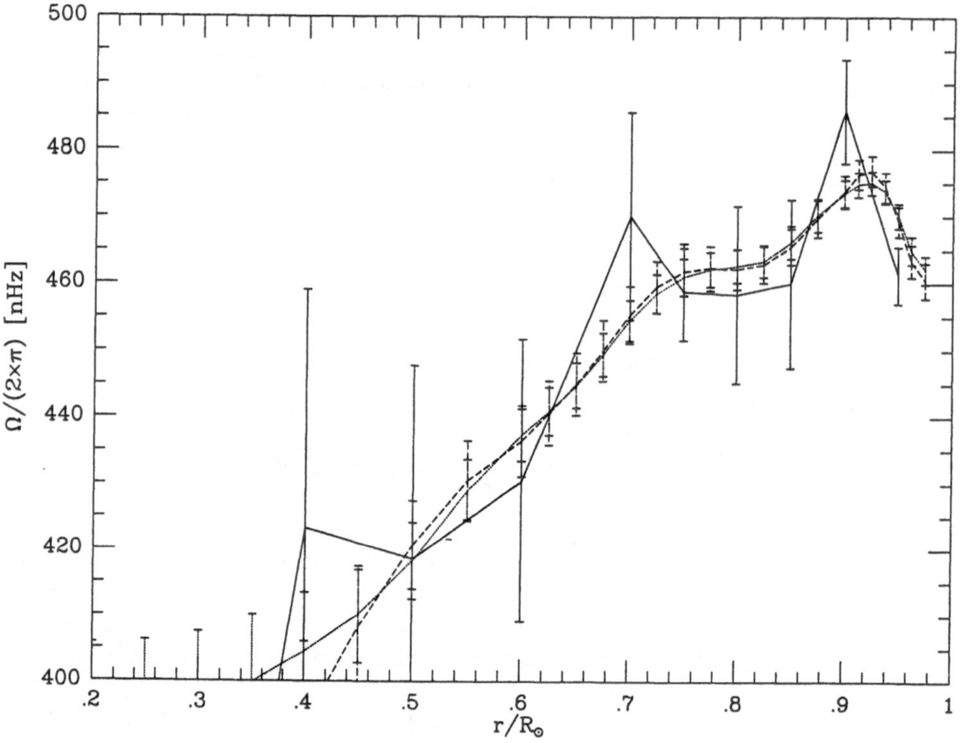

Fig. 1. Inverted rotation rate using respectively an iterative version of the spectral expansion method (solid), a constrained least-squares method using the first derivative constraint (short dashes) and a constrained least-squares method using the second derivative constraint (long dashes).

to the outermost 15% of the solar radius. While, in the deeper regions, no significant differences in the profiles can be noticed, the addition of high degree splitting does lead to a reduction in the uncertainty of the solution in those regions.

The main feature observed in the deeper regions, is a decrease of the rotation rate with depth down to 420 nHz around 0.5 of the solar radius. The present level of accuracy of the low degree splittings prevents any inversion method from providing significant results deeper than 0.5 R.

The potential presence of a discontinuity at the base of the convection zone is indicated by a sharp increase in the rotation rate around 0.7 R in the spectral expansion solution and by the presence of an instability around that depth when the smoothness constraint is relaxed in the constrained least-squares based inversion procedures (not shown).

Between 0.77 and 0.82 of the solar radius, the inverted profiles present a plateau around 462 nHz in the lower part of the convection zone, consistent with the observed surface magnetic feature rotation rate of 462 nHz (Snodgrass 1983). This suggests that

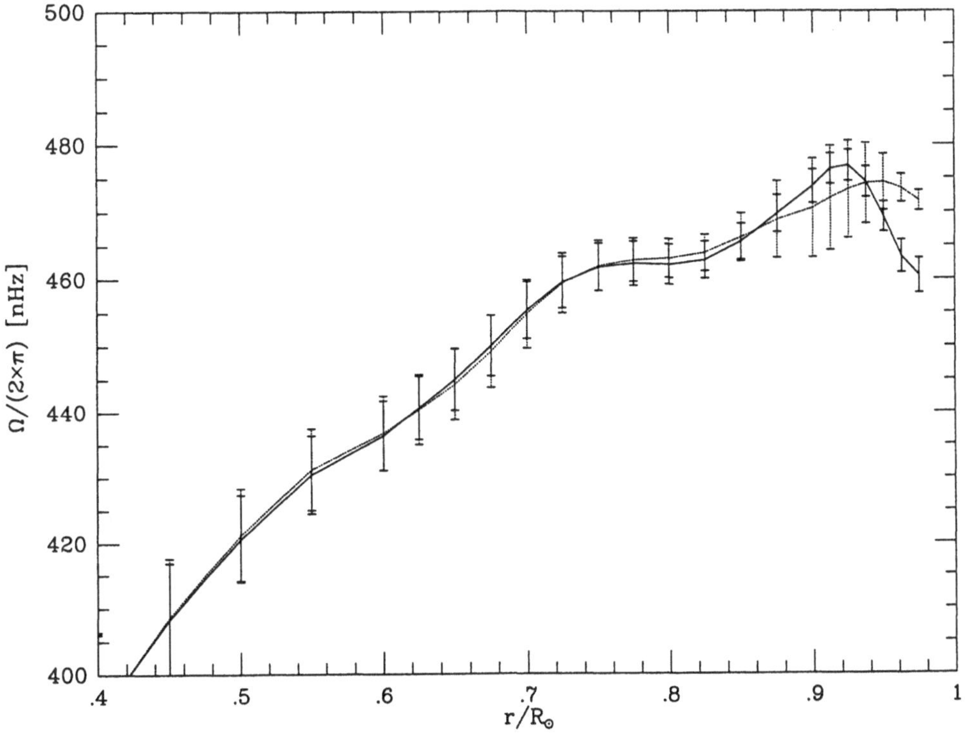

Fig. 2. Inverted rotation rate obtained with (solid) and without (dash) the high degree mode splittings ($120 < \ell \leq 500$), both using the second derivative constrained least-squares method.

there may be a relationship between magnetic structures and this region of the solar interior.

A robust feature of the inverted curves is a maximum in the rotation rate of about 475 nHz near $0.92\,R$. Indeed, this feature persists when the smoothness constraint is increased in the constrained least-squares inversion procedures (not shown). This feature is present only when the high degree modes are included in the inversion. The uncertainty on the solution obtained without the high degree modes, for $r \leq 0.94\,R$ is large enough to include this feature. Above $0.94\,R$ the two solutions diverge, but recall that the inner turning point for a typical 5 minute p-mode, at ℓ of 120 is around $0.91\,R$ and therefore the inverted profile computed without the high degree n-averaged splittings is not significant anymore. This higher-than-surface rotation rate is consistent with the 473.0 nHz rotation rate determined from cross-correlation of doppler measurements (Snodgrass and Ulrich 1989) and attributed to the supergranulation network.

Finally the rotation rate in the outermost 6% of the Sun decreases with radius, with a slope which extrapolates to a surface rotation rate consistent with the spectroscopic determination of 451.7 nHz (Snodgrass 1984). While the inner turning point for a typical

5 minute p-mode, at ℓ of 500, is around 0.989 R, the present level of uncertainty of the high degree splittings prevents the inversion of a significant rotation rate in the outermost 2.5% of the Sun.

5 Conclusions

While our results confirm a slower-than-surface rotation rate in the deeper part of the Sun and a potential discontinuity at the base of the convection zone, the addition of high degree mode splittings in the inversion procedure gives inverted profiles that lead to an attractive picture of the solar rotation rate in the regions below the surface that would reconcile different non-seismic estimates of the "surface" rotation rate.

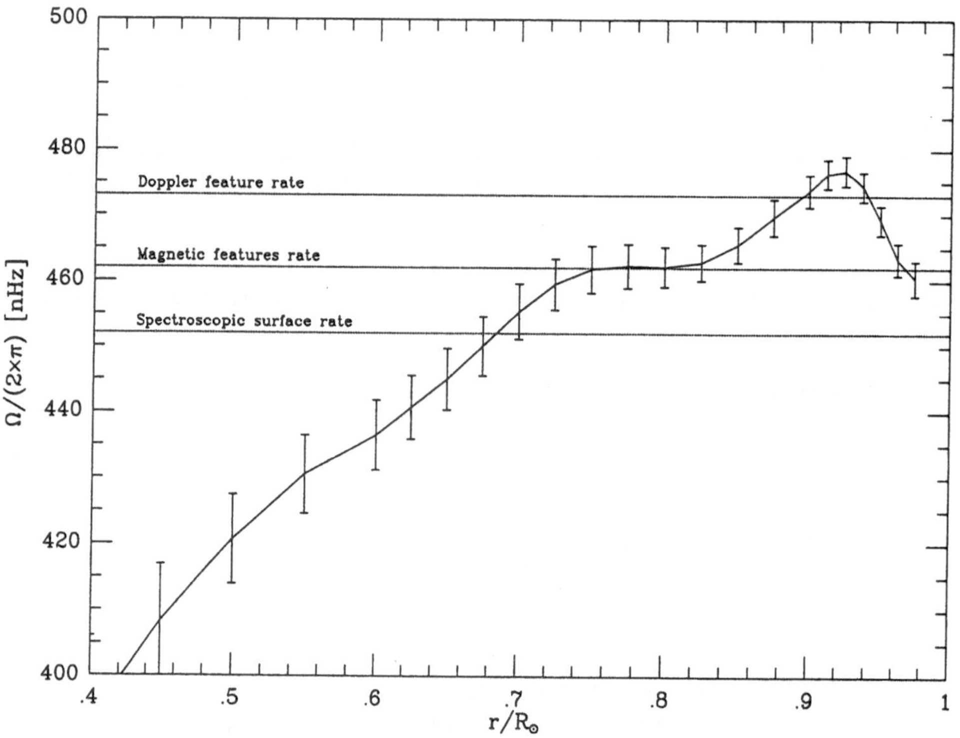

Fig. 3. Inverted rotation rate obtained using the high degree modes, and the second derivative constrained least-squares method compared to the non seismic determinations of the "surface" rotation rate.

This new picture, as presented in Fig. 3, would indicate, on one hand, that the rotation rate of the magnetic features may be driven by a layer rooted in the lower part of the convection zone where strong magnetic fields would be concentrated.

On the other hand, it indicates that the supergranulation network rotation rate observed at the surface would be representative of a layer centered around $0.92\,R$.

Finally, the outermost part of the inverted profile indicates a negative radial gradient consistent with the spectroscopic determination of the surface rotation rate.

Some caution has to be exerted when interpreting these results. It may not be valid to combine data from two different epochs (1984 and 1988) if significant temporal variations turn out to have been present. This potential problem will be alleviated as soon as the reduction of the new 1988 CCD data is completed.

Acknowledgements

The research described here was supported in part by NASA grant NAGW-13 at USC and NAGW-472 at UCLA, and by NSF grant INT-8400213 at USC. Part of the work was done using JPL's Cray X-MP/18 supercomputer. A. Cacciani also acknowledges support from the Italian Consiglio Nazionale delle Richerche and from the Ministerio della Publica Intruzione.

References

Christensen-Dalsgaard, J., and Schou, J. 1988, in *Seismology of the Sun and Sun-like Starts*, ed. E. Rolfe, ESA SP-286, (ESA Publication Division, Noordwijk), p.149.

Duvall, T.L., Jr., Dziembowski, W.A., Goode, P.R., Gough, D.O., Harvey, J.W., and Leibacher, J.W. 1984, *Nature*, **310**, 22.

Dziembowski, W.A., Goode, P.R., and Libbrecht K.G. 1989, *Astrophys. J. Letters*, **337**, L53.

Goode, P.R., Dziembowski, W.A., Korzennik, S.G., and Rhodes, E.J., Jr. 1989, submitted to *Astrophys. J.*

Gough, D.O. 1985, *Solar Phys.*, **100**, 64.

Korzennik, S.G., Cacciani, A., Rhodes, E.J., Jr., Tomczyk, S., and Ulrich, R.K. 1988, in *Seismology of the Sun and Sun-like Starts*, ed. E. Rolfe, ESA SP-286, (ESA Publication Division, Noordwijk), p.117.

Parker, R. 1977, *Ann. Rev. Earth Planet. Sc.*, **5**, 35.

Rhodes, E.J., Jr., Cacciani, A., and Korzennik, S.G. 1990, These proceedings.

Snodgrass, H.B. 1983, *Astrophys. J.*, **270**, 288.

Snodgrass, H.B. 1984, *Solar Phys.*, **94**, 13.

Snodgrass, H.B., and Ulrich, R.K. 1989, submitted to *Astrophys. J.*

Tomczyk, S. 1988, Ph.D. disseration, University of California, Los Angeles.

Woodard, M.F. 1989, *Astrophys. J.*, **347**, 1176.

Has the Sun's Internal Rotation Changed Through this Activity Cycle ?

P. R. Goode [1], W. A. Dziembowski [2], E. J. Rhodes, Jr. [3], and S. Korzennik [4]

[1]Physics Department, New Jersey Institute of Technology, Newark, NJ 07102, USA
[2]N. Copernicus Astronomical Center, Polish Academy of Sciences, ul. Bartycka 18, 00-716 Warszawa, Poland
[3]Department of Astronomy, University of Southern California, Los Angeles, CA 90089, USA
[4]Department of Astronomy, University of California, Los Angeles, CA 90024, USA

Abstract: The internal rotation of the Sun is determined from each of the six available sets of solar oscillation splitting data. These data span this activity cycle and best sample the region near the base of the convection zone. Going inwards through the convection zone into the outer radiative interior, the robust results are a decrease in the rotation rate in the equatorial plane and a trend away from the surface-like differential rotation toward solid body rotation. In the equatorial plane of the radiative interior, the rotation rate seems to systematically increase through the solar cycle. If true, this suggests that the interior has a role in the activity cycle.

In recent years, seismological techniques have enabled us to learn about the Sun's internal rotation. Duvall *et al.* (1984) used the solar oscillation data of Duvall and Harvey (1984) to determine that the rotation near the Sun's equatorial plane is fairly constant, at the surface rate, through the convection zone. Duvall *et al.* (1984) also report that beneath that zone the rotation rate declines going toward the deep interior. Duvall, Harvey and Pomerantz (1986) have confirmed the result in the convection zone and determined that the entire convection zone mimics the differential rotation observed for the solar surface. This report of differential rotation throughout the convection zone has been confirmed by Brown *et al.* (1989) using the data of Brown and Morrow (1987), Dziembowski, Goode and Libbrecht (1989) and Christensen-Dalsgaard and Schou (1988) using the data of Libbrecht (1989), and Rhodes *et al.* (1990) using the data of Rhodes *et al.* (1987, 1990). As well the investigators using the latter three data sets argued for a roughly 10% radial gradient in rotation and a transition away from differential rotation toward solid body rotation through the convection zone into the radiative interior. In fact, there are six available sets of solar oscillation splitting data from which one may determine the internal rotation rate of the Sun. We determine this rate using precisely the same approach so that we may determine the rate's confirmed properties. Since these six data sets span the current activity cycle, we can look for time dependent trends in rotation.

The available data imply a rotation law of the form

$$\Omega(r,\Theta) = \Omega_0(r) + \Omega_1(r)\cos^2\Theta + \Omega_2(r)\cos^4\Theta \quad (1)$$

Regularization is our chosen inversion method. In this least squares method one minimizes α_s^2 rather than χ_s^2 , where

$$\alpha_s^2 = \chi_s^2 + \eta_s M P_s \quad (2)$$

and where $s = 0$, 1 and 2, M is the number of splitting multiplets and

$$P_s = \int \mid d\Omega_s(r)/dr \mid^2 dr \quad (3)$$

The choice of the integrand reflects the aforementioned conclusions that the rotation rate is a fairly weak function of radius. The quantity η_s is not a free parameter, rather it is varied until χ_s^2 is one. Then, the 1σ errors in the inversion are consistent with the 1σ errors in the data. In practice, the value of η_s is further reduced until short wavelength features almost begin to appear in the calculated rotation law. The value of α_s^2 is not very sensitive to our de facto approach because the χ_s^2-space is relatively flat. The chosen value of η_s is 10^{-6} for each data set except that of Duvall and Harvey (1984). This latter data set is the only one in which the sampling is restricted to the region of equatorial plane. The results for $\Omega_0(r)$ are shown in the Fig. 1. The decrease in $\Omega_0(r)$ with r is the clear robust trend confirming the results reported by various investigators. Since the inversion constrains against a radial gradient, the actual gradient is sharper than reported here. The six data sets span nearly the range in l, the angular degree. This means that the oscillations best sample nearly the same region of the solar interior – between 0.5 and 0.8 of the radius where the gradient is sharpest. Eliminating all oscillations above $l = 60$, so that all six sets have same high-l cutoff, the discrepancies between the values near the solar surface disappear. The rotation laws in the Fig. 1 include no $l < 10$. If all $l < 20$ are eliminated, all sets have the same low-l cutoff and there are negligible changes in the relative values of $\Omega_0(r)$ near 0.4 of the radius. There is an apparent time dependent trend in $\Omega_0(r)$. The trend is that the smallest value of $\Omega_0(r)$ at 0.4 of the radius comes from the earliest data set from the current solar activity cycle and subsequent sets result in rates at 0.4 which increase through the cycle. The largest value of $\Omega_0(r)$ corresponds to the end of the last cycle. The least square errors in $\Omega_0(r)$ tend to be large near 0.4. However, sizeable differences are indicated in the Fig. 1 between the results from the data sets of Duvall and Harvey (1984) and Libbrecht (1989) in particular. The hint of a role for the deep interior in the Sun's activity cycle will be tested with subsequent splitting data.

Acknowledgements

P.R.G. is partially supported by AFOSR-89-0048. S.K and E.J.R are partially supported by NASA grant NAGW-13 and NSF grant INT-8400213.

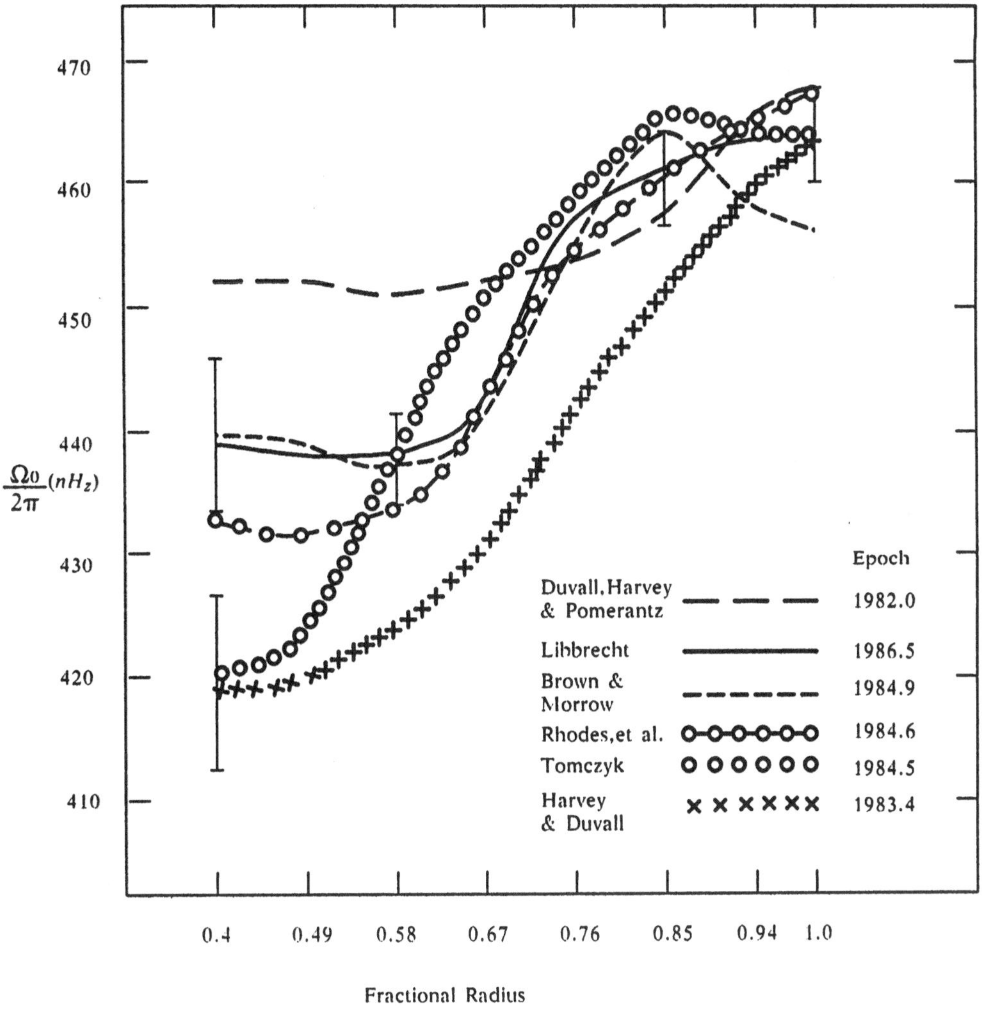

Fig. 1. The six $\Omega_0(r)/2\pi$ functions vs. fractional radius. The error bars are on the inversions from Libbrecht's (1989) data and at $0.4R$ for the Duvall and Harvey (1984) data.

References

Brown, T.M., Christensen-Dalsgaard, J., Dziembowski, W.A., Goode, P.R., Gough, D.O., and Morrow, C.A., 1989, *Astrophys. J.*, **343**, 526.

Brown, T.M., and Morrow, C.A., 1987, *Astrophys. J. Letters*, **314**, L21.

Christensen-Dalsgaard, J., and Schou, J., 1988, in *Seismology of the Sun and Sun-like Stars*, ed. E.J. Rolfe, ESA SP-286 (ESA Publication Division, Noordwijk), p.149.

Duvall, T.L., Dziembowski, W.A., Goode, P.R., Gough, D.O., Harvey, J.W., and Leibacher, J.W., 1984, *Nature*, **310**, 22.

Duvall, T.L. and Harvey, J.W., 1984, *Nature*, **310**, 19.

Duvall, T.L., Harvey, J.W. and Pomerantz, M.A., 1986, *Nature*, **321**, 500.

Dziembowski, W.A., Goode, P.R. and Libbrecht, K.G., 1989, *Astrophys. J. Letters*, **337**, L53.

Libbrecht, K.G., 1989, *Astrophys. J.*, **336**, 1092.

Rhodes, E.J.,Jr., Cacciani, A., Korzennik, S., Tomczyk, S., Ulrich, R.K., and Woodard, M.F., 1990, *Astrophys. J.*, in press.

Rhodes, E.J., Jr., Cacciani, A., Woodard, M., Tomczyk, S., Korzennik, S., and Ulrich, R.K., 1987, in *The Internal Solar Angular Velocity*, eds. B.R. Durney and S. Sofia (Reidel, Dordrecht), p. 75.

Tomczyk, S., 1988, Ph.D. dissertation, University of California at Los Angeles.

Internal Solar Rotation and the Boundary Layer Non-linear Dynamo

G. Belvedere [1], M.R.E. Proctor [2], and G. Lanzafame [1]

[1]Istituto di Astronomia, Università di Catania, Italy
[2]Department of Applied Mathematics and Theoretical Physics, University of Cambridge, England

Abstract: Most recent helioseismological results suggest the location of dynamo action in the boundary layer between the radiative and convective zones. This is confirmed by some preliminary results of a non-linear dynamo model in a spherical shell (0.05 $R_\odot$ thick), with full time and latitude resolution. The internal solar rotation profile deduced by helioseismological data is assumed. Periodic symmetric regime is found for negative helicity and absolute values of the dynamo number of order unity. These small values are compatible with the steep radial angular velocity gradient in the boundary layer.

1 Introduction

At the present moment the relevance of helioseismology to solar physics is really outstanding. Indeed, the ever more sophisticated helioseismological results are giving us the possibility of an increasingly deeper insight into the physics of the solar interior, and improving our knowledge of the Sun's internal structure and dynamics, which is fundamental in order to test our global knowledge of stellar structure and evolution. For instance, inverting solar oscillations data allows to test the validity of the solar standard model, to determine the depth of the solar convection zone to the highest degree of accuracy, to infer the solar rotation rate as a function of latitude and depth down to the radiative core and to explore the strength and topology of magnetic fields in and below the convection zone.

Therefore, the up to date indirect observational evidence of what happens in the Sun — even if some caution is necessary since the results are somewhat provisional — poses serious and precise constraints to theoretical work.

The impact of helioseismology on solar physics results also in a new challenge to dynamo theory, which is currently considered as the most plausible theoretical framework to understand both solar and stellar activity, even if several uncertainties do still exist, first of all, about the location of dynamo action.

Three possible locations have been suggested in modelling the solar dynamo:

(i) the whole convection zone

(ii) the base of the convection zone (1st scale height)
(iii) the boundary (overshoot) layer between the convective zone and the radiative one.

The last possibility seems to be the most realistic on the basis of the following argument. Spatial separation of the ω-effect and the α-effect is not plausible in so far as it would imply problems of upward and downward field transport in a turbulent medium. On the other hand, since stability of magnetic flux tubes against magnetic buoyancy suggests that the ω-effect is located deep in the convection zone or better in the boundary layer (Spiegel and Weiss 1980, Spruit and Van Ballegooijen 1982), also the α-effect is expected to operate mainly at deep levels.

The location of dynamo in the boundary layer is indirectly supported by the helioseismic results. The most recent helioseismological data (Harvey 1988; Dziembowski *et al.* 1989; Brown *et al.* 1989; Rhodes *et al.* 1990) seem to agree as to the following (provisional) scenario:

(i) the surface angular velocity $\omega(R_\odot, \theta)$ persists throughout the convection zone ($1R_\odot$ to $0.75R_\odot$): *i.e.* $(d\omega/dr)_{\mathrm{c.z.}} \sim 0$ or slightly > 0.
(ii) beneath the convection zone ($0.7R_\odot$ to $0.65R_\odot$), rigid rotation dominates with $\omega_0 \sim 2.7 \cdot 10^{-6}\,\mathrm{rad\,s^{-1}}$, which is the surface value at latitude $\sim 37°$.

These imply that the equatorial and polar rotation rates do converge to the intermediate value ω_0 below the convection zone.

Let us see what the implications to solar dynamo are:

(i) radial shear driven dynamo, operating in the convection zone, might be unrealistic
(ii) radial shear driven dynamo, operating in the boundary layer, is still supported by helioseismological data.

The argument is the following (with reference to the northern hemisphere). Since $\alpha < 0$ in the boundary layer (Yoshimura 1975; Glatzmaier 1985a,b; Gilman *et al.* 1989) and interpolating the helioseismological results in the boundary layer, $\partial\omega/\partial r < 0$ at higher latitudes ($\geq 37°$) and $\partial\omega/\partial r > 0$ at lower latitudes ($\leq 37°$), we get poleward ($\alpha\partial\omega/\partial r > 0$) and equatorward ($\alpha\partial\omega/\partial r < 0$) migration respectively, in agreement with the observational evidence shown by different tracers of the solar cycle (polar faculae and prominences on the one side, spots and faculae on the other side). Therefore, dynamo can still work and reproduce the observations if is located in the boundary layer.

2 The Model

This has very recently been confirmed by preliminary results obtained in the context of a boundary layer ($0.05R_\odot$ thin) nonlinear model.

We retain the α-effect formalism and the simplest non linear interactions, but use a representation of the spatial structure only in the radial direction (radial truncation), thus allowing full latitudinal dependence of fields and flows (this means the possibility of constructing butterfly diagrams).

The mean field dynamo equation for the time evolution of the magnetic field $\boldsymbol{B} = B(r,\theta)\boldsymbol{\Phi} + \nabla \times [A(r,\theta)\boldsymbol{\Phi}]$ where r, θ are spherical polar coordinates, $\boldsymbol{\Phi}$ is the unit vector in the azimuthal direction, $B\boldsymbol{\Phi}$ is the toroidal, and $\boldsymbol{B}_p \equiv \nabla \times A\boldsymbol{\Phi}$ is the poloidal part of $\boldsymbol{B}$, are:

$$\frac{\partial A}{\partial t} = \alpha F(r,\theta)B + \eta_T \left[\nabla^2 - \frac{1}{r^2 \sin^2\theta}\right] A \qquad (1)$$

$$\frac{\partial B}{\partial t} = r\sin\theta \boldsymbol{B}_p \cdot \nabla \left[\frac{U(r,\theta)}{r\sin\theta}\boldsymbol{\Phi}\right] + \eta_T \left[\nabla^2 - \frac{1}{r^2\sin^2\theta}\right] B. \tag{2}$$

Here αF is the usual α-effect, with F representing its spatial structure and α its magnitude. The quantity η_T is a turbulent diffusivity.

The dynamical influence of the magnetic field enters the model through its effect on the differential rotation $U(r,\theta) = u_0 + u$, where u_0 is a prescribed velocity field and u is a perturbation driven by the mean Lorentz force and subject to viscous damping. The simplest equation that encompasses these features is:

$$\frac{\partial u}{\partial t} = \frac{1}{\mu_0}\left[(\nabla \times \boldsymbol{B}) \times \boldsymbol{B}\right]_{\boldsymbol{\Phi}} + \rho\nu_T \left[\nabla^2 - \frac{1}{r^2\sin^2\theta}\right] u \tag{3}$$

where ν_T is a turbulent viscosity.

The model equations have been solved in a spherical shell representing the boundary layer, with suitable boundary conditions. For full details about the method we address the reader to Belvedere *et al.* (1989), which worked out a similar model in the solar convection zone.

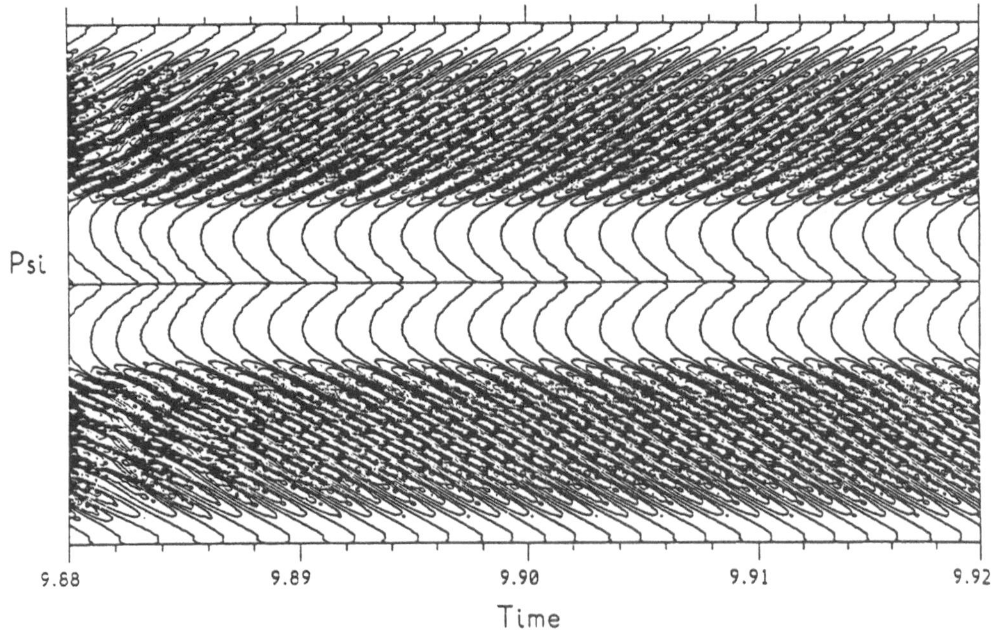

Fig. 1. Butterfly diagram with equatorial and polar branches for $d = 0.05R_\odot$ and $D = -1$.

Assuming the differential rotation profile u_0 in the boundary layer as given by interpolating the most recent helioseismological data, the results show the existence of periodic (dynamo wave-like) stable solutions with both equatorward and poleward migrating branches, as shown by related butterfly diagrams (Fig. 1). These solutions are found for dynamo number $D = \alpha\omega_0 d^3/\eta_T^2 \approx -1$, where d is the thickness of the boundary layer.

The relatively small value of $|D|$ is a consequence of the very steep radial angular velocity gradient in the thin layer, since, for the onset of dynamo action, $|\alpha\partial\omega/\partial r|$ must exceed a critical value.

However, these results are preliminary: only the linear regime has been investigated yet and some numerical instabilities do still exist. We have to improve the quality of results, especially concerning the relative amplitudes of polar and equatorial branch toroidal fields (see Fig. 1).

Further, the non-linear regime is to be investigated.

References

Belvedere, G., Pidatella, R.M., and Proctor, M.R.E. 1989, *Geophys. Astrophys. Fluid Dyn.*, in press.

Brown, T.M., Christensen-Dalsgaard, J., Dziembowski, W., Goode, P.R., Gough D.O., and Morrow, C.A. 1989, *Astrophys. J.*, **343**, 526.

Dziembowski, W.A., Goode, P.R., and Libbrecht, K.G. 1989, *Astrophys. J. Letters*, **337**, L53.

Gilman, P.A., Morrow, C.A., and De Luca, E.E. 1989, *Astrophys. J.*, **338**, 528.

Glatzmaier, G.A. 1985a, *Geophys. Astrophys. Fluid Dyn.*, **31**, 137.

Glatzmaier, G.A. 1985b, *Astrophys. J.*, **291**, 300.

Harvey, J.W. 1988, in *Seismology of the Sun and Sun-Like Stars*, ed. E.J. Rolfe, ESA SP-286 (ESA Publication Division, Noordwijk), p.55.

Rhodes, E.J., Cacciani, A., Korzennik, S., Tomczyk, S., Ulrich, R.K., and Woodard, M.F. 1990, *Astrophys. J.*, in press.

Spiegel, E.A., and Weiss, N.O. 1980, *Nature*, **287**, 616.

Spruit, H.C., and Van Ballegooijen, A.A. 1982, *Astron. Astrophys.*, **106**, 58.

Yoshimura, H. 1975, *Astrophys. J. Suppl.*, **29**, 467.

VI

Seismology of Stars

Toward Seismology of δ Scuti Stars

W. Dziembowski

Nicolaus Copernicus Astronomical Center, Bartycka 18, 00-716 Warsaw, Poland

Abstract: Many among δ Scuti variables were shown to be multiperiodic and all of them probably are. Being in various stages of early evolutionary phase these stars seem excellent candidates for seismic testing the basic assumptions and physics of the stellar evolution theory. What hinders any major progress in the field is an unresolved problem of connecting sparse spectra of the observed oscillation frequencies to dense spectra calculated for the models. Determination of the rotation rate in the interior is probably the most important goal of asteroseismology. In the case of δ Scuti stars there is a complication following from the high density of the spectrum that the standard formula for the rotational splitting may not be valid. To understand mode selection mechanism we have to go beyond the linear adiabatic theory of stellar oscillations. There are still uncertainties in determining mode stability, but the real difficulty lies in prediction which of many unstable models may reach detectable amplitudes. Trapping in the envelope is a possible mechanism of mode selection, but it is not clear yet whether it finds support in observational data. Observational information on spherical harmonics associated with the observed periodicities is crucial for mode identification. This was obtained with use of both photometric and spectroscopic data for number of δ Scuti stars. There are, however, uncertainties which must be clarified with new data and improved methods of their analysis. Efforts in obtaining periodograms for unevolved objects having simple theoretical frequency spectra are encouraged.

1 Introduction

The progress in the field of δ Scuti star seismology is embarrassingly slow. On the side of accomplishments we have perhaps only the finding due to Andreasen and Petersen (1988) that the discrepancy between the observed and calculated period ratios for radial modes may be removed by allowing a drastic – factor 2.5 – upward revision of opacity in the temperature range $\log T = 5.2 - 5.9$. A similar revision, as first suggested by Simon (1982), allows to reconcile the period ratios in the double mode Cepheids. Furthermore results of helioseismology also indicate that the opacity is higher than currently assumed (Christensen-Dalsgaard *et al.*, 1988). This, however, refers to higher temperature and the required enhancement is much smaller.

It is possible that these results through encouraging research on opacities will have an important impact on the theory of stellar evolution, but much more is expected from asteroseismology. In particular, we would like to subject to an observational test our assumption concerning chemical composition changes. We would also like to measure internal rotation and magnetism. These are important quantities which are likely responsible for diversification of star properties in the δ Scuti domain of the H-R diagram and the evolution of these quantities is very poorly understood.

For these goals we need data on nonradial modes. There is no doubt that significant fraction of periodicities found in δ Scuti variables is due to such mode excitation. The difficulty lies in connecting measured frequencies to specific eigenmodes of stellar oscillations. Spherical harmonic degree, l, and azimuthal order, m, of a mode are subject, at least in principle, of an observational determination. Its radial order may only be inferred with help of the theory of stellar oscillations. This is quite difficult problem for these stars and in this review I will focus primarily on the routes leading to its resolution.

2 Theoretical Frequency Spectra

2.1 Effects of the Evolution on Star Pulsation Properties

Changes in stellar oscillation properties during the evolution are a consequence of changes in the behavior of the Brunt-Väisälä frequency, N. In Fig. 1 taken from Królikowska's (1988) thesis the $N(r)$ function is shown for three models representing various stages of evolution of $1.6M_\odot$ star within the δ Scuti instability strip. Determination of this function, which records the evolutionary history of the stellar core, is certainly most exciting task of asteroseismology. One may see that the radial pulsations of low order cannot probe $N(r)$ in the region where most of the evolutionary changes take place because such modes are trapped in the envelope. Higher order pulsations penetrate deeper region but their frequencies are sensitive primarily to the sound speed in the outer layers.

Table 1. Parameters of selected models of a 1.6 $M_\odot$ star.

Model	X_c	$L/L_\odot$	$T_{\rm eff}$ [K]	$R/R_\odot$	Π_0 [d]
A	0.700	8.49	8760	1.36	0.0371
B	0.360	10.78	8330	1.58	0.0514
C	0.155	11.79	7830	1.87	0.0661
D	0.007	12.99	7770	2.03	0.0743
E	0.000	17.11	7830	2.25	0.0859
F	0.000	18.11	7300	2.69	0.1100
G	0.000	17.83	7100	2.76	0.1190

For nonradial modes the inner region characterized by large values of N develops into a gravity-wave propagation zone. This gives rise to a set of intermediate frequency modes of dual character: acoustic in the envelope and gravitational in the core. With the progress of evolution the g-mode order increases quickly and the frequency separation between consecutive modes at the same degree, $\Delta\omega$, becomes approximately given by the following formula

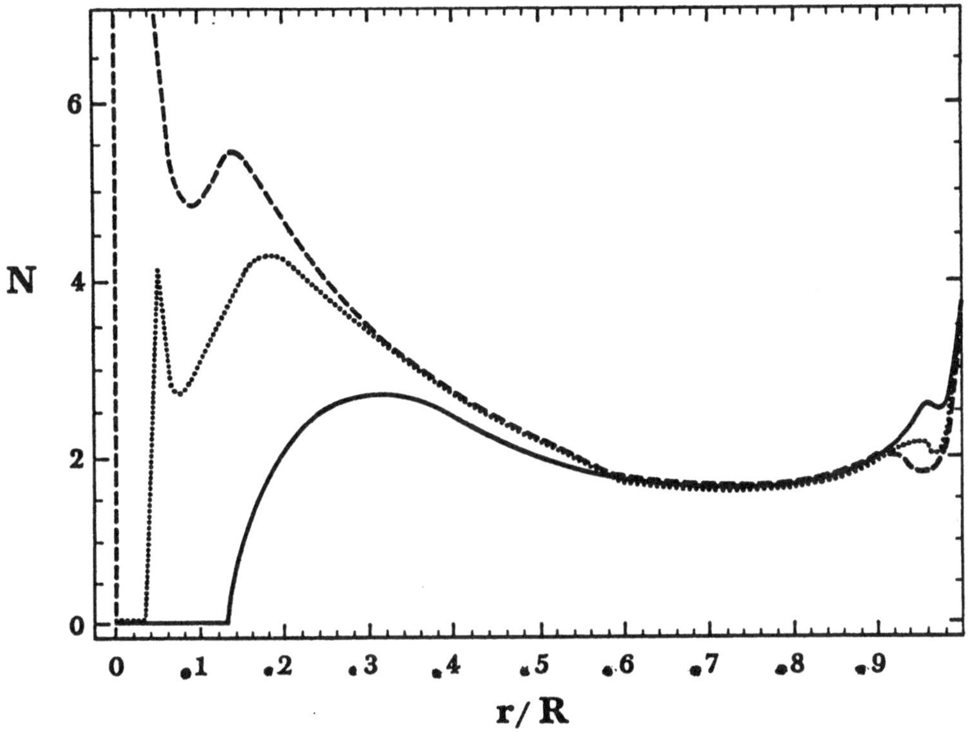

Fig. 1. The Brunt-Väisälä freguency in units of $\sqrt{4\pi G < \rho >}$ in three models of 1.6 $M_\odot$ star: at ZAMS (solid line), at the end of the central hydrogen burning (dotted line), at the shell burning phase (dashed line). In these units the frquencies of the first three radial modes are approximately : 2.05, 2.65 and 3.15. Plots do not show complicated patterns in near $r = R$; at the boundary N is between 20 and 30 in these models.

$$\frac{\Delta\omega}{\omega} \approx \frac{\pi}{\sqrt{l(l+1)}} \left(\int \frac{N}{\omega} \frac{dr}{r} \right)^{-1} , \tag{1}$$

where the integral should be evaluated in the inner region where $N > \omega$. To get this simple expression one actually has to assume $N \gg \omega$.

Complete surveys of nonradial mode frequencies in selected ranges for chosen models of δ Scuti stars were published by Dziembowski (1977a) and Lee (1985). Here in Figs.2 and 3 we reproduce frequency spectra for modes corresponding to $l \leq 3$ in the range of first five radial modes are shown for selected models of a $1.6 M_\odot$ star as calculated by Dziembowski and Królikowska (1990). Some parameters of these models are given in Table 1. The quantity E_G/E plotted as the ordinate is the fraction of the mode energy contained in the gravity propagation zone. It thus a good indicator of the physical nature of a mode. The figures cover essentially whole evolutionary phase of relevance to δ Scuti

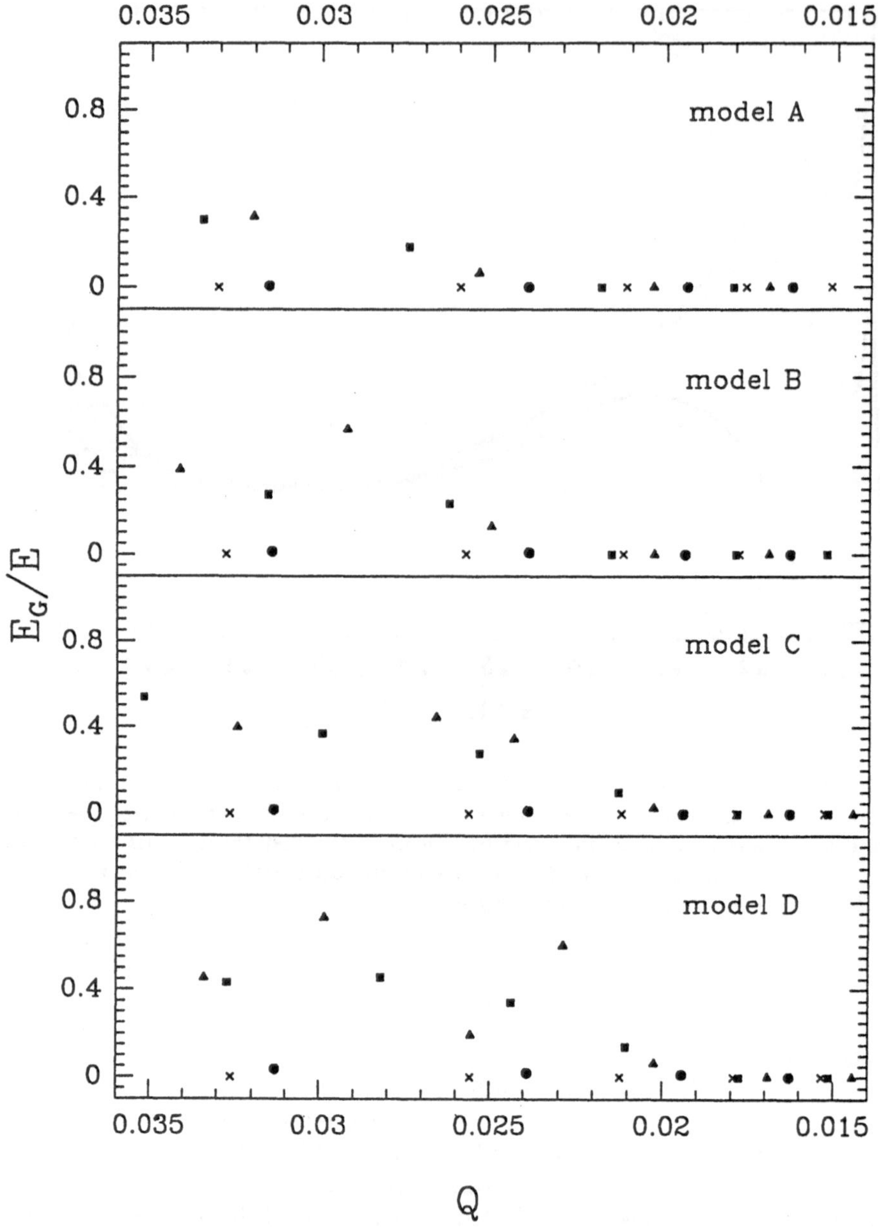

Fig. 2. Modes of spherical harmonic degrees $l = 0$ (crosses), $l = 1$ (filled circles), $l = 2$ (filled squares), $l = 3$ (filled triangles), in models of $1.6 M_\odot$ star representing various stages of the Main Sequence evolutionary phase, Q is the pulsation constant, E_G/E is the fraction of the oscillation energy coming from the g-mode propagation zone.

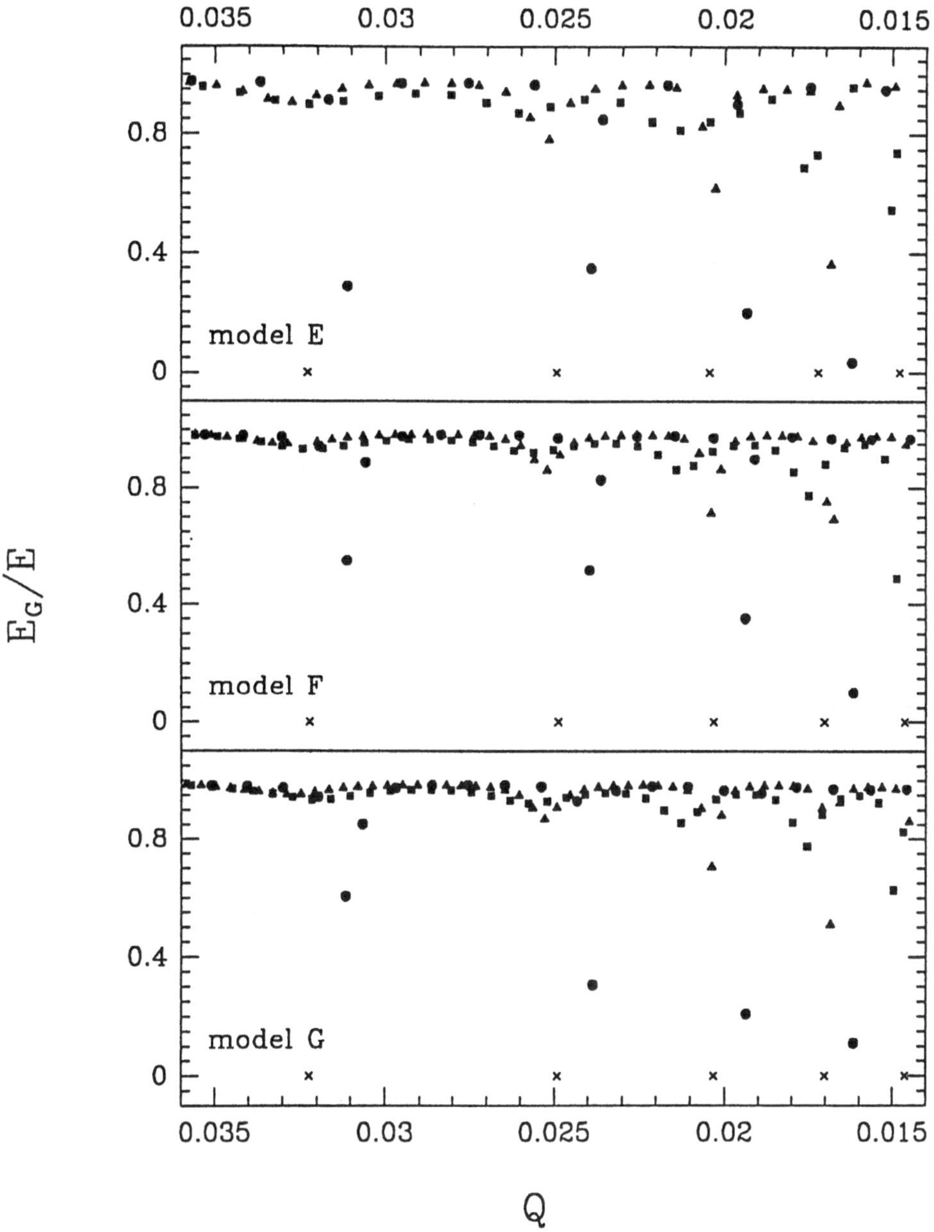

Fig. 3. The same as Fig. 2 but for the models of $1.6M_\odot$ in post-Main Sequence evolutionary phase.

stars. One may follow there appearance of new modes with the progress of the evolution and note that for $l > 0$ these are predominantly g-modes.

2.2 Modes Trapped in the Envelope

Significant trapping in the envelope occurs only for nonradial oscillations with $l = 1$. It may be noted that the trapped mode have their frequencies somewhat higher than their radial counterparts. Occasionally, as a result of an internal resonance a pair of partially trapped modes occurs. For $l = 2$ and $l = 3$ only traces of trapping are visible especially at the high frequency. Similar trapping pattern were found in an evolved 2.0 $M_\odot$ star models, though the effect for the dipole modes was noticeably weaker.

The fact that in evolved δ Scuti star models the modes corresponding to $l = 2$ and 3 do not exhibit the trapping effect to the same degree as those with $l = 1$ or $l = 4$ could be seen in results obtained for sequence of 1.5 $M_\odot$ star models by Dziembowski (1977a). For still higher l-values the trapping effect is increasingly more pronounced, but luminosity amplitudes of such mode suffering large cancellation are not likely to reach a measurable level.

It would greatly facilitate the task of the asteroseismology if the possibility of reaching observable amplitude is restricted to the trapped modes. We would have then an easy method of selecting from the dense spectra few candidates which can be associated with observed periodicities. What is important this selection principle relies only on the linear adiabatic theory of stellar oscillation. In an already mentioned paper (Dziembowski and Królikowska, 1990) we discuss observational and theoretical arguments supporting such a conjecture. I will return to this problem in the course of present review.

2.3 Frequency Splitting by Rotation

Multiplet structures in the frequency spectra are only sources of observational information about the internal rotation and magnetic fields. Although in periodograms for some δ Scuti stars some nearly equidistant patterns were found and they were interpreted as a rotational splitting, we have to regard an actual probing of these quantities as a matter of distant future. So far the observed structures were used only to infer the l-value.

The well-known Ledoux formula,

$$\omega(m) = \omega(0) - m(1 - C)\Omega + O(\Omega^2), \tag{2}$$

describes the structure for a slow uniform rotation. For acoustic oscillations the value of C is of the order of 0.1 for the lowest order modes and decreases with frequency, asymptotically $C \sim \omega^{-2}$. For high-order g-modes $C \approx 1/l(l + 1)$. For the oscillations observed in δ Scuti stars we, thus, expect $C \sim 0.1$ except if $l = 1$ untrapped modes are excited, when we have $C \approx 0.5$.

Allowing the $r-$dependence in Ω does not change the linearity of the $\omega(m)$ function, but the $\theta-$ dependence does. In the latter case (see e.g. Brown *et al.*, 1989) $\omega(m)$ may be represented as an odd-order polynomial. With the differential rotation like that found in the sun's surface the $\omega(m)$ function is still strongly dominated by the linear term and, in fact, we expect much weaker latitudinal dependence in predominantly radiative interiors of these stars.

Leaving aside possibility of very strong magnetic fields one should consider nonlinear effects of rotation as the most likely cause of braking the equidistant structure of the multiplets. For evolved star models these effects are much more important than it follows from the values calculated by Saio (1981) for a polytropic model. For validity of Eq.(2) we have to assume, in particular that

$$\left(\frac{m\Omega C}{\omega}\right)^2 \ll 2\frac{\Delta\omega}{\omega}, \tag{3}$$

where $\Delta\omega$ denotes now the frequency distance to the nearest mode of the same l or l differing by 2. For Main Sequence or polytropic stars the r.h.s. may be much smaller than 1 only due to an occasional degeneracy. For evolved stars, due to small frequency separation between consecutive modes inequality (3) may not be satisfied for any mode. In such instances functional forms of $\omega(m)$ may be found only by means of numerical calculations (see e.g. Berthomieu *et al.*, 1978).

3 Mechanism of Mode Selection

3.1 Linear Instability

Results of stability analyses for radial and nonradial modes in δ Scuti star models are available in several publications (Dziembowski, 1975, 1977a; Stellingwerf, 1979; Fitch, 1981; Lee, 1985). Usually, mode stability is characterized by its growth rate

$$\gamma = \frac{W}{2\omega E}, \tag{4}$$

where W is the rate of mode energy gain, called often the work integral.

For modes of low spherical harmonic degree W is virtually l-independent, because contribution to it comes almost solely from the model envelope where the eigenfunctions are only weakly l-dependent. Thus, the instability ranges shown in Fig. 4 are valid for all such modes. These plots combined with those shown in Figs. 2 and 3 reveal the multitude of unstable modes. The spectrum is still denser for models with $M = 2M_\odot$ better corresponding to luminosity class III objects.

At $l \leq 3$ the light amplitude reduction due to averaging positive and negative contributions over stellar disc is not very large. The growth rates for nonradial modes are reduced relative to the radial modes at the same frequency by factor $1 - E_G/E$, but their values are not reflected in any simple way in the terminal amplitudes. We, thus have to conclude that the linear nonadiabatic calculation are far insufficient to set useful restrictions enabling mode identifications in evolved objects.

The plots shown in Fig. 4 show all well-known properties of the opacity mechanism in δ Scuti stars. Its calculated efficiency depends in a sensitive way on the treatment of subphotospheric convection and, especially on the helium content. This latter property makes nonadiabatic calculations a potential tool for probing composition in the outer layers. Cox *et al.* (1979), who first studied this problem, suggested that occurrence of nonvariable objects in the δ Scuti domain of the H-R diagram is due to an inward He diffusion. Unfortunately, since only a small fraction of the unstable modes is observed, the instability boundaries cannot yet be regarded as measurable quantities.

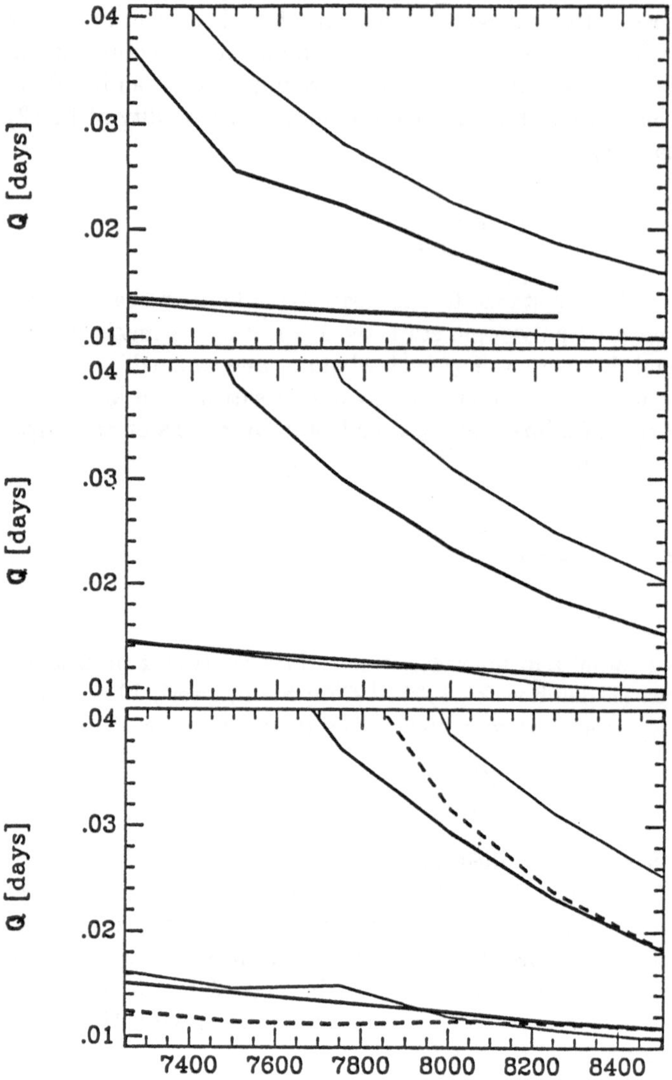

Fig. 4. Ranges of pulsation constant values where the instability driven by the opacity mechanism occurs plotted as function of the effective temperature. The thick solid lines give the boundaries obtained assuming a standard fractional helium content $Y = 0.28$, the thin lines give the boundaries obtained with $Y = 0.38$. In these two cases the convective flux in subphotospheric layers was ignored. The dashed lines were obtained upon including this flux in the equilibrium models, but ignoring its Lagrangian perturbation. Constant masses and luminosities were assumed at each sequence: $M = 1.6M_\odot$ and $L = 11L_\odot$ for lower plot, $M = 1.6M_\odot$ and $L = 18L_\odot$ for middle plot, $M = 2M_\odot$ and $L = 40L_\odot$ for upper plot. The effect of the convective flux inclusion in more luminous models is smaller than that shown for $L = 11L_\odot$.

The influence of convection on mode stability is the main source of uncertainty on the level of the linear theory. The two distinct approximations made to obtained the results shown in Fig. 4 are very crude, but I hope they correctly show where the results are most sensitive to treatment of the convective flux.

3.2 Amplitude Limitation

Linear instability does not guarantee that the mode attains an observable amplitude and, in fact, that it is at all excited. There are two physically distinct nonlinear effects that may limit the amplitude growth. First is the saturation of the opacity mechanism which is the dominant effect in large amplitude pulsators such as RR Lyrae stars or Cepheids. Second is the resonant mode coupling to modes that are linearly stable and therefore are dissipating energy.

These two mechanisms may be conveniently discussed with help of the following equation for the complex amplitude, $\tilde{A}$,

$$\frac{d\tilde{A}_j}{dt} = \tilde{A}_j\left(\gamma_j - \sum_k \alpha_{j,k}A_k^2\right) - \sum_p H_p s_p \tilde{A}_{p,1}\tilde{A}_{p,2}\exp[-i(\omega_j - \omega_{p,1} - \omega_{p,2})t]$$

$$+\sum_q H_q \tilde{A}_{q,1}\tilde{A}^*_{q,2}\exp[i(\omega_{q,1} - \omega_{q,2} - \omega_j)t], \tag{5}$$

where complex coefficients $\alpha_{j,k}$ describe the saturation, $A_k =| \tilde{A}_k |$ is the real amplitude, $s = 1/2$ for the two-mode resonance $(p,1 \equiv p,2)$ and $s = 1$ for the three-mode case, H_p are coupling coefficients which integrals of products of eigenfunctions for the involved modes, * denotes complex conjugation. Frequency differences occurring in the formula are assumed to be small. In the first of the terms describing resonant coupling the summation covers triplets in which mode j occurs as the highest frequency mode. In the second it occurs as one of the lower frequency modes.

Let us consider effect of the saturation alone. A simple system of equations for real amplitudes, A, follows then from Eq. (5):

$$\frac{dA_k^2}{dt} = 2A_k^2\left[\gamma_k - \sum_j \Re(\alpha_{j,k})A_k^2\right]. \tag{6}$$

In our application the difficulty consists in large number of modes that has to be simultaneously considered. Both observations and results of numerical modelling show that in most cases of the Cepheid-type pulsators only one mode is present at the finite amplitude development of the instability even if there are more unstable modes. We do not know whether the same applies to the δ Sct type variables, where the number of such modes is so much larger. I have argued against possibility that in these stars the saturation is due to the excitation of a large number of mostly invisible modes (Dziembowski, 1980), but problem demands more careful examination by means of looking for solutions of Eq. (6).

A simple case of resonant interaction between an unstable mode low-l (subscript 0) and two stable g-modes (subscripts 1 and 2) that are excited as a result of the parametric instability occurring if $\omega_1+\omega_2 \approx \omega_0$ was studied by Dziembowski and Królikowska (1985).

The saturation was ignored. Thus, nonlinear development of the modes driven by the opacity mechanism was treated as independent. The system of Eq. (5) written for the three coupled modes allows always a constant amplitude solution. In particular, we have

$$A_0 = \frac{1}{H}\sqrt{\gamma_1\gamma_2(1+q^2)}, \tag{7}$$

where

$$q = \frac{\omega_0 - \omega_1 - \omega_2}{\gamma_0 + \gamma_1 + \gamma_2}. \tag{8}$$

Note that $H \neq 0$ only if $l_2 = l_1 + l_0 - 2j$, where $j = 0, 1, ... \leq l_0/2$, and $m_0 = m_1 + m_2$. We showed that the g-modes are most likely of high degree, which for low l_0 implies $l_1 \approx l_2$, and such that $\omega_1 \approx \omega_2$.

Numerical calculations conducted for a 1.4 $M_\odot$ ZAMS star revealed that only few modes have chance to reach the amplitude level of 0.01 mag. Neglect of the saturation effects seems thus justified. The g-mode amplitudes are by many orders of magnitude lower. In this investigation effects of rotation were ignored. In a subsequent paper (Dziembowski *et al.*, 1988) we showed that significant reduction of the expected amplitudes occurs if the equatorial velocity exceeds some 50 km/s.

The constant amplitude solutions may represent a terminal development of the parametric instability only if it is stable. We found that in the frequency range corresponding to the first three radial modes such solutions are in most cases stable and that in the remaining cases periodic limit cycles are most likely solutions. In the latter cases Eq. (5) gives a good estimate of the mean amplitudes (Moskalik, 1986). For the higher order modes the constant amplitude solutions are unstable and in most cases the instability implies an unbound amplitude growth. This occurs if $\gamma_0 > | \gamma_1 + \gamma_2 |$. The inequality is fulfilled for such modes when they interact with g-modes trapped in the inner cavity which is true in most of the cases. The latter modes are very adiabatic; *i.e.* they have very low values of $| \gamma |$. Królikowska (1988) found that in evolved objects this is true for all modes that may be driven by the opacity mechanism because with the progress of evolution the center of gravity of all relevant g-modes moves toward the star center. The growth of amplitudes will be ultimately halted by interactions with parametrically excited new generation(s) of gravity modes.

A quantitative theory of the coupling involving larger number of modes is not yet available, but it seems likely that this effect determines amplitudes of the unstable modes, at least in those cases when the three mode coupling cannot prevent their growth. In evolved models the parametrically excited g-modes are essentially deep interior oscillations. One may therefore expect that the effect is weaker for the modes trapped in the envelope and, consequently, they are allowed to attain larger amplitudes. Thus, there is a physical justification for the conjecture that these modes are preferentially excited.

4 Nonradial Modes Observed in δ Scuti Variables

Naturally, asteroseismologists are mainly interested in objects with possibly large number of excited modes. Kurtz (1988) reviewed recently observational aspects of multimodal δ Scuti stars. The record holders are 78 Tau and 4 CVn with five modes detected in each case. A preliminary analysis of new observations of 63 Her (Mangeney *et al.*, 1988) suggests a possibility that up to seven modes are seen in this object.

The case of 78 Tau deserves special attention because short periods of the excited modes indicate that the object is still in the Main Sequence phase of evolution and therefore its spectrum should be relatively easy to interpret. Furthermore, we have additional information about the star following from the fact that it is in a binary system. Recent analysis by Kovács and Paparó (1988) provide accurate values of the frequencies which are all close to 0.014 c/d and apparently associated with modes of the same p-mode order. Unfortunately, we do not have observational information about l and m values.

Breger *et al.* (1990) interpret five frequencies they found in 4 CVn as a sequence corresponding to a single value of l equal 2 or 3 and three radial orders. Two pairs are interpreted as rotational split components. This identification, however, is not well justified. Firstly, because it relies on model frequencies provided by Fitch (1981) who applied an incorrect boundary condition eliminating artificially most of the frequencies. Secondly, validity of Eq. (2), which is essential for this identification, is questionable the case of this evolved object.

Modes reported in 63 Her occur in an unusually wide frequency range. There is more than factor five between the extreme values. If confirmed these data constitute a serious challenge to the theory. It is particularly difficult to understand excitation of very long period modes in this relatively hot object.

Information about spherical harmonics of the observed modes is extremely important both as a clue to mode identification and as a help in clarifying mechanism of mode selection. Certainly best source of it is line profile behavior as first noted by Ledoux (1951). Subsequently, several researchers used a comparison of observed and simulated profiles as a tool for mode identification. Smith (1982) determined in this way l's for the doublet at $f \approx 5.2\,\mathrm{day}^{-1}$ in δ Sct itself finding the values of 0 and 2. He also deduced $l = 2$ for two modes excited in 28 Aql and two modes in 14 Aur. Balona (1986a,b) developed a direct algorithmic method of deducing l and m values from the observed profile variations. It would be very important to apply his method to δ Scuti stars which, to my best knowledge, has not yet been done.

There is a simple test of the hypothesis that in evolved (long period) δ Scuti stars only trapped modes are preferentially excited. This follows from the fact that the trapping does not occur for modes with l =2 and 3. For discrimination between $l \leq 1$ and $l > 1$ cases a method of Balona and Stobie (1979) using only photometric data should be sufficient. In this method the crucial quantity is the phase difference between the color and luminosity variations, $\Delta\psi$. A simple expression for the flux changes caused by nonradial oscillations (Dziembowski, 1977b; Balona and Stobie, 1979) implies $\Delta\psi > 0$ for $l = 0$, $\Delta\psi = 0$ for $l = 1$, and $\Delta\psi > 0$ for $l > 1$ Improvements of the expression for the flux introduced by Watson (1988) result in some decrease of $\Delta\psi$.

Most values of $\Delta\psi$ compiled by Watson (1988) are within the errors consistent with the $l = 0$ or $l = 1$ identification. The exceptions are one mode in δ Sct itself and four modes in BD+28°1494. While for the latter object there is a reason for taking the large negative values with reservation the case of δ Sct must be regarded as a challenge to the idea that only trapped modes can be excited, because the results in this case are consistent with the line profile analysis. However, even the case of δ Sct cannot be regarded as a compelling evidence against our selection rule. All the spectroscopic mode identification must be regarded as tentative and more work is required to rule out possibility that $l = 1$ or perhaps 4. The method developed by Balona (1986a,b) appears best suited for the task. Returning to the photometric method, we would like to point out that among the values of $\Delta\psi$ listed by Watson (1988) there two large positive values which cannot be interpreted in terms of any l-value and therefore may indicate there some problems with the method validity. Furthermore, the method cannot easily distinguish between the cases $l = 2$ and 4, and in the latter case a partial trapping may occur.

5 Discussion

Seismology of δ Scuti stars is the field where close collaboration between observers and theoreticians is particularly important. Sparse frequency spectra obtained so far from data analyses are confronted with *embarras de richesse* of very large number of modes that are found unstable in appropriate stellar models. The nonlinear theory has not yet reached the level enabling selection of few candidates for association with observed periodicities. Observational information on the l-values is an important hint for the theory.

We have seen that with available information on spherical harmonic degrees of the modes excited in δ Scuti variables we cannot yet support or disprove the conjecture that all this modes are at least partially trapped in the acoustic cavity. There is a need for a new careful examination of both photometric and spectroscopic data and methods of extracting values of l. We must contemplate, thus, a possibility that the mode selection is not simply related to the trapping phenomenon and that it depends on complicated nonlinear and, perhaps, random processes. Therefore, it is probably better to focus current observational efforts on Main Sequence (short period) objects whose simple theoretical frequency spectra give better prospect for mode identification.

Situation may improve drastically if with an amplitude resolution such as expected from space observations we will see a significant fraction of the unstable low-degree modes. Mode identification will be again possible and observed dense spectra will become a source of detailed information about the deep interiors of evolved stars. However, before undertaking expensive space programs both ground-based observational and theoretical work is needed aimed at estimating probability that indeed this large number of modes will be detected.

References

Andreasen, G.K., and Petersen, S.O. 1988, *Asrtron. Astrophys.*, **192**, L4.
Balona, L.A. 1986a, *Monthly Notices Roy. Astron. Soc.*, **219**, 111.
Balona, L.A. 1986b, *Monthly Notices Roy. Astron. Soc.*, **220**, 647.
Balona, L.A., and Stobie, R.S. 1979, *Monthly Notices Roy. Astron. Soc.*, **189**, 649.

Berthomieu, G., Gonczi, G., Graff, P., Provost, J., and Rocca, A. 1978, *Asrtron. Astrophys.*, **70**, 597.
Breger, M., McNamara, B.J., Kerschbaum, Ling, H., Shi-yang, J., Zi-he, G., and Poretti, E. 1990, *Asrtron. Astrophys.*, in press.
Brown, T.M., Christensen-Dalsgaard, J., Dziembowski, W.A., Goode, P., Gough, D.O., and Morrow, C.A. 1989, *Astrophys. J*, **343**, 526.
Christensen-Dalsgaard, J., Gough, D.O., and Thompson, M.J. 1988, in *Seismology of the Sun and Sun-like Stars*, ed. E.J. Rolfe, ESA SP-286 (ESA Publication Division, Noordwijk), p.493.
Cox, A.N., King, D.S., and Hodson, S.W. 1979, *Astrophys. J*, **231**, 798.
Dziembowski, W. 1975, *Mem. Soc. Roy. Sci. Liège*, **8**, 289.
Dziembowski, W. 1977a, *Acta Astron.*, **27**, 95.
Dziembowski, W. 1977b, *Acta Astron.*, **27**, 203.
Dziembowski, W. 1980, in *Nonradial and Nonlinear Stellar Pulsation*, eds. H.A. Hill and W.A. Dziembowski (Springer, Berlin), p.22.
Dziembowski, W., and Królikowska, M. 1985, *Acta Astron.*, **35**, 5.
Dziembowski, W., and Królikowska, M. 1990, *Acta Astron.*, in press.
Dziembowski, W., Królikowska, M., and Kosovichev, A. 1988, *Acta Astron.*, **38**, 5.
Fitch, W.S. 1981, *Astrophys. J*, **249**, 218.
Kovács,G., and Paparó, M. 1989, *Monthly Notices Roy. Astron. Soc.*, **237**, 355.
Królikowska, M. 1988, Ph.D. Thesis, (Copernicus Astronomical Center, Warsaw).
Kurtz, D.W. 1988, in *Multimode Stellar Pulsations*, eds G. Kovacs, L. Szabados, and B. Szeidl (Kultura, Budapest), p.95.
Ledoux, P. 1951, *Astrophys. J.*, **114**, 373.
Lee, U. 1985, *Publ. Astron. Soc. Japan*, **37**, 279.
Mageney, A., Cheverton, Belmonte, J.A., Däppen, W., Saint-Pé, O., Praderie, F., Roca Cortès, Fuensalinda, J, and Alvarez, M. 1988, in *Seismology of the Sun and Sun-like Stars*, ed. E.J. Rolfe, ESA SP-286 (ESA Publication Division, Noordwijk), p.551.
Moskalik, P. 1986, *Acta Astron.*, **36**, 333.
Saio, H. 1981, *Astrophys. J*, **249**, 299.
Simon, N.R. 1982, *Astrophys. J. Letters*, **260**, L87.
Smith, M.A. 1982, *Astrophys. J*, **254**, 242.
Stellingwerf, R.F. 1979, *Astrophys. J*, **227**, 935.
Watson, R.D. 1988, *Astrophys. Space Sci.*, **140**, 255.

Oscillations in roAp Stars

D. W. Kurtz

Department of Astronomy, University of Cape Town,
Rondebosch 7700, South Africa

Abstract: A brief introduction to the rapidly oscillating Ap (roAp) stars and a review of their properties is presented. Evidence which strongly suggests that the roAp stars pulsate obliquely in high-overtone, low-degree modes is shown. This review then concentrates on four problems of particular interest to asteroseismologists: 1) The interpretation of the solar-like amplitude spectrum of HR 1217; 2) Mode lifetimes in roAp stars; 3) The nature of the harmonics in HR 3831; and 4) Evolutionary period changes in roAp stars.

1 Introduction

The rapidly oscillating Ap (roAp) stars are interesting objects for asteroseismological investigation. I have recently given an extensive general review of these stars (Kurtz 1990), so this review will concentrate on selected observations of interest to seismologists; see Kurtz (1990) for literature references not given in this paper.

The roAp stars are high-overtone, low-degree p-mode pulsators with masses around $2\,M_\odot$ and main sequence to subgiant luminosities. They have basically dipolar global magnetic fields of a few hundred to a few thousand gauss which have axes inclined to the rotation axis of the star. The pulsation modes are aligned with the magnetic axis of the star which puts the observer in the unique position of viewing the pulsation from varying aspect. This condition makes the identification of the degree and order of the pulsation possible in many cases. The amplitude modulation of the pulsation with rotational aspect constrains the pulsation geometry, yielding information about the rotational inclination, i, and the magnetic obliquity, β. In principle, stars which pulsate in more than one mode with different ℓ may allow the determination of i and β without reference to any other data.

Figure 1 shows a plot of the pulsation phase of the principal mode in HR3831 versus magnetic phase. This star has a well determined rotational period from its mean light variations of P_{rot}=2.851962±0.000014 day; the magnetic phase is calculated without reference to the pulsation. The pulsation phase, φ, has been calculated by fitting the principal frequency ν=1.42801257±0.00000009 mHz to sections of the data 2 cycles long (23.34 min) with a function $A\cos(2\pi\nu t+\varphi)$. The horizontal lines are separated by π radians. The vertical lines are the calculated times of magnetic quadrature; these are very near phases 0.25 and 0.75 because $i > 80°$. This diagram clearly shows the phase reversal characteristic of an oblique dipole pulsator. The φ error bars are $\pm 1\sigma$; they are dependent inversely on amplitude, hence amplitude can approximately be seen in this diagram also. Note the low errors (high amplitude) near the magnetic extrema and the large errors near quadrature where the amplitude goes through zero.

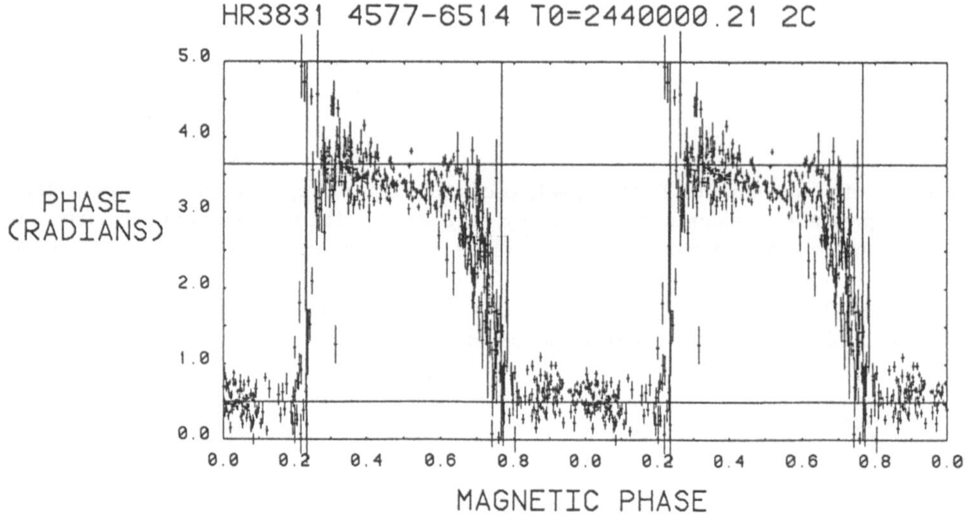

Fig. 1.

Some of the roAp stars are apparently single-mode pulsators; these stars seem to have modes with frequencies and amplitudes stable over a time-span of years. Other roAp stars are multiple mode pulsators; some of them have modes with slightly variable amplitudes; some have modes with lifetimes as short a days. In several cases $\Delta\nu_0$, the spacing between consecutive overtones, has been determined for roAp stars: $\Delta\nu_0$=34 or 68 μHz for HR 1217; $\Delta\nu_0$=52 μHz for HD 60435; $\Delta\nu_0$=57.8 μHz for HD 101065; $\Delta\nu_0$=68 μHz (perhaps) for HD166473; $\Delta\nu_0$=50.6 μHz (perhaps) for HR 7167; and $\Delta\nu_0$=58 μHz (perhaps) for HR8097(γ Equ). These $\Delta\nu_0$ are consistent with the late-A spectral types and IV-V luminosity classes of the roAp stars. It has not yet been possible to determine with certainty $\delta\nu$, the second order term which lifts the degeneracy between modes of (n,ℓ), $(n-1,\ell+2)$, etc., for any roAp star, although there are some indications that $\delta\nu \simeq$ 3 μHz, which is somewhat smaller than expected theoretically.

2 The Amplitude Spectrum of HR 1217

Kurtz *et al.* (1989) obtained 365 hr of observations of HR 1217 at 8 observatories during three months in 1986. They found that the 6-min light variations are amplitude modulated in phase with the magnetic variations, with amplitude maximum and magnetic maximum coinciding (as in HR 3831), but with mean light minimum occurring at a slightly, but probably significantly, different time. A frequency analysis of 324 hr of those observations over a 46-day time-span yielded 6 principal frequencies, all of which are amplitude modulated with the rotation of the star. Figure 2 shows a schematic amplitude spectrum which bears a striking resemblance to the solar p-mode spectrum. The spacing of the principal frequencies is to scale; the separations are in μHz. The rotational sidelobes are marked Ω; their separations and the separations of the secondary frequencies from the principal frequencies have been exaggerated for clarity (from Kurtz *et al.* 1989).

From Fig. 2 and other data Kurtz *et al.* (1989) concluded that

1) HR 1217 is an oblique rotator with a centered dipole magnetic field. The geometry of the field requires that tan i tan β=0.52±0.03.

2) HR 1217 is an oblique pulsator with the pulsation axis and the magnetic axis aligned. This is required by the coincidence of the times of amplitude maximum and magnetic maximum and the phases of the frequency triplets.

3) There are six principal frequencies of pulsation in HR 1217; ν_2 and ν_4 are dipole modes (ℓ=1, m=0); from the form of the amplitude modulation, $\tan i \tan\beta = 0.68 \pm 0.04$ for ν_2 and $\tan i \tan\beta = 0.51 \pm 0.03$ for ν_4, in reasonable agreement with the magnetic field measurements. The frequencies ν_3 and ν_5 cannot be described by single spherical harmonics — they look similar to dipole modes, but they have amplitude minima which are lower than expected.

4) The frequencies ν_1 to ν_5 have an alternating frequency spacing: $\Delta_1 = \nu_2 - \nu_1 = \nu_4 - \nu_3 = 33.27 \pm 0.08\mu$Hz; $\Delta_2 = \nu_3 - \nu_2 = \nu_5 - \nu_4 = 34.75 \pm 0.08\mu$Hz.

5) It is not possible to discriminate between the hypothesis that $\nu_1, \nu_2, \nu_3, \nu_4$ and ν_5 are basically due to alternating even and odd ℓ-modes with $\nu_0 = 68\,\mu$Hz, and the hypothesis that they are all basically due to dipole modes with $\Delta\nu_0 = 34\,\mu$Hz. The hypothesis of alternating even and odd ℓ-modes seems incorrect because the amplitude ratios of the rotational sidelobes to the central frequencies are incorrect for ℓ=even, although theoretically the appearance of modes with $\ell \neq 1$ is expected to be altered by the magnetic field. The hypothesis of ℓ=1 for all modes seems incorrect because the alternating Δ_1, Δ_2 frequency spacing is then difficult to explain and seems more consistent with alternating even and odd ℓ-modes.

6) If $\Delta\nu_0 = 68\,\mu$Hz, then HR 1217 lies 0.7 mag above the ZAMS and has a radius of $R = 1.9R_\odot$; if $\Delta\nu_0 = 34\,\mu$Hz, then HR 1217 lies 1.9 mag above the ZAMS and has a radius of $R = 3.4R_\odot$ (assuming that $R_{\rm ZAMS} = 1.4R_\odot$ at F0).

7) The sixth principal frequency, ν_6, could be an ℓ=2 mode if $\Delta\nu_0 = 34\,\mu$Hz, but then the alternating Δ_1, Δ_2 frequency spacing is difficult to explain. If $\Delta\nu_0 = 68\,\mu$Hz, then the Δ_1, Δ_2 spacing makes sense but the $\nu_6 - \nu_5 = 1.503\Delta_1$ spacing is inexplicable; ν_6 is an enigma.

8) There are secondary frequencies which indicate that the principal frequencies are amplitude modulated on a time-scale of months. This is not likely to be due to pulsation in higher degree modes (which would produce amplitude modulation on a much shorter

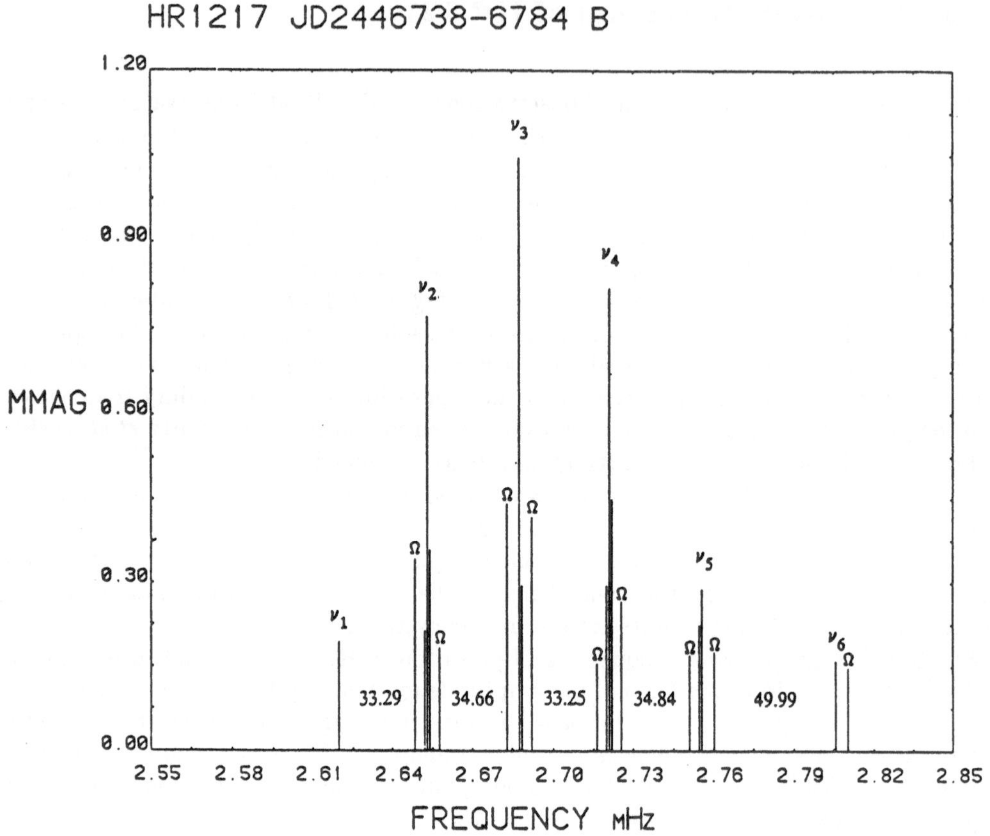

Fig. 2.

time-scale), and hence indicates that the amplitudes of the principal frequencies are not completely stable.

3 Mode Lifetimes in the roAp Stars

There is now strong evidence that the mode lifetimes in some of the roAp stars are much shorter than the time-span of the observations. HD 60435 has the most complex frequency spectrum of any of the roAp stars. It pulsates with many frequencies which have a basic separation of 25.8 μHz and appear to be alternating even and odd ℓ-modes, but the lifetimes of at least some of the modes seem to be as short as a few days (Matthews *et al.* 1986a; 1987).

A rotation period of 7.6793±0.0006 day has been determined for HD 60435 from its mean light variations (Kurtz *et al.* 1990). The mean light curves show a double wave of unequal depth, so both magnetic poles are seen, probably from different aspects, indicating that neither i nor β are near 90°. The time of mean light extremum coincides with pulsation maximum, indicating that HD 60435 is an oblique pulsator. Although HD 60435 is only a mild spectrum variable, a study of the line strengths made simultaneously with high-speed photometric observations suggests that Sr II maximum coincides with pulsation maximum (Matthews *et al.* 1986b). Since Sr II maximum is usually coincident with magnetic maximum in Ap stars, this also suggests that HD 60435 is an oblique pulsator. Two upper-limit measurements of the magnetic field strength (Landstreet and Bohlender, reported by Matthews *et al.* 1987) give $|B_{\rm eff}| < 1000$ G using the rotation ephemeris of Kurtz *et al.* (1990).

Amplitude modulation of the pulsation frequencies over one rotation cycle can be seen in many of the data sets, but during the next rotation cycle, the same frequencies may not be present. If they do return, they do so at their original frequencies. This behavior is qualitatively similar to the sun's; it indicates that HD 60435 is pulsating in many high-overtone p-modes with lifetimes which can be shorter than one rotation cycle. This conclusion is based on over 400 hr of high-speed photometric observations, much of which was obtained contemporaneously from Chile and South Africa; frequency resolution and aliases do not confuse the analysis.

Very recently Kreidl, Kurtz, Bus, and Birch (in preparation) have found that in 1989 HD 217522 was pulsating with multiple frequencies near 1.20 mHz and 2.01 mHz; for 97 hr of data obtained in 1982, Kurtz (1983a) found only frequencies near 1.21 mHz. It is clear that the 2.01-mHz frequencies have been excited in the interval between the data sets.

Other roAp stars also show evidence of mode lifetimes on various time-scales. Libbrecht (1988) suggests from his radial velocity measurements and the photometry of Kurtz (1983b) that the lifetimes of the pulsation modes in γ Equ may also be as short as those in HD 60435, but γ Equ is not a well-studied roAp star. HR 1217 has six pulsation frequencies which have constant amplitudes over one rotation period of 12.4572 days, but which show variable amplitudes on a time-scale of four rotation periods (Kurtz *et al.* 1989). HD 101065 has one large-amplitude (about 10 mmag peak-to-peak in B) pulsation mode and at least one much smaller amplitude pulsation mode (Martinez and Kurtz 1990); the large amplitude mode shows evidence of some variation in amplitude over the 10 yr it has been observed, but on a shorter time-scale it appears stable. In HR 3831 the principal dipole oscillation is stable with a lifetime on the order of the 5.3 yr for which it has been observed. HR3831 is singly periodic (with harmonics).

There is a qualitative pattern here: It appears that the lifetimes of the pulsation modes are short in those roAp stars that pulsate in many modes, whereas the lifetimes are longer in the stars with few or only one mode. This is not a problem of resolution; the separation between modes, both theoretically and observationally, is in the range 20 to 80 μHz, so that the modes can, in principle, be resolved with a single night of observation. On the other hand, the number of stars from which this conclusion is made is few.

Many fundamental data are lacking for the roAp stars, so it is not yet possible to say whether pulsation amplitudes, mode lifetimes, number of modes excited, or anything else is correlated with fundamental physical parameters. Only a few roAp stars have luminosity estimates, and most of these are from secondary methods; accurate space-based parallaxes for these stars are needed. Temperatures are somewhat uncertain, in most cases due to the spectral peculiarities; in the worst case, HD 101065, the temperature dispute has ranged over 1500 K. Only the magnetic fields of HR 1217 and γ Equ can be described as well-studied, and γ Equ has not yet gone through one magnetic cycle in the 40 yr it has been observed — there is some dispute whether the magnetic field variations arise from oblique rotation with a period of about 70 yr, or whether they are due to a stellar magnetic cycle. Rotation periods are also not known for most roAp stars.

These fundamental data are needed for asteroseismology to be used to its fullest extent to probe the interior structure and interior magnetic field configuration of these stars, to determine their masses and ages, and to discover what governs the mode lifetimes.

4 Problems in Understanding the First and Second Harmonics of the Principal Frequency in HR 3831

Table 1. A non-linear least-squares fit of the seven frequencies, $\nu_1 - \nu_7$, determined for the entire 1980-1986 data set for HR 3831

	frequency mHz		amplitude mmag	phase radians
ν_1	1.42395445±0.00000002	$A^{(1)}_{-1}=$	2.095±0.017	1.166±0.011
ν_2	1.42801257±0.00000009	$A^{(1)}_{0}=$	0.403±0.017	−0.830±0.060
ν_3	1.43207097±0.00000002	$A^{(1)}_{+1}=$	1.707±0.017	1.168±0.014
ν_4	2.84790845±0.00000019	$A^{(2)}_{-2}=$	0.189±0.017	1.147±0.126
ν_5	2.85602534±0.00000009	$A^{(2)}_{0}=$	0.419±0.017	1.046±0.057
ν_6	2.86414166±0.00000019	$A^{(2)}_{+2}=$	0.188±0.017	1.353±0.126
ν_7	4.28809652±0.00000035	$A^{(3)}_{+1}=$	0.104±0.017	1.764±0.230

σ=1.6150 mmag

$\nu_2 - \nu_1$=4.05812±0.00009 μHz
$\nu_3 - \nu_2$=4.05840±0.00009 μHz
$\nu_5 - \nu_4$=8.11689±0.00021 μHz=2(4.05845±0.00011 μHz)
$\nu_6 - \nu_5$=8.11632±0.00021 μHz=2(4.05816±0.00011 μHz)
ν_{rot}=4.05829±0.00002 μHz

$(\nu_2 - \nu_1)-(\nu_3 - \nu_2)$=−0.28±0.13 nHz
$[(\nu_5 - \nu_4)-(\nu_6 - \nu_5)]/2$=0.29±0.15 nHz
$\nu_5 - 2\nu_2$=0.20±0.20 nHz
$\nu_7 - 3\nu_2$=4.05881±0.00044 μHz
$<\Delta\nu>=[(\nu_2 - \nu_1)+(\nu_3 - \nu_2)+(\nu_5 - \nu_4)/2 +(\nu_6 - \nu_5)/2]/4$=4.05828±0.00005 μHz

Note to Table 1: These parameters fit the relation $\Delta B = \sum A_i \cos[2\pi\nu_i(t - t_0) + \varphi_i]$ where t_0=JD2444598.96±0.03.

A frequency analysis of HR 3831 yields seven frequencies which are given in Table 1 (Kurtz, Shibahashi and Goode, in preparation). Interpretation of the high frequency triplet, ν_{456}, in HR 3831 is an interesting problem. To very high accuracy $\nu_4=2\nu_1$, $\nu_5=2\nu_2$, and $\nu_6=2\nu_3$. In spite of this, Kurtz (1982) suggested that ν_{456} is not the first harmonic of ν_{123}, but rather is due to pulsation in an oblique quadrupole ($\ell=2$, $m=0$) mode which is driven by a small, undetectable first harmonic of ν_{123}, hence forcing the exact 2:1 ratio between the frequencies. In general, an oblique eigenmode will have $(2\ell+1)$ observable frequencies. Thus an ($\ell=2, m=0$) mode will show five frequencies split by the rotation frequency (see Shibahashi 1986). Assuming $\ell=2$, $m=0$ for ν_{456}, the ratios of its amplitudes, along with the constraint provided by the dipole interpretation of ν_{123} of $(A_{+1}^{(1)} + A_{-1}^{(1)})/A_0^{(1)} \equiv x = 9.4 \pm 0.4$, require that $i=86°$, $\beta=36°$, or vice versa (Kurtz 1982). The $A_{+1}^{(2)}$ and $A_{-1}^{(2)}$ components of the quintuplet have amplitudes below the observed noise level for these i and β and hence are not observed.

The radius of HR 3831 is about $R = 2.0R_\odot$. For $P_{\rm rot}=2.851962$ day this gives an equatorial rotational velocity of $v = 36\,{\rm km\,s^{-1}}$. Carney and Peterson (32) measured $v \sin i = 33 \pm 3\,{\rm km\,s^{-1}}$ for HR 3831 which requires that $i \simeq 90°$; this was taken to support the quadrupole interpretation of ν_{456} which gave $i=86°$.

Shibahashi (private communication) pointed out that the quadrupole interpretation of ν_{456} is not correct. From Table 1 the phases of $A_{\pm 2}^{(2)}$ and $A_0^{(2)}$ are all equal within the errors at magnetic maximum, thus ν_{456} cannot be due to an $m=0$ quadrupole mode. This is easily seen geometrically: the amplitudes of ν_{456} sum to zero at quadrature, yet an $m=0$ quadrupole has a maximum at quadrature (equal to $-1/2$ its amplitude when seen pole-on). This means that the only constraint on i and β for HR 3831 comes from ν_{123}. That requires i or $\beta > 80°$; Carney and Peterson's (1985) measurement of v sin i removes the ambiguity: It is $i > 80°$.

Kurtz, Shibahashi, and Goode (in preparation) have considered this problem and find the following.

4.1 The Observed Form of ν_{456}

The three frequencies ν_{456} are in phase at magnetic maximum so their amplitudes sum to a maximum at that time. At quadrature $A_{+2}^{(2)}$ and $A_{-2}^{(2)}$ are again in phase with each other, but π radians out of phase with $A_0^{(2)}$. Since $A_{+2}^{(2)} + A_{-2}^{(2)} = A_0^{(2)}$, this means that the amplitude of ν_{456} is zero at quadrature. This is the behavior of a mode which is modulated by $\cos^2\alpha$ (where α is the angle between the pulsation pole and the line-of-sight [see Shibahashi 1986; Kurtz and Shibahashi 1986]): it is zero at quadrature and maximum at both magnetic extrema, but without the phase reversal of the dipole ($\cos\alpha$-modulated) mode.

We can derive this formally as follows. Let us assume for ν_{456} that

$$\Delta L/L \propto \cos^2\alpha \cos(\omega t + \varphi). \tag{1}$$

It is easy to see that

$$\cos^2\alpha \propto \{2P_2(\cos\alpha) + P_0(\cos\alpha)\}. \tag{2}$$

This means that $\cos^2\alpha$ modulation gives rise to a frequency quintuplet, just as a quadrupole mode does, but the central component will have P_2 and P_0 contributions. With a bit of manipulation we can find that

$$(A_{+2}^{(2)} + A_{-2}^{(2)})/A_0^{(0+2)} = x^2/(x^2+2) \equiv y \qquad (3)$$

where $x = \tan i \tan \beta$. We know that $x = 9.4 \pm 0.4$ from ν_{123}. The above equation gives $y = 0.98 \pm 0.08$ calculated from x (which depends on the amplitudes of ν_{123}); $y = 0.90 \pm 0.07$ calculated from the amplitudes of ν_{456} in Table 1; these are in excellent agreement. Similar arguments lead to

$$(A_{+1}^{(2)} + A_{-1}^{(2)})/A_0^{(0+2)} = 4x/(x^2+2) \equiv z \qquad (4)$$

where $z = 0.42$ for $x = 9.4$. Hence $(A_{+1}^{(2)} + A_{-1}^{(2)})/2 = 0.08\,\mathrm{mmag}$ which is lost in the noise of our analysis and hence explains why those components are not seen. The conclusion from this argument is that the high frequency triplet in HR 3831 is amplitude modulated by $\cos^2\alpha \propto \{2P_2(\cos\alpha) + P_0(\cos\alpha)\}$.

4.2 Theoretical Expectation

The 2:1 ratio between the frequencies ν_{456} and ν_{123} strongly suggests that ν_{456} is the first harmonic of ν_{123}. Since ν_{123} is a dipole mode with a surface distortion proportional to $\cos\theta$, where θ is the colatitude, the first harmonic should show a surface distortion of the form $\cos^2\theta$ since the harmonic arises from the second order terms in the wave equation. Hence the surface distortion of the first harmonic of a dipole mode should look like $\cos^2\theta \propto \{2P_2(\cos\theta) + P_0(\cos\theta)\}$; that is, the harmonic looks to be two parts quadrupole plus one part radial, *in the reference frame of the star*.

When seen from an observer's point of view the relative amplitudes of the quadrupole and radial components of the distortion must be weighted by an integration over the visible surface; the quadrupole component will have a lower apparent amplitude due to cancelling between out-of-phase surface areas. It is easy to show that the observed amplitudes of the radial to quadrupole modes must be weighted by

$$R = \int I(\theta) P_0(\cos\theta) \cos\theta d\cos\theta \Big/ \int I(\theta) P_2(\cos\theta) \cos\theta d\cos\theta, \qquad (5)$$

where $I(\theta)$ is the limb-darkening function. If we ignore limb-darkening ($I(\theta) = 1$), then the observed amplitude of the quadrupole component must be reduced by $1/R = 1/4$ relative to the radial component.

The result of this is that we expect the first harmonic to be modulated as $\{1/2P_2(\cos\alpha) + P_0(\cos\alpha)\}$; however, the observations clearly show that ν_{456} is modulated as $\{2P_2(\cos\alpha) + P_0(\cos\alpha)\}$.

4.3 Conclusion about ν_{456}

The modulation of ν_{456} goes as $\cos^2\alpha$ as seen by the observer, whereas we expect that it should go as $\cos^2\theta$ in the reference frame of the star. Ignoring limb-darkening, these two differ by exactly a factor of 4 in the quadrupole contribution. The reason for this is not known.

4.4 The Second Harmonic Frequency ν_7

The second harmonic frequency lies at $3\nu_2 + \nu_{\rm rot}$. It should arise from third order terms and hence, for a dipole fundamental oscillation, should have a surface distortion proportional to $\cos^3\theta$. This can be written as a linear sum of $P_1(\cos\theta)$ and $P_3(\cos\theta)$, but the P_3 term disappears as seen from the observer's viewpoint when $I(\theta) = 1$; otherwise the P_3 term contributes very little, so the second harmonic should look like a dipole. This means that the amplitude of $A_{-1}^{(3)}$ should be greater than the amplitude of $A_{+1}^{(3)}$ as is the case for the low frequency triplet.

It is, therefore, a surprise that the $A_{+1}^{(3)}$ component is observed and not the $A_{-1}^{(3)}$ component. This may be due to observational noise: the amplitude of ν_7 is only 0.10 ± 0.02 mmag in a region of the amplitude spectrum where there are noise peaks as high as 0.08 mmag (the probability of a false alarm at exactly $\nu_7 = 3\nu_2 + \nu_{\rm rot}$ is low, so ν_7 is real). It is therefore possible that $A_{-1}^{(3)}$ is hidden by a negative (out-of-phase) noise peak in the amplitude spectrum of the available data; only new data will determine if that is true. It is also possible that ν_7 is not part of the second harmonic of ν_{123} since we cannot show that ν_{456} is the first harmonic of ν_{123}.

4.5 Conclusions about HR 3831 and the Oblique Pulsator Model

HR 3831 is pulsating in an (ℓ=1, m=0) dipole mode with the pulsation axis and magnetic axis aligned. The pulsation amplitude modulation and the magnetic field strength modulation are in phase and have the same form; they require the same i, β geometry: $\tan i \tan\beta = 9.4 \pm 0.4$, so either i or $\beta > 80°$. The observed value of $v \sin i$ removes this ambiguity, requiring that $i > 80°$. From photometry of its companion, HR 3831 lies 0.5 mag above the ZAMS and has a radius of $R = 2.0 R_\odot$. Pulsation maximum, magnetic extremum, and mean light extremum coincide.

The oblique pulsator model accounts for the observed splitting of the pulsation frequencies, which is equal to $\nu_{\rm rot}$ (or $2\nu_{\rm rot}$) to high accuracy. It accounts for the relative amplitudes of ν_{123} and the phases of ν_1 and ν_3; the phase of ν_2 is still unexplained. The natures of the (apparent) first harmonic frequencies, ν_{456}, and second harmonic frequency, ν_7, are not yet known.

5 Period Changes in the roAp Stars

Observations of period changes, $\dot{P} \equiv dP/dt$, in pulsating stars provide a powerful tool. (*a*) Secular $\dot{P}$s are caused by changes in stellar structure and hence measure stellar evolution in real time. Winget *et al.* (1985) detected an evolutionary $\dot{P}$ in the hot pre-white dwarf star PG 1159-035 which they suggest was caused by gravitational contraction of the star during the 4.4 yr of observations. (*b*) Measurement of a periodic variation in $\dot{P}$ due to Doppler shifts is an accurate way to measure binary orbits when the pulsation frequencies are accurately known. Winget *et al.* (1990) have found a variable $\dot{P}$ for one of the pulsation frequencies of the ZZ Ceti star G29-38 which they suggest is caused by orbital motion of G29-38 about an unseen companion which probably has a mass in excess of the Chandrasekhar limit, indicating a neutron star or black hole.

The pulsation frequencies in the roAp stars can be determined to very high accuracy for the stars which pulsate in modes with long lifetimes. For example, the frequencies of highest amplitude in HR 3831 are accurate to 20 pHz (Table 1). A preliminary examination of the HR 3831 data yields $\dot{P} = (3.4 \pm 3.8) \times 10^{-11}$ s/s. The determination of $\dot{P}$ for HR 3831 is slightly complicated by the phase reversal it undergoes (see Fig. 1); a more rigorous handling of that phase reversal will lower the above limit on $\dot{P}$, as will new data.

Hence measurement of $\dot{P}$s in the roAp stars potentially can determine much about them. Heller and Kawaler (1988) have calculated $\dot{P}$s for evolutionary A star models. They show that for HR 1217 $\dot{P}$ should be detectable with the 6 yr of observations available if $\Delta\nu_0 = 34\,\mu$Hz and $\Delta M_V = 1.9$ mag, whereas it should not be detectable if $\Delta\nu_0 = 68\,\mu$Hz and $\Delta M_V = 0.7$ mag. Unfortunately, alias problems in the early single-site data for HR 1217 preclude making this test yet, but in a few years time it may distinguish between $\Delta\nu_0 = 34\,\mu$Hz and $68\,\mu$Hz and hence distinguish whether the pulsation modes in HR 1217 are basically all dipole modes, or whether they are alternating even and odd ℓ-modes.

Martinez and Kurtz (1990) have found that $\dot{P}$ in the principal frequency of HD 101065 is much larger than expected from evolutionary changes in a H core-burning A star. This could be due to a very late evolutionary state for HD 101065 if it is a unique object, rather than an extreme example of an Ap star, or it could be due to a Doppler shift. In the latter case the secondary has a mass less than $0.1\,M_\odot$ unless the orbit is within 2° of pole-on. Future observations will test this possibility.

It is interesting to note that the roAp stars which seem least interesting asteroseismologically, *i.e.*, the singly periodic stars, are the ones which are most amenable to determination of $\dot{P}$. After an initial accurate determination of the pulsation frequency in such a star, a modest yearly observing effort could yield a detectable $\dot{P}$ in 5 to 10 years. HD 137949 (33 Lib) and HD 134214 are the best candidates for this sort of study (see Kurtz 1990).

References

Carney, B.W., and Peterson, R.C. 1985, *Monthly Notices Roy. Astron. Soc.*, **212**, 33p.

Heller, C.H., and Kawaler, S.D. 1988, *Astrophys. J.*, **329**, L43.

Kurtz, D.W. 1982, *Monthly Notices Roy. Astron. Soc.*, **200**, 807.

Kurtz, D.W. 1983a, *Monthly Notices Roy. Astron. Soc.*, **205**, 3.

Kurtz, D.W. 1983b, *Monthly Notices Roy. Astron. Soc.*, **202**, 1.

Kurtz, D.W. 1990, *Ann. Rev. Astr. Astrophys.*, **28**, in press.

Kurtz, D.W., Matthews, J.M., Martinez, P., Seeman, J., Cropper, M., Clemens, J.C., Kreidl, T.J., Sterken, C., Schneider, H., Weiss, W.W., Kawaler, S.D., Kepler, S.O., van der Peet, A., Sullivan, D.J., and Wood, H.J. 1989, *Monthly Notices Roy. Astron. Soc.*, **240**, 881.

Kurtz, D.W., and Shibahashi, H. 1986, *Monthly Notices Roy. Astron. Soc.*, **223**, 557.

Kurtz, D.W., van Wyk, F., and Marang, F. 1990, *Monthly Notices Roy. Astron. Soc.*, in press.

Libbrecht, K.G. 1988, *Astrophys. J. Letters*, **330**, L51.

Martinez, P., and Kurtz, D.W. 1990, *Monthly Notices Roy. Astron. Soc.*, in press.

Matthews, J., Kurtz, D.W., and Wehlau, W. 1986a, *Astrophys. J.*, **300**, 348.

Matthews, J., Kurtz, D.W., and Wehlau, W. 1987, *Astrophys. J.*, **313**, 782.

Matthews, J.M., Slawson, R.W., and Wehlau, W.H. 1986b, in *Hydrogen Deficient Stars and Related Objects*, ed. K. Hunger, D. Schönberner, and N. Rao Kamesnara (Reidel, Dordrecht), p.313.

Shibahashi, H. 1986, in *Hydrodynamic and Magnetohydrodynamic Problems in the Sun and Stars*, ed. Y. Osaki (University of Tokyo, Tokyo), p.195.
Winget, D.E., Kepler, S.O., Robinson, E.L., Nather, R.E., and O'Donoghue, D. 1985, *Astrophys. J.*, **292**, 606.
Winget, D.E., Nather, R.E., Clemens, J.C., Provencal, J., Bradley, P.A., Wood, M.A., Claver, C.F., Robinson, E.L., Hine, B.P., Grauer, A.D., Fontaine, G., Achelleos, N., Marar, T.M.K., Seetha, S., Ashoka, B.N., O'Donoghue, D., Warner, B., Kurtz, D.W., Vauclair, G., Chevreton, M., Kanaan, A., Kepler, S.O., Augustein, T., van Paradijs, J., Hansen, C.J., and Liebert, J., 1990, *Astrophys. J.*, in press.

In Search of the Ap Instability Strip

Jaymie M. Matthews

Department of Geophysics and Astronomy, University of British Columbia,
Vancouver, Canada

Abstract: The lack of a definite excitation mechanism for the rapidly oscillating Ap (roAp) stars has hampered both the understanding of these p-mode pulsators and the photometric search for new members of the class. In an effort to identify an empirical 'instability strip' for the class, Strömgren colours ($b-y$, $u-b$, $[m_1]$, $[c_1]$, and β) of the roAp stars have been compared to colour-colour distributions of large samples of Ap stars (Vogt and Faundez 1979) and spectroscopically normal δ Scuti stars (Breger 1979). The loci of roAp stars are best defined in the $[c_1]$–$[u-b]$, $[c_1]$–$(b-y)$ and $[c_1]$–$[m_1]$ planes. The $[c_1]$–$(b-y)$ diagram suggests that roAp stars cover a range in luminosity comparable to the δ Scuti stars and share a common spectroscopic peculiarity more specific than the SrCrEu subclass to which they belong.

1 Introduction

The rapidly oscillating Ap (or roAp) stars are now recognized as excellent objects for the techniques of stellar seismology and, to date, remain the only main sequence stars other than the Sun available for such analysis. In fact, seismology of roAp stars offers a unique aspect not anticipated even for other solar-type oscillators: a probe of stellar magnetic field geometries and *interior* field strengths. The reader will find an excellent review by Kurtz of the latest observations and theory of roAp stars elsewhere in this volume.

The oscillation frequency spectra of several of the roAp stars bear striking similarities to the solar "five-minute" oscillations as seen in integrated light; both are attributed to p-mode pulsations of low degree and high radial overtone. On one hand, the strong chemical peculiarities and magnetic fields of the roAp stars represent new and powerful diagnostic tools in stellar seismology. On the other, these phenomena seem to simultaneously define *and* obscure the underlying pulsation mechanism. The roAp stars may occupy the δ Scuti instability strip, based on the similar ranges in $(b-y)$ for the two types of pulsator. But the high overtones of the roAp modes require a major difference in the pulsational characteristics of the two classes, even though Ap and δ Scuti stars should have similar masses and radii. The consequences of chemical diffusion in Ap stars make the He II ionization mechanism more difficult to invoke (Dolez *et al.* 1988) and might even introduce other κ mechanisms (Matthews 1988). The magnetic field itself may drive the pulsations (Shibahashi 1983). To make matters worse, tests of any of these ideas face the added complication that normal photometric calibrations of stellar temperature and luminosity are not valid for the line-blanketed flux distributions of cool Ap stars.

For these reasons, the theories of roAp pulsation are still too vague and numerous to guide the difficult search for new members of the class, while the few known oscillators place only weak constraints on the models. Comparison of the individual photometric indices of the roAp stars to normal-star calibrations (e.g., Kurtz 1990) does set some empirical limits on the phenomenon. So too does the possibility that all roAp stars belong to the late-A – F0 SrEuCr subclass. But neither has proven to be an unambiguous indicator of pulsation among cool Ap stars.

Therefore, I have examined samples of the known roAp variables, SrCrEu stars, other Ap stars and δ Scuti variables in various Strömgren colour-colour diagrams to (i) determine which photometric indices best isolate the roAp stars from other chemically peculiar stars, and (ii) look for similarities between the roAp and δ Scuti pulsators.

2 The Data Samples

Strömgren $b-y$, m_1, c_1, and β indices for the 14 known roAp stars were taken from Kurtz's (1990) compilation. Indices (except β) have been published by Vogt and Faundez (1979) for 341 Ap stars brighter than $V \simeq 9.5$; eight of these belong to the set of roAp stars and were removed from the 'Ap' sample. Also, a subset of 51 SrCrEu stars were identified among the remaining stars, based on spectral classifications by Bidelman and MacConnell (1973). Breger's (1979) list of δ Scuti stars was the source of the Strömgren photometry for these variables. In this list, complete sets of Strömgren indices are not available for 37 of the stars, while 16 have δ Delphini, Am(:), Ap, or "metal-deficient" spectral types. This leaves a total of 75 δ Scuti variables which should be spectroscopically (and chemically) normal.

From the observed Strömgren colours in all three samples, the dereddened indices $[c_1]$, $[m_1]$ and $[u-b]$ were calculated via the standard photometric relations (cf. Golay 1974).

The reader will observe that, in the following colour-colour planes, one roAp star is consistently set apart from the pack. Ironically, this loner is the prototype for the class: HD 101065 or Przybylski's star, the first to be discovered (Kurtz 1980). Spectroscopically, this star is peculiar even among peculiar stars. Its complex spectrum (Przybylski 1982, Wegner *et al.* 1983) suffers from exceptionally severe line blanketing that completely distorts its photometric indices. Even so, many of the arguments presented below still apply to HD 101065.

3 The Loci of roAp Stars

3.1 The $[c_1]$–$[u-b]$ Plane

Most of the stars in all three samples lie in the spectral range where both $[c_1]$ and $[u-b]$ serve as luminosity indicators for normal stars. (These indices are not independent, tied together by the relation $[u-b] = [c_1] - [m_1]$.) Figure 1 shows the $[c_1]$ vs. $[u-b]$ distributions of the different groups. The majority of the Ap stars form a fairly narrow linear envelope and the ('normal') δ Scuti stars define an even tighter parallel sequence at the lower edge of that envelope. This is not surprising given the expected correlation between the two indices. In fact, the increased scatter in the Ap sequence compared to the δ Scuti stars is almost certainly due to the range of chemical peculiarity spanned by the former and colour variations during the oblique rotation of those stars.

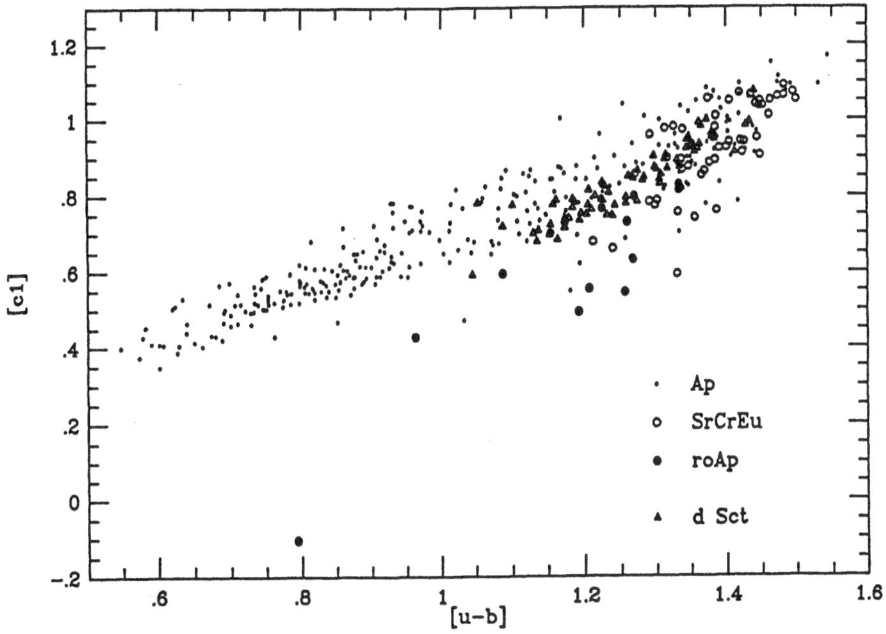

Fig. 1.

All of the roAp stars, and a few of the Ap and SrCrEu stars, fall below the main band of Ap stars. This reflects Kurtz's (1990) observation that a negative δc_1 index (relative to Crawford's (1979) calibration) is a good indicator of the roAp phenomenon. However, as Kurtz himself has noted, a serious problem with this criterion is that the more luminous roAp stars can have c_1 indices depressed by line blanketing which still result in positive δc_1. Two of the known oscillators (HD 60435 and 10 Aql) fall into this category, and other roAp stars may have escaped detection because of this. The location of cool Ap stars in the $[c_1]$ vs. $[u-b]$ plane resolves this ambiguity. Both indices are about equally sensitive to the true Balmer jump caused by surface gravity effects (i.e., in the absence of line blanketing). Therefore, actual luminosity differences between stars are not very prominent in this plane; witness the small scatter of the A-F V-III stars in the δ Scuti sample. Instead, deviations from the mean distribution should measure severe line blanketing (hence, extreme abundance anomalies) independent of luminosity. For example, in this plane, the roAp star HD 60435 ($\delta c_1 = +0.035$) coincides with another oscillator, HD 6532, whose δc_1 of -0.050 implies a lower luminosity.

3.2 The $[c_1]$–$(b - y)$ Plane

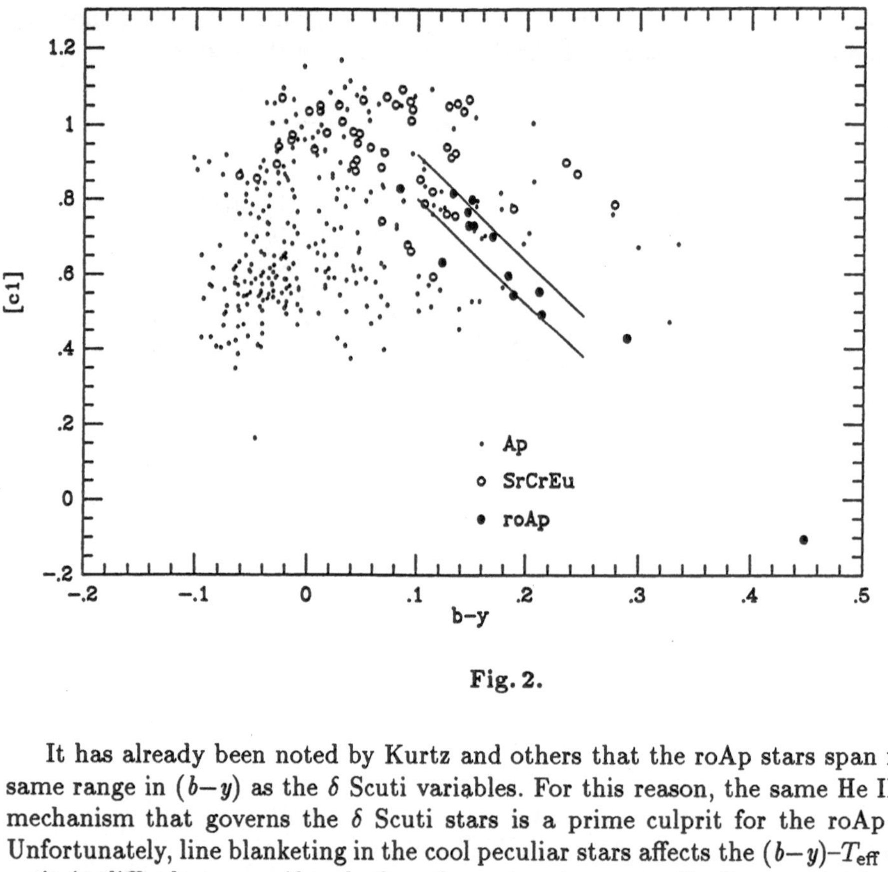

Fig. 2.

It has already been noted by Kurtz and others that the roAp stars span roughly the same range in $(b-y)$ as the δ Scuti variables. For this reason, the same He II ionization mechanism that governs the δ Scuti stars is a prime culprit for the roAp instability. Unfortunately, line blanketing in the cool peculiar stars affects the $(b-y)$–$T_{\rm eff}$ calibration, so it is difficult to specify whether the roAp stars actually lie completely within the classical instability strip. Even so, a comparison of the roAp and δ Scuti pulsators in the $[c_1]$ vs. $(b-y)$ plane – shown in Fig. 2 – is instructive. (A plot of $[c_1]$ vs. β yields similar results and is not reproduced here.) Both groups lie along parallel sequences in such a plot, which for normal stars is a luminosity vs. temperature diagram. This implies that the differences in chemical peculiarity from one roAp star to another are not so large as to destroy their intrinsic luminosity–$T_{\rm eff}$ distribution. Rather, a common type of peculiarity among these stars appears to introduce primarily systematic shifts in $[c_1]$ and $(b-y)$ relative to normal stars.

However, we already know that the known roAp stars are essentially confined to the SrCrEu subclass of peculiar stars. Does Fig. 2 simply reiterate the fact that all roAp variables show the characteristic peculiarity of SrCrEu stars? Not according to Fig. 3, in which the roAp, SrCrEu and other Ap stars are plotted in the $[c_1]$ vs. $(b-y)$ plane. Note that the Ap stars are spread across a wide distribution, reflecting their range in temperature and peculiarity. Significantly, the subset of SrCrEu stars shows a range much wider than the roAp band, particularly at the cool end of the distribution. Even allowing for a few possible spectral misclassifications and the limited size of the sample, this suggests that *the roAp stars share a common spectroscopic peculiarity even more specific than the SrCrEu subclass.*

The solid lines in Fig. 3 are models of main sequence stars with normal composition (Hauck and Mermilliod 1975, Relyea and Kurucz 1978), plotted where they overlap the $(b-y)$ range of the roAp stars. The upper line represents $\log g = 4.0$; the lower, 4.5. The surface gravities from these models cannot be safely applied to the roAp stars, but the agreement between the slope of the models and that of the roAp distribution reinforces the argument above.

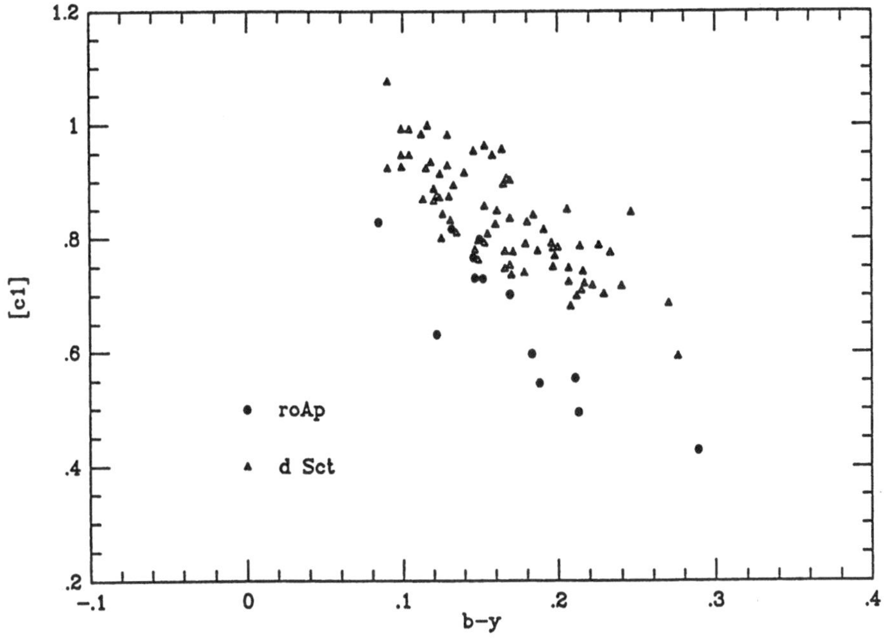

Fig. 3.

3.3 The $[c_1]$–$[m_1]$ Plane

Cameron (1967) first investigated the distribution of chemically peculiar stars (about 300 Ap and 100 Am) in the c_1 vs. m_1 plane. Figure 4 shows the Ap, SrCrEu and roAp samples discussed here plotted in the dereddened colour-colour plane. The "c_1–m_1 gap" found by Cameron centred near $(c_1,m_1) \simeq (0.8,0.17)$ is not very noticeable in Fig. 4; it was probably an artifact of Cameron's reddened sample. Both Cameron's and my diagrams show that most of the hotter Ap stars lie to the upper left of the diagonal sketched in Fig. 4. The SrCrEu stars cut across this distribution almost perpendicularly in the upper right of the figure. The known roAp stars appear confined to the lower right of the diagonal. This reflects in part the already-established fact that roAp variables have some of the smallest $[c_1]$ indices found in SrCrEu stars. However, the apparent roAp cutoff at the 'elbow' of the general Ap distribution in $[c_1]$ vs. $[m_1]$ may serve as a useful selection criterion and hold some physical significance for the pulsation mechanism.

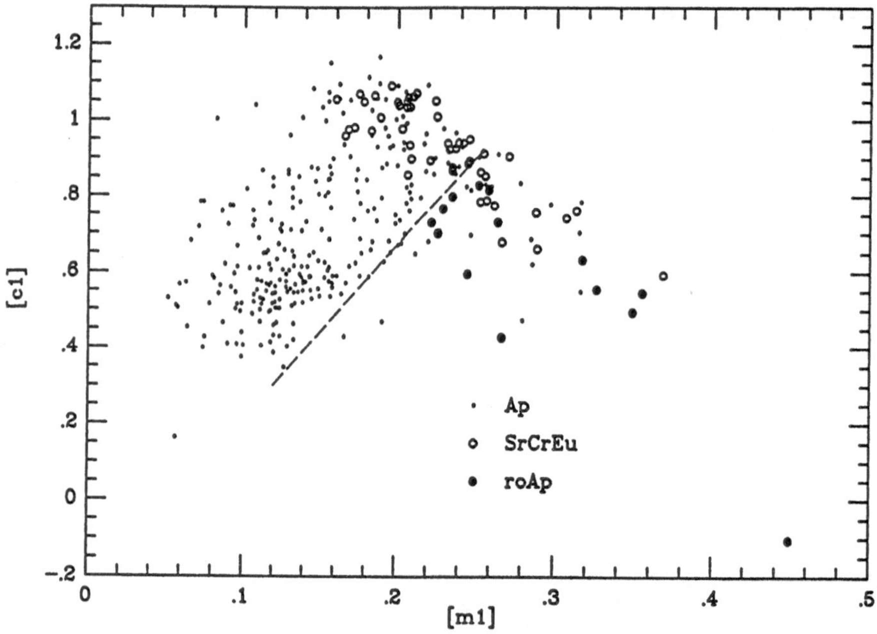

Fig. 4.

4 Conclusion

The location of cool Ap stars in both the $[c_1]$–$[u-b]$ and $(b-y)$ planes is a better indicator of the roAp phenomenon than the δc_1 index alone. The cutoff of roAp variables in the $[c_1]$–$[m_1]$ plane may be an important clue as to the physical cause of the pulsations.

References

Bidelman, W.P., and MacConnell, D.J. 1973, *Astron. J.*, **78**, 687.

Breger, M. 1979, *Publ. Astron. Soc. Pacific*, **91**, 5.

Cameron, R.C. 1967, in *The Magnetic and Related Stars*, ed. by R.C. Cameron (Mono Book Corp, Baltimore), p.471.

Crawford, D.L. 1979, *Astron. J.*, **84**, 1858.

Dolez, N., Gough, D.O., and Vauclair, S. 1988, in *Advances in Helio- and Asteroseismology*, eds. by J. Christensen-Dalsgaard and S. Frandsen (Reidel, Dordrecht), p.291.

Golay, M. 1974, *Introduction to Astronomical Photometry* (Reidel, Dordrecht), p.183.

Hauck, B., and Mermilliod, M. 1975, *Astron. Astrophys. Suppl.*, **22**, 235.

Kurtz, D.W. 1980, *Monthly Notices Roy. Astron. Soc.*, **193**, 29.

Kurtz, D.W. 1990, *Ann. Rev. Astron. Astrophys.*, in press.

Matthews, J.M. 1988, *Monthly Notices Roy. Astron. Soc.*, **235**, 7p.

Przybylski, A. 1982, *Astrophys. J.*, **257**, L83.

Relyea, L.J., and Kurucz, R.L. 1978, *Astrophys. J. Suppl.*, **37**, 45.

Shibahashi, H. 1983, *Astrophys. J. Letters*, **275**, L5.

Vogt, N., and Faundez, M. 1979, *Astron. Astrophys. Suppl.*, **36**, 477.

Wegner, G., Cummins, D.J., Byrne, P.B., and Stickland, D.J. 1983, *Astrophys. J.*, **272**, 646.

α Circini: Variability in the Infrared and Visible

Werner W. Weiss [1], Hartmut Schneider [2], Rainer Kuschnig [1],

and Patrice Bouchet [3]

[1]Institut für Astronomie der Universität Wien,
Türkenschanzstr. 17, A-1180 Wien, Austria
[2]Universitäts-Sternwarte, Geismarlandstr. 11, D-3400 Göttingen, F.R.Germany
[3]European Southern Observatory, La Silla,
Karl-Schwarzschild-Str. 2, D-8046 Garching, F.R.Germany

Abstract: Photometric observations of the rapidly oscillating Ap (CP2) star α Cir, covering the wavelength range from the UV to the near IR, indicate the detection of pulsation also at IR-wavelengths.

1 Introduction

For some of the cooler magnetic CP2 stars high overtone ($n \gg 1$), low-degree ($\ell \leq 3$) nonradial pulsation modes are observed. The periods range from about 4 minutes to 16 minutes. Up to now, fourteen of these pulsating stars, also called "rapidly oscillating Ap stars (roAp)," are known. The most recent list is given in Table 1 of Kurtz, 1990. Several reviews have been published on this subject: Kurtz (1986, 1988, 1990), Weiss (1986), Shibahashi (1987), and Matthews (1988).

One serious observational problem is posed by the very small light amplitude of the order of a few millimagnitudes or less which reflects the nonradial pulsation of these stars. With the exception of α Cir (Weiss and Schneider, 1984; Schneider and Weiss, 1989), the information on the wavelength dependence of the light amplitudes is scarce and restricted to occasional observations in Johnson V which were performed to determine the phase lag between temperature and luminosity variation. The general picture, however, fits to what is known for example for δ Sct type variables: the photometric amplitude decreases with increasing wavelengths.

Recently, Leifsen and Maltby (1988), have corroborated that for the sun too, the amplitude decreases with increasing wavelength. However, at 2.23 μm a sharp and unexpected increase of the power by a factor of about 50 was observed by them in the power spectrum in the 5-minutes frequency domain. In addition, the authors find evidence for pulsation periods around 3.9 minutes (4.3 mHz) in their whole-disk data, well beyond the known 5-minutes oscillations of our sun.

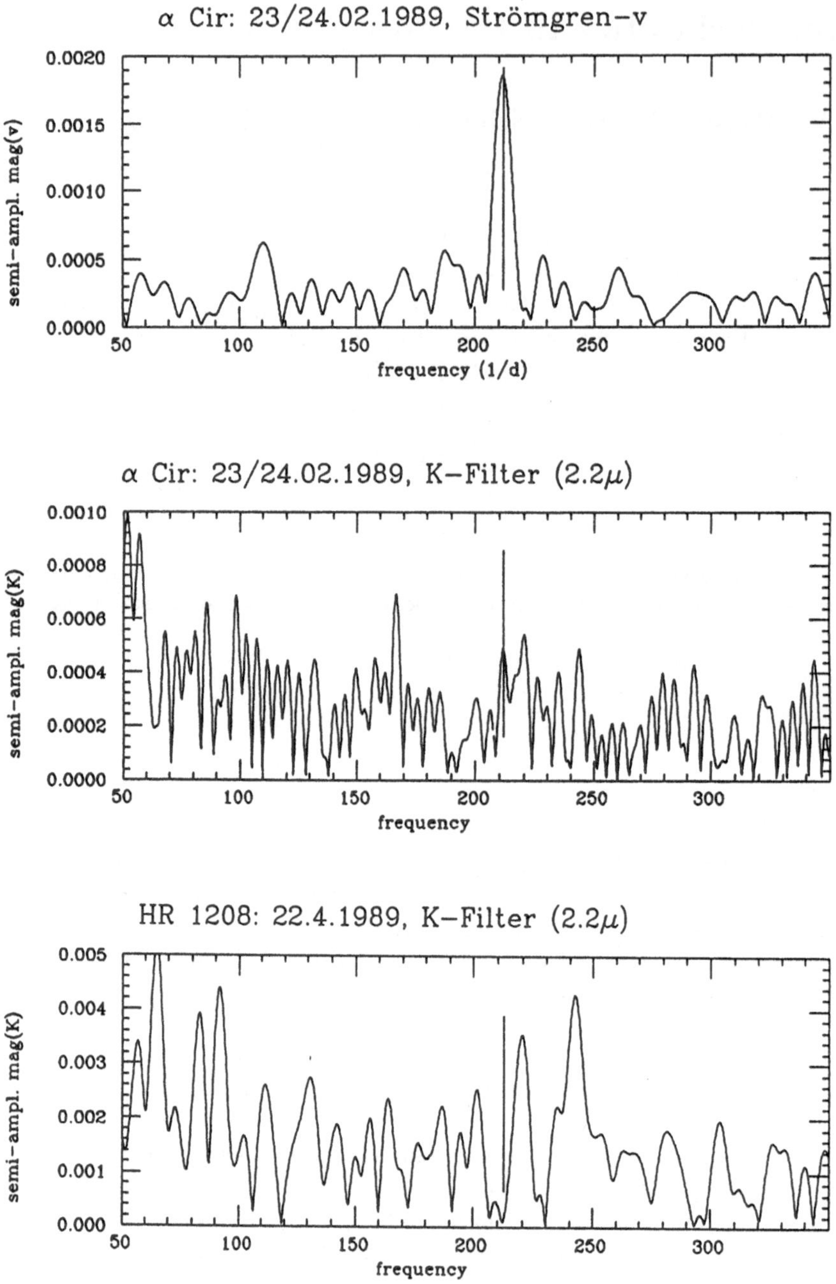

Fig. 1. Power spectrum of the Strömgren v photometry, obtained with the Bochum 0.62m telescope (upper panel) and of the IR photometry at 2.2 μm with the ESO 1.0 m telescope (middle panel) for α Cir. The Power spectrum of the IR photometry for the reference star HR 1208 is presented in the lower panel.

2 Observations

Line opacities due to molecular features cannot be expected to be relevant for CP2 star atmospheres and hence a significant photometric amplitude in the IR can not be expected for roAp stars. However, for curiosity, we observed α Cir on February 24, 1989, at La Silla with the ESO 1.0m telescope and contemporaneously with the Dutch 0.91m and Bochum 0.62m telescope. The analysis of these data determined observe the presumably constant star HR 1208 one month later (April 22, 1989) in order to check the instrumental stability. The standard ESO IR-photometer with the K-filter and a 15" diaphragm was used. A guide star was selected in the photometer offset to check regularly the tracking of the telescope while we were continuously monitoring our program star.

Power peaks at 6.54 and 5.91 minutes period are detectable in the IR photometry of α Cir. They can probably be attributed to instrumental effects, because both frequencies can be found also in our data for the reference star HR 1208 (6.54 and 5.94 min), but not in contemporaneous data in the visible region.

Contemporaneous observations in the Walraven system (V, B, L, U, W) were obtained with the Dutch 0.91m telescope (Fig. 2) during our IR observations of α Cir. The stability of the Walraven photometer was checked by observations of HD 89393 (Fig. 3) which were obtained in the same night, but have preceeded the photometry of α Cir.

The observations started with an airmass of 1.78 and ended with 1.24. Despite the rather large airmass, we obtained excellent data with a noise-level of about 0.4 mmag (Fig. 1). The data, obtained one month later for HR 1208, however, are five times as noisy, because the observations had to be done during daytime.

The Bochum 0.62m telescope was used to obtain additional contemporaneous data with a Strömgren v filter. We observed α Cir continuously during about 3.5 hours, again without using a comparison star, but with regularly checking the centering of α Cir in the diaphragm.

3 Conclusion

With a least squares sine fit we found an upper limit for the semi-amplitude of 0.49 mmag in the infrared data with a period of 6.8 minutes. This frequency is well known from photometry in the visible (Kurtz and Balona, 1984; Weiss and Schneider, 1984) and is clearly observed in our present Walraven and Strömgren photometry. In Figs. 1, 2, and 3 we have marked this pulsation frequency. But the power in our IR data exceeds the noise only marginally. It seems to be therefore justified to speak of an upper limit rather than of a true detection of the pulsation in the IR.

The strongest signal in the infrared, however, was obtained with a period of 8.63 minutes with a semi-amplitude of 0.70 mmag. Such a frequency has not yet been observed for α Cir in the visible and we can only speculate about the reality of this mode. An interference with the spectral window of our data set can be excluded, since the frequency in the interval from 100 to 300 cycles/d with the strongest power (5% of the peak magnitude) does not coincide. Furthermore, we note that no peak is present at the 8.63 minutes period for the data of our reference star HD 89393 (Fig. 3). In addition, no power at this frequency (166.9 c/d), exceeding the noise level, is present in the reference

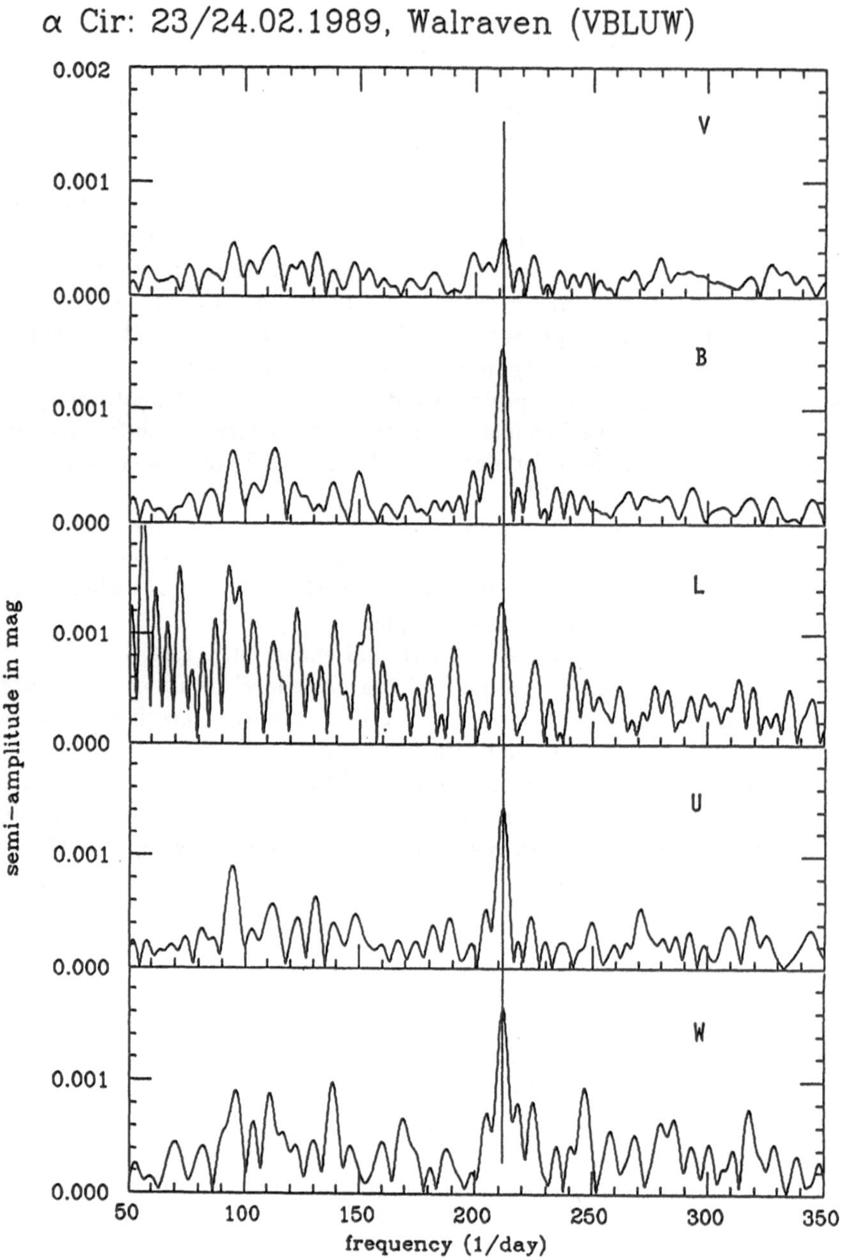

Fig. 2. Power spectrum of the Walraven photometry for α Cir, obtained with the Dutch 0.91m telescope.

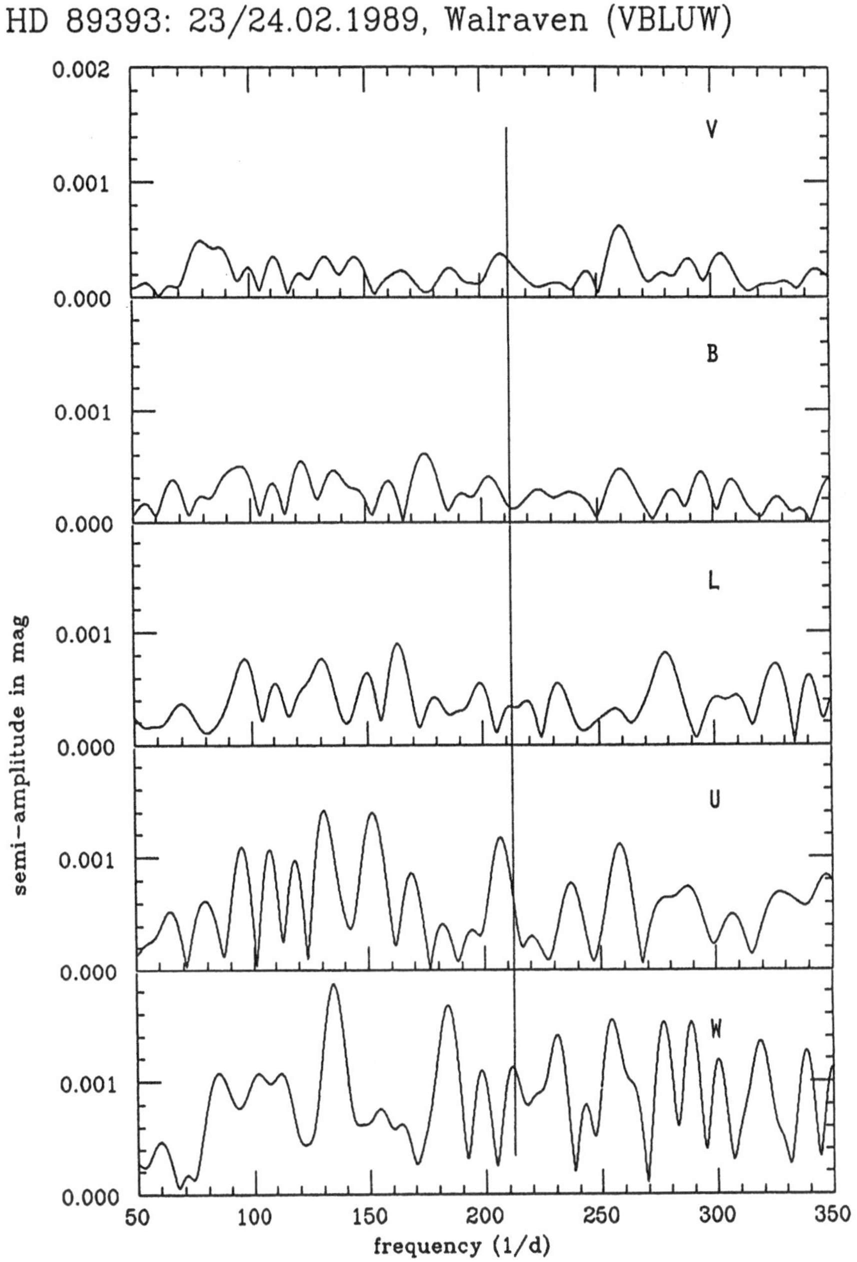

Fig. 3. Power spectrum of the Walraven photometry for the reference star HD 89393, obtained with the Dutch 0.91m telescope.

data for HR 1208 which were taken with the identical equipment during the first half of the same night. However, we cannot exclude a jet unidentified instrumental effect and we are therefore planning supplementary IR observations in the near future.

Acknowledgements

The authors gratefully acknowledge the generous support granted by ESO, by the Austrian "Fonds zur Förderung der wissenschaftlichen Forschung," project No. 6927, and by the "Österreichische Mineralöl-Verwaltung," which financed the travel expenses for two of the authors. It is a pleasure for WWW to thank the organizers for their hospitality during this excellent workshop.

References

Kurtz, D.W. 1986, in *Seismology of the Sun and Distant Stars*, ed. by D.O. Gough (Reidel, Dordrecht), p.131.

Kurtz, D.W. 1989, in *Multimode Stellar Pulsations*, ed. by G. Kovács, L. Szabados, and B. Szeidl (Konkoly Obs., Budapest), p.107.

Kurtz, D.W. 1990, *Ann. Rev. Astron. Astrophys.*, in press.

Kurtz, D.W., and Balona, L.A. 1984, *Monthly Notices Roy. Astron. Soc.*, **210**, 779.

Leifsen, T., and Maltby, P. 1988, in *Seismology of the Sun and Sun-Like Stars*, ed. E. Rolfe, ESA SP-286 (ESA Publication Division, Noordwijk), p.169.

Matthews, J. 1988, IAU General Assembly in Baltimore, Comm. 29 (WG on Ap Stars), preprint.

Schneider, H., and Weiss, W.W. 1989, *Astron. Astrophys.*, **210**, 147.

Shibahashi, H. 1987, in *Stellar Pulsation*, eds. A.N. Cox, W.M. Sparks, and S.G. Starrfield (Springer, Berlin), p.112.

Weiss, W.W. 1986, in *Upper Main Sequence Stars with Anomalous Abundances, IAU Colloq. No. 90*, eds. C.R. Cowley, M.M. Dworetsky, and C. Megessier (Reidel, Dordrecht), p.219.

Weiss, W.W., and Schneider, H. 1984, *Astron. Astrophys.*, **135**, 148.

Pulsations and Chemical Composition of Main Sequence Magnetic Stars

Sylvie Vauclair [1] and Noël Dolez [1 2]

[1]Observatoire Midi-Pyrénées, Toulouse, France
[2]C.E.R.F.A.C.S., Avenue G.Coriolis, 31100 Toulouse, France

Abstract: This paper is a discussion of the present situation concerning the abundance anomalies in Ap Stars as possible catalysts of pulsations through the κ-mechanism. A preliminary model is suggested, which is part of a work in progress, in collaboration with D. Gough.

1 Observational Constraints and Model Suggestions

The rapidly oscillating Ap stars discovered by Kurtz (1982) lie among the coolest magnetic chemically peculiar stars. They oscillate with periods between 4 and 15 minutes and the amplitude of the light variations vary, at least in some of them, in phase with the magnetic field. These observations lead Kurtz (1982) to suggest an explanation of these stars in terms of an oblique pulsator model, namely nonradial pulsations aligned with the magnetic axis, this magnetic axis itself being inclined with respect to the rotation axis. Since 1982, many observations have been pursued by Kurtz and co-workers, leading to more and more precise constraints on the periods and amplitudes of the pulsation modes: see the reviews by Kurtz (1985, 1990).

The most intriguing among pulsating Ap stars may be HR 1217 (HD 24712), for which six modes can be seen with nearly uniformly spaced frequencies of average separation 33.3 μHz, showing however a slight difference of about 1 μHz in adjacent frequency spacings (Kurtz and Seeman 1983). New results about this star are given by Kurtz (1990). The frequency spacings seem difficult to reconcile with mode frequencies computed with the asymptotic theory. Another puzzle in HD 24712 was pointed out by Shibahashi and Saio (1985). They computed the maximum frequency for which a wave could be reflected in the atmosphere of such a star and found $\nu_{\max} = 1.897$ mHz while frequencies up to $\nu = 2.4$ mHz are observed.

Thus, apart from giving an excitation mechanism for the pulsations, a model of pulsating Ap stars should be able to solve the following questions :

1) why are the amplitudes of the pulsations maximum at the magnetic pole and minimum at the magnetic equator, or why is the pulsation axis aligned with the magnetic axis ?

2) how can waves with higher frequencies than the maximum one computed by Shibahashi and Saio (1985) be reflected in the stellar atmospheres ?

3) is it possible to reconcile observed and computed frequency spacings ?

Several models have been suggested to explain these pulsations: Dolez and Gough (1982) proposed an excitation mechanism related to the abundance anomalies which are observed in these stars, and which are presumably due to element segregation. They discussed the fact that if helium was overabundant at the magnetic poles, it could trigger the pulsations in the direction of the magnetic axis, and lead to unstable modes with high n as observed. They fell across an apparent inconsistency, as element segregation leads to helium depletion at the magnetic poles while helium accumulation should be necessary for triggering the pulsations. Shibahashi (1983) suggested magnetic overstability at the poles as an excitation mechanism for the oscillations. Mathys (1985) proposed that, contrary to Kurtz's (1982) oblique pulsator model, the oscillations could be aligned with the rotation axis instead of the magnetic axis, and that the amplitude modulations could be due to abundance spots on the star, which would block part of the flux. Matthews (1988) suggested that the pulsations could be triggered by a silicon accumulation at the magnetic poles. None of these ideas lead to satisfactory explanations of the observations.

Dolez and Gough (1982) suggestion has recently been revisited (Dolez *et al.* 1986), with the added hypothesis of a stellar wind at the magnetic poles. Segregation in a stellar wind, as proposed by Vauclair (1975) to account for helium rich main-sequence stars and discussed more recently by Michaud *et al.* (1987), can lead to helium accumulation in the magnetic polar regions of pulsating Ap stars. These helium rich regions would lie too deep in the star to be observed at the surface, but they could possibly trigger the pulsations. The magnetic equator regions should be helium depleted and never show any helium accumulation, even inside the star, due to the reduction of the wind velocity by the horizontal field lines. This model looks quite promising although, as discussed below, many more computations and comparison with the observations are needed.

2 The Helium Model

Helium gravitational settling in a stellar wind was invoked by Vauclair (1975) to account for He-rich stars, which are observed at the hot end of the magnetic star sequence. These 20000 K to 25000 K stars exhibit a helium enrichment by a factor 2 to 3 which cannot be explained by diffusion alone, the upward radiative force on helium being smaller than the downward gravitational force [see Montmerle and Michaud (1976)]. Vauclair (1975) showed that an overall mass loss flux of order $10^{-12} M_\odot \,\mathrm{yr}^{-1}$ could convect helium up to the observed layers and leave it there due to the larger effect of gravitational settling when helium is neutral: then the collisions are about 50 times less effective than when it is once ionized, and helium is no more pulled out by the hydrogen flux.

Evidences of the occurrence of outflowing winds controlled by large magnetic fields in chemically peculiar stars were obtained later by Barker *et al.* (1982) for the He-rich star HD 184927 and by Brown *et al.* (1985) for the CP2 star HD 21699 (B6V). IUE spectra present a blueshifted component of CIV lines in both of these stars. The line profiles vary with time, in phase with the magnetic field, as expected [see also the discussion on the magnetically controlled winds by Casinelli and Lamers (1987)]. Using the model developed by Castor *et al.* (1981), Michaud *et al.* (1987) deduced from these observations a mass loss rate of about $5 \times 10^{-12} M_\odot \, yr^{-1}$ at the magnetic poles.

A similar process can apply for cooler magnetic stars. In case of helium gravitational settling in a mass loss flux, the helium outward motion may be reversed in the region where helium becomes neutral. However, for smaller effective temperatures, this region lies inside the star and not at the surface as in He-rich stars. The outer layers may appear He-poor, while helium accumulation occurs deeper in the star.

Computations of helium abundance variations have been done in a $1.6 \, M_\odot$ star. As 2D codes are not yet available, two different models have been computed and compared: a "polar star" which behaves like the polar regions of real stars (vertical magnetic lines) and an "equatorial star" which behaves like the equatorial regions of real stars (horizontal magnetic lines). Convection was supposed inhibited in the "polar star", not in the "equatorial star".

An example of helium abundance variations in a "polar star" is given in Fig. 1a.

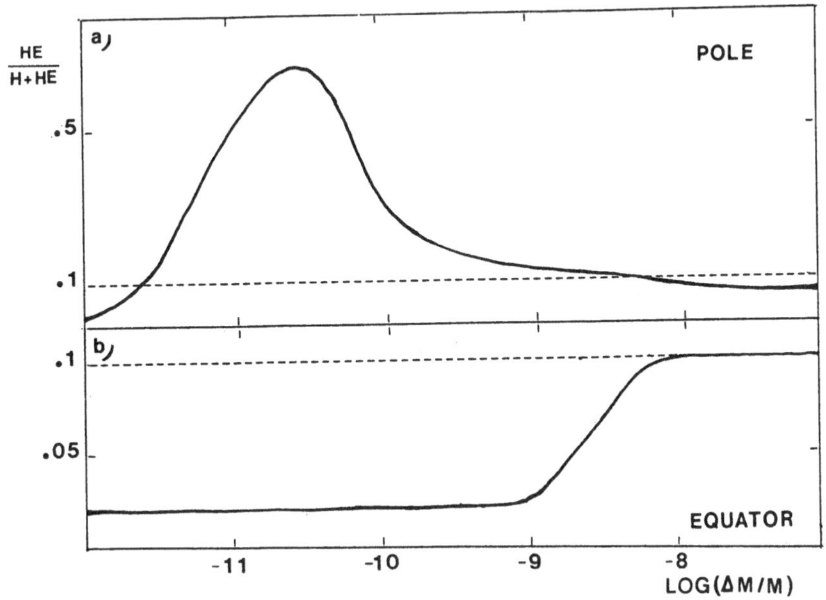

Fig. 1. Examples of helium abundance profiles at the poles and at the equator of a $1.6 \, M_\odot$ star, obtained after 10^{-6} yrs. The mass loss rate at the poles is taken as $2 \times 10^{-14} \, M_\odot \, yr^{-1}$.

A helium accumulation is obtained mostly in the first helium ionization region. The helium abundance shape depends on the mass loss flux, but also on the importance of turbulence below the helium peak. In the absence of magnetic field, such a situation is highly unstable and helium would rapidly be mixed to underneath layers. This "thermohaline convection" must be somewhat inhibited in the presence of the magnetic lines, as helium rich regions do exist in the so-called helium rich stars. To what extend it is inhibited is unfortunately an unknown parameter. The situation is quite different and more simple in the "equatorial star" where helium is depleted in the outer layers (Fig. 1b).

3 Discussion: Pulsations in Ap Stars

Preliminary computations show that a helium accumulation as given in Fig. 1a can trigger pulsations in the frequency range 1 to 1.9 mHz in our "polar stars". The instability is basically due to the small helium accumulation in the second ionization zone, i.e. around $\log(\Delta M/M) = -9.5$. The large helium peak in the first ionization zone has a much smaller effect on the destabilization of the modes. Modeling of the energy loss at the equator has to be introduced in further computations.

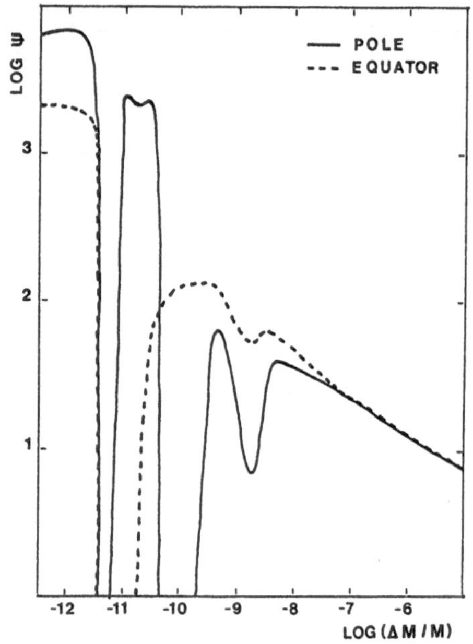

Fig. 2. Variation with depth of the function $\psi = \phi \left(GM/R^3\right)^{-1}$ in a 1.6 $M_\odot$. ϕ is defined as in Shibahashi and Saio (1985). The critical frequency is obtained as $\nu_{\rm crit} = \sqrt{\phi}/(2\pi)$. Solid lines are for the polar regions, dashed lines for the equatorial ones.

The large μ gradient induced by the helium decrease above the peak leads to an increase of the boundary in the dispersion diagram and may explain why we see frequencies larger than the "critical frequency" in HD 24712. Figure 2 displays the ϕ function as defined by Shibahashi and Saio (1985). We can see two characteristic features in the ϕ behavior at the magnetic pole compared to the magnetic equator. First the ϕ value is larger in the polar atmosphere than in the equatorial one. We have approximately :

$$\phi_{\rm pole}/\phi_{\rm eq} \simeq 1 + 2\nabla_\mu \tag{1}$$

where $\nabla_\mu = d\ln\mu/d\ln P$ [see Shibahashi and Saio (1985)].

Second there is a peak in the ϕ function in the middle of the superadiabatic region. This ϕ peak is due to the large μ gradient in the helium peak.

More sophisticated models are needed to answer the question of the frequency spacing. However, one can see from the preceding discussion that magnetic stars cannot be considered as spherical stars so that the asymptotic behavior of the modes is not applicable here. Observations of frequency spacings can be used as constraints on the non spherical models.

In summary, the helium model for pulsating magnetic stars seems quite promising, but more computations (may be with 2D codes) and comparison with observations are needed in these non spherical objects.

References

Barker, P.K., Brown, D.N., Bolton, C.T., and Landstreet, J.D. 1982, in *Advances in Ultraviolet Astronomy: Four years of IUE research* (NASA CP-2238), p.589.

Brown, D.N., Shore, S.N., and Sonneborn, G. 1985, *Astron. J.*, **90**, 1354.

Casinelli, J.P., and Lamers, H.J.G.L.M. 1987, in *Exploring the Universe with the IUE Satellite*, ed. Y. Kondo (Reidel, Dordrecht), p.139.

Castor, J.I., Lust, J.H., and Seaton, M.J. 1981, *Monthly Notices Roy. Astron. Soc.*, **194**, 547.

Dolez, N., Gough, D.O. 1982, in *Pulsations in Classical and Cataclysmic Variable Stars*, ed J.P. Cox and C.J. Hansen (JILA, Boulder), p.248.

Dolez, N., Gough, D.O., and Vauclair, S. 1986, in *Advances in Helio- and Asteroseismology*, ed. J. Christensen-Dalsgaard and S. Frandsen (Reidel, Dordrecht), p.291.

Kurtz, D.W. 1982, *Monthly Notices Roy. Astron. Soc.*, **200**, 807.

Kurtz, D.W. 1985, in *Seismology of the Sun and Distant Stars*, ed. D.O. Gough (Reidel, Dordrecht), p.441.

Kurtz, D.W. 1990, these proceedings.

Kurtz, D.W., and Seeman, J. 1983, *Monthly Notices Roy. Astron. Soc.*, **205**, 11.

Mathys, G. 1985, *Astron. Astrophys.*, **151**, 315.

Matthews, J. 1988, *Monthly Notices Roy. Astron. Soc.*, **235**, 7p.

Michaud, G., Dupuis, J., Fontaine, G., and Montmerle, T. 1987, *Astrophys. J.*, **322**, 302.

Montmerle, T., and Michaud, G. 1976, *Astrophys. J. Suppl.*, **31**, 489.

Shibahashi, H. 1983, *Astrophys. J. Letters*, **275**, L5.

Shibahashi, H., and Saio, H. 1985, *Publ. Astron. Soc. Japan*, **37**, 245.

Vauclair, S. 1975, *Astron. Astrophys.*, **45**, 233.

The Seismology of Sun-like Stars

P. Demarque and D. B. Guenther

Center for Solar and Space Research, Yale University,
P.O.Box 6666, New Haven, CT 06511, U.S.A.

Abstract: By observing the p-mode oscillation spectrum of a star and determining its first and second order frequency spacings, $\Delta\nu$ and δ, it is possible to obtain constraints which are useful in modeling the star. We show, as an example, how the positions of the frequencies of the p-modes changes from the familiar "picket fence" or regular spacing to a more random irregular spacing as the sun evolves from the ZAMS to the base of the giant branch. For α Cen, ε Eri, Procyon, Arcturus, and μ Cas we describe, in specific terms, how the frequency spacings can be used to improve our knowledge about each of these stars. We discuss the important advantages of observing the p-modes of stars in star clusters. We also discuss how rotational splittings could be used along with other stellar diagnostics to infer the rotational history of the star's interior.

1 Introduction

It is still unclear at this point what impact seismic studies will have on the future development of stellar structure and evolution for sun-like stars. This is because no conclusive observations are as yet available. The hopes and prospects are, however, still very great. In this paper, we describe several promising lines and areas of research in which seismology could become a major tool of research in this area.

2 Seismology of the Evolving Sun

As an illustration of the evolutionary behavior of the oscillation properties of sun-like stars, we present a preliminary analysis of a data set, taken from the work of D. Guenther and S. Cersosimo, which contains all non-radial oscillation modes for $\ell = 0$, 1, 2, and 3, and $n = 1$ to 49, for an evolving model of the sun. The purpose of this work is to explore the pulsation properties of sun-like stars are different phases of evolution, with an eye for possible useful diagnostics beyond the quantities derived from asymptotic theory and used by Christensen-Dalsgaard (1986, 1988) and others since then.

For the purpose of reference, Fig. 1 shows the evolutionary track of the sun in the theoretical HR-diagram. Figure 2 shows the variation with time of the quantity $\Delta\nu$, used by Christensen-Dalsgaard. It is inversely proportional to the sound travel time through

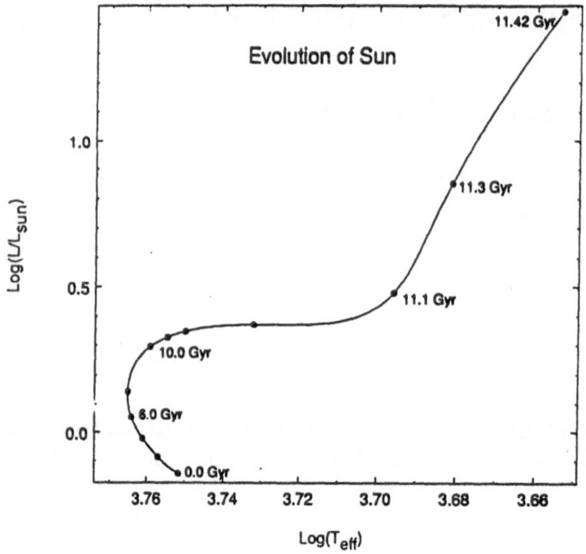

Fig. 1. Post-main sequence evolution of the sun in the theoretical HR-diagram.

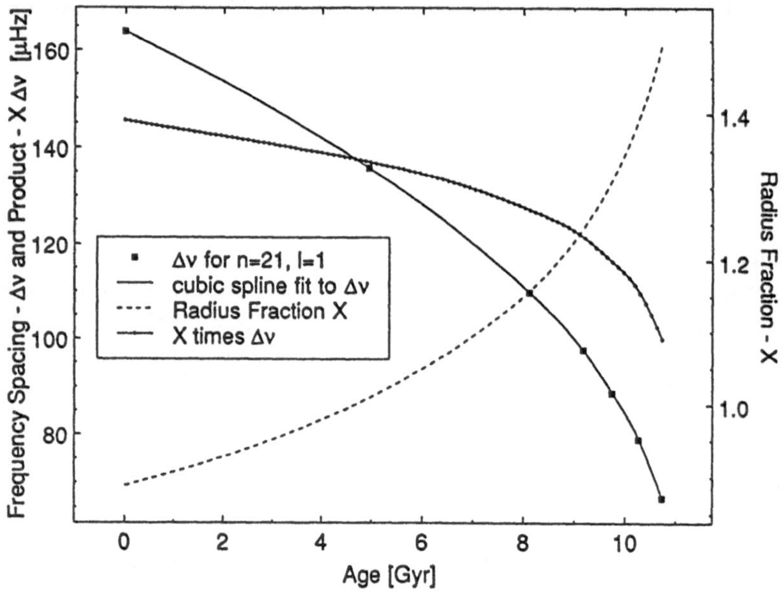

Fig. 2. Effect of the changing radius of the evolving sun on the asymptotic spacing of the p-modes.

the radius of the star. The product (radius × $\Delta\nu$) is seen to vary slowly near the main sequence, only to decrease sharply as the stellar radius begins to increase rapidly around the turnoff.

Figure 3 illustrates the behavior of the oscillation spectrum in two frequency ranges as the model evolves. The height of each frequency spike symbolizes the value of ℓ, shown here in four steps, corresponding to $\ell = 0$, 1, 2, and 3, respectively. We recognize the familiar regular "picket-fence" pattern in models near the main sequence. Also familiar is the evolving pattern of $\Delta\nu$ and δ. As the radius increases, $\Delta\nu$ decreases. As the central concentration increases, the second order δ is also seen to decrease. As the model approaches the giant branch, however, the stellar envelope expands, and a pure helium core develops, surrounded by a thin energy producing shell. As this less simple internal structure develops, the mode frequency pattern becomes more complex and loses its regularity, due to the splitting of modes into several multiplets.

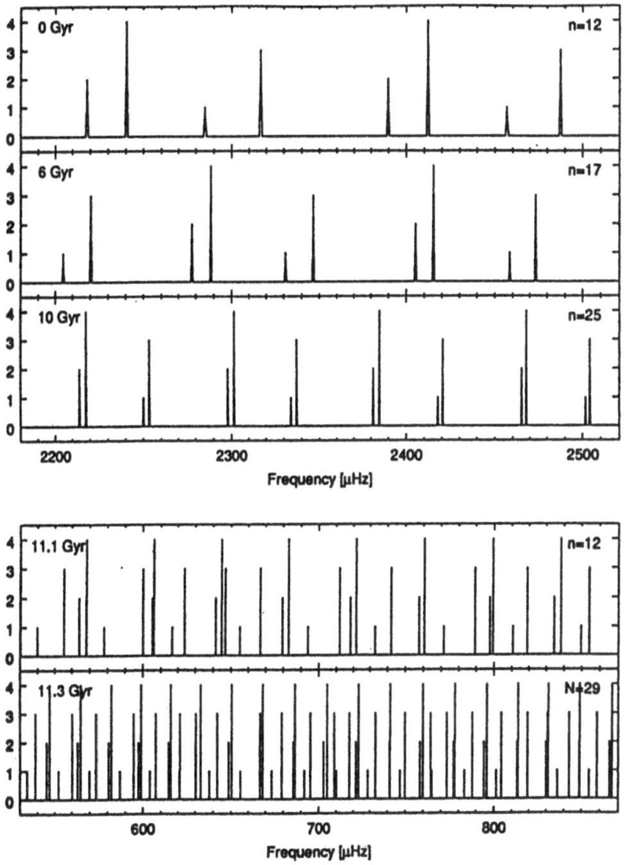

Fig. 3. Evolution of the sun's p-mode spectrum.

Indeed, Fig. 4, which introduces the changing behavior of the Brunt-Väisälä frequency as a function of time and position, shows that the nature of the modes themselves becomes more complicated, as mixing between pressure and gravity modes takes place.

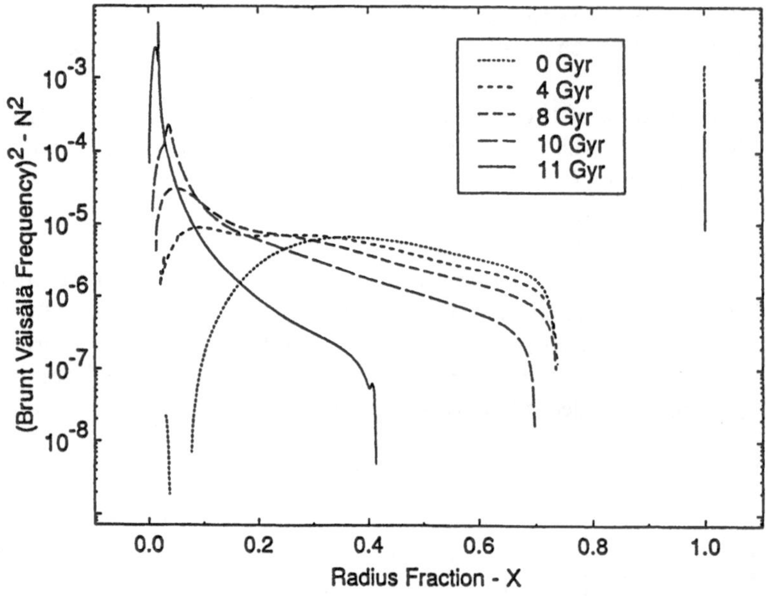

Fig. 4. Evolution of the Brunt-Väisälä frequency.

3 Some Interesting Stars

A few bright or nearby sun-like stars represent particularly interesting targets for seismic observations. Some of these stars have already been the object of intensive attempts to detect p-mode oscillations analogous to the solar oscillations, or perhaps even g-modes, as observed in PG-1159 type stars (Kawaler *et al.* 1985). Generally, the first objective is to determine $\Delta\nu$, which is a measure of the stellar radius. One could in principle also determine δ, which is a measure of central concentration. Although there is yet no convincing evidence that such oscillations have been observed in any of the cases listed below, they are listed here because they each have a special astrophysical significance. In some instances, the search for oscillations have already stimulated interesting theoretical studies.

3.1 α Centauri

Our closest neighbor is also a binary system, whose two components bracket the sun in mass (1.1 and $0.9\,M_\odot$, respectively), and are close analogues of the sun in terms of chemical composition and ages. Already, it has been noted that models constructed under solar assumptions require a smaller α than the sun to fit the radius of α Cen A (Demarque *et al.* 1986). Now that very detailed abundances (Furenlid and Meylan 1990) have been obtained, this system deserves more detailed modelling with the help of Los Alamos Opacity Library opacities specifically calculated for the purpose.

3.2 ε Eridani

This is a chromospherically active solar analogue, which suggests a young age (Noyes *et al.* 1984a). Its short rotation period is also consistent with an age of at most 1 Gyr. This has been confirmed by recent evolutionary sequences which include rotation by Guenther *et al.* (1989). These sequences, which were performed using the same rotational parameters as were needed to fit the sun (Pinsonneault *et al.* 1989) and sun-like stars in open clusters (Pinsonneault *et al.* 1990), yielded the surface rotational period of about 12 days observed by Hallam and Wolff (1981) (see Fig. 5). An interesting feature of the young models for ε Eri is the need to use a value of α much less than unity to fit the radius of the star (Guenther 1987; Soderblom and Däppen 1988), suggesting that the effective α may be affected by rotation and/or magnetic fields in this young star (Tayler 1986).

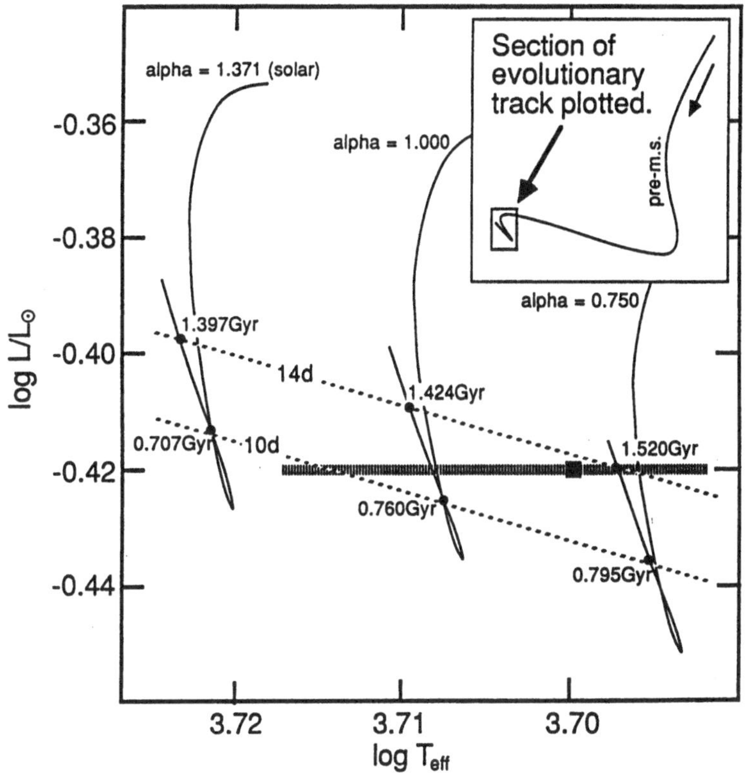

Fig. 5. Pre-main sequence evolution with rotation of ε Eri, in the HR-diagram for three values of the mixing-length parameter. Surface rotation periods and ages are marked (the observed period is 12 days). The error in the effective temperature is also shown.

3.3 Procyon

Procyon is another star whose oscillations have been the object of a great deal of attention, so far inconclusively. The theoretical work of Christensen-Dalsgaard and Frandsen (1983) had predicted that the p-mode oscillations should reach their maximum amplitude for F stars, which have very shallow convection zones. Detailed modelling motivated by these observations has already revealed some information about the extent of overmixing at the edge of the convective core due to overshoot, and has suggested seismic tests of its evolutionary status (Demarque and Guenther 1986, 1988a). Figure 6 illustrates the possible stages of evolution of Procyon. Pre-main sequence, labelled 0, which would assign an age less than 10 Myr, seems ruled out by the existence of Procyon's white dwarf companion. The core burning stage, labeled 1 is the most likely, since it corresponds to the longest time scale. Stages 2 and 3, core exhaustion and shell narrowing phases, respectively, could only be identified unequivocably by seismology. In this case, δ from the p-modes, and even more so the spectrum of g-modes, are strongly affected by the structure of the core.

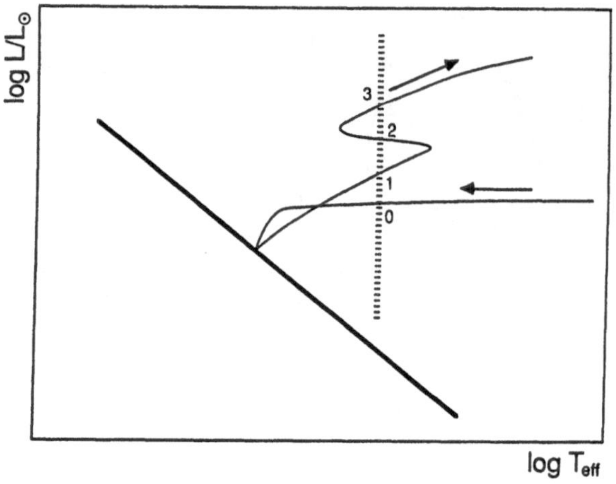

Fig. 6. Theoretical HR-diagram indicating the four possible phases of evolution of Procyon. The arrows show the direction of evolution.

3.4 Arcturus

Arcturus is a red giant with [Fe/H]=-0.7. This metallicity is found both among the metal-rich disk globular clusters (e.g. 47 Tuc) and the most metal-poor open clusters (e.g. NGC 2420), and therefore leaves the origin and therefore the age of Arcturus ambiguous. A determination of its mass through seismology would provide an important clue about the age of field stars in the galactic thick disk, which is currently a subject of controversy (Sandage 1986; Norris and Green 1989). The question is: is Arcturus like the red giants in the disk globular cluster 47 Tuc, with a mass near $0.8\,M_\odot$ and an age of about 13

Gyr? or should it be assigned to the same galactic population as NGC 2420, whose red giants have masses near $1.3\,M_{\odot}$, and an age closer to 5 Gyr?

3.5 μ Cas

A visual binary system of the halo population, μ Cas could provide the first direct mass estimates for halo stars, once its orbital parameters will have been determined with sufficient accuracy with HST (Hegyi *et al.* 1989). The aim, then, together with precise metallicity determination, is to derive the helium abundance. This would in turn place an upper limit on the primordial abundance of helium. Although in principle, a full solution for the system could be derived from the HST observations alone, uncertainties in the temperature scale for low metallicity stars cause an uncertainty in the age derived, which then translates to an uncertainty in the helium abundance. The additional constraint of p-mode observations to test the radii and central concentrations of both members of μ Cas would place very rigorous limits on the primordial helium abundance of this halo stars.

4 What We Can Learn from Seismic Observations in Star Clusters

4.1 Open Star Clusters in the Galactic Disk

The traditional advantages of studying stars in star clusters, which are practically coeval, and of uniform chemical composition, and which display a well understood range of stellar masses, apply also to seismology. We can find in our own galactic neighborhood a variety of clusters, such as α Persei, the Pleiades, Coma, the Hyades, NGC 752, NGC 188, NGC 6791, whose ages range from about 1 Myr to that of the sun or older, and with metallicities close to solar. These clusters offer as a group the opportunity to study the seismic properties of sun-like stars in considerable details as they evolve. Combined with other types of observations (see also Section 5), there are three traditional types of seismic data in well studied clusters which could provide unique information. The first two are derived from the asymptotic theory, the parameters $\Delta\nu$ and δ. They are:

(1) From the evolution of the $\Delta\nu$ parameter, which yields directly the evolution of the radius as a function of time, *i.e.*, $R(t)$. This is the weak link of stellar evolution theory, and in using theoretical data to interpret stellar populations (Demarque *et al.* 1988). This problem is particularly accute in sun-like stars, which have surface convection zones, with poorly understood superadiabatic layers in their outer parts. Also poorly understood are the outer radiative layers of cool stars, with a complex interaction of molecular opacities, convection, magnetic activity and winds.

(2) From the evolution of the δ parameter, the changes in the degree of central concentration caused by evolution. This is a test of the mass distribution in the core, and presents a direct test of stellar evolution theory.

The third parameter is simply the splitting of oscillation modes due to the Doppler effect, each mode probing the rotational state of different regions of the star's interior. Section 5 contains more discussion of rotation. Finally, star clusters provide a controlled testing ground for the kind of exploration for diagnostics of internal structure evolution that are described in Section 2.

4.2 The Special Case of Hyades

The Hyades star cluster provides a special opportunity to use seismology as a test of stellar structure theory. It is sufficiently nearby for its distance to be measured by several astrometric means. It is also relatively young (about 500 Myrs), with a well defined main sequence, with only few evolved stars, including a few red giants, and some white dwarf remnants. These reasons make the Hyades uniquely suited for a combined astrometric and seismic approach.

4.2.1 Astrometry

Astrometry provides three independent ways of estimating the distance to the Hyades: direct trigonometric parallaxes of individual cluster members, the moving cluster convergent point method, and the presence of binary systems with known orbital periods. Because the Hyades are near the limit of parallax determination by traditional techniques, the uncertainties of this method are large. But there are two other methods:

(a) the the convergent point method. This method has been refined over the years, and yield a distance modulus for the cluster (van Bueren 1952).

(b) the binary orbits yield a mass-luminosity relation, as well as a mass-radius relation, which can be used to derive the distance modulus. One requires consistency between (1) and (2) (McClure 1982). With the help of theoretical isochrones based on stellar evolution tracks, and using the value of Z derived from spectroscopy, one can then derive the helium abundance Y. At the same time, the luminosity-radius relation yield an estimate for $< \alpha >$, the mean effective mixing-length parameter within the cluster. With enough data, possible variations with mass of $< \alpha >$ could be explored.

4.2.2 Seismology

Note that along the main-sequence, the ratio M/R is nearly constant, and we have therefore:

$$\Delta\nu \propto \left[\frac{GM}{R^3}\right]^{1/2} \propto \frac{1}{R}. \tag{1}$$

From the stellar models, we know that:

$$R\Delta\nu = c(Y, Z, t). \tag{2}$$

It follows that from the observed $\Delta\nu$'s, one can derive R for individual stars, and therefore also derive the effective α for each individual star. It must be understood that the specific value of α is strictly a function of the details of the particular stellar structure code, and has therefore no fundamental significance in itself, except as a relative measure of convective efficiency in the outer layers of the convection zone.

5 The Role of Internal Rotation

There is observational evidence that internal rotation plays an important role in the evolution of sun-like stars. The following lines of evidence lead to this conclusion:

(a) the distribution of surface rotational velocities among pre-main sequence stars in young star clusters, and the changes in this distribution as a function of time, as evidenced by observations of clusters of different ages (Soderblom 1983; Stauffer and Hartmann 1986, 1987).

(b) the associated pre-main sequence distribution of the abundances of Li and Be as a function of mass and age among cluster stars (Boesgaard and Budge 1988a, b; Hobbs and Pilachowski 1988; Balachandran *et al.* 1988; Rebolo and Beckman 1988).

(c) the near constancy of the rotational periods of field subgiant stars, which reveal the progressive dredge-up of angular momentum from their deep layers, as their radii increase (Noyes *et al.* 1984b; Demarque and Guenther 1988b).

(d) the evidence that mixing takes place of CNO processed elements from the interior to the stellar surface. Although this mixing could be the result of processes other than rotationally induced currents or instabilities, rotation provides a natural explanation for these observations (Sweigart and Mengel 1979; Smith 1987). This is particularly so in the light of direct evidence for the presence of rotation in both earlier and later phases of evolution.

(e) the observations of rotational velocities in blue horizontal-branch stars, corresponding to rapid rotation in the cores of their main sequence progenitors. This is so in spite of a complex internal evolution, and likely opportunities for angular momentum loss during their giant evolution (Peterson *et al.* 1983; Peterson 1985).

(f) the evidence that some white dwarf stars still rotate at a substantial rate (Pilachowski and Milkey 1987).

It is possible, through a phenomenological approach to make progress in understanding the rotational history of sun-like stars (Endal and Sofia 1978, 1981; Pinsonneault *et al.* 1889). The most detailed recent studies reveal that it is possible to learn a great deal about the rotational history of stars, and the effects of rotation on evolution by constraining the models with the available data on surface rotation rates, the abundances of light elements and CNO cycled isotopes. The available models clearly show that the destruction of Li in particular is proportional to the magnitude of the initial total angular momentum J_0. Li variations thus are primarily a measure of the distribution in J_0. They also indicate that on the other hand, the mixing of CNO processed elements depends on the amount of angular momentum trapped in the deep interior by the presence of a gradient in mean molecular weight μ caused by the conversion of hydrogen into helium. Only later in the evolution, on the giant branch, is part of this angular momentum transported out as the original μ gradient has been removed by hydrogen burning. It is thus possible to constrain the initial distribution in J_0 and the form of the wind law, to set limits on the time-scale of angular momentum transfer in stellar interiors, to gain some insight on the efficiency of mixing as J is transferred inside stars and of μ gradients in isolating the core.

Seismic observations would add a new dimension in this area, in a direct way by providing estimates of current internal rotation through mode splitting measurements, and indirectly by testing the accumulated effects of the rotational history on the internal structure.

Acknowledgements

Financial support from NASA grants NAGW-777 and NAGW-778 to Yale University is gratefully acknowledged.

References

Balachandran, S., Lambert, D.L., and Stauffer, J.R. 1988, *Astrophys. J.*, **333**, 267.
Boesgaard, A.M., and Budge, K.G. 1988a, *Astrophys. J.*, **325**, 749.
Boesgaard, A.M., and Budge, K.G. 1988b, *Astrophys. J.*, **332**, 410.
Christensen-Dalsgaard, J. 1986, in *Seismology of the Sun and the Distant Stars*, ed. D.O. Gough (Reidel, Dordrecht), p.23.
Christensen-Dalsgaard, J. 1988, in *Advances in Helio- and Asteroseismology*, IAU Symp. No. 123, eds. J. Christensen-Dalsgaard and S. Frandsen (Reidel, Dordrecht), p.8.
Christensen-Dalsgaard, J., and Frandsen, S. 1983, *Solar Phys.*, **82**, 469.
Demarque, P., and Guenther, D.B. 1986, in *Cool Stars, Stellar Systems, and the Sun*, eds. M. Zeilik and D.M. Gibson (Springer, Berlin), p.187.
Demarque, P., and Guenther, D.B. 1988a, in *Advances in Helio- and Asteroseismology*, IAU Symp. No. 123, ed. J. Christensen-Dalsgaard and S. Frandsen (Reidel, Dordrecht), p.287.
Demarque, P., and Guenther, D.B. 1988b, in *Seismology of the Sun and Sun-like Stars*, ed. E. Rolfe (ESA-286, Noordwijk), p.99.
Demarque, P., Guenther, D.B., King, C.R., and Green, E.M. 1988, in *Calibration of Stellar Ages*, ed. A.G.D. Philip (L. Davis Press, Schenectady), p.101.
Demarque, P., Guenther, D.B., and van Altena, W.F. 1986, *Astrophys. J.*, **300**, 773.
Endal, A.S., and Sofia, S. 1978, *Astrophys. J.*, **220**, 279.
Endal, A.S., and Sofia, S. 1981, *Astrophys. J.*, **243**, 625.
Furenlid, I., and Meylan, T. 1990, *Astrophys. J.*, **350**, 827.
Guenther, D.B. 1987, *Astrophys. J.*, **312**, 211.
Guenther, D.B., Demarque, P., Tucker, D., and Pinsonneault, M.H. 1989, in preparation.
Hallam, K.L., and Wolff, C.L. 1981, *Astrophys. J. Letters*, **248**, L73.
Hegyi, D., Demarque, P., Kurucz, R., and Sneden, C. 1989, approved HST observing proposal.
Hobbs, L.M., and Pilachowski, C. 1988, *Astrophys. J.*, **334**, 734.
Kawaler, S.D., Winget, D.E., and Hansen, C.J. 1985, *Astrophys. J.*, **298**, 752.
McClure, R.D. 1982, *Astrophys. J.*, **254**, 606.
Norris, J.E., and Green, E.M. 1989, *Astrophys. J.*, **337**, 272.
Noyes, R.W., Baliunas, S.L., Belserene, E., Duncan, D.K., Horne, J., and Widrow, L. 1984a, *Astrophys. J. Letters*, **285**, L23.
Noyes, R.W., Hartmann, L.W., Baliunas, S.L., Duncan, D.K., and Vaughan, A.H. 1984b, *Astrophys. J.*, **279**, 763.
Peterson, R.C. 1985, *Astrophys. J. Letters*, **294**, L35.
Peterson, R.C., Tarbell, T.D., and Carney, B.W. 1983, *Astrophys. J.*, **265**, 972.
Pilachowski, C.A., and Milkey, R.W. 1987, *Publ. Astron. Soc. Pacific*, **99**, 836.
Pinsonneault, M.H., Kawaler, S.D., and Demarque, P. 1990, *Astrophys. J. Suppl.*, in press.
Pinsonneault, M.H., Kawaler, S.D., Sofia, S., and Demarque, P. 1989, *Astrophys. J.*, **338**, 424.
Rebolo, R., and Beckman, J.E. 1988, *Astron. Astrophys.*, **201**, 267.
Sandage, A.R. 1986, *Ann. Rev. Astron. Astrophys.*, **24**, 421.

Smith, G. 1987, *Publ. Astron. Soc. Pacific*, **99**, 67.
Soderblom, D. 1983, *Astrophys. J. Suppl.*, **53**, 1.
Soderblom, D., and Däppen, W. 1988, in *Advances in Helio- and Asteroseismology*, IAU Symp. No. 123, eds. J. Christensen-Dalsgaard and S. Frandsen (Reidel, Dordrecht), p.281.
Stauffer, J.R., and Hartmann, L.W. 1986, *Publ. Astron. Soc. Pacific*, **98**, 1233.
Stauffer, J.R., and Hartmann, L.W. 1987, *Astrophys. J.*, **318**, 337.
Sweigart, A.V., and Mengel, J.G. 1989, *Astrophys. J.*, **229**, 624.
Tayler, R.J. 1986, *Monthly Notices Roy. Astron. Soc.*, **220**, 793.
van Bueren, H.G. 1952, *Bull. Astr. Inst. Netherlands*, **11**, 385.

Acoustic Oscillations in Main-Sequence Stars: HD155543

J. A. Belmonte, F. Pérez Hernández, and T. Roca Cortés

Instituto de Astrofísica de Canarias, 38200 La Laguna, Tenerife, Spain

Abstract: High-speed photometric techniques have been found useful as a way to study the acoustic mode signature in low main sequence stars. In this work, the discovery of solar-like oscillations associated to the presence of acoustic modes of pulsation in the F2V star HD155543, located outside of the instability strip, is reported. This finding has been obtained through an analysis of a long series of data (184 hours) obtained in 20 nights of observation with two twin three channel photometers attached to two 1.5 m telescopes sited at two observatories, simultaneously: Teide (OT) at Tenerife (Spain) and San Pedro Mártir (SPM) at Baja California Norte (Mexico). The major results yielded have been: the range of frequencies where p-modes signal is present (1 to 3 mHz); an upper limit of 20 μmag for the amplitude of the modes; the mean spacing between modes of equal degree ℓ and consecutive order n, $\Delta\nu_0 = 97.3 \pm 0.6$ μHz and two possible values of D_0, 1.4 or 1.8 μHz. The values of these parameters agree, within the resolution, with those yielded by standard computed models of main sequence stars compatible with the luminosity and effective temperature already known for HD155543. These results open new perspectives for astero-seismology in the near future.

1 Introduction and Observations

As an extension of the discovery of the rich solar p-mode spectrum of oscillations (Claverie *et al.*, 1979), several attempts have been conducted pursuing the discovery of oscillations in stars other than the Sun. First observational attempts were conducted by Traub, Mariska and Carleton (1978), using an interferometric technique, and by Deubner and Isserstedt (1983) using wide band photometry; however, both failed to obtain a positive result. On the other hand, theoretical predictions of amplitudes and frequencies to be expected have been made by Christensen-Dalsgaard and Frandsen (1983), (hereafter CDF). Nevertheless, marginally positive results have been reported in literature using different spectrophotometric techniques, e.g. Noyes *et al.* (1984), on ϵEri (K2V); Gelly *et al.* (1986), on αCMi (F5IV) and αCen (G2V).

In this paper, we report the discovery of the presence of acoustic oscillations in a main sequence star, the F2 dwarf HD155543, by means of rapid continuous photometry.

The observations on this star were obtained during a three weeks observing run in May and June 1987, using two twin three-channel photometers attached to the Cassegrain foci of the 1.5m Carlos Sánchez Telescope at OT, and the 1.5m telescope at SPM. These observations of an F2 dwarf were part of a two-site campaign of observation of the δScuti star 63Her and the aforementioned HD155543, using the technique of rapid differential photometry. A general description of that campaign including: instrument description, general performance, reduction and analysis techniques, results on 63Her and preliminary results on HD155543 can be found in Mangeney *et al.* (1988) and Belmonte *et al.* (1989), whilst a fine description of the analysis performed on HD155543 and the results yielded can be found in Belmonte *et al.* (1989). In outline, we were able to obtained 184 hours of high quality data on both stars, which represents a duty cycle of ~38% of the possible coverage of ~20 days span of data available.

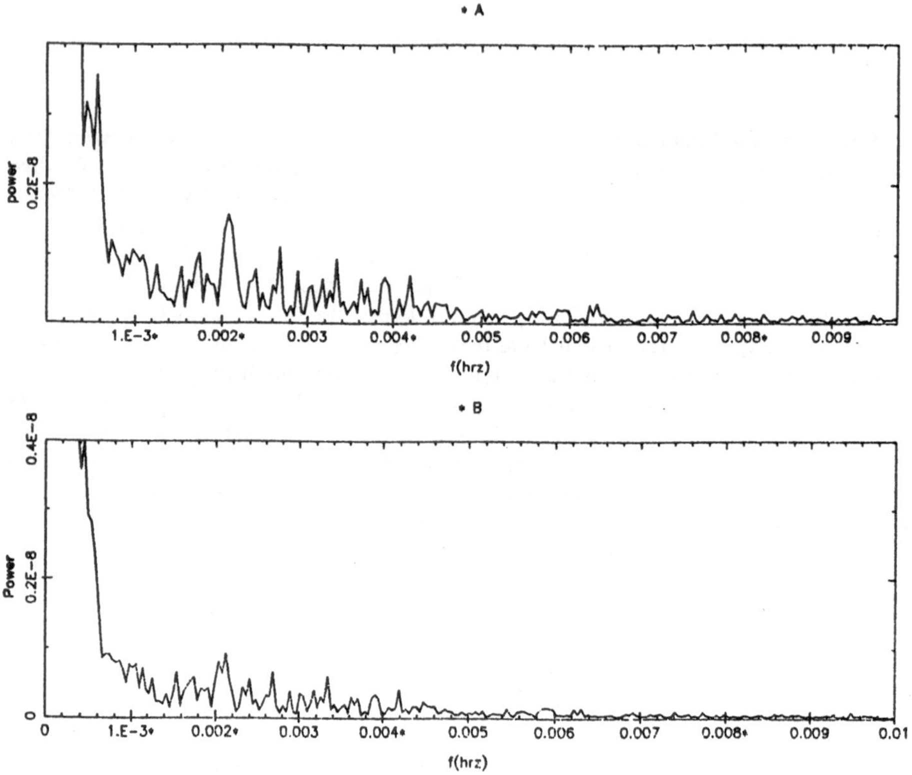

Fig. 1. Average power spectra of the data obtained at OT for HD155543 (* A) and 63Her (* B). Notice the excess of signal in the uppermost plot and its power level.

2 Data Analysis

The raw data consist of six second integrations obtained continuously throughout the night. These were converted to instrumental magnitudes and then corrected for the mean atmospheric extinction using standard least squares fit to airmass. Series of rapid photometric residuals were finally obtained. For an extensive description of these procedures see Belmonte (1989) and Belmonte *et al.* (1989). The power spectrum of each residual

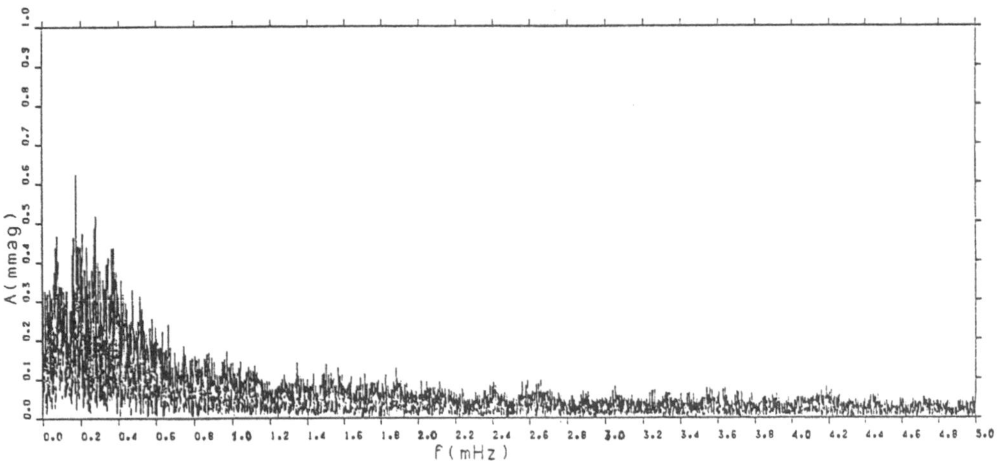

Fig. 2. Amplitude spectrum of the 184 hour time span series of observation of the F2V star HD155543. Data are reduced in a standard way and further corrected for long-term sky transparency variations by a 1 hour moving mean filter.

nightly series was computed and iteratively added and averaged. The results of this analysis, for data taken at OT, is shown in Fig. 1, where the average spectrum of HD155543 (* A) is plotted together with that obtained by the same procedure on 63Her (* B). Both spectra show a similar pattern of peaks and noise level; however, an excess of signal $\leq 10^{-9}$ in power (≤ 30 μmag in amplitude), in the frequency range from 1 to 3.5 mHz, is present in the uppermost panel (HD155543), perhaps related to the presence of actual stellar signal in the data taken on the F2V star.

In parallel, residual series of HD155543 were also corrected for long-term sky transparency variations through numerical filtering with a $\sim$1 hour moving mean. Series from different nights and observatories were then put together to produce a single time series, 184 hours long. This series was analyzed via an Iterative Sine Wave Fitting (hereafter ISWF) procedure for unequally spaced data, obtaining its amplitude spectra at a sampling frequency of 0.578 μHz, (equivalent to $\delta\nu$=20 d^{-1}). The amplitude spectrum of the data corrected with the moving mean filter is presented in Fig. 2. As it can be seen, atmospheric transparency still plays an important role at lower frequencies while, for frequencies higher than 2 mHz, the spectrum is slowly decreasing, with a noise level well below 100 μmag, that should be mainly ascribed to atmospheric scintillation effects.

Following CDF, assuming an excitation mechanism by turbulent convection, amplitudes of several tens of μmag would be expected for a star like HD155543; so, if present, the p-mode spectrum will be hidden amongst the noise. However, from the asymptotic theory, the acoustic modes for $n \gg \ell$ have the property of being roughly equally spaced in frequency (Tassoul, 1980), hence producing a remarkable signature in the frequency domain. In order to look for this, several spectra of the amplitude spectrum itself were computed (Pallé ,1986; Pérez Hernández, 1986; Gelly *et al.*, 1986). The range of frequencies from 0.2 to 5 mHz was considered and further divided into intervals of different width (~1.5 mHz) and its spectrum calculated again via ISWF. The only structure present in all the spectra was one around 12 μHz which is due to the 11.57 μHz separation of the daily sidebands, being a consequence of the observing window function. The other best signal appear in the intervals from 1 to 3 mHz, where a clear peak was present at 49 μHz. From theoretical predictions (see CDF) for an F2 Zero Age Main Sequence (hereafter ZAMS) star, the last value would, in principle, be possible for either $\Delta\nu_0$ (mean separation between acoustic modes of consecutive n and equal degree ℓ) or its half value, $\Delta\nu_0/2$.

Indeed, in order to estimate the range of frequencies where the presence of p-modes could be expected in the observed star, and its corresponding frequency spacing ($\Delta\nu_0$), p-mode frequencies have been calculated for several stellar models using the code of Christensen-Dalsgaard (1982).

The values reported in literature (see Belmonte *et al.*, 1989) for the absolute visual magnitude (hence the luminosity) and for the effective temperature of HD155543 are: M_v=3.32±0.29 ($L = 3.7 \pm 1L_\odot$), $T_{\rm eff} = 6700 \pm 200$ K. From a grid of models of mass from 0.7 to 2.0 $M_\odot$ main sequence stars, $M = 1.35M_\odot$ and $M = 1.4M_\odot$ provide the extreme values of the mass of the star in question (at any age) that match the above stated values for the luminosity and effective temperature; also a model with $M = 1.3M_\odot$ gives values of L and $T_{\rm eff}$ at the limit of the observational ones, at an age of ~1.4 Gyr. Values of X_s=0.7335 and α=1.635 have been considered for the hydrogen abundance by mass and the mixing length parameter respectively. These values are such that, for a model of $M = 1M_\odot$ and an age (hereafter τ) of 4.75 Gyr, the code gives the radius and the luminosity for the present Sun with a relative error below 10^{-5}. The frequencies have been calculated using the linear and adiabatic approximations. Only modes with $0 \leq \ell \leq 3$ (the only degrees one would expect to observe due to the absence of spatial resolution) have been computed. From the results yielded by the computations of stellar models, we would expect the range of highest amplitude p-modes to be $1.1 \leq \nu \leq 3.3$ mHz; while values between 40 to 51 μHz and from 1.1 to 1.9 μHz are expected for $\Delta\nu_0/2$ and D_0 (see Christensen-Dalsgaard, 1988), respectively.

3 Discussion

When performing the spectrum of the spectrum of helioseismological high quality data a large dominant peak is found at a frequency separation of 68 μHz (Pérez Hernández, 1986), which is actually $\Delta\nu_0/2$ for the Sun. Similarly, for the acoustic modes of an F2V star, their most significant feature would be the half value of the frequency spacing $\Delta\nu_0$. The most significant peak yielded by the previous analysis is found precisely for a frequency separation of 49 μHz. Moreover, this value was the dominant when the frequency intervals from 1 to 3 mHz were analyzed.

To choose the signals with highest amplitudes present, several frequency intervals of 0.8 mHz width, in the range between 1.4 and 3 mHz with a shift of 0.2 mHz between them, were analyzed looking for the p-mode signature. In all plots, the most significant peak was centred at $S = 49\,\mu$Hz; however, the spectrum of the spectrum for the interval 2.0-2.8 mHz was found to include the clearest signal. A value of $\Delta\nu_0/2$=49 μHz can thus be considered for the acoustic modes of oscillation of HD155543, with the maximum signal around 2.4 mHz. A cut-off frequency slightly higher than 3 mHz can be expected, since the frequency intervals over 3 mHz show a decrease in the amplitude of the 49 μHz peak.

As a further test, series of artificial data using the theoretical frequencies from the models computed in the preceding section were produced. Since the $1.4 M_\odot$ ZAMS model has a $\Delta\nu_0/2$=49 μHz, it was decided to use the frequencies computed with this model to create a series of artificial data. A sum of sinewaves over all modes with degree $0 \le \ell \le 3$ and frequency between 2.0 and 2.8 mHz were produced using random phases; no noise was added and their relative amplitudes of, 82, 98, 55 and 8 % for modes of ℓ=0, 1, 2, 3 respectively, were chosen to take into account the mode sensitivity of the observations, since we were observing photometrically disk-integrated light. The artificial time series was submitted to the same observational window as the measured one. These data were analyzed via ISWF as the observational data series did.

The amplitude squared spectrum of the amplitude spectrum between 2.0 and 2.8 mHz was recomputed for both, theoretical and observational series. A sampling interval of 416 seconds (corresponding to a third of the intrinsic resolution) has been used, from $\sim 6 \times 10^3$ to 10^6 seconds. The results of both analysis are plotted in Fig. 3(a,b). In a time axis, Fig. 3(a,b) should show a set of equidistant peaks as a consequence of the signature of the p-modes present in the data, a set of nearly equidistant peaks in frequency in the amplitude spectrum (Gelly *et al.*, 1986). Indeed in Fig. 3a, such set of peaks (numbered from # 1 to # 18) is clearly seen, modulated by the observational window function. The peak # 1 appears at a frequency separation of 98 μHz, while the dominant peak # 2 appears at $S = 49\mu$Hz. The peak marked with $*$ corresponds to the value of $\Delta\nu_{20}$ ($\nu_{n,0} - \nu_{n-1,2} \approx 6D_0$). Examining now Fig 3b, the observational results, a set of equidistant peaks can also be identified, being numbered from # 1 to # 19. Using the frequency separation of all identified peaks, a value of 97.3±0.6 μHz is found (if only the peaks up to # 10 are used the value yielded is of 97.4±0.7 μHz) which corresponds to the frequency of peak # 1. Consequently, following the above explanation this value has to be assigned to $\Delta\nu_0$ for the p-modes present in the data. Notice also that peak # 2, at 48.7 μHz, is the clearest signal as already expected.

In order to rule out some other possibilities which might cause the results to be artifacts of the data reduction and/or analysis, a final verification was performed. In order to destroy the phase information, the observational residual series was divided into three hours intervals. These intervals were then randomly redistributed to reproduce a new white noise series with the same time length, window function and total energy as the original, but where any actual stellar signal with phase coherency longer than three hours (as p-modes are expected to be) would be partially destroyed, as a fractional randomization of the data is done. This new series was submitted to the same analysis as the original and artificial series. The result obtained is plotted in Fig. 3c, where the amplitude squared spectrum of the spectrum is presented which has to be compared with the ones of the artificial (3a) and of the observational (3b) series. The peaks at 11.6 and 5.8 μHz, marked with crosses, are the ones due to the daily sidebands, while the peak at 49 μHz, present in a and b, has completely disappeared.

Hence, this analysis supports the assumption made that the peak at $S = 48.7 \pm 0.3 \mu$Hz should be considered as a feature associated with an actual stellar signal and not an artifact of the reduction and analysis procedures.

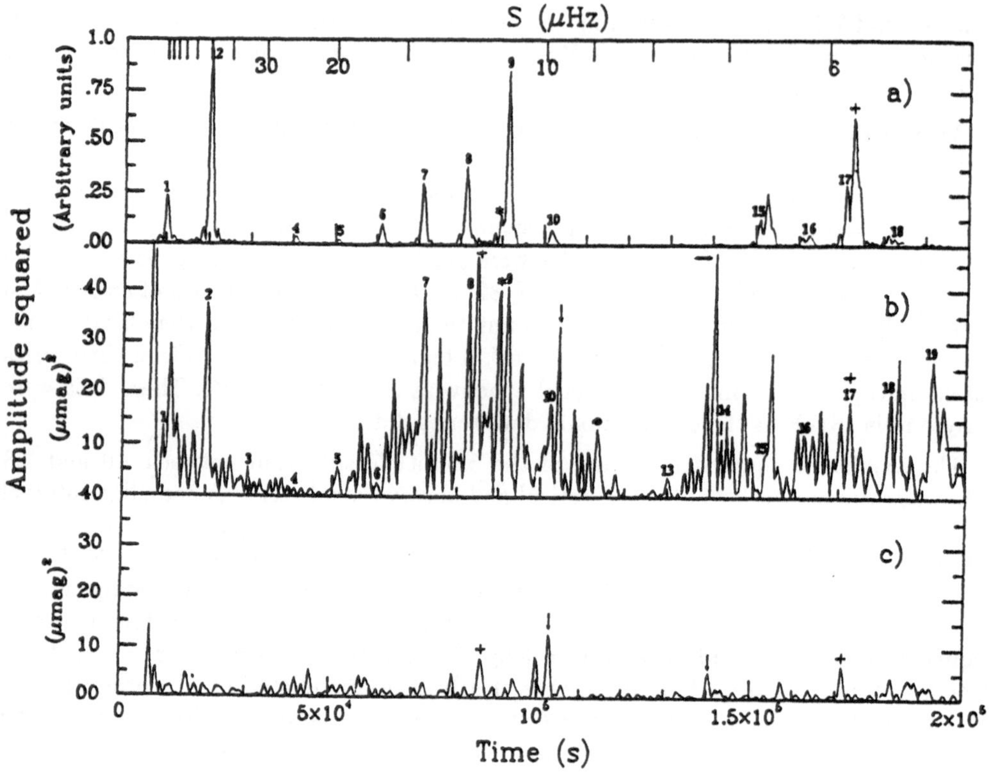

Fig. 3. Amplitude squared spectra of the HD155543 amplitude spectra (from 2 to 2.8 mHz, see text) of: **a)** theoretical artificial series; **b)** observational series; **c)** randomly redistributed observational series.

Assuming this value as correct ($\Delta\nu_0$=97.3 μHz), a folding analysis was preformed on the observational data (Gelly *et al.*, 1986; Jiménez *et al.*, 1987). The procedure is very simple: the squared amplitude spectrum of the series is cut into 97.7 μHz width intervals; afterwards, the ten central intervals within the range 1.85 to 2.83 mHz, where the clearest p-mode signature is seen, were folded onto the first one and averaged out (that is, to the frequencies of each interval an adequate integer number of times 97.7 μHz is subtracted in order to have all intervals with a frequency axis from 0 to 97.7 μHz). The performed folded spectrum has to be taken with caution because it can lead to a misinterpretation of the real features present in the spectrum, if the value considered for $\Delta\nu_0$ is not exactly the one present in the data. However, from this procedure we can speculate with the result of a value of $\Delta\nu_{20}$=8.7±0.5 μHz (D_0=1.4±0.1 μHz) for HD155543. Nevertheless, further evidence for values of $\Delta\nu_{20}$ could come from Figs. 3a and b, where the symbol $*$ represents a peak that could be assigned to this magnitude ($\Delta\nu_{20}$=11.1 μHz, D_0=1.8±0.1 μHz), although the value of 8.7 μHz yielded by the folding analysis is also present in Fig. 3b, a peak marked with a •. Moreover, we found an upper limit for the amplitude of the p-modes with degree ℓ=0,1 of the order of 20-30 μmag, in agreement with the signal excess given by the average spectrum of Fig. 1 (⋆ A).

4 Conclusions and Future Prospects

Strong evidence of the presence of global acoustic oscillations in a main sequence star, out of the instability strip, has been discovered through a good series of photometric data. The range of frequencies where p-modes seem to be present (1 to 3 mHz), together with the value of $\Delta\nu_0$=97.3 μHz and the possible values of D_0 (1.4 or 1.8 μHz) are in good agreement with the ones obtained by computing eigenfrequencies for MS standard models and also with previous determinations of $T_{\rm eff}$ and L for HD155543, thus giving more confidence to the interpretation of the data.

Multi-site ground based rapid photometry during long periods of observation has proven to be a powerful strategy for the detection of stellar oscillations and thus some important stellar parameters can be indirectly measured. However, due to the atmospheric noise, up to now it fails to identify the actual frequencies of the acoustic modes. Further improvements have been planned with future three site campaigns (OT, SPM, and some place at USSR, China, Japan, etc...), as well as with the construction of a new 4 channel photometer, provided with a chopping system to measure 3 stars and skybackground, simultaneously, through a pair of narrow-band filters. With such device, we hope to be able to correct most of the atmospheric contribution to noise, either transparency fluctuations or scintillation (Harvey, 1988), thus allowing to identify independent frequencies and amplitudes of the modes.

Acknowledgements

We want to acknowledge Dr. A. Mangeney, Prof. F. Praderie, Dr. E. Fossat and Dr. P.L. Pallé for valuable work and discussions which enriched the paper.

References

Belmonte J.A. 1989, Ph.D. Thesis. Universidad de La Laguna (Spain).

Belmonte J.A., Mangeney A., Chevreton M., Praderie F., Saint-Pé O., Puget P., Alvarez M., and Roca Cortés T. 1989, *Astron. Astrophys.*, submitted.

Belmonte J.A., Pérez Hernández F., and Roca Cortés T. 1989, *Astron. Astrophys.*, in press.

Christensen-Dalsgaard J. 1982, *Monthly Notices Roy. Astron. Soc.*, **199**, 735.

Christensen-Dalsgaard J. 1988, in *Proc. IAU Symp. No. 123, Advances in Helio- and Asteroseismology*, eds. J. Christensen-Dalsgaard and S. Frandsen (Reidel, Dordrecht), p.295.

Christensen-Dalsgaard J., and Frandsen S. 1983, *Solar Phys.*, **82**, 469.

Claverie A., Isaak G.R., McLeod C.P., van der Raay H.B., and Roca Cortés T. 1979, *Nature*, **282**, 591.

Deubner F.L., and Isserstedt J. 1983, *Astron. Astrophys.*, **126**, 216.

Gelly B., Grec G., and Fossat E. 1986, *Astron. Astrophys.*, **164**, 383.

Harvey J.W. 1988, in *Proc. IAU Symp. No. 123, Advances in Helio- and Asteroseismology*, eds. J. Christensen-Dalsgaard and S. Frandsen (Reidel, Dordrecht), p.497.

Jiménez A., Pallé P.L., Roca Cortés T., Domingo V., and Korzennik S. 1987, *Astron. Astrophys.*, **172**, 323.

Mangeney A., Chevreton M., Belmonte J.A., Däppen W., Saint-Pé O., Praderie F., Roca Cortés T., Fuensalida J.J., and Alvarez M. 1988, in *Seismology of the Sun and Sun-like Stars*, ed. E. Rolfe, ESA SP-286 (ESA Publication Division, Noordwijk), p.551.

Noyes R.W., Baliunas S.L., Belserene E., Duncan D.K., Horne J., and Widrow L. 1984, *Astrophys. J. Letters*, **285**, L23.
Pallé P.L. 1986, Ph.D. Thesis. Universidad de La Laguna (Spain).
Pérez Hernández F. 1986, Tesis de Licenciatura. Universidad de La Laguna (Spain).
Tassoul M. 1980, *Astrophys. J. Suppl.*, **43**, 469.
Traub W.A., Mariska J.T., and Carleton N.P. 1978, *Astrophys.J.*, **223**, 583.

Search for Five-Minute Oscillations in Late-Type Stars by Fabry-Perot Interferometer

Hiroyasu Ando

National Astronomical Observatory, Mitaka-shi, Tokyo 181, Japan

Abstract: A program of the accurate measurement of radial velocity variations has been started for the various stars using a Fabry-Perot interferometer at Okayama Astrophysical Observatory since 1986 to diagnose the internal structure of stars. In particular, we have accumulated such data for the late type star α CMi, which might have the largest amplitude of "Five-minute oscillations" according to the theoretical prediction (Ando 1976). Here we will give a brief explanation of the instrument and the data reduction procedure. The power spectrum for a time series of radial velocity variations in α CMi has been calculated. The noise level in this spectrum has been estimated to be about $10\,\mathrm{m\,s^{-1}}$. At this level, it is safe to say that any significant signal related to the stellar oscillations may not be detected. To confirm the "Five-minute oscillations" in this star, we should continue to reduce the noise level by improving the sensitivity of the instrument and/or by accumulating data moreover.

1 Introduction

In the past decade, much efforts have been dedicated to asteroseismology by many authors (Traub *et al.* 1978; Campbell and Walker 1979; Baliunas *et al.* 1981; Smith 1982, 1983; Noyes *et al.* 1984; Fossat *et al.* 1984; Gelly *et al.* 1986; Pietraszewski *et al.* 1986, 1988; Ando *et al.* 1988; Belmonte *et al.* 1990; Isaak *et al.* 1990; Schmider and Mosser 1990). The ultimate objective of asteroseismology is to look into the internal structure of stars and to give a great impact to the theory of the evolution and structure of stars. This impact is not only in stellar problems, but also in the cosmological problems; e.g., age of globular clusters, primordial helium abundance.

However, even now the detectability of oscillations in late-type (solar type) stars is an important issue and the asteroseismology is still on the way of the mature science.

In this talk, we present our attempt to detect "Five-minute Oscillations" in one of late-type stars (α CMi) at Okayama Astrophysical Observatory.

2 Instrumentation

The radial velocity measurement was taken using Fabry-Perot interferometer attached to the coude focus of 74 inches telescope at Okayama Astrophysical Observatory. Fabry-Perot interferometer has 2.5 Å as a free spectral range at 5000 Å, and resolving power of 10^5 ($3\,\mathrm{km\,s^{-1}}$). As it has very narrow passband, coude spectrograph was used as a pre-optics to make order sorting.

The blue and red wings were consecutively monitored at a proper absorption line of α CMi, in which the integration time for each wing is 4 seconds. To improve the signal to noise ratio, 8 consecutive data for each wing are summed up to be converted into a time series of the radial velocity variations.

3 Observations

The observed object (α CMi) was selected according to the criteria; it should be apparently bright, and has period as short as possible, amplitude as large as possible. α CMi is apparently bright (0 mag.), and may have period of 14 min. and amplitude of $1\,\mathrm{m\,s^{-1}}$ from theory (Ando 1976).

The observational journals are listed in Table 1. The data for early days above the midline have already been analysed and published (Ando *et al.* 1988). In the following, we discuss the data for the latter terms.

Table 1. Observational Journal for α CMi

Data		Number of Observation	Noise for each point(m/s)
1986 Feb	22	111	220
	23	72	240
	24	257	300
Dec	8	154	290
	10	197	300
1987 Dec	29	112	280
	30	61	230
	31	404	423
1988 Jan	1	115	139
	3	318	393
	5	147	225
	6	143	294
1989 Feb	26	262	221
	27	120	193

4 Results

Power spectra of α CMi for a time series given in Table 1 have been calculated using Deeming's (1975) technique. The power spectrum for the former term was shown in Ando *et al.* (1988). Any significant peak has not been found. The noise level is estimated to be $13\,\mathrm{m\,s^{-1}}$.

Figure 1 shows the power spectrum for the latter term. It has a significant peak around period of 20 minutes and also some peaks in the lower frequency region. To check the instrument noise, we observed the signal of Th lamp and found the long-term variations and quasi-sinusoidal variation with period of 20 minutes in the time series. At first we could not distinguish whether this short-term quasi-sinusoidal variation comes from the instrument or from Th lamp itself. To separate this, we made the radial velocity measurement of α Boo (see Fig. 2). Eventually we confirmed this signal comes from the instrument, since the significant peak with the same period as in the experiment has been identified in α Boo. Now it is better to carry out the radial velocity measurement using a proper calibration source.

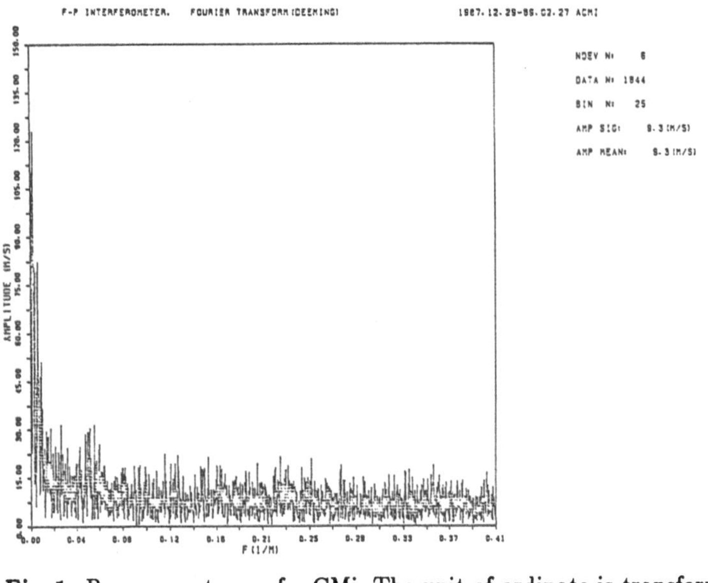

Fig. 1. Power spectrum of α CMi. The unit of ordinate is transformed in velocity unit ($\mathrm{m\,s^{-1}}$).

It is clear that we should accumulate the data of α CMi to reduce the noise level to the level of $1\,\mathrm{m\,s^{-1}}$.

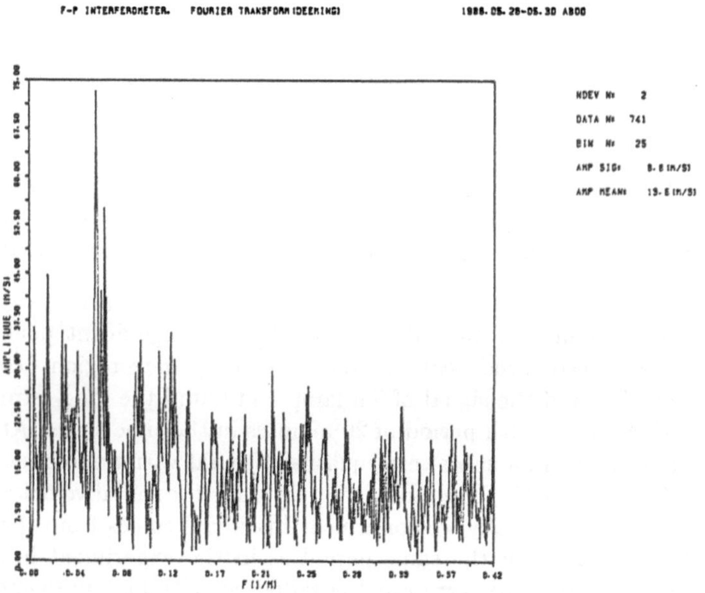

Fig. 2. Power spectrum of α Boo. The instrumental noise is clearly seen around frequency of 0.05.

5 Prospect

It is obvious that the constant improvement in efficiency and stability of the instruments should be pursued. However, it is certain from many reports in this workshop that the quality of the data particularly for the radial velocity measurement depends severely on the accessibility of the large telescopes. We need more photons, in other words, much larger aperture telescopes. In the next decade, we will be in a good position to access 8 m class telescopes to lead asteroseismology to more fruitful and mature science.

References

Ando, H. 1976, *Publ. Astron. Soc. Japan*, **28**, 517.

Ando, H., Watanabe, E., Yutani, M., Shimizu, Y., and Nishimura, S. 1988, *Publ. Astron. Soc. Japan*, **40**, 249.

Baliunas, S.L., Hartmann, L., Vaughan, A.H., Liller, W., and Dupree, A.K. 1981, *Astrophys. J.*, **246**, 473.

Belmonte, J.A., Pérez Hernández, F., and Roca Cortés, T. 1990, these proceedings.

Campbell, B., and Walker, G.A.H. 1979, *Publ. Astron. Soc. Pacific*, **91**, 540.

Deeming, T.J. 1975, *Astrophys. Space Sci.*, **36**, 137.

Fossat, E., Grec, G., Gelly, B., and Decanini, Y. 1984, *Compt. Rend Acad. Sci., Ser. 2*, **299**, 17.

Gelly, B., Grec, G., and Fossat, E. 1986, *Astron. Astrophys.*, **164**, 383.

Isaak, G.R., Innis, J.L., Brazier, R.I., and McLeod, C.P. 1990, these proceedings.

Noyes, R.W., Baliunas, S.L., Belserene, E., Duncan, D.K., Horne, J., and Widrow, L. 1984, *Astrophys. J. Letters*, **285**, L23.

Pietraszewski, K.A.R.B., Bell, C.R., Ring, J., Reay, N.K., and Leeper, M. 1988, in *Advances in the Helio- and Asteroseismology, IAU Symp. No.123*, eds. J. Christensen-Dalsgaard and S. Frandsen (Reidel, Dordrecht), p.517.

Pietraszewski, K.A.R.B., Reay, N.K., and Ring, J. 1986, in *Seismology of the Sun and the Distant Stars*, ed. D.O. Gough (Reidel, Dordrecht), p.377.

Schmider, F.-X., and Mosser, B. 1990, these proceedings.

Smith, M.A. 1982, *Astrophys. J.*, **253**, 727.

Smith, M.A. 1983, *Astrophys. J.*, **265**, 325.

Traub, W.A., Mariska, J.T., and Carlton, N.P. 1978, *Astrophys. J.*, **223**, 583.

Mode Trapping in Pulsating White Dwarfs

Steven D. Kawaler [1] and Philip Weiss [2]

[1]Department of Physics, Iowa State University, Ames, IA 50011, USA
[2]Department of Astronomy, Yale University, New Haven, CT 06511, USA

Abstract: White dwarfs undergo nonradial g-mode pulsation in three separate instability regions during their cooling history. The coolest pulsating white dwarfs are the ZZ Ceti stars, and the hottest correspond to the pulsating PG1159 stars. The adiabatic pulsation properties of stellar models accurately reproduce many of the observed properties of the pulsations. In particular, trapping of modes by composition transition regions near the surface of these stars is an adiabatic phenomenon. Here we present models of ZZ Ceti stars which show trapped modes with periods that are very close to those observed, and illustrate a technique for using these periods to determine the mass of the surface hydrogen layer.

1 Introduction

Research in helioseismology mainly concerns analysis of the oscillations of the present-day Sun. However, the observed pulsations of white dwarfs indicate that solar-mass stars also undergo nonradial oscillations during the final stages of their evolution. Data on the pulsations of white dwarfs are produced at a much more manageable rate than for solar oscillations; we can therefore obtain a more or less complete description of their observed pulsational behavior. Our knowledge about the interior of solar-mass stars during their tenure as white dwarfs is therefore becoming more thorough than for any other class of stars through the use of the techniques of stellar seismology [see, for example, the reviews by Winget (1988), and Winget and Fontaine (1982)].

The coolest class of pulsating white dwarfs are the DAV, or ZZ Ceti stars. The DAV stars have very pure hydrogen surface layers, and lie in a narrow strip in $T_{\rm eff}$ near 12 000 K. Below the surface hydrogen layer (with a thickness of less than 10^{-4} solar masses) lies a thicker layer of helium, which in turn surrounds a degenerate carbon/oxygen core. Variable white dwarfs pulsate with periods between 100 and 1000 seconds (10 to 100 times longer than the radial pulsation time scale); almost all are multiperiodic. It is clear that nonradial g-modes are responsible for the observed pulsations. Because of the nearly neutral stratification of the degenerate core, g-modes propagate with large amplitude in the outer nondegenerate layers; hence they are essentially envelope modes in white dwarfs.

The pulsations are driven by ionization of hydrogen in the DAV stars. This mechanism is quite potent, and would lead to destabilization of many more modes than are observed if not for the "trapping" of oscillations by the compositional stratification of the outer layers (Winget *et al.* 1981). If a normal mode has a node in the composition transition zone, then the amplitude of that mode will be greatly reduced in the interior. Modes which have nodes near the transition zone are "trapped", *i.e.* they do not penetrate below the region of changing composition. In a white dwarf, the high density within the core results in large kinetic energies of oscillation even though the amplitude of those oscillations are small in the core. Trapped modes, on the other hand, have essentially zero amplitude below the composition transition zones, and therefore have much smaller pulsational kinetic energies. Since the growth rate for unstable modes is inversely proportional to the kinetic energy of the mode, modes that are trapped by the composition transition region are selectively amplified in white dwarfs.

Mode trapping as a selection mechanism successfully explains why we see only a few modes in pulsating DAV stars, as shown by Winget *et al.* (1981) and Dolez and Vauclair (1981) (see also Bradley *et al.* 1988). This work shows qualitative agreement between the periods of trapped modes in models and the observations. More recently, Hansen (1987) and Brassard and Hansen (1990) reported asymptotic analyses of mode trapping in white dwarfs that allow more quantitative comparisons between theory and observation. Their work provides theoretical tools for determining the thickness of the surface hydrogen layer in DAV stars. In this paper, we report on a preliminary examination of the quantitative aspects of mode trapping in DAV stars. We use complete models of white dwarfs with various hydrogen layer thicknesses and effective temperatures to examine how the periods of the trapped modes depend on these parameters. With the analytic analysis of Hansen (1987) as a guide, we show how the periods of trapped modes relate to the thickness of the surface hydrogen layer. With this relationship, we use the observed periods in DAV white dwarfs to determine the thickness of their surface hydrogen layers

2 Models, Methods, and Results

To compute our white dwarf models, we integrate the equations of stellar structure from the surface downwards, following specification of the effective temperature and luminosity. Integration proceeds until 10^{-3} solar masses from the center. The core composition is set as pure carbon, the thickness of the helium layer is set at $10^{-2}M_*$, and the thickness of the hydrogen layer is a free parameter. In composition transition regions, the helium abundance changes linearly with mass through the transition zone. The luminosity at each point is proportional to the interior mass. Convective fluxes are computed using the mixing length set to the smaller of the pressure scale height and the distance to the top of the convection zone.

We iterate on the total stellar luminosity to ensure that the density of the innermost model point is within 2 percent, but less than, the mean density of the remaining central ball. This is an important point; if the inner boundary is not self consistent (for example if the density of the central ball was lower than the overlying layer) then the Brunt-Väisälä frequency becomes unphysical and adversely affects pulsation calculations. Using this procedure, our models are very consistent with those resulting from evolutionary calculations, but have the advantage that the surface hydrogen layer thickness can be varied systematically.

Most models are 0.60 solar masses, corresponding to the mean mass for field white dwarfs. We computed models with effective temperatures from 11 000 K to 9 500 K, bracketing the blue edge for pulsational instability in these models. For each $T_{\rm eff}$, we computed models with hydrogen layer thicknesses from 10^{-4} to $10^{-10} M_*$. The upper limit corresponds to models that lose no mass following the planetary nebula nucleus phase; models with thinner layers correspond to various degrees of mass loss at earlier phases of evolution. We computed the adiabatic pulsation calculations using the pulsation code described by Winget (1981).

We find clear evidence for trapping of nonradial g-mode oscillations in our models. As an example, consider Fig. 1 which shows the pulsational kinetic energy as a function of radial node number n for a 0.60 solar mass model at 9 900 K with a surface hydrogen layer of 10^{-8} stellar masses. The composition transition zone in this model spans about two pressure scale heights. Minima in the kinetic energy occur at n=2, 7, 11, 14, and 18 for l=2 modes, corresponding to trapped modes with 0, 1, 2, 3, and 4 nodes between the surface and the composition transition region. Periods for these modes are 185, 366, 512, 637, and 782 seconds, respectively. Trapped modes with l=1 have the same values for n. Also shown in Fig. 1 is the stability coefficient (in $\rm yr^{-1}$) for each mode. The negative values indicate unstable modes. Clearly, local minima in the growth times correspond to trapped modes.

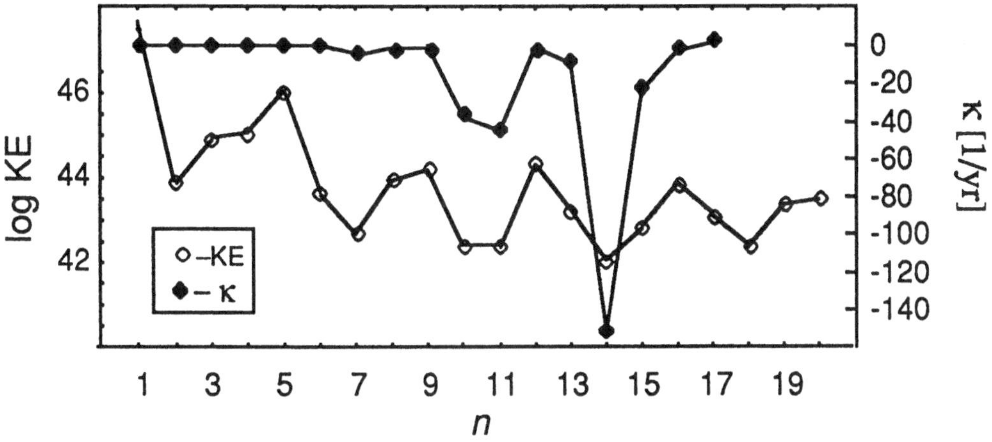

Fig. 1.

The thinner the hydrogen layer, the more pronounced the trapping. The primary reason for this is that the difference in period between trapped modes increases with decreasing thickness of the hydrogen layer. Trapping is favored when the period of a mode approaches those given by the relationship given by Hansen (1987, see below). Since models with thin hydrogen layers have several modes separating trapped modes, they have a larger probability of having a mode with a period near the favored periods. In models with thick surface hydrogen layers ($M_{\rm H} > 10^{-6} M_*$) the natural g-mode spacing approaches the spacing between trapped modes, so that the relative differences in kinetic energies are small. Hence trapping begins to fail as a mode selection mechanism for $M_H > 10^{-6} M_*$.

The degree to which modes are trapped also depends on the steepness of the composition gradient. Trapped modes in the model shown in Fig. 1 have kinetic energies that are reduced by a factor of about 100 as compared to untrapped modes; a model with a composition transition zone covering 4 pressure scale heights showed trapping by factors of less than 30.

Hansen (1987) showed that the periods for trapped g-modes in white dwarf envelopes should conform to the relation

$$P(i)^2 = 4\pi^2 \lambda_i^2 \left[Z_d l(l+1) \frac{GM}{R^3} \right]^{-1} . \tag{1}$$

where i=0, 1, 2, ... is the number of nodes between the surface and the composition transition region, Z_d is the fractional depth of the transition region, and λ_i is a series of constants of order 1 to 10 for i=0 to 5. With our models, we derive the values of λ_i, shown in Table 1, for more realistic input physics. With these coefficients, the Hansen (1987) relationship accurately gives the periods of trapped modes to within 3 percent .

Table 1. Trapping coefficients for white dwarfs.

i	0	1	2	3	4
λ_i	2.21 ± 0.06	4.38 ± 0.06	6.06 ± 0.20	7.58 ± 0.22	9.00 ± 0.20

The values of λ_i do not depend strongly on $T_{\rm eff}$, M_H, M_{He}, the mass of the model, or the thickness of the composition transition region. Figure 2 shows the periods of trapped l=2 modes as a function of the surface hydrogen layer mass in 0.60 solar mass models at 10 000 K computed using our values for λ_i. Note that these curves are insensitive to the effective temperature, and that to transform to l=1 one multiplies the periods by the square root of 3.

Figure 2 maps the trapped l=2 pulsation periods into the surface hydrogen layer thickness for 0.60 solar mass white dwarfs. While it is difficult to assign masses for individual pulsators, and to identify the value of l of the pulsations, several DAV stars show periods that are very close to those seen in the models. In Fig. 2, we indicate the observed periods for four DAV stars, placing them at our best estimate of M_H. This preliminary analysis indicates that these DAV stars have values of M_H ranging from a few $\times 10^{-5}$ to less than 10^{-10} stellar masses. Clearly, though, more general conclusions must await more careful determinations of the pulsation periods of these and other DAV stars, and more systematic application of the models presented here to the observational data.

Acknowledgements

This work was sponsored in part by NASA grant NAGW-1364 to Iowa State University and in part by NASA grant NAGW-778 to Yale University.

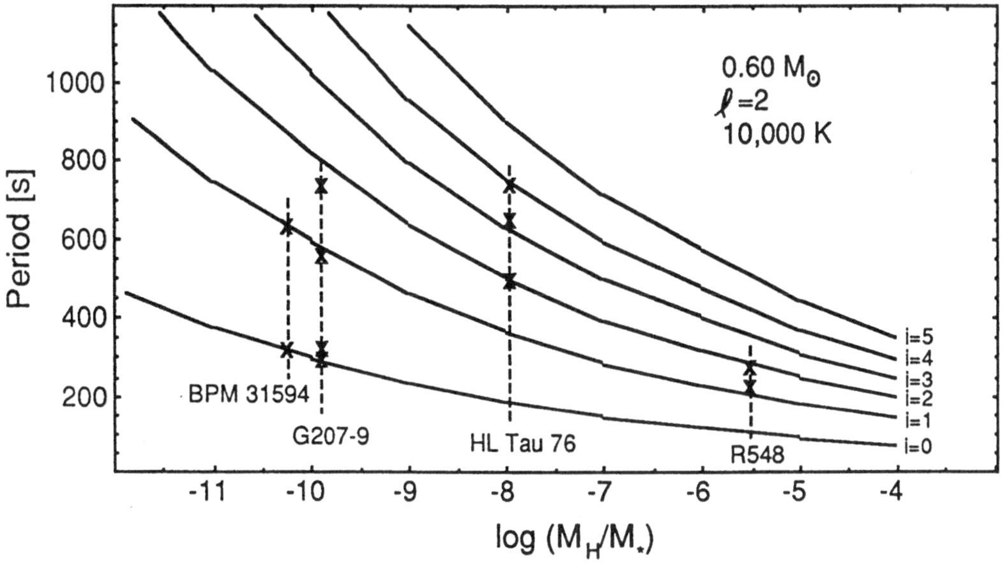

Fig. 2.

References

Bradley, P., Winget, D., and Wood, M. 1989, in *White Dwarfs, IAU Colloq. No. 114*, ed. G. Wegner (Springer, Berlin), p.286.

Brassard, P., and Hansen, C. J. 1990, preprint.

Dolez, N., and Vauclair, G. 1981, *Astron. Astrophys.*, **102**, 375.

Hansen, C.J. 1987, private communication.

Kawaler, S. 1988, in *The Second Conference on Faint Blue Stars, IAU Colloq. No. 95*, eds. A. Phillip, D. Hayes, and J. Liebert, (L. Davis Press, Schenectady), p.297.

Winget, D.E. 1981, Ph. D. thesis, University of Rochester.

Winget, D.E. 1988, in *Advances in Helio- and Asteroseismology, IAU Symp. No. 123*, eds. J. Christensen-Dalsgaard and S. Frandsen (Reidel, Dordrecht), p.305.

Winget, D.E., and Fontaine, G. 1982, in *Pulsations of Classical and Cataclysmic Variable Stars*, eds. J.P. Cox and C.J. Hansen, (JILA, Boulder), p.46.

Winget, D.E., Van Horn, H.M., and Hansen, C.J. 1981, *Astrophys. J. Letters*, **245**, L33.

Chemical Composition and Instability Mechanism in the PG1159 Stars

Gérard Vauclair

Observatoire Midi-Pyrénées, 14 av. E. Belin, 31400 Toulouse, France

Abstract: Our present understanding of the excitation mechanisms in the hot PG1159 white dwarfs is discussed. The identification of the correct excitation mechanism is made difficult by the uncertainties regarding the derivation of the chemical composition in these stars. A model is presented in which the C, N, O abundance reflects the diffusion equilibrium abundance of these elements in a helium dominated envelope. It is shown that the deep accumulation of these elements predicted by diffusion could play a role on the g-modes instability by the κ-mechanism, as the thermal time scale of the relevant layers does agree with the observed periods for the pulsating PG1159. Furthermore, such a model would naturally provide a filtering mechanism as a consequence of the inhomogeneous abundance distribution.

1 Introduction

Along the white dwarf cooling sequence, one finds three instability strips for the PG1159, the DB, and the DA variable stars respectively. The existence of these instability strips, at quite different stages of the cooling sequence, provides a unique opportunity to check our theory of white dwarf evolution and, more generally, our knowledge of stellar structure and evolution [see Kawaler and Hansen (1989) for a review]. In the two cooler instability strips, the instability mechanism is identified with the κ-γ mechanism due to partial ionization of hydrogen in the DA variables (Dolez and Vauclair 1981; Winget *et al.* 1982) and of helium in the DB variables (Winget *et al.* 1983). In the PG1159, the instability mechanism is not yet entirely understood, mainly as a consequence of the difficulty encountered in deriving the chemical composition of these very young, hot white dwarfs.

The nonradial gravity modes in PG1159 stars may be unstable by the usual two main mechanisms: the ε-mechanism or the κ-γ mechanism. The ε-mechanism may be efficient in white dwarfs which have retained enough mass in their outer layers so that a fraction of the total luminosity could be due to residual nuclear burning at the bottom of the envelope. In PG1159 stars which do not show any hydrogen in their spectrum, the 3α reaction at the bottom of the helium envelope could destabilize g-modes. This mechanism, first proposed by Kawaler (1986), met two difficulties: (1) the unstable g-modes are those which have large amplitude in the nuclear burning region; they are necessarily of low radial order; these modes have too short periods compared to the observations; (2) the planetary nebula nuclei, thought to be the progenitors of most of the white dwarfs, should also show nonradial modes instabilities, as they have more fuel than their descendants (Kawaler *et al.* 1986). A survey of planetary nebula nuclei by Hine and Nather (1987) has not confirmed this prediction. The κ-γ mechanism was proven to be responsible for the instability in the DB and DA variable white dwarfs. At the temperature of the PG1159 (10^5 K), the partial ionization of heavy elements could provide the instability mechanism. Starrfield *et al.* (1983, 1984, 1985) and Starrfield (1987) have shown that partial ionization of carbon and oxygen may trigger g-modes instability with the right periods. However, this explanation also encounters some difficulties. The homogeneous envelope models should be dominated by C and O (with He less than 20 % in mass) for the κ-γ mechanism to work. Such an extreme chemical composition is not excluded at the present time; it should however be consistent with what is known about the composition of progenitors and descendants of these stars. The evolutionary stage of the PG1159 should fit a global description in which pure hydrogen and/or pure helium white dwarfs could have PG1159 carbon-oxygen rich, helium-hydrogen poor progenitors. Whether such a consistent evolutionary scheme may be obtained is not demonstrated presently. Furthermore, while these models show unstable modes with the correct periods, they do not provide any filter mechanism and so they predict much more unstable modes than observed.

As far as the outer layers chemical composition is concerned, the spectral analyses conclude to contradictory results. On one hand, the line profiles analyses applied to absorption lines in the visible wavelengths, using state of the art NLTE, do conclude to carbon rich, helium poor composition (Werner *et al.* 1989). On the other hand, the analyses of UV and soft X-ray observations do support a helium dominated composition, with C/N/O abundance at most solar (Barstow and Holberg 1989). These contradictions illustrate the difficulty in modelling stellar atmospheres and line profiles at these extreme effective temperatures.

In the present paper, an alternative model is proposed, with the following assumptions : (1) the PG1159 have He dominated envelopes and (2) the outer layers are stable enough against turbulent mixing to allow element diffusion. This model has been presented elsewhere (Vauclair 1987, 1989). It has been shown to be in rough qualitative agreement with photospheric observed abundance. It was also suggested that the diffusion mechanism responsible for heavy element levitation in the photosphere could be relevant to the g-modes instability. The present paper is a progress report about this aspect of the alternative model for PG1159 stars.

2 Alternative Model

In a pure He envelope white dwarf, in the absence of any turbulent mixing (due to rotation, mass loss ...) diffusion has been proven to be an efficient mechanism (Vauclair *et al.* 1979; see Vauclair 1987, 1989 for reviews). At the hot end of the white dwarf cooling sequence, where are found the PG1159 stars (at $T_{\rm eff}$ = 100 000 K - 150 000 K), the radiation field is so intense as to support a small amount of heavy elements in the outer layers. The element abundance which can be supported is the one for which the upward radiative acceleration exactly balances the downward gravitational/thermal acceleration (Vauclair 1989). As a consequence, the equilibrium abundance reached by a given element varies with depth: the metal abundance in these models cannot be homogeneous. Figure 1 illustrates the equilibrium abundance achieved in a typical PG1159 star model (0.6 solar mass, $T_{\rm eff} = 100\,000\,{\rm K}$, $\log g = 7.6$, pure He). The deep minima in the C, O, and N abundance distribution do reflect the low radiative acceleration provided through the EUV resonance lines of the noble gas configuration. It is suggested that, while the outermost maxima are related to the abundance observed in the photosphere, the innermost ones may be related to the nonradial modes stability. The basic argument to support this idea is that in these deep layers the He opacity being very small, an admixture of trace amounts of impurities (the N abundance at the deep maximum just reaches the solar value) could produce a large variation of the opacity on a small scale height. This variation of opacity is typically the physical basis of the κ-mechanism.

Before proceeding further along this line, one must check whether such deep metal clouds may be relevant to the pulsation problem by comparing the thermal time scale to the pulsation periods. It is well known that any stellar layer will ignore the perturbations due to the pulsations as long as its characteristic thermal time scale differs too much from the pulsation period (Dziembowski and Koester 1981). A layer located at radius r has a chance to drive the pulsation if its thermal time scale $\tau_{\rm th}$ is comparable to the pulsation periods Π:

$$\tau_{\rm th} \equiv \int_{M_*}^{M(r)} \frac{C_v T\, dM}{L} \simeq \Pi . \tag{1}$$

Figure 1 shows the variation of $\tau_{\rm th}$ in the model. A comparison with typical pulsation periods for PG1159 stars (400 s – 1000 s) reveals that the potential driving zones are just at the depth where diffusion theory does predict the accumulation of C, N, and O. More specifically, they do coincide in this model with the nitrogen cloud. At this layer temperature ($T = 10^6\,{\rm K}$) partial ionization of C, N, and O may provide the instability mechanism. Note that in the Starrfield *et al.*'s (1983, 1984, 1985) model, the g-modes instability is due to partial ionization of C and O at this temperature. In the present model, the coincidence of the potential driving zone with the maximum of the nitrogen cloud is a strong indication that the deep clouds predicted by the diffusion theory could play a role in the excitation of the gravity modes. In the following, the nitrogen cloud will be regarded as the potential driving zone and will be referred to as the N-shell.

At this stage, one may ask whether the g-modes which could be excited at the level of the N-shell do have periods in agreement with observations. One may also ask whether the narrow metallic cloud which could drive the pulsations could also efficiently filter them. It is clear that in order to become unstable, a given mode should have an amplitude as large as possible in the relevant driving zone. An illustration of this is given on Fig. 2. A series of g-modes has been calculated for ℓ= 1, 2, and 3. On Fig. 2, the amplitude of the radial displacement is shown for 2 modes with ℓ=2. The amplitude is normalized at the bottom of the photosphere (Rosseland optical depth =1.). The g_4^2 mode, with period 305 s, has a node in the middle of the N-shell and has very little chance to get excited. On the contrary, the mode g_9^2, with a period of 543 s has a maximum almost coinciding with the N-shell which makes this mode a good candidate for instability. The N-shell will trigger instability of those modes which have large amplitude in the shell. If one considers the series of increasing radial order g-modes, for a given ℓ, one may follow the variation of their amplitude in the N-shell. For low order radial modes, the amplitude decreases with increasing order until one gets to the mode which has its first node at the N-shell (as the

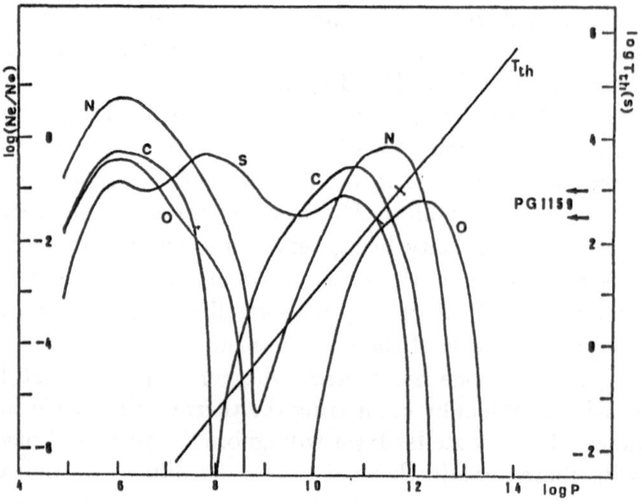

Fig. 1. Diffusion equilibrium and thermal time scale. C, N, O, and Si abundance stratification in a model of PG1159 star (T_{eff} = 10^5 K). The abundance (left scale), in solar units, and on a logarithmic scale, is plotted versus the logarithm of the pressure. The thermal time scale (right scale) is of the order of the PG1159 typical pulsation periods in the middle of the 2nd nitrogen diffusion cloud.

g_4^2 discussed above). For increasing radial orders, the amplitude will then increase and go to a maximum as the first lobe of the eigenfunction coincides spatially with the N-shell. As the radial order is further increased, the lobes of the eigenfunction will concentrate further towards the stellar surface and the amplitude at the N-shell will decrease. Figure 3 illustrates this variation of the amplitude in the N-shell. The results presented here are preliminary. The nonradial eigenfunctions have been computed for ℓ=1, 2, 3 (the most probably observable azimuthal numbers) and for only the 11 or 12 first radial orders. One clearly sees the increase of the amplitude when the first lobe of the eigenfunctions passes through the N-shell. For ℓ=2, the amplitude starts to decrease after this first lobe has passed the N-shell. This indicates that there should be an efficient filtering mechanism which favors the g-modes with periods around 1300 s (for ℓ=1), 750 s (for ℓ=2) and 500 s (for ℓ=3). This filter mechanism depends mainly on the depth of the driving zone, which in turn depends on the effective temperature and gravity. This should be explored across the PG1159 instability strip.

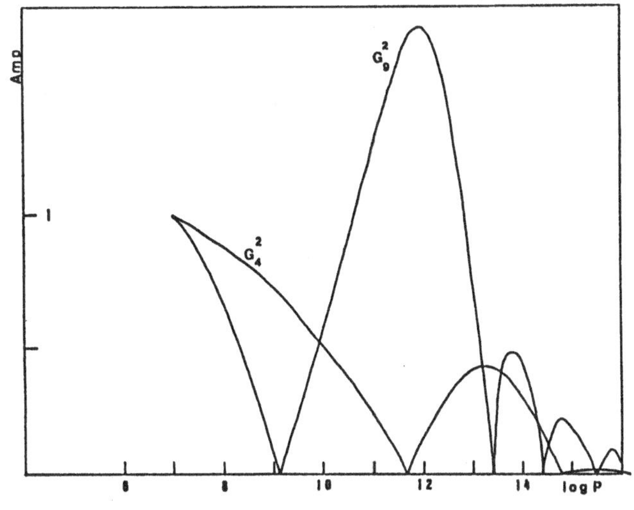

Fig. 2. Amplitude of the radial displacement (normalized at unity at the photosphere) for two $\ell = 2$ g-modes : the g_4^2 has a node in the N-shell and has no chance to be excited by κ-mechanism, while the g_9^2 has its maximum amplitude in the N-shell and is a good candidate for instability.

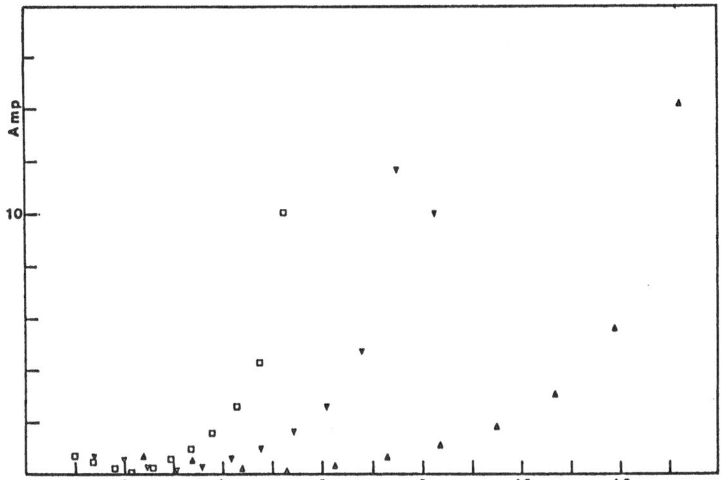

Fig. 3. Amplitude of g-modes in the N-shell. The amplitudes of g-modes at the level of the 2nd N-shell, normalized at the bottom of the photosphere, are plotted as a function of the periods. The modes are represented by upward triangles for ℓ=1, downward triangles for ℓ=2 and squares for ℓ=3. The modes amplitude shows maxima for periods around 1300 s (ℓ=1), 750 s (ℓ=2), and 500 s (ℓ=3). The N-shell acts as a filter for the corresponding g-modes.

3 Conclusions

In a model of PG1159 stars in which the assumptions made are: (1) that helium dominates the composition in the envelope, and (2) that diffusion is responsible for the levitation of heavy elements, it is suggested that the metal clouds in diffusion equilibrium may play a determinant role in the pulsation problem. It is found here that the deep nitrogen cloud concentrates at the depth where the thermal time scale equals the typical pulsation periods observed in PG1159. This strongly suggests that such metal clouds could potentially drive g-modes pulsations. It is also argued that the narrow metallic accumulation clouds, if they are able to destabilize g-modes, would provide a powerful filtering mechanism. The fully nonadiabatic nonradial linear stability analysis for PG1159 models, taking into account the inhomogeneous composition derived from diffusion equilibrium considerations is in progress (Vauclair, Dolez, and Turck-Chièze, in preparation).

References

Barstow, M.A., and Holberg, J.B. 1989, in *Proc. of Berkeley Colloquium on Extreme Ultraviolet Astronomy*

Dolez, N., and Vauclair, G. 1981, *Astron. Astrophys.*, **102**, 375.

Dziembowski, W., and Koester, D. 1981, *Astron. Astrophys.*, **97**, 16.

Hine, B.P., and Nather, R.E. 1987, in *The Second Conference on Faint Blue Stars, IAU Colloq. No. 95*, eds. A.G.P. Philip, D.S. Hayes, and J. Liebert (L. Davis Press, Schenectady), p.627.

Kawaler, S.D. 1986, Ph.D. Thesis, University of Texas at Austin.

Kawaler, S.D., and Hansen, C.J. 1989, in *White Dwarfs, IAU Colloq. No. 114*, ed. G. Wegner (Springer, Berlin), p.97.

Kawaler, S.D., Winget, D.E., Hansen, C.J., and Iben, I. 1986, *Astrophys. J. Letters*, **306**, L41.

Starrfield, S.G. 1987, in *The Second Conference on Faint Blue Stars, IAU Colloq. No. 95*, eds. A.G.P. Philip, D.S. Hayes, and J. Liebert (L. Davis Press, Schenectady), p.309.

Starrfield, S.G., Cox, A.N., Hodson, S.W., and Pesnell, W.D. 1983, *Astrophys. J. Letters*, **268**, L27.

Starrfield, S.G., Cox, A.N., Kidman, R., and Pesnell, W.D. 1984, *Astrophys. J.*, **281**, 800.

Starrfield, S.G., Cox, A.N., Kidman, R., and Pesnell, W.D. 1985, *Astrophys. J. Letters*, **293**, L23.

Vauclair, G. 1987, in *The Second Conference on Faint Blue Stars, IAU Colloq. No. 95*, eds. A.G.P. Philip, D.S. Hayes, and J. Liebert (L.Davis Press, Schenectady), p.341.

Vauclair, G. 1989, in *White Dwarfs, IAU Colloq. No. 114*, ed. G. Wegner (Springer, Berlin), p.176.

Vauclair, G., Vauclair, S., and Greenstein, J.L. 1979, *Astron. Astrophys.*, **80**, 79.

Werner, K., Heber, U., and Hunger, K. 1989, in *Proc. of the Hot Star Workshop: Intrinsic Properties of Hot Luminous Stars*, Boulder, Colorado.

Winget, D.E., Van Horn, H.M., Tassoul, M., Hansen, C.J., and Fontaine, G. 1983, *Astrophys. J. Letters*, **268**, L33.

Winget, D.E., Van Horn, H.M., Tassoul, M., Hansen, C.J., Fontaine, G., and Carroll, B.W. 1982, *Astrophys. J. Letters*, **252**, L65.

Mode Identification in a Slowly Rotating Star from Line Profile Variations

L. A. Balona

South African Astronomical Observatory,
P. O. Box 9, Observatory 7935, South Africa

Abstract: A new quantitative least-squares method of identifying the spherical harmonic order from the moments of the line profile is presented. The method is applicable to stars in which the projected rotational velocity is of the same order or less than the pulsational amplitude. Together with a previous method which applies only to stars in which the projected rotational velocity exceeds the pulsational amplitude, it is now possible to obtain an objective mode identification and least-squares estimates of the pulsational parameters for any multiperiodic star.

1 Introduction

The problem of mode identification is crucial if stellar pulsation is to be used as a probe of stellar properties. When a large number of modes are observed, as in the sun, this is easily accomplished by comparing observed and theoretical pulsation frequencies. For the vast majority of stars in which only a few modes are observed, this method cannot be used. Yet important information on the gross stellar properties could be extracted if the pulsation modes could be identified. As an example we need only mention the sensitivity of the period ratio in double-mode Cepheids to the adopted mass and the consequent mass discrepancy problem which is still unresolved.

Mode identification using relationships between the amplitudes and phases of light and colour variations has been considered by Balona and Stobie (1980) and Watson (1988). But most of the information concerning stellar pulsation resides in the variation of spectral line profiles. The advent of high-resolution digital spectroscopy has made a study of the line profile variations eminently suitable as a means of mode identification. This is done by comparing the observed variations with a grid of computed profiles for a range of modes and pulsation parameters. Osaki (1971) and Kambe and Osaki (1988) have described the appearance of line profile variations for a large range of pulsation parameters.

This method is adequate when only one mode is present, but even in this case there is no simple way of taking into account the effect of temperature modulation (which becomes very important in rapidly-rotating stars). Furthermore, there is not indication of the uniqueness of a given solution. The method is impractical for multiperiodic stars since it is impossible to separate the effect of different modes. To overcome these problems, Balona (1986b, 1987) developed a method which is based on the time variation of the line-profile moments. This leads to an algorithm which enables mode identification to be made for any number of frequencies and gives a least-squares estimate of the pulsation parameters. However, the method is only applicable to stars where the projected rotational velocity is significantly larger than the pulsational velocity.

A discussion of the time-variation of the moments for slowly-rotating stars is given by Balona (1986a). This work is of a qualitative nature and aims to obtain mode identification criteria from an inspection of the relative amplitudes and phases of the Fourier components of the moments. In this note we present an extension of this method which allows a fully quantitative solution of the problem. The algorithm presented here complements that discussed in Balona (1986b, 1987). The two methods allow quantitative mode identification for multiperiodic stars with arbitrary projected rotational velocities.

2 Variation of the Line-profile Moments

The velocity component in the direction of the observer at a point on the surface is the sum of the contributions due to rotation, v_{rot}, and pulsation, v_{pul}. If the intrinsic line profile is infinitely narrow, the n-th moment is given by:

$$M_n = \int (v_{\mathrm{rot}} + v_{\mathrm{pul}})^n \, dL \tag{1}$$

where dL is the component of luminosity of the surface element in the direction of the observer. The integral is taken over the visible hemisphere of the star. The luminosity varies with the pulsation frequency ω, but its effect on the line profile is negligible unless the projected rotational velocity is large compared with the pulsational velocity amplitude (Balona 1987).

It is evident that the n-th moment will vary with frequencies as high as $n\omega$. Since the intensity of the line profile at a given wavelength is a function of these moments, it follows that the first and higher harmonics may have significant power. In fact, it is easy to show that for axisymmetric modes the variation of the intensity of the line profile at the central wavelength does not have a frequency equal to the pulsation frequency, but one which is twice as large. If photometry is unavailable, it is possible to obtain the frequencies of oscillation from an analysis of the time variation of line profile intensities at fixed wavelengths (e.g. Gies and Kullavanijaya 1988). Care must be taken to ensure that the frequencies obtained in this way are not harmonics or cross-coupling of one frequency with another. In fact, it seems likely that the 2.26 hour period found by Gies and Kullavanijaya (1988) in ϵ Per is not a real eigenfrequency, but the first harmonic of the mode at 4.47 hours. Analysis of the time variation of the moments is probably a better way of obtaining the pulsation frequencies since these effects can be more easily controlled.

Since $(v_{\mathrm{pul}})^r$ contains a time dependence of the form $\sin^r \omega t$, it is clear that only odd powers of v_{pul} contribute to a variation with frequency ω and that even powers will contribute to frequency 2ω. Provided that the projected rotational velocity is significantly larger than the pulsational velocity, the effect of the second and higher powers of v_{pul}

can be neglected and consideration of the first power of $v_{\rm pul}$ is sufficient to obtain the Fourier components of frequency ω accurate to second order. This is the case discussed by Balona (1986b, 1987).

When the projected rotational velocity is comparable to, or less than $v_{\rm pul}$, this approximation cannot be used. A full solution of the problem requires evaluation of integrals over the visible hemisphere of several products of spherical harmonics. It is not possible to evaluate these integrals in a simple way. However, the first and second moments require an evaluation of integrals involving not more than the product of two spherical harmonics. Furthermore, the time variation is unaffected by the form of the intrinsic profile or the projected rotational velocity (Balona 1986b). Comparison of the observed Fourier components with calculated Fourier components of the first and second moments allows a least-squares estimation of the pulsation parameters together with a discriminant which quantifies the goodness of fit of a particular solution. In this note we describe how this can be done and supply a numerical example.

3 Variation of the First and Second Moment

We use a coordinate system $(r,\ \theta,\ \phi)$ in which $\theta = 0$ points towards the observer and $\mu = \cos\theta$. The adopted limb darkening law is

$$h = 1 - u + u\cos\theta. \tag{2}$$

When the projected rotational velocity is small, we can neglect the effects of temperature modulation etc. and write the n-th moment, normalized to unit area, as:

$$m_n = b\int_0^{2\pi}\int_0^1 (v_{\rm rot} + v_{\rm pul})^n(\mu + \beta\mu^2)d\mu\, d\phi \tag{3}$$

where $\beta = u/(1-u)$ and $v_{\rm rot} = v\sin i(1-\mu^2)^{1/2}\sin\phi$. The normalizing factor is $1/b = \pi(1 + 2\beta/3)$. For a spheroidal mode of order (ℓ, m) we have:

$$v_{\rm pul} = N_{\ell m}V_r\sum_{k=-\ell}^{\ell} a_{lmk}\left\{\mu P_{\ell k} + \alpha(1-\mu^2)\frac{dP_{\ell k}}{d\mu}\right\}\sin(\omega t + k\phi + \chi_r) \tag{4}$$

where $N_{\ell m}$ is the normalization factor for spherical harmonics, V_r is the vertical component of pulsational velocity, α is the ratio of horizontal to vertical velocity amplitude and χ_r is a phase constant. The transformation between a coordinate system defined by the pulsation axis to one defined by the direction of the observer is made by the summation. The matrix elements of $a_{\ell mk}$ are given in Balona (1986a).

The first moment (centroid) is:

$$m_1 = -2\pi b\, N_{\ell m}\, V_r\, \sin(\omega t + \chi_r)\, a_{\ell m0}[J^0_{\ell 0} + \alpha(\ell+1)(J^0_{\ell 0} - j^0_{\ell+1,0})]. \tag{5}$$

where the J- and j-integrals are discussed in Balona (1987). For a given mode at a particular orientation, V_r and χ_r can be determined from the Fourier components of m_1.

The normalized second moment is:

$$m_2 = m_2' + b\int_0^{2\pi}\int_0^1 (v_{\rm rot})^2(\mu+\beta\mu^2)d\mu d\phi + 2b\int_0^{2\pi}\int_0^1 v_{\rm rot}v_{\rm pul}(\mu+\beta\mu^2)d\mu d\phi$$
$$+ b\int_0^{2\pi}\int_0^1 (v_{\rm pul})^2(\mu+\beta\mu^2)d\mu d\phi, \qquad (6)$$

where m_2' is the second moment of the intrinsic line profile. The second term has been evaluated by Balona (1986b, 1987) and represents the contribution of rotational velocity to the line broadening. The third term varies with frequency ω and has been evaluated using the J- and j-integrals (Balona 1987). It vanishes for axisymmetric modes.

The last term has a component which varies with frequency 2ω; We will denote it by $m_2(2\omega)$. Defining

$$B_{\ell k}^{nm} = \int_0^1 \mu^n P_{\ell m} P_{km} d\mu, \qquad (7)$$

we have:

$$m_2(2\omega) = -\pi b(N_{\ell m}V_r)^2 \cos 2(\omega t + \chi_r) \sum_{k=-\ell}^{\ell} (-1)^k a_{\ell m k} a_{\ell m,-k}$$
$$[\alpha_r^2(B_{\ell\ell}^{3k} + \beta B_{\ell\ell}^{4k}) + \alpha_h^2(B_{\ell+1,\ell+1}^{1k} + \beta B_{\ell+1,\ell+1}^{2k})$$
$$- 2\alpha_r\alpha_h(B_{\ell,\ell+1}^{2k} + \beta B_{\ell,\ell+1}^{3k})] \qquad (8)$$

where $\alpha_r = 1 + \alpha(\ell+1)$ and $\alpha_h = \alpha(\ell+1)$. The B-integrals may be evaluated using recurrence relations (see appendix). There is an additional component which evaluates to a constant and contributes to the overall line broadening. This can be calculated in a similar manner.

Given the ratio of horizontal to vertical velocity, α, the summation can be evaluated for any given mode (ℓ, m) at a particular inclination. From the Fourier decomposition of the second moment, the value of V_r and χ_r can be determined and the quantities $V_r \sin\chi_r$, $V_r \cos\chi_r$ calculated. The value of $V_r \sin\chi_r$ and $V_r \cos\chi_r$ can also be determined from the Fourier decomposition of the first moment (see above). The two independent estimates are then used to form the discriminant:

$$\sigma^2 = [(V_r \sin\chi_r)_1 - (V_r \sin\chi_r)_2]^2 + [(V_r \cos\chi_r)_1 - (V_r \cos\chi_r)_2]^2. \qquad (9)$$

The most probable values of ℓ, m, and inclination are found when σ reaches a minimum value.

The algorithm is not modified when more than one mode is present, since the Fourier coefficients can be calculated separately for each frequency of oscillation if the observations are sufficiently numerous and well-phased. In principle this algorithm can be applied to rapidly rotating stars as well, except that the Fourier amplitude of $m_2(2\omega)$ will be small relative to $m_2(\omega)$. The method proposed by Balona (1987) will be numerically more accurate and also provides estimates of the effect of temperature modulation. Moreover it will give an estimate of the projected rotational velocity and the width of the intrinsic profile as well.

4 Numerical Example

As an illustration, we generated line profiles for a mode with the following parameters: $\ell = 2$, $m = -1$, $V_r = 50\,\mathrm{km\,s^{-1}}$, $\alpha = 0.1$, $\chi_r = 300°$, $u = 0.36$, $v \sin i = 30\,\mathrm{km\,s^{-1}}$, $i = 60°$. We used a Gaussian with standard deviation of $10\,\mathrm{km\,s^{-1}}$ as an intrinsic profile. Normalizing the line profile to unit area we obtain the following Fourier decomposition for the first and second normalized moments:

$$m_1 = -0.516 + 3.13 \sin \omega t - 5.427 \cos \omega t; \tag{10}$$

and

$$m_2 = 365.9 - 94.4 \sin \omega t - 46.98 \cos \omega t - 26.86 \sin 2\omega t + 15.48 \cos 2\omega t. \tag{11}$$

The discriminant, σ, was calculated for all values of m between $\ell = 0$ and $\ell = 4$ at intervals of one degree of inclination. Table 1 lists the first few solutions in order of increasing σ.

Table 1. Values of the smallest discriminant, σ, and the resulting pulsational parameters for a numerical example. The inclination, i and phase angle, χ_r are in degrees. The vertical component of pulsational velocity amplitude, V_r, the r.m.s. intrinsic line width, W_i, and projected rotational velocity, $v \sin i$, are in $\mathrm{km\,s^{-1}}$.

ℓ	m	i	σ	V_r	χ_r	W_i	$v \sin i$
2	-1	60	0.020	49.98	300.0	9.88	30.1
1	0	54	0.125	37.91	300.0	-	-
2	0	0	1.115	25.99	300.0	-	-
3	-1	26	1.468	88.86	300.0	9.28	11.1
3	-2	66	5.225	113.8	300.0	6.45	11.2

The projected rotational velocity, $v \sin i$, and r.m.s. width of the intrinsic line, W_i, can be obtained from the zero points of the second and fourth moments, as described by Balona (1986b, 1987). However, these zero points include contributions from integrals involving the pulsational velocity but which evaluate as constants (see above). The resulting estimates will be inaccurate unless $v \sin i \gg V_r$. Consequently, in this example we determined $v \sin i$ from the term $m_2(\omega)$ in the second moment and obtained W_i from the corrected zero point of the second moment. Notice that $m_2(\omega)$ vanishes for axisymmetric modes, so that an estimate of the projected rotational velocity and intrinsic line width cannot be obtained from the first two moments alone.

References

Balona, L.A. 1986a, *Monthly Notices Roy. Astron. Soc.*, **219**, 111.
Balona, L.A. 1986b, *Monthly Notices Roy. Astron. Soc.*, **220**, 647.
Balona, L.A. 1987, *Monthly Notices Roy. Astron. Soc.*, **224**, 41.
Balona, L.A., and Stobie, R.F. 1980, *Monthly Notices Roy. Astron. Soc.*, **192**, 625.
Gies, D.R., and Kullavanijaya, A. 1988, *Astrophys. J.*, **326**, 813.
Kambe, E., and Osaki, Y. 1988, *Publ. Astron. Soc. Japan*, **40**, 313.
Osaki, Y. 1971, *Publ. Astron. Soc. Japan*, **23**, 485.
Watson, R.D. 1988, *Astrophys. Space Sci.*, **140**, 255.

Appendix: Evaluation of the B-integrals

To calculate $B_{\ell k}^{nm}$, we proceed as follows. By simple integration we obtain:

$$B_{\ell\ell}^{2n,\ell} = \frac{2[(2\ell)!]^2}{(\ell!)^3} \frac{(2n)!(n+\ell+1)!}{n![2(n+\ell+1)]!} \tag{A.1}$$

and

$$B_{\ell\ell}^{2n+1,\ell} = \frac{[(2\ell)!]^2}{2^{2\ell+1}(\ell!)^3} \frac{n!}{(n+\ell+1)]!} \tag{A.2}$$

Starting with $\ell = 0$, we obtain B_{00}^{n0}. Using

$$(\ell+1)B_{\ell,\ell+1}^{n\ell} = (2\ell+1)B_{\ell\ell}^{n+1,\ell}, \tag{A.3}$$

we obtain B_{01}^{n0}. Using

$$(\ell+1)^2 B_{\ell+1,\ell+1}^{n\ell} = (2\ell+1)^2 B_{\ell\ell}^{n+2,\ell}, \tag{A.4}$$

we obtain B_{11}^{n0}. Using

$$\ell^2(\ell+1)^2 B_{\ell+1,\ell+1}^{nm} = \ell^2(2\ell+1)^2 B_{\ell\ell}^{n+2,m} - 2\ell(2\ell+1)(\ell^2-m^2)B_{\ell-1,\ell}^{n+1,m} + (\ell^2-m^2)^2 B_{\ell-1,\ell-1}^{nm} \tag{A.5}$$

we obtain B_{22}^{n0}. Using

$$\ell(\ell+1)B_{\ell,\ell+1}^{nm} = \ell(2\ell+1)B_{\ell\ell}^{n+1,m} - (\ell^2-m^2)B_{\ell-1,\ell}^{nm} \tag{A.6}$$

we obtain B_{21}^{n0} etc. In this way all the required B-integrals can be evaluated using recurrence relations.

Long-Term Variations of Nonradial Oscillations in a Rapidly Rotating Early-Type Star (ζ Oph)

Eiji Kambe [1], Hiroyasu Ando [2], and Ryukou Hirata [3]

[1]Department of Astronomy, University of Tokyo, Bunkyo-ku, Tokyo 113, Japan
[2]National Astronomical Observatory, Mitaka-shi, Tokyo 181, Japan
[3]Department of Astronomy, Kyoto University, Sakyo-ku, Kyoto 606, Japan

Abstract: We have been monitoring the line profile variations of a Be star, ζ Oph, since 1986 to investigate their connection to the Be-activity. From the period analysis of the variations, we have found that two modes detected in August 1989 are identical to the modes dominated in April 1987 and in May 1988 (Kambe *et al.* 1989). The amplitudes and the ratio of horizontal-to-radial velocity amplitudes, of oscillations k, seem to vary from season to season, but it must be confirmed by further observations. The relation of the line profile variations to the Be-activity in this star is not yet clear.

1 Introduction

Recently the relation between nonradial oscillations in Be stars and its Be-activity has been suspected (Abbot *et al.* 1986, Smith 1986). To clarify the connection between them, we need the extensive observations concentrated on one or two of these stars over the periods of Be phenomenon (Baade 1987, Percy 1987, Harmanec 1990). So far, however, only a few results have been reported which show the long-term variations of the nonradial oscillations in Be stars (Penrod 1987, Smith 1989).

In Japan, we have been extensively monitoring the line profile variations of some Be stars (ζ Oph, λ Eri, 28 Cyg, etc.) since 1986. In this paper, we show the results of the mode identification of the nonradial oscillations in ζ Oph and discuss its variation in recent three years. A part of our results has already been published (Kambe *et al.* 1989).

2 Observations and Techniques of Analysis

From 1987 to 1989, we have obtained about 130 high resolution spectra with high signal-to-noise ratio of a Be star, ζ Oph (O9.5Ve). The spectra were obtained with the RCA-CCD camera system attached to the Coude spectrograph of the 1.88 m reflector at the Okayama Astrophysical Observatory. The HeI λ 6678 line has been mainly monitored, but some H_α profiles have been also obtained as a diagnosis of Be phenomenon. The star has been in its quiescent phase since 1982 and no emission feature has been observed.

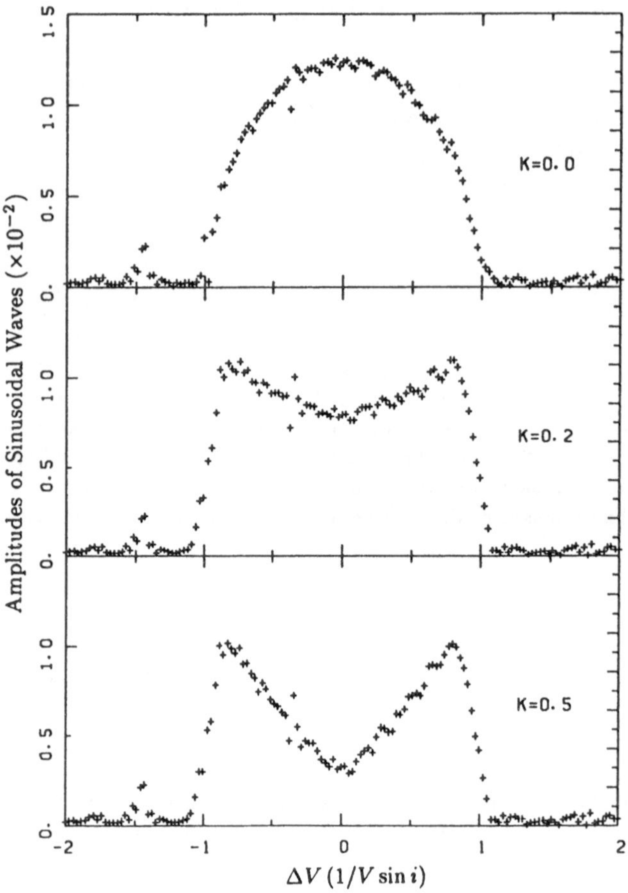

Fig. 1. Wavelength (or velocity displacement) dependence of the amplitude of sinusoidal waves to which data is fitted at each wavelength point. The dependence is shown for three different sectoral modes which have the same values of parameters except k values (shown in the figure). See Kambe *et al.* (1989) in detail.

The spectra were analyzed with the method developed by Gies and Kullavanijaya (1988) and extended by Kambe *et al.* (1989). The periods and the amplitudes of oscillations, the number of azimuthal nodes, m, and k can be estimated by using this method. Here it should be noted that the estimated k values are reliable only in such a case that the line profile variations are mainly caused by the velocity fields of oscillations. In Fig. 1, we plotted the wavelength (or velocity displacement) versus the amplitudes of the sinusoidal waves, to which the temporal variation of line depth at each wavelength is fitted, for three different sectoral modes which have the same l, m, ($l = -m = 8$) and amplitudes but different k (In these models, only velocity fields of oscillations are considered). As seen in this figure, the dependence of the amplitudes of sinusoidal waves on wavelength is different according to k values (for the discussion in detail, see Kambe *et al.* 1989). Thus, we can estimate k values from this figure. However, this method may not works well in a case when temperature variations cannot be neglected. In such a case, small k value might be derived even for g-modes as pointed out by Lee and Saio (1990), though their results are preliminary.

3 Results and Discussion

The results of mode identification are tabulated in table 1. Periods of oscillations as well as values of m and k are shown in the table. The maximum amplitude of the fitted sinusoidal waves among wavelength points (except noise) are also estimated in the last column of the table in units of continuum level. Of course, the amplitudes of oscillations might be estimated in velocity units from the line profile fitting method. However, we dare not to do so here, because further assumptions, such as limb darkening law, the extent of gravitational darkening and line strength, must be done to estimate the velocity amplitudes of oscillations, which make them ambiguous.

Table 1. Results of mode identification.

Observing run	m	Period (hr)	k	Amplitude of sinusoidal wave (10^{-3})
	-4 ± 1	3.89 ± 0.15	0.2 ± 0.1	~ 3.0
1987 Apr.	-7 ± 1	2.72 ± 0.08	0.4 ± 0.2	~ 3.5
	-2 ?	low	?	~ 1.2
1988 May	-7 ± 1	2.44 ± 0.08	0.2 ± 0.1	~ 3.8
	-4 ± 1	3.33 ± 0.1	0.2 ± 0.1	~ 3.2
	-9 ± 1	1.86 ± 0.05	0.2 ± 0.1	~ 2.4
1989 Aug.	-7 ± 1	2.44 ± 0.08	0.1 ± 0.1	~ 5.5
	-4 ± 1	3.33 ± 0.1	0.4 ± 0.1	~ 3.5

Two modes have been detected as identical in these seasons, *i.e.*, an $l = -m = 4$ NRP with a period of 3.33 hr and an $l = -m = 7$ NRP with a period of 2.44 hr. (The periods of main two modes in 1987 correspond to the one day alias periods in 1988.) These periods are compared with other results (Vogt and Penrod 1983, Balona and Engelbrecht 1987, Harmanec 1989) in detail in the paper of Kambe *et al.* (1989). These modes are prograde in the rotating frame with periods about 8 hours (here, we assume that $R = 8R_\odot$, $v \sin i \sim 400\,\mathrm{km\,s^{-1}}$ and $i \sim 90°$), which correspond to low order g-modes. The k values almost correspond to the theoretically expected ones ($M = 18M_\odot$; $k = 0.2 \sim 0.4$). Since k values are not very close to zero, the velocity displacements might be dominant as the cause of line profile variations in ζ Oph.

Our data indicate some long-term variations of nonradial oscillations in this star.

Especially, the amplitude of the mode with a period of 2.44 hr seems to increase considerably over these seasons. The confirmation by further observation is needed, however, because it is possible that the amplitudes of the modes estimated here are largely affected by the cycle-to-cycle variations of the amplitudes of oscillations, which has been extensively studied by Smith and his colleagues (Smith *et al.* 1987, Smith 1989) for some line profile variables. Here, we emphasize that the monitoring of long-term variations of the amplitudes of oscillations is important, because they might represent the energy of oscillations, while cycle-to-cycle variations might be largely affected by the transient atmospheric phenomenon at the surface of the star.

The k values seem to change over these observing terms, though it must also be confirmed using more abundant data. The utilization of some different lines is desirable, because ambiguities due to some blended lines, such as atmospheric lines of the earth, exist in the estimated k values.

We have found some indications of the long-term variations of nonradial oscillations. However, the relation of such long-term variations of nonradial oscillations and Be-activity has not yet been clear. Such retrograde modes that are inferred by Ando's mechanism for the Be phenomenon (Ando 1986) were not detected yet. But we cannot say at present whether the modes exist or not, because our data still have very limited time coverage. Therefore, more accumulation of data is needed to understand more deeply the property of nonradial oscillations in this star.

References

Abbot, D.C., Garmany, C.D., Hansen, C.J., Henrichs, H.F., and Pesnell, W.D. 1986, *Publ. Astron. Soc. Pacific*, **98**, 29.

Ando H. 1986, *Astron. Astrophys.*, **163**, 97.

Baade, D. 1987, in *Physics of Be Stars, IAU Colloq. No. 92*, eds. A. Slettebak and T. P. Snow (Cambridge University Press, Cambridge), p. 361.

Balona, L.A., and Engelbrecht, C.A. 1987, in *Physics of Be Stars, IAU Colloq. No. 92*, eds. A. Slettebak and T. P. Snow (Cambridge University Press, Cambridge), p.87.

Gies, D.R., and Kullavanijaya, A. 1988, *Astrophys. J.*, **326**, 813.

Harmanec, P. 1989, *Bull. Astron. Inst. Czech.*, **40**, 201.

Harmanec, P. 1990, *Bull. Astron. Inst. Czech.*, in press.

Kambe, E., Ando, H., and Hirata, R. 1989, submitted to *Publ. Astron. Soc. Japan*.

Lee, U., and Saio, H. 1990, *Astrophys. J.*, in press.

Penrod, D.G. 1987, in *Physics of Be Stars, IAU Colloq. No. 92*, eds. A. Slettebak and T. P. Snow (Cambridge University Press, Cambridge), p.463.

Percy, J. 1987, in *Physics of Be Stars, IAU Colloq. No. 92*, eds. A. Slettebak and T. P. Snow (Cambridge University Press, Cambridge), p.49.

Smith, M.A. 1986, in *Hydrodynamic and Magnetohydrodynamic Problems in the Sun and Stars*, ed. Y. Osaki (University of Tokyo, Tokyo), p.145.

Smith, M.A. 1989, *Astrophys. J. Suppl.*, **71**, 357.

Smith, M.A., Fullerton, A.W., and Percy, J.R., 1987, *Astrophys. J.*, **320**, 768.

Vogt, S.S., and Penrod, D.G. 1983, *Astrophys. J.*, **275**, 661.

Low-Frequency Oscillations of Rotating Massive Stars

Hideyuki Saio and Umin Lee

Department of Astronomy, University of Tokyo, Bunkyo-ku, Tokyo 113, Japan

Abstract: Stellar rotation significantly modifies the property of nonradial oscillations when the frequencies (in the co-rotating frame) are comparable to or less than the frequency of rotation. The angular dependence of the amplitude of such an oscillation cannot be expressed by a single spherical harmonic, $Y_l^m(\theta, \phi)$. The amplitude of a g-mode tends to be confined to a narrow equatorial region compared to the non-rotating case. In addition to g-modes, which exist in a radiative equilibrium region, *inertial* (oscillatory convective) modes exist in a convective region. Half of the inertial modes have *negative* energy, while all the g-modes have positive energy. When the oscillation frequency of a positive energy mode is close to that of a negative energy mode, resonance coupling between the two modes occurs and energy flows from the negative energy mode to the positive one to increase the amplitude of both modes; *i.e.*, to lead to the *overstability* of the oscillations. Therefore, overstable g-mode oscillations are possible for all rotating stars with inner convective and outer radiative regions. The resonance excitation of g-modes in the radiative envelope by inertial oscillations in the rotating convective core give natural explanation for rapid variations of early type stars. From observed periods of a variable early type star we can obtain information on the superadiabatic temperature gradient and the angular rotation frequency of the convective core.

Recent observations have revealed that rapid light and line profile variations on time scales of ~ 0.2 to ~ 2 days are ubiquitous in early type stars. Recent reviews on these objects are given by, e.g., Baade (1987), Percy (1987), Smith (1988), and Harmanec (1989). There are two competent models to explain the light and line profile variations; *i.e.*, nonradial oscillations and rotation with inhomogeneous surface brightness. Many of these stars show light variations with double-wave light curves, which are argued in preference of the rotation origin (e.g., Balona and Engelbrecht 1986; Harmanec 1989). The double-wave light curves are consistent with the existence of "superperiod"(= azimuthal order m times observed oscillation periods), which are obtained from line profile variations (e.g., Smith 1985, Gies and Kullavanijaya 1988). A double-wave (or multi-wave) light curve can be reproduced by superposing nonradial oscillations (with sinusoidal light curves) which obey the rule of superperiod; *i.e.*, for each oscillation m times the period makes the superperiod (= the period of the multi-wave light curve). The superperiod can be explained by nonradial oscillations which have very low frequencies with respect to a rotating body, because in such a case observed frequencies of oscillations are written as

$$\sigma_{obs} = \omega - m\Omega \simeq -m\Omega, \tag{1}$$

where ω is the oscillation frequency seen in the frame rotating with angular frequency Ω. The rotating star spots model is the limiting case of $\omega = 0$ and Ω is the angular frequency of rotation at the stellar surface. In other words, nonradial oscillation model with $\omega \ll |m\Omega|$ and Ω being the surface rotation frequency is indistinguishable from the star spots model. However, we note that the superperiod of ϵ Per ($\simeq$ 13.5 hour) seems to be too short for the surface rotation period (Gies and Kullavanijaya 1988; *cf.* Harmanec 1989). This difficulty is avoided by regarding the superperiod as the rotation period of the *core*, which is possible when envelope oscillation modes are resonantly excited by core oscillations which have very low frequencies in the co-rotating frame of the core.

The properties of low-frequency nonradial oscillations are significantly modified from those of a non-rotating star. In order to understand such oscillations we have to take an approach significantly different from that familiar for nonradial oscillations of non-rotating stars. The angular dependence of the amplitude of such an oscillation mode cannot be expressed by a single spherical harmonic, $Y_l^m(\theta,\phi)$. We need to sum all possible terms proportional to $Y_l^m(\theta,\phi)$ with $l \geq |m|$ for a given m. Then the nonradial oscillations of rotating stars are described by infinite number of equations. Oscillations are divided into two groups; even and odd modes (Berthomieu *et al.* 1978). Perturbation of a scalar variable for an even (odd) mode is represented by a sum of the terms proportional to Y_l^m with $l = |m|, |m|+2, \ldots$ ($l = |m|+1, |m|+3, \ldots$), and symmetric (anti-symmetric) to the equator.

Asymptotic analyses using WKBJ approximation are possible (Berthomieu *et al.* 1978; Lee and Saio 1987a, 1989). The square of the radial wave number for low-frequency oscillations is written as (Lee and Saio 1987a)

$$k_r^2 r^2 \simeq N^2 \lambda / \omega^2, \tag{2}$$

where N is the Brunt-Väisälä frequency, and λ is an eigenvalue of a symmetric matrix $\boldsymbol{W}$, which gives the relation between the pressure perturbation and the horizontal displacement. [The matrix $\boldsymbol{W}$ tends to a diagonal matrix whose diagonal elements are given by $l_j(l_j+1)$ with $l_j \equiv |m| + j$ (j represents an integer including zero).] Note that equation (2) is reduced to the relation for nonrotating stars when λ is replaced by $l(l+1)$. However, since λ's can take negative values (see Figs. 34.3 and 34.4 of Unno *et al.* 1989 or Fig. 6 of Lee and Saio 1987a), different things happen. Equation (2) shows that k_r^2 is positive in a convective zone ($N^2 < 0$) if λ is negative; *i.e.*, there exist waves propagative in a convective zone, which we call *inertial* or oscillatory convective waves.

Frequencies of inertial (even) modes with $n = 1$ ($n \equiv$ radial order) are shown in Fig. 1 as functions of the angular frequency of rotation of the core, $\Omega_{\rm core}$ for various values of m. These are obtained by solving truncated equations including only two spherical harmonic components for the convective core of a $10\,M_\odot$ ZAMS model. In this paper we only consider prograde waves, which have positive ω's for negative m. Since the relation between ω and $\Omega_{\rm core}$ is proportional to the square root of the superadiabatic temperature gradient, $\sqrt{\nabla - \nabla_{\rm ad}}$, in the convective core, the x- and y-axes are normalized by this quantity. For given n and m two frequencies (ω_H and ω_L; $\omega_H > \omega_L$) exist at each rotation frequency $\Omega_{\rm core}$. Lee and Saio (1990b) have shown that the wave energy of inertial modes on the low-frequency branches (*i.e.*, having ω_L) are negative, while the wave energy of the other inertial modes and g-modes are positive. Oscillation frequencies for high order g-modes, which are propagative in the radiative envelope, are located in the same frequency range of Fig. 1. When a positive energy mode has a oscillation frequency close to that of a negative energy mode, resonance coupling between the two modes allows energy to flow from negative to positive energy modes, which results in a single overstable mode (Lee and Saio 1986, 1989, 1990b). This overstability exists even when nonadiabatic effect of radiation is included (Lee and Saio 1987b). The same mechanism works for differentially rotating stars (Lee 1988), in which a inertial mode with frequency ω_L in the rotating core couples with a g-mode with frequency $\omega_e = \omega_L - m(\Omega_{\rm core} - \Omega_{\rm env})$ in the frame co-rotating with the envelope. These overstable modes are consistent with observed properties of rapid variations of early type stars. Since ω_L is much smaller than $|m|\Omega$ (Fig. 1), the frequency in observer's frame is $|m|\Omega_{\rm core}$ (eq. [1]), which shows the existence of a superperiod originated by the rotation period of the core. The amplitude of an oscillation on the surface is strongly confined in a narrow equatorial belt if $\omega_e \ll |m\Omega_{\rm env}|$. Line profile variations caused by such oscillations are discussed in Lee and Saio (1990a).

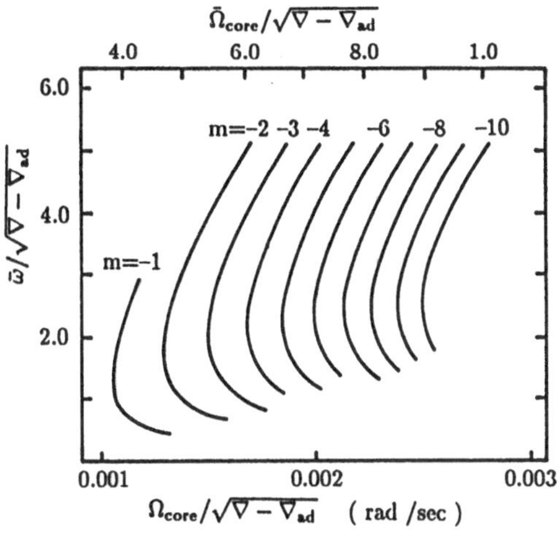

Fig. 1. Loci of inertial (oscillatory convective) modes in the plane of oscillation frequency (in the co-rotating frame) versus rotation frequency of the core rotation $\Omega_{\rm core}$ for various azimuthal order m. The bars on $\bar{\omega}$ and $\Omega_{\rm core}$ indicate that these are normalized by $\sqrt{GMR^{-3}}$.

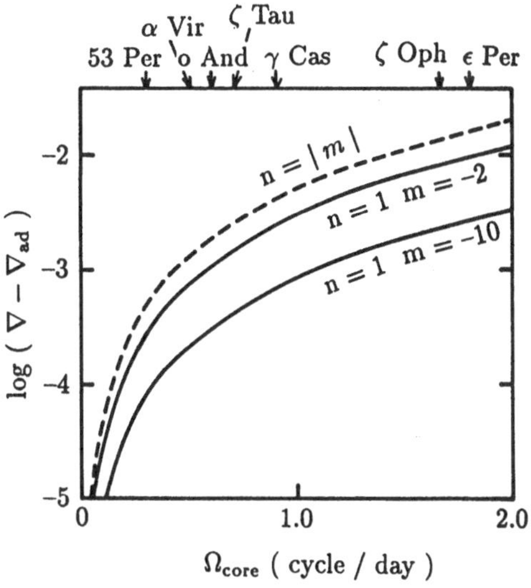

Fig. 2. The range of the superadiabatic temperature gradient in the convective core inferred from superperiods of variable early type stars. The horizontal axis is the angular frequency of rotation of the convective core.

For $n = 1$ modes we can read from Fig. 1 approximate values of $\Omega/\sqrt{\nabla - \nabla_{\rm ad}}$ at which resonant excitations are expected to occur. From these values we can calculate the superadiabatic temperature gradient for each rotation frequency of the core Ω. These relations for $m = -2$ and -10 are shown in Fig. 2. Since the curve for a given m shifts leftward as n increases (Lee and Saio, 1987a), the curves for the modes with $n > 1$ are located above the curve for $n = 1$ in Fig. 2. It is interesting to note that all the curves for the modes with $n = |m|$ $(-1 > m > -10)$ are located close to the curve for $m = -1$ in Fig. 1. Therefore, if we assume that inertial modes whose horizontal scales are comparable to radial ones (*i.e.*, $n \simeq |m|$) are responsible for rapid variations of early type stars, we have a relation of $\Omega/\sqrt{\nabla - \nabla_{\rm ad}} \sim 10^{-3}$ rad/sec. This relation is shown by a dashed curve in Fig. 2. Also shown in this figure are approximate rotation frequencies of the cores of some variable early type stars, which are obtained from the superperiod of each variable (53 Per, Balona 1987; α Vir, Smith 1985; o And, Hill *et al.* 1989; ζ Tau, Yang *et al.* 1989; γ Cas, Yang *et al.* 1988; ζ Oph, Kambe *et al.* 1989; ϵ Per, Gies and Kullavanijaya 1988).

This figure shows that in order for this mechanism to be responsible for these variables, the superadiabatic temperature gradient in the convective core must be much larger than that expected for a non-rotating star ($\nabla - \nabla_{\rm ad} \sim 10^{-6}$; Osaki 1974). Since rotation tends to inhibit convective motion, a large superadiabatic temperature gradient is probably necessary to transfer heat energy generated by nuclear reactions in the core.

References

Balona, L.A. 1987, in *Stellar Pulsation*, eds. A.N. Cox, W.M. Sparks, and S.G. Starrfield (Springer, Berlin), p.83.

Balona, L.A., and Engelbrecht, C.A. 1986, *Monthly Notices Roy. Astron. Soc.*, **219**, 131.

Baade, D. 1987, in *Proc. IAU Colloq. No. 92, Physics of Be Stars*, eds. A. Slettebak and T.P. Snow (Cambridge Univ. Press, Cambridge), p.361.

Berthomieu, G., Gonczi, G., Graff, Ph., Provost, J. and Rocca, A. 1978, *Astron. Astrophys.*, **70**, 597.

Gies, D.R. and Kullavanijaya, A. 1988, *Astrophys. J.*, **326**, 813.

Harmanec, P. 1989, *Bull. Astron. Inst. Czech.*, **40**, 201.

Hill, G.M., Walker, G.A.H., Dinshaw, N., Yang, S., and Harmanec, P. 1989, *Publ. Astron. Soc. Pacific*, **101**, 258.

Kambe, E., Ando, H., and Hirata, R. 1989, preprint.

Lee, U. 1988, *Monthly Notices Roy. Astron. Soc.*, **232**, 711.

Lee, U., and Saio, H. 1986, *Monthly Notices Roy. Astron. Soc.*, **221**, 365.

Lee, U., and Saio, H. 1987a, *Monthly Notices Roy. Astron. Soc.*, **224**, 513.

Lee, U., and Saio, H. 1987b, *Monthly Notices Roy. Astron. Soc.*, **225**, 643.

Lee, U., and Saio, H. 1989, *Monthly Notices Roy. Astron. Soc.*, **237**, 875.

Lee, U., and Saio, H. 1990a, *Astrophys. J.*, in press.

Lee, U., and Saio, H. 1990b, *Astrophys. J.*, submitted.

Osaki, Y. 1974, *Astrophys. J.*, **189**, 469.

Percy, J. 1987, in *Proc. IAU Colloq. No. 92, Physics of Be Stars*, eds. A. Slettebak and T.P. Snow (Cambridge Univ. Press, Cambridge), p.49.

Smith, M.A. 1985, *Astrophys. J.*, **297**, 206.

Smith, M.A. 1988, in *Pulsation and Mass Loss in Stars*, eds. R. Stalio and L.A. Willson (Kluwer, Dordrecht), p.251.

Unno, W., Osaki, Y., Ando, H., Saio, H., and Shibahashi, H. 1989, *Nonradial Oscillations of Stars* (2nd ed.) (University of Tokyo Press, Tokyo), Chap. 6.

Yang, S., Ninkov, Z., and Walker, G.A.H. 1988, *Publ. Astron. Soc. Pacific*, **100**, 233.

Yang, S., Walker, G.A.H., Hill, G.M., and Harmanec, P. 1989, preprint.

The MUSICOS Network for MUlti-SIte COntinuous Spectroscopy

Bernard Foing [1], Claude Catala [2] and the MUSICOS team

[1]Space Science Department of ESA, ESTEC,
2200AG Noordwijk, The Netherlands

[2]Institut d'Astrophysique Spatiale, BP 10, 91371 Verrieres le Buisson, France
and
DESPA, Observatoire Paris-Meudon, 92195 Meudon, France

Abstract: MUSICOS, (for MUlti SIte COntinuous Spectroscopy) is a project for a multisite network of high resolution spectrometers around the world partly dedicated to continuous spectroscopy, under study in France, with European and extra-European collaborators. This network aims to serve the solar/stellar community (for the study of asteroseismology, stellar rotational modulation, surface structures, Doppler imaging, variable winds, coordinated multi-frequency observations with space satellites...) specially for programs requiring a continuous spectroscopic coverage around the clock during several days. A major scientific goal of MUSICOS is to allow the study of non radial pulsations (e.g. of OB, Be, delta Scu, fast rotating B stars) and ultimately asteroseismological studies of solar type stars, using the most complete continuity. This goal drives the highest constraints on the spectrometer (high efficiency, high S/N, wavelength range over 500A, spectral stability). The first MUSICOS campaign dedicated to 3 targets in december 1989, involved sites and telescopes in Mauna Kea (CFH and 2.2m UH), Kitt Peak Mc Math, La Silla 1.4m CAT, France 1.5m OHP, Crimea 2.6m Shajn and China 2.16m Xinglong with two fiber-fed spectrographs transported for this campaign at Hawaii and Xinglong. The current MUSICOS instrument design (fiber-fed cross-dispersed echelle spectrograph with a resolution 30000 coupled to a CCD detector, for duplication) are described; the expected performances for asteroseismology and the strategy for a multisite network of similar spectrographs are presented.

1 Introduction

MUSICOS is an international project started in 1988 which aims at developing a multisite network of high resolution spectrographs. This project has received support from many scientists, from European countries, America and Asia so far. The goal of Musicos is to facilitate multi-site spectroscopic observations, first by setting up an organisation helping the coordination of observations from existing instruments at different sites, then by designing, developing and installing similar spectrometers in well-chosen sites around the world, for which part of the time would be devoted to multi-site observations.

2 The 1st MUSICOS Workshop in 1988

We organised in June 1988 a national workshop in France about the scientific use of multi-site spectroscopy (see Catala and Foing 1988). The workshop was attended by 50 participants, from different fields of solar and stellar physics. The interest for continuous spectroscopy was recognised for various topics (for the study of asteroseismology, stellar rotational modulation, surface structures, variable winds, for stellar activity programs where short term phenomena occurs, or fast rotation modulation allows application of Doppler imaging method. Also the study of wind variability, flare patrol, or eclipse imaging was stressed. Additionally, a joint network of high resolution spectroscopy and photometry would give a simultaneous support to continuous satellite observations. The need for multi-site continuous coverage (taking as example networks of solar seismology) was especially recognised for the asteroseismology programs e.g. of OB, Be, delta Scu, fast rotating B stars and solar type stars (see Däppen 1988, Catala and Foing 1988).

Given the positive impact of this workshop and the interest shown by several other European teams, we have decided to get started on the practical and technical aspects of the project as soon as 1989. In a first phase, we intend to organise multi-site campaigns using resident instruments on various telescopes around the world and a transportable fiber-fed mono-order spectrometer (ISIS, developed at Meudon Observatory and used until 1988 at the 1.93 m telescope at OHP). In a second phase, we shall continue in 1990 the design and development of an cross echelle fiber-fed spectrometer, and perform tests in the laboratory and at telescope on this prototype to assess the scientific performances of this fiber-fed spectrograph. In a third phase, we shall develop an instrument that can be proposed for duplication to different foreign collaborators. This will allow at long term multi-site campaigns with identical instruments, and to reach the limits in quality, sensitivity and homogeneity of ground based continuous spectroscopy that are set by asteroseismology requirements.

3 Asteroseismology Requirements

For asteroseismology, the need for a high S/N, and the possibility of observing a large number of photospheric lines, was stressed, together with the specific requirement of a high spectral spectral stability and possibility of very accurate velocity calibration. An observing continuity better than 60 percent in order to decrease the effect of parasite sidelobes and aliasing in the power spectrum, and to increase the signal to noise and resolution on the oscillation modes was requested, which argues for multi-site network. The need for prereduction of data at each site, in order to control in real time the instrumental parameters (quality of night, drifts) was made clear at the workshop.

The application domain would concern first the beta CMa stars, which oscillations can be observed in radial velocity, and for which Doppler imaging effect due to fast rotation allows to have access to information on modes with degree l higher than 4.

Other techniques can make use of line intensities oscillations. For the delta Scuti stars, continuous multi-site photometry has permitted the detection of non -radial modes. The Doppler rotation image information and the effect of large scale inhomogeneity can be studied spectroscopically. The same apply to Bp and Ap stars.

For solar-type oscillations, the velocity signal is much reduced (of the order of 15 cm/s per stronger solar mode). Classical slit spectrographs using one or a few lines are not stable and efficient enough. Resonance cell techniques have a good stability but can apply only to 0 or 1st magnitude stars with 4m class telescopes, which are oversubscribed and make difficult a continuous observing from several sites.

For having access to fainter stars from later-type closest stars or from other classes in the H-R diagram, the multi-line concept may be relevant in the future. For seismology velocity measurements, the MUSICOS spectrometer at 30000 resolution only with a 2m telescope would give on a 3 magnitude star a S/N of 100 in 1 mn, which coupled to the measurement of 400 lines relatively free of blending available on the spectral domain in the cross echelle mode would in principle allow to reach an oscillation detection of 3m/s per mn exposure, and thus a noise of 5cm/s for a continuous 72 hours observation.

The fact that solar oscillations have been detected with classic spectrographic instrument by measuring scattered or reflected light, but that the noise on stellar observations do not go below a rms velocity error of 40 m/s, suggests errors arising from point sources, with the beam geometry being sensitive to depointing of the stellar beam for these measurements. Tests on such fiber-fed cross spectrometer have been achieved using Th lamps for accurate calibration (e.g. by T. Brown, 1990). It is thus very important to improve the wavelength stability of the spectrograph down to the level allowed by the photon noise, and to reduce the sensitivity to guiding errors in particular, for instance using a fiber-fed spectrograph. Stability tests will be achieved, and accurate procedures for wavelength calibration are under study.

4 Technical Design of MUSICOS Instrument

From the scientific requirements after the MUSICOS workshop it was realised that a large range of scientific programmes require a spectral resolution 10000, 30000, and 80000. The main requirements from asteroseismology is the availability of large spectral domain in order to measure a number of photospheric lines in solar-type stars, and the stringent constraint for spectral stability down to the 10m/s range. The spectrograph can be fed by an optical fiber (Felenbok, 1988) which makes the instrument transportable and adaptable on another telescope.

The current design of the instrument include the following specifications: a spectral resolution of 30000 minimum, a spectral domain 3900-8700 A that can be covered in two successive exposures (blue and red), an efficiency allowing a S/N better than 200 in 1 hour for a star of 6th magnitude with a 2m telescope and an astronomical CCD; a Thomson CCD 1024 × 1024 (19 μ pixel) proposed as a detector; a fiber feeding; the possibility to acquire 2 spectra simultaneously. Initially, until october 1989, the selected fiber (spectran SG820) had a core of 110 μ, which allowed to input 1.5 arcsec with an efficiency of 85 percent in the spectrograph for a 2m telescope opened at f/5.5. An image slicer, of Bowen-Walraven type, would thus have been needed to give finally on the detector 2.5 layers of width 50 μ, superposed on a total height of 315 μ, with an efficiency better than 90 percent. At present, the selected fiber has a core of 50 μ, which corresponds to 2 arcsecs for an entrance aperture f/2.5. Its chromatic transmission is near 90 percent for a large domain in the visible and in the red, and remains better than 70 percent at 3900 A. Thus, an image slicer is not anymore necessary. The cross dispersion will be given by a FLINT rectangle prism working in double pass at deviation minimum. The whole is articulated around an objective (possibly from the market). A slight tilt of the grating, and a lateral mirror will allow a comfortable mounting of the detector cryostat. This set up has the advantage of being very compact, luminous and inexpensive.

5 The MUSICOS December 1989 Observing Campaign

For this first broad MUSICOS campaign at the end of 1989, three programs were chosen among those which really require multi-site spectroscopic observations:
1) Short-period spectroscopic variations in Be stars
2) Corotating stream structures in the winds of PMS Herbig Ae stars
3) Doppler Imaging and flare monitoring of RSCVn-type active stars

Initially, the sites, telescopes and observers to be involved in this campaign, were the following: Mauna Kea 2.2m UH (T. Simon), 3.6m CFHT (P. Felenbok, B.H. Foing), Kitt Peak MacMath (J. Neff), La Silla 1.4m CAT (S. Char, S.Jankov), France 1.5m OHP (J. M. LeContel, A.M. Hubert), Crimea 2.6m Shajn (I. Tuominen, J. Huovelin), China 2.16m Xinglong (Huang Lin, Jiang Shi-Yang, Li QiBin, C. Catala).

After the general announcement sent by J. Butler through the Multi wavelength IAU Working group, and also the IAU circular for our multi-site continuous spectroscopy campaign from 8 to 17 december 1989 (MUSICOS 89 campaign), we received a large observing support (ground based spectroscopic or photometric, or satellite observations) and finally 17 telescopes, including IUE, were pointed towards our targets. In addition to resident spectrographs at OHP, Crimea, CFH, ESO, Lick, Mc Math, we used the two existing versions of the ISIS fiber-fed spectrograph in mono-order mode (one of them used until 1988 at OHP), and specially transported for the MUSICOS campaign on the 2.2m University of Hawaii telescope and on the 2.16m newly installed telescope at the Chinese XingLong station.

Final participants for the december 89 MUSICOS campaign included: B.H.Foing, (ESA/IAS), co-PI MUSICOS, coordination HR1099; C.Catala (DESPA), co-PI MUSICOS, coordination AB Aur; and the participants in the SPECTROSCOPIC campaign at Hawaii: T. Simon (UH), J. Baudrand, J.G Cuby, (Meudon), Installation of ISIS bis spectrograph on 2.2m U. Hawaii; P.Felenbok, J.Czarny (Meudon) for observations at Canada France Hawaii 3.6m; U.S.A.:J. Neff GSFC and J.Avellar, for NSO observations at MacMath Kitt Peak ; Chile: S.Jankov, IAS CES/CAT for remote control observations from ESO Garching S.Char (IAS), A.M. Lagrange et al (IAP) for 1m Ca spectrophotometry at ESO La Silla; IUE: T. Ayres (Boulder) IUE observations Brazil: R. de la Reza, C.Torres (Brazil National Observatory) France: A.M. Hubert, H. Hubert, M. Floquet (Meudon) with spectro Aurelie OHP; Austria: J. Hron,H.Maitzen (Wien) for 1.5m Schoepfl; USSR: I. Tuominen, Huovelin (Helsinki) P.Petrov, A. Scherbakov, I. Savanov (Crimea Obs), for observations with Coude Spectrograph on Crimea Shajn telescope China: Li Qi Bin, Huang Lin,Jiang Shi-Yang, Zhao Disheng (Beijing AO), C.Catala, J.Guerin, M.Dreux,

B.Foing installation of ISIS instrument at Xinglong Obs.2.1 m telescope India: K.K. Ghosh Vainu Bappu Observatory.

The participants in the PHOTOMETRIC campaign were: Chile, G. Cutispoto (Catania), photometry ESO La Silla; Mexico, San Pedro martir: Alvarez (UNAM), photometry ; Turkey :V. Keskin, C. Akan, S. Evren, Ege University UBV 48cm; Greece: J.Seidarakis, S.Mavridis; Australia: Dr. Page (Mt Tambourine observatory); US: Dr Brown (Manitoba Glenlea Obs.).

6 MUSICOS Project Group Organisation

There is in France a MUSICOS project group composed of C. Catala (Meudon) and B.H.Foing (ESA/IAS) as Principal Investigators, P. Felenbok, A.M. Hubert, J. Czarny (Meudon), J.M. Le Contel and E. Fossat (Nice). J. Baudrand (Meudon) is the instrument project manager. A group of 30 associated scientists in France is involved in the project. Also contacts were established with European groups in Italy (in Trieste and Catania), in Scandinavia (in Upssala, Helsinki, and Aarhus), in Spain, in United Kingdom (at Armagh). Associate countries such as US (in Hawaii, in Boulder, Goddard, and with the SYNOP group), in USSR (in Crimea) and in China (at Beijing Astonomical observatory) wish to participate to the multi-site project. A group of international co-investigators from institutes providing funding, hardware, observing site or manpower support to the project is being set, together with an international group of Associated scientists interested in the further scientific use of the network.

7 MUSICOS Planning

Since 1982, with our collaborators, we have participated to Multi-Site Multi-Wavelength Observing campaigns. Next MUSICOS campaign with existing instruments would include sites in Hawaii, ESO, OHP, Pic du Midi, Canarias, Crimea, Kitt Peak, AAT, China etc). The joint analysis and collaborations have started already from these observing campaigns. At the end of 1989, we organised an observing campaigns including transport of ISIS and ISIS-bis instruments in complementary sites (Xing Long, Hawaii). In january-mars 1990, we worked on the preliminary reduction and analysis of the results from this campaign. A MUSICOS workshop organised at Meudon on 27-30 March is to provide presentation of these results, further discussion and future organisation and realization of the MUSICOS project (see Catala and Foing 1990). 1989 was devoted also to the design of the spectrograph. In 1990, the current ISIS bis spectrograph will be transformed into a cross echelle spectrograph. At the end of 1990, instrument tests and qualification of this MUSICOS prototype are to take place at Pic du Midi Observatory. 1991 should see the final development of the MUSICOS spectrometer model to be duplicated by the participant countries. In 1991-1992, we plan the installation in remote sites (such as Hawaii, Xinglong, Canarias), and the start of the full network operations. When the network is operational, a Multi-Site Guest Observer program will be offered to the community.

The philosophy of the MUSICOS project is to associate the groups interested also on the scientific return, and this is the reason why we start already collaborative campaigns to learn how making, reducing, analysing the results from multisite data. Also, some programs require special observational strategy and might be interleaved with other programs or require service observing. Asteroseismology will require continuous coverage and a good control on the stability and the quality of the measurements. The MUSICOS project will be open to any input and collaboration during the development phase, and through the observing proposals to the community during the operational phase.

References

Brown, T. 1990, in *IAU Colloq. No. 121, Inside the Sun*, eds. G. Berthomieu and M. Cribier (Kluwer, Dordrecht), in press.

Catala, C. and Foing, B. H. (ed.) 1988, *Proceedings 1st MUSICOS Workshop on MUlti SIte COntinuous Spectroscopy* (Meudon Observatory Publications).

Catala, C. and Foing, B. H. (ed.) 1990, *Proceedings 2nd MUSICOS Workshop on MUlti SIte COntinuous Spectroscopy* (Meudon Observatory Publications).

Däppen, W. 1988, in *Proceedings 1st MUSICOS Workshop on MUlti SIte COntinuous Spectroscopy*, eds. C. Catala and B.Foing (Meudon Observatory Publications), p.11.

Felenbok, P. 1988 in *Proceedings 1st MUSICOS Workshop on MUlti SIte COntinuous Spectroscopy*, eds. C. Catala and B. Foing (Meudon Observatory Publications), p.123.

Seismology of Jupiter

Francois-Xavier Schmider [1] and Benoit Mosser [2]

[1]Astro Unit, Queen Mary College, Mile End Road, London E1 4NS, UK
[2]Observatoire de Meudon, 5 pl. J. Jansen, 92195 Meudon Pal Cedex, France

Abstract: The detection of p-modes on stars or planets is a powerful instrument for study of the internal structure of the object. The difficulty of the detection is due to the required stability for the detector. A stellar seismometer has been built in Nice University and tested on the planet Jupiter in November 1987. The observations show a power spectrum similar to p-modes spectrum in the range of 10 to 20 minutes, with amplitude of a few meters per second, well above the noise level of the instrument. A tentative has been made to identify the peaks as low degree p-modes. The implication of this identification is discussed. Future observations are also envisaged.

1 Introduction

The success of helioseismology has incited some teams to try to identify similar acoustic modes on other objects, mainly on solar-type stars. The Nice University group, using a stellar seismometer based on a resonance sodium cell, was successful in the detection of stellar oscillations on Procyon and α Centauri in 1983 and 1984 (Gelly *et al.* 1986). A new instrument, based on the same principle, has been conceived and realized in order to improve the performance (Schmider 1989). This instrument has been tested at Haute-Provence Observatory in October and November 1987. We first observed Jupiter, in order to calibrate the instrument sensitivity on solar oscillations in the light reflected by the planet. The sensitivity of the instrument is determined by the maximum slope in the flange of the sodium line, so that it is degraded, in the case of Jupiter, by the fast rotation of the planet. For this reason, the solar oscillations have not been detected. The noise level for frequencies higher than $0.5\,\mu$Hz is of about $50\,\mathrm{cm\,s^{-1}}$ for 5 nights of observations, well above the amplitude of solar oscillations. On the other hand, strong variations have been observed in the 10 minutes range, where the possible acoustic modes of Jupiter were expected. A Fourier analysis of the signal shows a set of peaks (Fig. 1) with amplitude of about $4\,\mathrm{m\,s^{-1}}$ between 0.6 and 2 mHz.

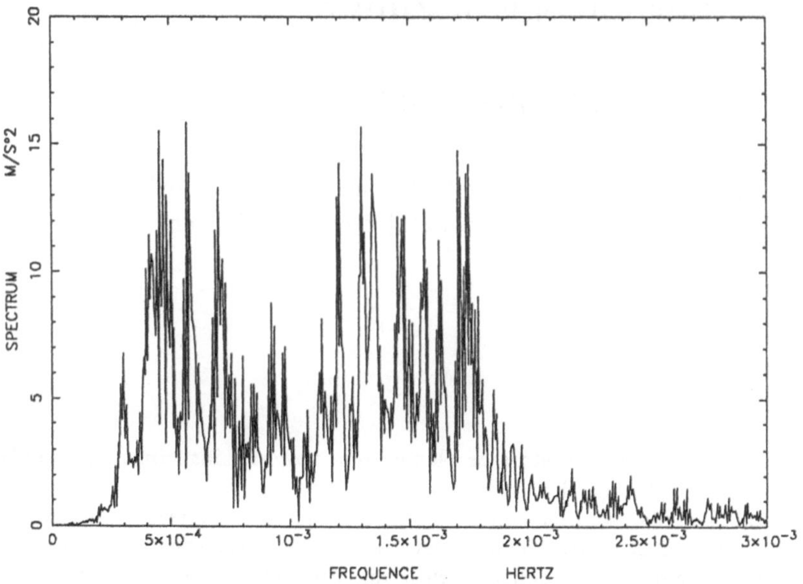

Fig. 1. Spectrum of the radial velocity of Jupiter observed during 5 nights with a sodium cell seismometer. A set of peaks with amplitude of about 4 m s^{-1} are visible in the range of 0.5 to 2 mHz.

2 Identification

The possibility of studying the internal structure of Jupiter by the measurement of global oscillations has been envisaged for a long time (Vorontsov *et al.* 1976), but it has never been detected in the past. It was then very important to look if the detected variations could be interpreted in term of global oscillations.

We tried to identify these peaks as Jovian acoustic modes, looking for a regular pattern in the power spectrum. The well known Tassoul's (1980) asymptotic equation predicts equidistant peaks in the case of low l, high n quantum numbers. A spectrum of the spectrum has been computed, showing first the presence in the spectrum of peaks separated by twice the jovian frequency rotation equal to $2 \times 28\,\mu$Hz (Fig. 2), and an equidistance of the peaks of 137 μHz. The fine structure of the power spectrum is visible in the echelle-diagram of Fig. 3. The presence of peaks separated by twice the jovian frequency rotation proves the jovian origin of the signal and essentially rules out all interpretation of the signal as resulting from atmospheric or instrumental effects. In term of global oscillations, this frequency spacing can be interpreted as the rotational splitting of the modes of degree $l = 1$.

A set of theoretical frequency has been computed in order to fit the data. Calculations were based on the interior structure model from Hubbard and Marley (1989), which fits the gravitational moments measurements. The interior model and the frequency computation take into account the discontinuity due to the presence of a dense core at the

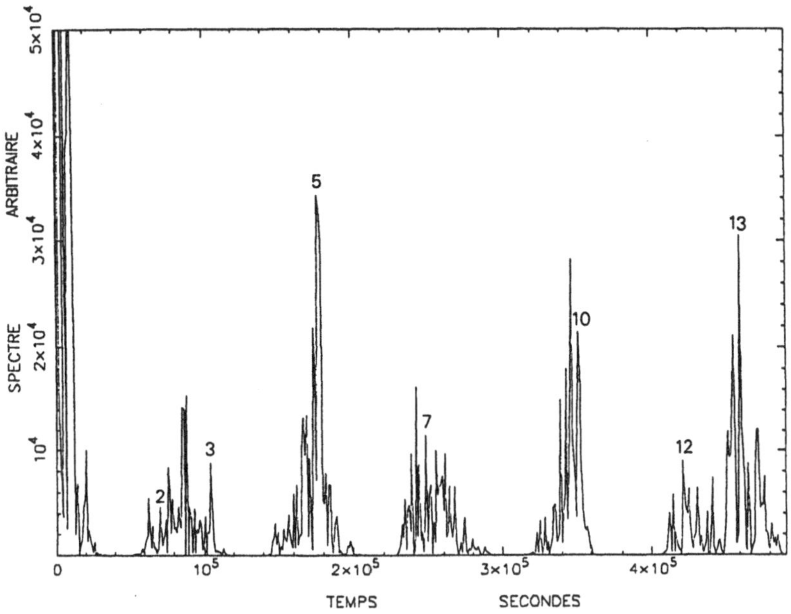

Fig. 2. Spectrum of the spectrum of the Fig. 1, which has the dimension of an autocorrelation. The gaps in the data due to the day time are clearly visible. The figure shows the harmonics of the jovian rotation period, here numbered, almost equal to 10 hours. This tends to prove the presence in the spectrum of peaks separated by 28 μHz, inverse of 10 hours.

center of the planet; the displacements of the modes due to the asphericity and the rotation (Mosser *et al.* 1989) are also introduced; finally, assuming an adiabatic temperature gradient, we recalculate the temperature and density profiles from the 1-bar to the 1-kbar pressure level, while in models of Hubbard and Marley (1989) atmospheric parameters are interpolated between the 1 to 700-bar pressure levels.

The observed discontinuity in the spacing of modes of degree $l = 1$ provided a value of 5500 km for the radius of the dense core, in good agreement with the theoretical prediction. On the other hand, the observed value (137 μHz) of the fundamental tone ν_0 of the Tassoul's (1980) expression disagrees from the theoretical one (154 μHz). The conflict may result from the used distribution of density in the upper levels of the molecular envelope, to which the gravitational moments are insensitive and whose contribution in ν_0 is very great. The assumption of Stevenson (1985), which proposes some enhancement of heavy materials in the outer layers, could explain a part of the observed disagreement. Moreover, substantial sources of uncertainty may come from the poorly known equations of state in the molecular and metallic hydrogen envelopes.

Assuming a value of ν_0 equal to the observed one, the computed frequencies fit very well the data, for all modes of order n from 3 to 15 and degree $l = 0$ and 1. The error bars are lower than the influence of the day-night alternation (11.6 μHz). The atmospheric cut-off frequency seems to be about 2 mHz, a little bit less than expected by the standard models of the jovian atmosphere.

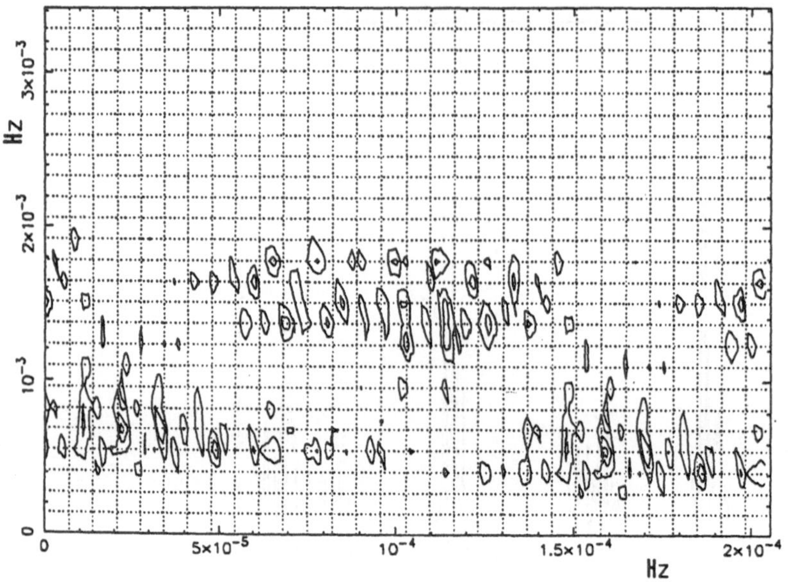

Fig. 3. Echelle-diagramme of the previous spectrum with an equi-distance of 137 μHz. This diagram is obtained by superposition of lines of spectrum having a length of 137 μHz. The presence of equidistant peaks in the spectrum is clearly visible as vertical alignments in this figure.

3 Future Observations

It appears that acoustical oscillations of Jupiter have been detected. Previous IR observations by Deming *et al.* (1989) report no detection of high degree sectoral p-mode oscillations with a level greater than 0.07 K. It has to be noticed that these observations are sensitive to a higher levels part of the jovian troposphere, so the amplitudes of the oscillations should be substantially decreased at this level. Moreover it is not evident to deduce an amplitude in intensity from the amplitude observed in velocity, and some very crude approximations have to be made. The calculation made by Deming *et al.* (1989), which gives a value of $1\,\mathrm{m\,s^{-1}}$ as an upper limit for the amplitude is only valid in the case of an isothermal atmosphere, which is obviously not the case. So the observed amplitude is not in contradiction with the observations of Deming *et al.* (1989).

A theoretical prediction of jovian oscillations amplitude has given a value of about $50\,\mathrm{cm\,s^{-1}}$ (Bercovici and Schubert 1987). The observed amplitude seems to contradict this expectation. A possible explanation comes from the convection in the deep interior of the planet which has not been taken into account in the calculations.

Given the importance of the results which can be derived, it is fundamental to confirm this observation and to try to detect modes of higher degree, which can provide us with the location of the molecular-metallic hydrogen transition and new informations on the atmosphere. New observations will be made independently using the Fourier Transform

Spectrometer at the CFH telescope in Hawaii. From ground-based observations, it is possible to detect modes of degree up to 30, assuming they have amplitude similar to those of low degree. This could be done by using a CCD camera after the sodium cell. The photon flux required to get a sensitivity better than $1\,\mathrm{m\,s^{-1}}$ on an image of Jupiter on 64×64 pixels can be achieved by using a 2-meter telescope. It will also permit to disentangle the observations from the effects of the contrasted structure of the surface of the planet in the velocity signal.

However, such ground-based observations, will suffer the day-night alternation which produces side-lobes in the spectrum and prohibits a direct determination of the frequencies of the modes. An investigation of Jovian seismology from space should then be urgently undertaken. This can be an important research domain for the Cassini mission.

4 Conclusion

Assuming the interpretation of the observations as p-modes oscillations is correct, the implications on the physics of Jupiter are multiple. Standard models should be refined to account for a lower fundamental tone. The cut-off frequency is also slightly lower than predicted by the theory, which suggests that the physics of the atmosphere should be improved. May be the most important, the amplitude of the oscillations is somewhat greater than predicted. It may be that the excitation theory developed for the Sun (Goldreich and Kumar 1988) cannot apply to Jupiter: it is not surprising since these two objects exhibit completely different interior structures. Any element should be considered as a part of a puzzle, the solution of which can only come from observations. It is thus of fundamental importance to get better measurements, namely continuous observations on a long period of time. Space observations are necessary, as well as networks for stellar and planetary seismology.

References

Bercovici, D., and Schubert, G. 1987, *Icarus*, **69**, 557.

Deming, D., Mumma, M.J., Espenak, F., Jennings, D.E., Kostiuk, Th., Wiedemann, G., Loewenstein, R., and Piscitelli, J. 1989, *Astrophys. J.*, **343**, 456.

Gelly, B., Grec, G., and Fossat, E. 1986, *Astron. Astrophys.*, **164**, 383.

Goldreich, P., and Kumar, P. 1988, *Astrophys. J.*, **326**, 462.

Hubbard, W.B., and Marley, M.S. 1989, *Icarus*, **78**, 102.

Mosser, B., Schmider, F.-X., Gautier, D., and Delache, Ph. 1990, submitted to *Astron. Astrophys.*

Schmider, F.-X., 1989, Thèse de Doctorat, Université Paris 7.

Stevenson, D.J. 1985, *Icarus*, **62**, 4.

Tassoul, M. 1980, *Astrophys. J. Suppl.*, **43**, 469.

Vorontsov, S.V., Zharkov, V.N., and Lubimov, V.M. 1976, *Icarus*, **27**, 109.

Lecture Notes in Mathematics

Vol. 1236: Stochastic Partial Differential Equations and Applications. Proceedings, 1985. Edited by G. Da Prato and L. Tubaro. V, 257 pages. 1987.

Vol. 1237: Rational Approximation and its Applications in Mathematics and Physics. Proceedings, 1985. Edited by J. Gilewicz, M. Pindor and W. Siemaszko. XII, 350 pages. 1987.

Vol. 1250: Stochastic Processes – Mathematics and Physics II. Proceedings 1985. Edited by S. Albeverio, Ph. Blanchard and L. Streit. VI, 359 pages. 1987.

Vol. 1251: Differential Geometric Methods in Mathematical Physics. Proceedings, 1985. Edited by P.L. García and A. Pérez-Rendón. VII, 300 pages. 1987.

Vol. 1255: Differential Geometry and Differential Equations. Proceedings, 1985. Edited by C. Gu, M. Berger and R.L. Bryant. XII, 243 pages. 1987.

Vol. 1256: Pseudo-Differential Operators. Proceedings, 1986. Edited by H.O. Cordes, B. Gramsch and H. Widom. X, 479 pages. 1987.

Vol. 1258: J. Weidmann, Spectral Theory of Ordinary Differential Operators. VI, 303 pages. 1987.

Vol. 1260: N.H. Pavel, Nonlinear Evolution Operators and Semigroups. VI, 285 pages. 1987.

Vol. 1263: V.L. Hansen (Ed.), Differential Geometry. Proceedings, 1985. XI, 288 pages. 1987.

Vol. 1265: W. Van Assche, Asymptotics for Orthogonal Polynomials. VI, 201 pages. 1987.

Vol. 1267: J. Lindenstrauss, V.D. Milman (Eds.), Geometrical Aspects of Functional Analysis. Seminar. VII, 212 pages. 1987.

Vol. 1269: M. Shiota, Nash Manifolds. VI, 223 pages. 1987.

Vol. 1270: C. Carasso, P.-A. Raviart, D. Serre (Eds.), Nonlinear Hyperbolic Problems. Proceedings, 1986. XV, 341 pages. 1987.

Vol. 1272: M.S. Livšic, L.L. Waksman, Commuting Nonselfadjoint Operators in Hilbert Space. III, 115 pages. 1987.

Vol. 1273: G.-M. Greuel, G. Trautmann (Eds.), Singularities, Representation of Algebras, and Vector Bundles. Proceedings, 1985. XIV, 383 pages. 1987.

Vol. 1275: C.A. Berenstein (Ed.), Complex Analysis I. Proceedings, 1985–86. XV, 331 pages. 1987.

Vol. 1276: C.A. Berenstein (Ed.), Complex Analysis II. Proceedings, 1985–86. IX, 320 pages. 1987.

Vol. 1277: C.A. Berenstein (Ed.), Complex Analysis III. Proceedings, 1985–86. X, 350 pages. 1987.

Vol. 1283: S. Mardešić, J. Segal (Eds.), Geometric Topology and Shape Theory. Proceedings, 1986. V, 261 pages. 1987.

Vol. 1285: I.W. Knowles, Y. Saitō (Eds.), Differential Equations and Mathematical Physics. Proceedings, 1986. XVI, 499 pages. 1987.

Vol. 1287: E.B. Saff (Ed.), Approximation Theory, Tampa. Proceedings, 1985–1986. V, 228 pages. 1987.

Vol. 1288: Yu. L. Rodin, Generalized Analytic Functions on Riemann Surfaces. V, 128 pages, 1987.

Vol. 1294: M. Queffélec, Substitution Dynamical Systems – Spectral Analysis. XIII, 240 pages. 1987.

Vol. 1299: S. Watanabe, Yu.V. Prokhorov (Eds.), Probability Theory and Mathematical Statistics. Proceedings, 1986. VIII, 589 pages. 1988.

Vol. 1300: G.B. Seligman, Constructions of Lie Algebras and their Modules. VI, 190 pages. 1988.

Vol. 1302: M. Cwikel, J. Peetre, Y. Sagher, H. Wallin (Eds.), Function Spaces and Applications. Proceedings, 1986. VI, 445 pages. 1988.

Vol. 1303: L. Accardi, W. von Waldenfels (Eds.), Quantum Probability and Applications III. Proceedings, 1987. VI, 373 pages. 1988.

Lecture Notes in Physics

Vol. 344: M. Rascle, D. Serre, M. Slemrod (Eds.), PDEs and Continuum Models of Phase Transitions. Proceedings, 1988. VII, 229 pages. 1989.

Vol. 345: H. Holden, A. Jensen (Eds.), Schrödinger Operators. Proceedings, 1988. V, 458 pages. 1989.

Vol. 346: D. Lüst, S. Theisen, Lectures on String Theory. VII, 346 pages. 1989.

Vol. 347: Ph. Blanchard, J.-P. Dias, J. Stubbe (Eds.), New Methods and Results in Non-linear Field Equations. Proceedings. 1987. VII, 133 pages. 1989.

Vol. 348: A. Bohm, M. Gadella, Dirac Kets, Gamow Vectors and Gel'fand Triplets. VII, 119 pages. 1989.

Vol. 349: Y.A. Kubyshin, J.M. Mourão, G. Rudolph, I.P. Volobujev, Dimensional Reduction of Gauge Theories, Spontaneous Compactification and Model Building. X, 80 pages. 1989.

Vol. 350: G. Tenorio-Tagle, M. Moles, J. Melnick (Eds.), Structure and Dynamics of the Interstellar Medium. Proceedings, 1989. XXI, 537 pages. 1989.

Vol. 351: A. Dervieux, B. Larrouturou (Eds.), Numerical Combustion. Proceedings, 1989. VII, 481 pages. 1989.

Vol. 352: K.B. Wolf (Ed.), Lie Methods in Optics II. Proceedings, 1988. XII, 197 pages. 1989.

Vol. 353: M. Barthes, J. Léon (Eds.), Nonlinear Coherent Structures. Proceedings, 1989. X, 280 pages. 1990.

Vol. 354: J. De Coninck, F. Dunlop (Eds.), Wetting Phenomena. Proceedings, 1988. VII, 112 pages. 1990.

Vol. 355: R. Lima, L. Streit, R. Vilela Mendes (Eds.), Dynamics and Stochastic Processes. Proceedings, 1988. V, 298 pages. 1990.

Vol. 356: R. Wehrse (Ed.), Accuracy of Element Abundances from Stellar Atmospheres. Proceedings, 1988. V, 109 pages. 1990.

Vol. 357: J.L. Lumley (Ed.), Whither Turbulence? Turbulence at the Crossroads. Proceedings, 1989. V, 535 pages. 1990.

Vol. 358: J. Kovalevsky, Astrométrie moderne. VI, 294 pages. 1990.

Vol. 359: K. Kirchgässner (Ed.), Problems Involving Change of Type. Proceedings, 1988. XII, 207 pages. 1990.

Vol. 360: Y. Mellier, B. Fort, G. Soucail (Eds.), Gravitational Lensing. Proceedings 1989. XV, 315 pages. 1990.

Vol. 361: P. Gaigg, W. Kummer, M. Schweda (Eds.), Physical and Nonstandard Ganges. Proceedings, 1989. IX, 310 pages. 1990.

Vol. 362: N.E. Kassim, K.W. Weiler (Eds.), Low Frequency Astrophysics from Space. Proceedings. XII, 280 pages. 1990.

Vol. 363: V. Ruždjak, E. Tandberg-Hanssen (Eds.), Dynamics of Quiescent Prominences. Proceedings, 1989. XI, 304 pages. 1990.

Vol. 364: T.T.S. Kuo, E. Osnes, Folded-Diagram Theory of the Effective Interaction in Nuclei, Atoms and Molecules. V, 175 pages. 1990.

Vol. 365: M. Schumacher, G. Tamas (Eds.), Perspectives of Photon Interactions with Hadrons and Nuclei. Proceedings. IX, 251 pages. 1990.

Vol. 366: M.-J. Goupil, J.-P. Zahn (Eds.), Rotation and Mixing in Stellar Interiors. Proceedings, 1989. XIII, 183 pages. 1990.

Vol. 367: Y. Osaki, H. Shibahashi (Eds.), Progress of Seismology of the Sun and Stars. Proceedings, 1989. XIII, 467 pages. 1990.

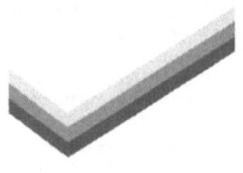